Engineering Solutions for Modern Challenges in Advanced Materials Science

Deepanraj Balakrishnan
https://orcid.org/0000-0001-8986-6692
Prince Mohammad Bin Fahd University, Saudi Arabia

Ratnasunil Buradagunta
https://orcid.org/0000-0001-9855-7808
Prince Mohammad Bin Fahd University, Saudi Arabia

Wattala Fernando
University of Dundee, UK

IGI Global
Scientific Publishing
Publishing Tomorrow's Research Today

Vice President of Editorial Melissa Wagner
Director of Acquisitions Mikaela Felty
Director of Book Development Jocelynn Hessler
Production Manager Mike Brehm
Cover Design Jose Rosado

Published in the United States of America by
IGI Global Scientific Publishing
701 East Chocolate Avenue
Hershey, PA, 17033, USA
Tel: 717-533-8845 | Fax: 717-533-7115
Website: https://www.igi-global.com E-mail: cust@igi-global.com

Library of Congress Cataloging-in-Publication Data

LCCN: 2025019251 (CIP Data Pending)
ISBN13: 9798337335568
Isbn13Softcover: 9798337335575
EISBN13: 9798337335582
British Cataloguing in Publication Data
A Cataloguing in Publication record for this book is available from the British Library.

Table of Contents

Preface xiii

Chapter 1
Artificial Intelligence and Machine Learning in Advanced Materials Science: A New Era of Innovation 1
Rituraj Jain, Marwadi University, India
Sitesh Kumar Singh, National University of Science and Technology, Muscat, Oman
Kamal Upreti, Christ University, India
Yohannes Bekuma, Wollega University, Ethiopia

Chapter 2
Beyond Algorithms: Real-World Constraints in Energy Materials Modeling and Design 35
Mona Albalawi, University of Tabuk, Saudi Arabia

Chapter 3
Machine Learning as a Catalyst in Advanced Materials Research and Development 77
Muhammad Akram, Abertay University, UK
Waqas Uddin Choudhry, Wien University of Applied Sciences, Hochschule, Austria

Chapter 4
Nanomaterials and Nanotechnology Applications 111
Patrick Ehi Imoisili, University of Johannesburg, South Africa

Chapter 5
Addressing Energy, Environment, and Housing Challenges Through Nanotechnology: Materials of the Future 147
Shehzad Ahmad, University of the Punjab, Lahore, Pakistan
Zainab Maqsood, University of the Punjab, Lahore, Pakistan
Ali Hasan, University of Agriculture, Faisalabad, Pakistan
Shahid Mahmood, University of the Punjab, Lahore, Pakistan
Ali Hussain, University of the Punjab, Lahore, Pakistan

Chapter 6
Sustainable Nanofluid Engineering Synthesis Techniques and Emerging Applications in Smart Materials 187
M. Amin Mir, Prince Mohammad Bin Fahd University, Saudi Arabia
Syed M. Hasnain, Prince Mohammad Bin Fahd University, Saudi Arabia
Noushi Zaidi, King Faisal University, Saudi Arabia
Mohammad Ayaz Ahmad, University of Guyana, Guyana
Ercan Bursal, Prince Mohammad Bin Fahd University, Saudi Arabia

Chapter 7
AI-Driven Modeling and Optimization of 3D Graphene–MXene Hybrid Nanocomposites 217
Cağdaş Kızıl, Antalya Chamber of Commerce and Industry, Project Development and State Support Office, Antalya, Turkey
Ali İhsan Kömür, Faculty of Science, Department of Chemistry, Akdeniz University, Antalya, Turkey
Onur Karaman, Vocational School of Health Services, Medical Imaging Techniques, Akdeniz University, Antalya, Turkey
Ceren Karaman, Vocational School of Technical Sciences, Department of Electricity and Energy, Akdeniz University, Antalya, Turkey

Chapter 8
Developing Nanostructured Calcium-Based Bio-Ceramic Materials From Natural Source: Potential Applications and Future Directions 251
Ratna Sunil Buradagunta, Prince Mohammad Bin Fahd University, Saudi Arabia
Ravikumar Dumpala, Visvesvaraya National Institute of Technology, India

Chapter 9
Creating a Sustainable Future: The Emergence of Eco-Friendly Materials 265
Gaydaa Al Zohbi, Prince Mohammad Bin Fahd University, Saudi Arabia

Chapter 10
Cradle to Cradle: Engineering the Ecological Renaissance with Advanced Sustainable Materials 321
Moatassim Raoof Shindy, Nano Cure Industrial Consulting Company, Egypt

Chapter 11
Biomass-Derived Activated Carbon: A Sustainable Approach for CO_2 Capture 359
Jesús Botello-González, Universidad Autónoma de Nuevo León, Mexico
Eduardo Soto-Regalado, Universidad Autónoma de Nuevo León, Mexico
Margarita Loredo-Cancino, Universidad Autónoma de Nuevo León, Mexico
Nancy Elizabeth Dávila-Guzmán, Universidad Autónoma de Nuevo león, Mexico
Jacob Josafat Salazar-Rábago, Universidad Autónoma de Nuevo León, Mexico

Chapter 12
Innovative Pavement Materials and Sustainable Engineering Practices for Next-Generation Infrastructure 395
Bader Eddin Alasali, National University of Science and Technology, Muscat, Oman
Mohanad Talal Alfach, School of Mathematics, Computer Science, and Engineering, City St. George's, University of London, UK

Compilation of References 437

About the Contributors 539

Index 547

Detailed Table of Contents

Preface xiii

Chapter 1
Artificial Intelligence and Machine Learning in Advanced Materials Science:
A New Era of Innovation 1
Rituraj Jain, Marwadi University, India
Sitesh Kumar Singh, National University of Science and Technology, Muscat, Oman
Kamal Upreti, Christ University, India
Yohannes Bekuma, Wollega University, Ethiopia

The increase of artificial intelligence (AI) and machine learning (ML) in the discovery, design, and optimization of new materials is causing a rapid acceleration of the field of materials science. This chapter addresses the principles for how artificial intelligence and machine learning can enable the predictive modeling, high-throughput screening, and smart production of polymers, alloys, ceramics, and nanomaterials. Emphasized are some of the techniques, such as hybrid approaches to artificial intelligence and physics, generative models, and reinforcement learning. Some important problems in the chapter are spoken about: lack of standards, incoherence of data, and unfeasibility of explaining the models. Along with that, this also attempts to investigate how the advantages of upcoming malware such as quantum computing, edge synthetic intelligence, and open information facilities can reinforce the research and innovation in the forthcoming age of materials.

Chapter 2
Beyond Algorithms: Real-World Constraints in Energy Materials Modeling
and Design 35
Mona Albalawi, University of Tabuk, Saudi Arabia

The world stands at a critical energy crossroads. While renewable energy capacity grows exponentially, our ability to store, convert, and utilize this energy efficiently still depends on breakthroughs in materials science. This chapter isn't just about promising lab results – it's about real-world solutions that must balance scientific innovation with engineering pragmatism and industrial realities. Furthermore, the world is racing toward an energy revolution, but the path isn't as simple as swapping fossil fuels for solar panels and batteries. Behind every headline about a "breakthrough" material, there's a messy, often untold story of trial, error, and real-world constraints. This chapter isn't just about the science of energy materials—it's about the people behind them, the frustrations they face, and the hard-won lessons that separate lab-scale promise from industrial reality.

Chapter 3
Machine Learning as a Catalyst in Advanced Materials Research and Development 77
Muhammad Akram, Abertay University, UK
Waqas Uddin Choudhry, Wien University of Applied Sciences, Hochschule, Austria

Machine learning is changing how materials are designed, tested, and improved. It does this by learning from data to guess properties, suggest new recipes, and pick better experiments. There are three main types of machine learning: supervised learning, which learns from labelled examples; unsupervised learning, which finds hidden groups in unlabeled data; and reinforcement learning, which learns by trying steps and getting feedback. Machine learning works on data from lab experiments, computer simulations, and past papers, but this data needs to be cleaned, features picked, and dimensions reduced, or models can get confused. With good data, ML can predict things like bandgap, hardness, stability, or lifetime. It can also do inverse design, which means starting from a target and working backward to a material or recipe, and optimize processes to get better results with fewer trials across polymers, alloys, nanomaterials, and composites. Some new directions are autonomous labs that loop measure-learn-decide, AI-guided design that ties simulations and experiments together more closely, and work that tries to make materials greener and less wasteful. There are still issues, such as a lack of high-quality data, labels that are noisy, and models that don't work with new chemistries or conditions. Because of this, it's important to combine physical rules and domain knowledge with machine learning and to use explainable methods so that decisions make sense. It's also important to keep an eye on ethical issues like bias, safety, and reproducibility. Overall, this review tries to give a clear and simple look at where machine learning helps and where it struggles in advanced materials and sustainability. It also tries to share ideas that will help future studies and encourage teams from different fields to work together to build and test new ideas.

Chapter 4
Nanomaterials and Nanotechnology Applications 111
Patrick Ehi Imoisili, University of Johannesburg, South Africa

Nano-materials (NMs) and nano-technology (NT) are some of the most amazing things that have happened in science, technology, and business in the last few years. They let us make materials that can do many things. Nanotechnology has also been materially advanced by various industrial sectors, encompassing quantum confinement, enhanced surface area-to-volume ratios, and properties that are tunable optically, mechanically, and electronically. This chapter provides a comprehensive history of the growth of nanoscience and nanotechnology, with a focus on the types and properties of (NMs). The possible uses of NMs were discussed, with a focus on how nanotechnology can benefit the economy and society in the construction and manufacturing industries. Information and communication technologies, sustainable energy, ballasting technology, environmental sustainability, health, and medical uses. The problems and chances that NMs and NT present were also discussed.

Chapter 5
Addressing Energy, Environment, and Housing Challenges Through Nanotechnology: Materials of the Future 147
Shehzad Ahmad, University of the Punjab, Lahore, Pakistan
Zainab Maqsood, University of the Punjab, Lahore, Pakistan
Ali Hasan, University of Agriculture, Faisalabad, Pakistan
Shahid Mahmood, University of the Punjab, Lahore, Pakistan
Ali Hussain, University of the Punjab, Lahore, Pakistan

The worldwide demand of energy and change of the climate highpoint the advanced need of sustainable energy storage. During on wind reliance solar is requires for energy storage. The major source for emissions and energy use is a building sectors. Advances nanostructured materials, alloys, ceramics and polymers are enable such next generation solar cells, flexible electronics and effective devices. In this chapter, explores economic aspects, application of advance nanostructured material in energy storage and in environment. Furthermore, explores challenges in building a sustainable future and fabrication strategies.

Chapter 6
Sustainable Nanofluid Engineering Synthesis Techniques and Emerging Applications in Smart Materials 187
M. Amin Mir, Prince Mohammad Bin Fahd University, Saudi Arabia
Syed M. Hasnain, Prince Mohammad Bin Fahd University, Saudi Arabia
Noushi Zaidi, King Faisal University, Saudi Arabia
Mohammad Ayaz Ahmad, University of Guyana, Guyana
Ercan Bursal, Prince Mohammad Bin Fahd University, Saudi Arabia

Colloidal suspensions of nanoparticles in regular base fluids, or nanofluids, have become attractive materials for improving heat transfer, thermal conductivity, and overall energy efficiency in a variety of engineering applications. Often referred to as "smart fluids," their ability to dynamically control heat transfer makes them useful in fields like biomedical engineering, microelectronics, transportation, nuclear energy, and food processing. This chapter provides a thorough review of current advancements in the study of nanofluids, covering synthesis strategies, colloidal stability maintenance tactics, and methodologies for assessing thermophysical characteristics. Innovative applications that aim to further optimize system performance, like hybrid nanofluids, are given particular attention. The existing difficulties in simulating the behavior of nanofluids, especially the absence of a universal predictive framework, are also covered in this chapter. The chapter concludes with a summary of future research possibilities, emphasizing the need for future practical applicability, long-term stability, and future studies on the nanoparticle-fluid interactions. For researchers, engineers, and business people trying to drive nanofluid technology, this chapter is a great resource.

Chapter 7
AI-Driven Modeling and Optimization of 3D Graphene–MXene Hybrid Nanocomposites 217
Cağdaş Kızıl, Antalya Chamber of Commerce and Industry, Project Development and State Support Office, Antalya, Turkey
Ali İhsan Kömür, Faculty of Science, Department of Chemistry, Akdeniz University, Antalya, Turkey
Onur Karaman, Vocational School of Health Services, Medical Imaging Techniques, Akdeniz University, Antalya, Turkey
Ceren Karaman, Vocational School of Technical Sciences, Department of Electricity and Energy, Akdeniz University, Antalya, Turkey

This chapter focuses on the simulation and optimization of electrochemical energy storage systems (EES) employing 3D graphene-MXene hybrid nanocomposites as electrode materials through the application of artificial intelligence (AI) and advanced materials design. In this regard, the overview of the possibilities of using different AI models to predict basic physicochemical structure-property relationships of materials and enhance performance parameters like specific capacity and cycling stability is provided. The chapter also explores fundamental themes of materials engineering including interlayer spacing, material porosity, and interface engineering, and evaluates the role that AI-based optimization can have in the design and production of high-performance hybrid nanostructures in EES. In addition, the fact that AI models can be used to design the material in the material design process is also mentioned as the means of ensuring that the structural integrity and stability of the electrode material are maintained and that its electrochemical performance is improved. The basic procedures of the AI model development and verification, such as preprocessing data, training the model, experimental verification, and feedback, are also addressed in detail. Moreover, the synthesis step that requires fundamental materials engineering issues like scalability, are also discussed along with other emerging issues and trends of designing EES. Finally, this section provides critical information on how AI can be used in the creation of next generation advanced hybrid nanostructures in high-performance EESs.

Chapter 8
Developing Nanostructured Calcium-Based Bio-Ceramic Materials From Natural Source: Potential Applications and Future Directions 251
Ratna Sunil Buradagunta, Prince Mohammad Bin Fahd University, Saudi Arabia
Ravikumar Dumpala, Visvesvaraya National Institute of Technology, India

Synthesizing calcium based ceramic materials from natural source such as fish bones, sea corals and egg shells is cost effective route to produce nanocrystalline bio-ceramic materials. These nano-ceramic phases contain special ions which help to induce additional benefits when used as biomaterials. Since, the size of the produced phases is in nano-level, applications including local drug delivery, surface coatings on implants and producing nano-composites can be benefited due to the promising properties of nano-crystalline materials. The present chapter provides a comprehensive summary of the developments in synthesizing nano-crystalline calcium based ceramic materials from the natural sources for biomaterial applications. The characterization, performance evaluation, and potential applications of using these nano-phases in biomedical applications are discussed in detail. Furthermore, future perspectives and the potential applications in medical fields are briefly presented.

Chapter 9
Creating a Sustainable Future: The Emergence of Eco-Friendly Materials 265
Gaydaa Al Zohbi, Prince Mohammad Bin Fahd University, Saudi Arabia

The growing environmental issues of the 21st century demand a paradigm shift in our material choices. This chapter investigates the emerging field of sustainable materials, analyzing their definition, classification, and different applications across industries. It explores the lifecycle evaluation of materials, emphasizing the significance of circular economy principles and minimized environmental impact. From bio-based polymers and recycled composites to advanced low-carbon concrete and sustainably sourced timber, the chapter features the advanced materials leading to a more sustainable future. Furthermore, it addresses the obstacles to broad implementation and highlight strategies for promoting sustainable material integration in design and manufacturing. This chapter aims to provide a comprehensive overview for researchers, practitioners, and policymakers seeking to embrace a material-conscious approach to building a resilient and sustainable world.

Chapter 10
Cradle to Cradle: Engineering the Ecological Renaissance with Advanced Sustainable Materials 321
Moatassim Raoof Shindy, Nano Cure Industrial Consulting Company, Egypt

The prevailing linear "take-make-dispose" industrial model has unleashed severe environmental and economic consequences, demanding an urgent paradigm shift towards truly circular systems. This chapter champions the Cradle to Cradle [C2C] philosophy as the foundational framework for this transformation, meticulously detailing how Advanced Sustainable Materials are engineering an ecological renaissance. We delve into pioneering advancements from 2020 to 2024, exploring the transformative potential of bio-based polymers, the circular economy's alchemy in material upcycling, breakthroughs in sustainable energy solutions [including lead-free perovskite solar cells and bio-derived battery electrolytes], and the burgeoning field of bio-inspired materials that seamlessly merge living and synthetic entities. By linking good academic ideas with those found in industry, this book goes beyond a simple overview. It offers an in-depth study of both the successes and challenges of developing sustainable materials. Their potential, performance, and longevity always need to be assessed on a large scale. Researchers and industry professionals rely on this book to analyze economic factors and regulations, focusing on the entire life cycle of materials. This chapter serves as a useful guide and a powerful call for change. We argue that it reveals paths to a future where excellent material performance is coupled with care for our planet. We emphasize that pioneering materials design is not only essential for the environment, but is the absolute foundation of the next industrial revolution.

Chapter 11
Biomass-Derived Activated Carbon: A Sustainable Approach for CO_2 Capture 359
Jesús Botello-González, Universidad Autónoma de Nuevo León, Mexico
Eduardo Soto-Regalado, Universidad Autónoma de Nuevo León, Mexico
Margarita Loredo-Cancino, Universidad Autónoma de Nuevo León, Mexico
Nancy Elizabeth Dávila-Guzmán, Universidad Autónoma de Nuevo león, Mexico
Jacob Josafat Salazar-Rábago, Universidad Autónoma de Nuevo León, Mexico

The release of carbon dioxide into the atmosphere has increased in recent years and represents a global concern for greenhouse warming. Over the past six decades, global temperature has risen significantly, which could represent an irreversible trend; consequently, technologies to capture carbon dioxide are required. Agro-wastes can be used to produce biochar or activated carbons, which serve as adsorbents for CO_2. These wastes are by-products of industrial processes and are widely available. Adsorbents are characterized to determine their CO_2 capture capacities, including textural properties and surface chemistry. Their performance can be enhanced by surface modifications, such as textural changes or heteroatomic doping. These materials can capture CO_2 in batch and continuous systems; however, powdered activated carbons are unsuitable for continuous processes due to pressure drop. To overcome this, they can be structured via additive manufacturing, improving strength and rheology. Recent studies explore alternatives like zeolites, silica, bentonite, and 3D-printed monoliths.

Chapter 12
Innovative Pavement Materials and Sustainable Engineering Practices for Next-Generation Infrastructure 395
Bader Eddin Alasali, National University of Science and Technology, Muscat, Oman
Mohanad Talal Alfach, School of Mathematics, Computer Science, and Engineering, City St. George's, University of London, UK

Because people rely heavily on durable surface transportation, there is now a strong demand for new pavement solutions. For decades, conventional asphalt and concrete were reliable, but now climate change, more vehicles and new regulations often make them less effective. It gives a review of new pavement engineering techniques and materials using polymer- and nano-modified asphalts, advanced concrete and recycled industrial waste. Showing what improves a material's toughness, helps the planet and increases its usefulness, we present engineers with practical choices for building strong and useful products. These materials are shown and highlight how connected and responsive building materials can transform infrastructure management. A lifecycle assessment acts as an approach designed to measure the sustainability of materials. Eventually, current problems and future research directions in building sustainable pavements are covered.

Compilation of References 437

About the Contributors 539

Index 547

Preface

INTRODUCTION

The accelerating pace of global technological change has placed unprecedented demands on materials science and engineering. From energy security and environmental sustainability to digital transformation and biomedical innovation, modern challenges increasingly hinge on the development of advanced materials that are intelligent, resilient, scalable, and sustainable. *Engineering Solutions for Modern Challenges in Advanced Materials Science* emerges from this critical context, bringing together diverse perspectives that reflect the evolving role of materials engineering in addressing complex, real-world problems.

Recent advances in artificial intelligence, nanotechnology, sustainable material design, and computational modeling have fundamentally reshaped how materials are discovered, engineered, and deployed. No longer confined to incremental improvements, contemporary materials research is increasingly interdisciplinary—integrating data science, physics-based modeling, chemical engineering, and systems-level thinking. This volume reflects that transformation by highlighting both methodological innovation and application-driven engineering solutions across energy, infrastructure, healthcare, and environmental systems.

As editors, we envisioned this book as a bridge between foundational scientific understanding and engineering practice. Rather than focusing solely on theoretical advances or laboratory-scale breakthroughs, the chapters emphasize scalability, feasibility, and societal relevance. Contributors from academia, industry, and applied research institutions provide insights that collectively underscore a shared goal: to translate advanced materials science into solutions that are technically sound, economically viable, and environmentally responsible.

OVERVIEW OF THE CHAPTERS

The volume opens with **Chapter 1**, which sets the conceptual foundation by exploring the transformative role of artificial intelligence and machine learning in advanced materials science. It highlights predictive modeling, generative design, and hybrid physics–AI approaches while critically addressing challenges such as data quality, model interpretability, and standardization.

Chapter 2 shifts the focus toward energy materials, emphasizing the often-overlooked gap between laboratory innovation and industrial deployment. By foregrounding real-world constraints, this chapter underscores the importance of engineering pragmatism in achieving meaningful energy transitions.

In **Chapter 3**, machine learning is further examined as a catalyst for materials research and development, with detailed discussions on data sources, model development, inverse design, and ethical considerations. The chapter provides a balanced view of both the opportunities and limitations of ML-driven materials engineering.

Chapters 4 and 5 explore nanomaterials and nanotechnology as enablers of solutions to energy, environmental, housing, and industrial challenges. These chapters highlight the multifunctionality of nanostructured materials and their socio-economic implications, while also addressing fabrication challenges and sustainability concerns.

The role of smart fluids and thermally responsive materials is examined in **Chapter 6**, which provides a comprehensive review of nanofluid synthesis, stability, characterization, and emerging applications across engineering domains.

Chapter 7 presents a focused case study on AI-driven modeling and optimization of graphene–MXene hybrid nanocomposites, illustrating how advanced computational tools can accelerate the design of next-generation energy storage materials.

Biomaterials and bio-derived ceramics are the focus of **Chapter 8**, which discusses sustainable synthesis routes using natural sources and highlights applications in biomedical engineering, coatings, and drug delivery systems.

Sustainability takes center stage in **Chapters 9 and 10**, which explore eco-friendly materials, circular economy principles, and cradle-to-cradle engineering frameworks. These chapters emphasize lifecycle thinking and the integration of environmental stewardship with material performance.

Carbon capture and climate mitigation strategies are addressed in **Chapter 11**, which examines biomass-derived activated carbon materials and advanced manufacturing approaches for scalable CO_2 adsorption technologies.

The volume concludes with **Chapter 12**, which explores innovative pavement materials and sustainable engineering practices for next-generation infrastructure, highlighting lifecycle assessment, smart materials, and resilience under evolving environmental conditions.

CONCLUSION

Collectively, the chapters in this volume demonstrate that advanced materials science is no longer defined solely by material performance, but by its capacity to respond to global challenges in an integrated and responsible manner. Engineering solutions must now balance innovation with sustainability, intelligence with explainability, and performance with long-term societal impact.

This book also underscores the growing importance of data-driven and AI-enabled methodologies in materials engineering. While these tools offer unprecedented opportunities for acceleration and optimization, the contributors consistently emphasize the need for transparency, validation, and alignment with physical principles. Such balance is essential for ensuring that technological progress remains robust and trustworthy.

Finally, we hope this volume serves as both a reference and a catalyst—for researchers seeking interdisciplinary insights, for engineers designing practical solutions, and for policymakers and industry professionals navigating the transition toward sustainable and resilient material systems. By bringing together diverse applications and methodologies under a unified engineering perspective, *Engineering Solutions for Modern Challenges in Advanced Materials Science* aims to contribute meaningfully to the future of materials innovation.

Chapter 1
Artificial Intelligence and Machine Learning in Advanced Materials Science: A New Era of Innovation

Rituraj Jain
https://orcid.org/0000-0002-5532-1245
Marwadi University, India

Sitesh Kumar Singh
https://orcid.org/0000-0002-7108-0808
National University of Science and Technology, Muscat, Oman

Kamal Upreti
https://orcid.org/0000-0003-0665-530X
Christ University, India

Yohannes Bekuma
https://orcid.org/0000-0003-4707-208X
Wollega University, Ethiopia

ABSTRACT

The increase of artificial intelligence (AI) and machine learning (ML) in the discovery, design, and optimization of new materials is causing a rapid acceleration of the field of materials science. This chapter addresses the principles for how artificial intelligence and machine learning can enable the predictive modeling, high-throughput screening, and smart production of polymers, alloys, ceramics, and

DOI: 10.4018/979-8-3373-3556-8.ch001

nanomaterials. Emphasized are some of the techniques, such as hybrid approaches to artificial intelligence and physics, generative models, and reinforcement learning. Some important problems in the chapter are spoken about: lack of standards, incoherence of data, and unfeasibility of explaining the models. Along with that, this also attempts to investigate how the advantages of upcoming malware such as quantum computing, edge synthetic intelligence, and open information facilities can reinforce the research and innovation in the forthcoming age of materials.

INTRODUCTION

The paradigm shift in materials science is the transition to the previous approach of trial and error in the analysis of the results into the models of data supported by Artificial Intelligence (AI) and Machine Learning (ML). In line with the need of the aerospace, energy and biomedicine industries to have high-tech materials, there should be a speedy process of new material discovery at a low cost. The use of AI/ML in materials science is presented in this chapter to accelerate innovation and address the issues of traditional processes. It reviews key AI/ML applications in materials design, processing and property prediction on the case study basis. In the field of materials science, as it is now, the hunt and optimization of novel materials have been based on experimental and theoretical models. Even though these approaches have brought about improvements in most sectors including electronics and aerospace, they are costly and time-consuming as they entail trial and error activities (Ninduwezuor-Ehiobu, et al., 2023). As the ability of AI and ML to enable discovery using data driven methods, a substitute of integration using AI and Machine Learning has emerged. It is possible to promote predictive modeling, pattern recognition, and data analysis on the design of materials based on these technologies with new and improved properties (Badini et al., 2023).

Specifically, AI/ML tools tend to be highly efficient when it comes to large datasets mining to estimate promising material candidates and optimizing production (Badini et al., 2023). Alloy design, catalyst development and next generation energy materials are included in such uses. The chapter is a critical review of AI and ML application and future opportunities in materials science, covering its creation (Ninduwezuor-Ehiobu et al., 2023).

The most important ones are the AI-ladder framework (simplest models to autonomous experimentation and related challenges like the quality of the data, the explainability of the model, and the ethical issues involved (Bappy, 2024). The chapter explains the importance of interdisciplinary cooperation and AI/ML use in the conscientious manner in case of the complete utilization of AI/ML in materials domain.

The primary contributions of this chapter include but are not limited to: (i) telling the reader how AI/ML is transforming the situations of material discovery and optimization, (ii) a critical assessment of the recent research and knowledge gaps, (iii) the introduction of the AI frameworks to material areas, and (iv) The Interdisciplinary roadmap to intelligent materials engineering.

The chapter provides a critical synthesis of the current state of the art of the applications of AI and ML in advanced materials science beyond mere overviews up to deep analytics. Rather than just a precis of the past research we focus our attention on evaluating their methodological coverage, shortcomings in practice, and how they might be improved. Such contributions include the emergence of domain-specific instrumentations such as DeepChem and GT4SD (Generative Toolkit for Scientific Discovery) that are playing an increasingly significant role in the data-rich molecular design, bio-virtual screening, and high-throughput experimentation. We also emphasize the interdisciplinary paradigms which integrate AI/ML with the knowledge of materials science (e.g., physics-informed learning and hybrid simulation-driven modelling) which, in sum, enhance the robustness and relevance of the prediction system. Finally, but certainly not least, the chapter outlines an end-to-end strategy of addressing a few acute issues regarding data sparsity, model generalization as well as computational scalability, in support of intelligent integration strategies, such as multimodal learning, open data platforms as well as physics-guided optimization. These will help to position AI/ML not as a tool but as empowering agents of the next generation of intelligent material engineering.

FOUNDATIONS OF AI AND ML IN MATERIALS SCIENCE

The advances in AI and ML have rendered them as potent instruments that contribute to the emergence of patterns and forecasted connections out of vast aggregates of material, like ML algorithms like the supervised, unsupervised and reinforcement learning can handle the quantity of information retrieved by methods of characterization, such as XRD (X-ray Diffraction), XPS (X-ray Photoelectron Spectroscopy), TEM (Transmission Electron Microscopy), SEM (Scanning Electron Microscopy) and more, and improve the accuracy and speed of interpretation (Lau et al., 2 ML is faster than other discovery methods, as it finds new relations that are concealed in the composition of material, structure and properties in a repeatable systematic way that cannot be found in other methods (trial and error experiment by hand).

In materials science, where inputs have many dimensions and little data exists in experiments, it is necessary to use domain knowledge in the AI/ML models. The combination of the data-driven approach with intuition of science reduces data sparse, extrapolates well and increases the relevance and reliability of a model (Liu et al.,

2023). It has led to the popularity of physics informed and physics-based learning model, which uses available scientific principles, and algorithmic reasoning, on tasks in solid state mechanics and prediction properties (Lau et al., 2023).

Different ML paradigms (supervised, unsupervised, and reinforcement learning) demonstrated domain-specific advantages in the context of the property prediction, materials classification, and process optimization application, which we take into consideration (Ali and Mashwani, 2023). You can give labelled data in supervised learning to predict an outcome, which is based on mechanical properties of alloys based on their composition (Alnuaimi & Albaldawi, 2024). Supervised learning is a form of learning in which a model is learned on labelled data to learn input feature to known output mappings - it is well suited to learning such things as prediction of mechanical strength as a function of material composition. Alternatively, the hidden patterns or structure in unlabelled data is discovered by unsupervised learning and might be applied to cluster similar material microstructures or discover the new material families. (Yadav et al., 2024). In dynamic conditions, reinforcement learning proves to be so effective to discover optimal synthesis protocols and manufacturing policies under the condition of trial-based feedback (Ninduwezuor-Ehiobu et al., 2023). Nevertheless, both approaches are involved in various fields of materials science and nonetheless, both can create breakthroughs in the strategic combination of the other in aiding to predict material, classify materials and design material using experiment(s). Table 1 introduces the main types of machine learning, namely, supervised, unsupervised, and reinforcement learning, their peculiarities, and real-life implementation in the context of materials science research.

Table 1. "Key Machine Learning Methods Categorized by Learning Type and Applications in Materials Science

ML Category	Definition	Common Algorithms	Materials Science Applications
Supervised Learning	Learning from labeled data to map input features to known outputs.	Linear Regression, Random Forest, SVM, Neural Networks	Predicting tensile strength, corrosion resistance, and electrical conductivity from material composition.
Unsupervised Learning	Analyzes unlabeled data to discover hidden patterns, groupings, or structures.	K-Means, PCA, DBSCAN, Autoencoders	Clustering materials based on microstructure, discovering unknown material families.
Reinforcement Learning	Learns through interaction with an environment using reward signals to optimize decisions.	Q-Learning, Deep Q-Networks (DQN), Proximal Policy Optimization (PPO)	Tuning synthesis parameters for alloy development, autonomous experimentation for optimal material properties.

Role of big data in materials research

The relevance of big data is the application of AI and ML in the materials research. As it has introduced the large-scale data resource databases in steel materials, it has boosted the research and development of materials. It is among the core strategies in designing and discovering steel material of machine learning models using experimental data, industrial manufacturing data and simulation data (Geng et al., 2023). The concept of the AI ladder is employed as a demonstrating tool to realize that, among the simplest data fitting methods to the more complicated functions, including the semi-autonomous experimentation and hypothesis, the AI is working around in materials science. Using this data-driven method, the researchers can predict the material properties, find novel materials, and streamline the process more efficiently and correctly (Al-Akayleh et al. 2024). Specifically, real time measuring is so interesting that it can be integrated with the AI models, which is what makes the dynamic determination of the most favorable or the parameters to adjust in various processes, such as drying of the food. It means that there is a possibility that AI can not only analyze current data but also participate in real-time in real-time decision-making in materials research.

AI/ML FOR MATERIAL DISCOVERY AND DESIGN

Data-driven material screening and prediction

To such an extent, AI and ML have completely transformed the process of data-based material screening and prediction that completely upsets the form of material discovering and optimization that has been occurring over decades. These computational methods can easily sample large chemical space - or large structural space where new materials that can be found with these desired properties are sought. On the other hand, large collections of known compounds enable us to train ML models predicting the properties of new compounds with a high level of solid accuracy, and without the need to spend money and time on the costly and extremely time-consuming experimental procedures. Although the ML based scoring functions have been shown to be applicable to both virtual screen as well as ligand optimization problems in the pharmaceutical research field, broader systematic classification and validation (Harren et al., 2024) is required to achieve broader false ground real-world applicability. This combined with AI utilized in high throughput experimentation and quantum computing open a new horizon of advancements in the energy storage lines, as well as the advanced manufactures (Muhammad et al., 2023).

Generation models and evolutionary algorithms to design novel materials

However, as will be shown later, recent advancements of the use of generative models and evolutionary algorithms are offering new abstractions of material design. They offer rapid hypothesis creation as well as broad research through a large space of designs. Deep learning and simulation methods have been utilized in these examples to identify relationship between structure responses (Park et al., 2023). The example of GT4SD (Manica et al., 2023) is a demonstration of the improvement of speed of material ideation (through data-driven representations). This is replaced by transformer-based models like ChatGPT now, which add chemicals to chemical space when performing material generative modeling (Mok & Back, 2024). More to the point, the GANS-based evolutionary computation is employed to more effectively train the models such as the generative adversarial networks, in addition to optimizing the material property settings in real-life applications (Wang et al., 2024).

Case studies on discovery of high-performance polymers, alloys, ceramics and nanomaterials

High performance materials have sort transformed industries in all areas of Aerospace and biomedical engineering. It gives a summary of case studies in the context of developing advanced polymers, alloys, ceramics and nanomaterials since each has its distinct properties and uses. It is a breakthrough to include nanocellulose in high performance polymers field. Shen et al. (2024) have proved that the cellulose nanocrystal (CNC) form, cellulose nanofibril (CNF) form, and bacterial nanocellulose (BNC) form of nanocellulose possesses superior mechanical characteristics because it surpasses conventional implant materials. These nanocellulose based composites possess properties very akin to those required in the case of orthopedic implants and are highly high strength, stiffness and biocompatible. Nanocellulose has demonstrated to aid the osteogenic differentiation and growth of stem cells in vitro as well as in vivo perforations in animal bone regeneration models in critical sized defects (Shen et al., 2024). The use of alloys in aerospace, especially magnesium alloys, is a field that has been spectacular to apply in recent years. Although magnesium (Mg) alloy has excellent mechanical and electrochemistry characteristics suitable to biodegradable implants, it has not been embraced by the medical field yet due to its low service strength and low corrosion rate in the body fluids (Bai et al., 2023). High emissivity oxidation and high anti-corrosion electroplating are just some of the surface treatment measures employed in producing these alloys to enhance their aerospace performance (Bai et al., 2023). E2209 filler would be best used in applications that demand high strength, toughness and corrosion resistance, especially within the marine, chemical and power generation settings (S. Sengottaiyan, et. al., 2025). Artificial intelligence (AI) fuzzy logic inference system presents the exploration on the drilling parameters effect on multi-walled carbon nanotubes (MWCNTs) (N. Senthilkumar, et. al., 2024). Composite materials can produce these nanotubes. The choice of composite material is determined by the purpose of use as well as the better qualities of the fibre (Senthilkumar, N., 2024).

It is also known that titanium alloys have also benefited in the laser additive manufactured (LAM) technology, which enables the creation of a new alloy composition and microstructure (Su et al., 2024). This is since the integration of auxiliary energy fields, such as thermal, acoustic, mechanical deformation, and magnetic fields, at the LAM process results in the improved dynamics of the melt pool and solidification behavior and microstructures and mechanical properties (Su et al., 2024). Ceramics engineering and its composites have gained prominence in the ceramics engineering sector due to their characteristics of great wear, corrosion, and heat (Almansoori et al., 2024). Aluminum Nitride (AlN) and Silicon Nitride (Si3N4), as well as zirconia (ZrO2), Silicon Carbide (SiC), and Alumina (Al2O3) have been studied a lot in terms of automotive, aerospace and tribological applications. Studies sought to address the inherent brittle nature of ceramics by embracing nanomaterials, including graphene and carbon nanotube to enhance the physics mechanical properties (Almansoori et al., 2024).

It is through nanomaterials that smart materials have provided new opportunities in a wide range of different fields. Such photo switchable supramolecular nanomaterials possessed special features (Younis et al. 2023). They were designed using azobenzene that includes small molecules and polymers. The origin of these materials is weak molecular interactions, and it was proved to find application in pH sensing and in capturing CO2 (Younis et al., 2023). Other advances in materials science have been on the development of biodegradable smart polymers. The biodegradable polymer matrices like Polylactic Acid (PLA) are the most widely used matrices in these polymers (Chan et al., 2023). These smart materials have undergone copolymerization, combination, or creation into composite material, and the fundamental properties and functionality have been altered through these methods, to achieve desired properties and functionality (Chan et al., 2023). In general, these case studies indicate that there are several applications and derivations of computational discovery that can successfully result in the identification of high-performance material.

Although there have been some significant improvements in materials design, there remains a lack of scalability, interpretability and generalizability of models across conduct in different material families in past case studies. As an example, AI applications in ceramics do not have such datasets compared to polymers or alloys. Moreover, much of the generative methodology does not make use of integration with physics-based simulation. The solution to these gaps in generic addressing requires the presence of Hybrid AI frameworks and aggressive validation pipelines to respond to physical plausibility and domain relevance.

Role of AI in virtual screening before experimental verification

The second application that is significant is the AI powered virtual screening used to greatly accelerate pre-experimental evaluation of compounds. AI based ligand based virtual screening is rapid, making compounds with high binding affinity with the target protein (Roggia et al., 2023) accessible using docking algorithms and QSAR models. Today, such tools have been capable of exceeding the state-of-the-art methods up to over order of magnitude with the help of Graph Convolutional Networks and feature fusion (Liu et al., 2024). On the platform, predictive pharmacokinetics and synthetic accessibility come together to scale up to platforms such as DrugFlow that offer lower cost and time in early drug discovery in the Shen et al (2024) and Kim et al (2023).

AI Frameworks and Libraries in Materials Science

Even other libraries such as DeepChem, Matminer, and ASE (Atomic Simulation Environment) offer domain adapted libraries in molecular featurization, dataset and model training. DeepChem enables the model of cheminformatics using a network of graph-based networks and quantum data sets to facilitate the design of drug-like materials. Matminer helps to accelerate Property prediction ML pipelines using feature extractors. Such libraries combined with simulation tools enhance reproducibility and time savings during the formulation that results in rapid translation of models into lab workflows.

PREDICTION OF MATERIAL PROPERTIES

ML techniques for predicting mechanical, thermal, electrical, and chemical properties

ML techniques have been used to predict several material properties. Ensemble techniques like random forests are actually very high accuracy predictors of tensile strength and elasticity. Another feasible alternative that has been adopted is artificial neural networks (ANN) in thermal property prediction of higher heating value of biomass (Brandic et al., 2024). Other ML algorithms along with support vector machines (SVMs) have been effective in predicting thermal conductivity of nanofluids. With very high accuracy, the shape of solvation free energy has been predicted with the help of ML models (Ferraz-Caetano et al., 2023). In addition, the ML has been applied to forecast the corrosion behavior (Ma et al. 2023) and tablet characteristics of interest in the material science sector. To some extent, these studies have discovered that feature selection and preprocessing of data are extremely significant in the accuracy of prediction. An example is (Onyiriuka, 2023), which introduced an improved feature selection algorithm in predicting the thermal conductivity of nanofluids and performed better than the available alternatives. Thus, ML methodology offers potent forecasting devices of numerous material qualities and leads to alternative algorithms exhibiting personal power. The material property prediction has also been driven to the limit by integration of the domain knowledge, and the sophisticated ML. Table 2 summarizes different machine learning methods and their specific domains of materials property and demonstrates the value that each method brings to the issue of the prediction of material properties in materials science.

Table 2. Machine Learning Techniques for Predicting Diverse Material Properties

Property Type	ML Technique	Description	Use Cases
Mechanical	Random Forest, GBM	Used to predict yield strength, ductility, and hardness from alloy composition and microstructure.	High-strength steel design, polymer blend optimization.
Thermal	ANNs, SVR	Applied in forecasting thermal conductivity and diffusivity in nanomaterials and composites.	Heat shielding ceramics for aerospace, cooling nanofluids.
Electrical	SVM, Decision Trees	Employed to model charge transport, dielectric behavior, and resistivity in semiconductors and nanofluids.	Flexible electronics, energy storage materials.
Chemical	Deep Learning, GCNs	Useful for modeling solvation energy, chemical reactivity, oxidation states, and corrosion rates.	Corrosion-resistant alloys, battery electrolyte prediction.

Deep learning for microstructure-property correlation

Deep learning methods have recently emerged as one of the potent methods to determine the relationship between microstructure and property in materials science. According to these approaches, these intricate features can be obtained with acceptable performance in microstructural images, and can be linked with the relevant properties, therefore, making the predictions more accurate and better insights into the structure property relations (Wijaya et al., 2024). Convolutional Neural Networks (CNNs) have been discovered to be promising in this sphere. As an example, using CNNs predicting ultimate tensile strength of Al-Si alloys with and without compositional and/or microstructural characteristics would be able to achieve high accuracy (R2 = 0.94) (Zhu et al., 2024). Like CNNs, they have also been applied to forecast elastic behavior of fiber reinforced composites on a wide parameter space of materials (Nakka et al., 2023). Earlier research started investigation regarding the means of further enhancement of innovative models of deep learning. As an example, diffusion probabilistic models have demonstrated the capability of synthetically generating microstructures, potentially retaining physical information, potentially futuring the range of known properties that can be forecasted (Wijaya et al., 2024). Multimodal fusion learning is also created to use compositional and microstructural data as one to make more comprehensive predictions of properties (Zhu et al., 2024).

Supervised learning for property prediction

Supervised learning algorithms have also turned out to be a good business venture in predicting the prices of property, as it demonstrates a higher accuracy compared to the normal methods. Real estate price prediction using different methods including SVM, RF, GBM, and neural network are researched; each of which has been put to practice by researchers (Choy & Ho, 2023). Traditional statistical methods like hedonic price models do not perform as well in terms of terms of explanatory power and minimized error as compared to these models. It is also fascinating that different algorithms are suited in several ways to predict. Considering RF and GBM, the accuracy of the two models performs better than SVM, whereas SVM performs better when making fast and partially accurate predictions in the time-limited environment (Ho et al., 2020). When rent prediction is compared with eager and lazy learning algorithms, it has been discovered that lazy learning algorithms, such as KNN and Locally Weighted Learning are more accurate and have less error in prediction (Heidari et al. 2021). Thus, supervised learning methods can be applied to the research undertaken on the valuation of properties and appraisal. However, when deciding regarding the choice of the best algorithm, the needs of every prediction task, i.e. accuracy, speed, and interpretability, should be considered carefully (Mathotaarachchi et al., 2024).

Unsupervised learning for material classification and trend discovery

Unsupervised learning techniques have been demonstrated to be usable in the discovery of material classification and trends in a variety of science domains. The large datasets in these methods could be automatically identified through pattern and structure without the use of labeled training data. In materials science, unsupervised learning algorithms are applied to analyze spectroscopic and nanoelectronics measurements in such a way that the unsupervised learning algorithms may be utilized to reveal patterns in the measurements and identify meaningful structures and characteristics (El Abbassi et al., 2021). Because the use of unsupervised learning has a huge potential of extracting valuable information to large unlabeled datasets, it is particularly useful in the discovery of new trends and relationships of material properties (Yan et al., 2015). The latter are, however, more adaptable to accommodate intricate and high-dimensional data with no knowledge of the data structure, although they are typically simpler to use with the assistance of supervised techniques. This makes them well adapted in discovery of concealed patterns and novel classes of materials. Lastly, unsupervised methods of learning may be employed as the formidable capabilities of material scientists to categorize materials, learn any trends, and even possibly make new discoveries and contributions to this area, otherwise (Eckhardt et al., 2022).

Challenges in model accuracy and generalization

It is not just that accuracy is a problem, but also generalization and even though machine learning models are very competent, which has potential, the generalization and accuracy of the machine is not feasible. The issue of loading the models performing very well on the training data and poorly on seeing new data (Xu et al., 2023) also is a frequently encountered problem. To address this issue, therefore, certain regularization techniques such as dropout and L2 norm penalty (Lu et al., 2021) are applied. Better generalization in the case of data augmentation is computationally costly, and it involves searching the best augmentation parameters (Mounsaveng et al., 2021). Another problem is the out of distribution prediction, particularly in case no data is available (Lu et al., 2021). In this situation, the generalization ability can be helped with the assistance of hybrid models that contain domain knowledge. To measure this uncertainty in the prediction, it is possible to understand and give credence through Bayesian methods (Lu et al. 2021). Research on machine learning has thus been placed on trading off the complexity of models versus their generalization capabilities.

AI IN MATERIALS PROCESSING AND SYNTHESIS OPTIMIZATION

Reinforcement Learning in Process Parameter Tuning

Recently, reinforcement learning (RL) has emerged as an effective AI technique in the optimization of process parameters to process materials and synthesize them. According to Epps et al. (2021), the RL algorithms can effectively search for the complex parameter space to learn the best conditions by synthesizing metal halide perovskite quantum dots. Based on the simulation of more than 600,000 experiments, researchers identified the optimal methods to make decisions to speed up materials development using multiple objectives. Like in other fields, including the field of computed tomography (CT) reconstruction, the RL is applied there: to the parameters automatic control in a human-like manner (Shen et al., 2018). This method compared to manual tuning in terms of quality of reconstructed CT images was of no use. The RL is also complemented with other AI methods to improve the effectiveness of parameter tune. Bayesian optimization and Gaussian process regression have been applied to optimize the hyper parameters of the RL algorithms in e.g., grid world example (Barsce et al., 2017). The meta-optimization approach can produce more efficient and effective parameter-tuning in materials processing and synthesis.

Figure 1 provides a systematic description of all the applications of Reinforcement Learning (RL) to materials science and engineering. Notably, the research and development diagram that involves RL is subdivided into four key branches associated with materials synthesis, CT imaging, meta-optimization, and alignment with other methods of AI, all of which are key research topics that have benefited by RL.

In materials synthesis, the optimization of the production of sophisticated materials, including metal halide perovskite quantum dots, has been accomplished to date, using RL. The RL helps to optimize conditions to synthesize a material and accelerate the process with the help of simulated experiments and algorithmic exploration. Due to this automated learning feature, the researchers can efficiently search for huge synthesis parameter space rapidly without incurring the classical trial and error mechanism.

In the case of CT imaging, RW makes it easier to improve medical and materials-based imaging. The scanning parameters can be automatically adjusted with the assistance of the parameter tuning policy network, and the image quality will be improved, and CT reconstruction will be more accurate. The RL applied in this application has been shown to perform better than manual means of tuning and provides an interesting illustration of the usefulness of RL in real time parameter adaptation.

The metameta optimization branch focuses on optimizing itself. The RL hyper parameters are optimized to maximize the learning efficiency with the help of Bayesian optimization and Gaussian Process Regression. Environments such as the simulations environment of Grid World have been executed to implement such techniques and come up with a more efficient and better reinforcement learning models.

The figure eventually leads to the correspondence between RL and other AI methods to enhance the effectiveness and quality of optimization processes. By integrating real learning (RL), with Bayesian techniques and surrogate modelling and search algorithms, researchers can obtain improved search performance, better optimization results and improved generalization.

Figure 1. Applications of Reinforcement Learning in Materials Science and Engineering

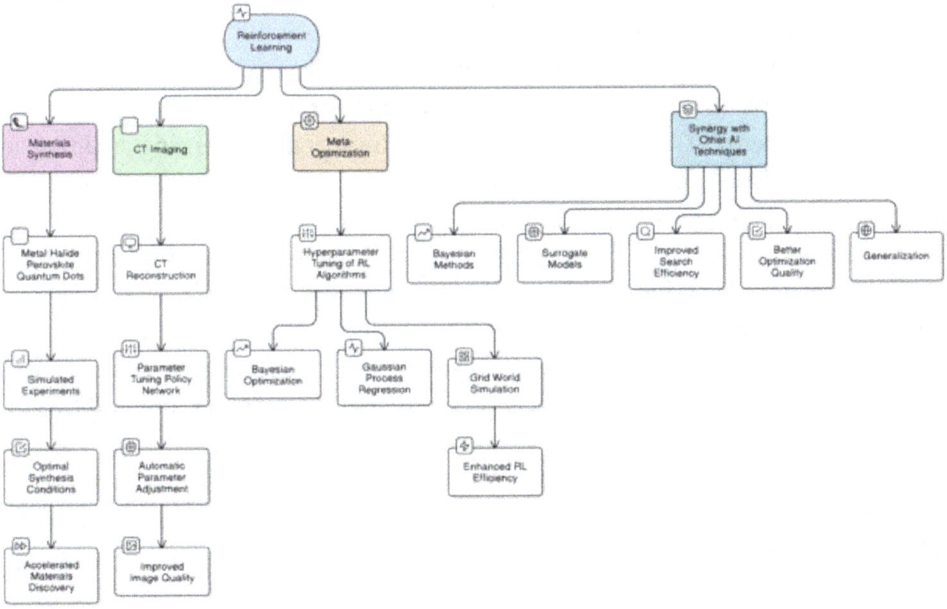

Simulation-Assisted Optimization

Due to its excellent results in materials research and design, simulation assisted optimization of materials, particularly using MD simulation and DFT, has recently been regarded as a very powerful tool. Based on MD simulations one develops the dynamic behavior of the system at the atomic level and DFT calculation provides correct electronic structure information. Important from the point of view of studying metal organic frameworks (MOFs) these methods allow the prediction of structural transformations, thermodynamic properties and catalytic behavior. However, surprisingly, ML techniques have been further integrated with the quantum and classical simulations to increase the efficiency and accuracy of these methods. Simulation setup can accelerate, and several key parameters can be predicted by ML (Formalik et al., 2023). The synergy of this between ML and simulation techniques has drastically increased the computational speed and stability of simulations, such as the case of simulations of ions near polarizable nanoparticles (Kadupitiya et al., 2020). As a result, simulation assisted optimization based on MD, DFT and ML opens a route to materials design and optimization. Using this multiscale computational approach, high throughput screening of materials, such as MOFs for gas separations, is possible as well as key insights in addition to experimental studies (Schaarschmidt et al., 2021).

Figure 2 depicts a complete flow diagram of the simulation assisted optimization in the materials research with ML integration. Starting with the initiation of the simulation assisted optimization, we then branch into two major original pathways, i.e.: machine learning integration and the traditional simulation methods.

Four major benefits are emphasized under the machine learning mainstream integration. By facilitating faster simulation set up, cheaper computation, better prediction of crucial parameters, and faster and more stable computation, ML can facilitate these. These advantages are important to enable the handling of these complex or materials systems and to accelerate the discovery cycle. In the figure, a particular case study focused on ion simulation around polarizable nanoparticles was highlighted showing practical use of ML in advancing as advanced simulation cases.

Simulation Methodologies: Molecular Dynamics (MD) simulations and Density Functional Theory (DFT) calculations, both fundamental methods in computational materials science, are the second pathways that represent simulation methodologies. They play a crucial role in simulating atomic scale interaction, energy landscape and structural behavior.

It also elaborates on how the diagram can be used in applications of Metal-Organic Frameworks (MOFs), where simulations are employed to predict structural transformations, thermodynamic properties, and catalytic behavior. For high coverage screening and materials design and optimisation efforts, these predictive insights orientatory and serve as a useful point of reference. These simulations are also used to complement experimental studies and validate and ensure practical relevance.

Finally, the framework results into generation of end insights for materials research, which allows for the data driven, accurately, and efficiently material design and material optimization strategies. This figure emphasizes the complementary role of ML and simulation leading to bringing about intelligent, predictive and scalable ways of carrying out the traditional research methodologies.

Figure 2. Simulation-Assisted Optimization Framework in Materials Research Enhanced by Machine Learning

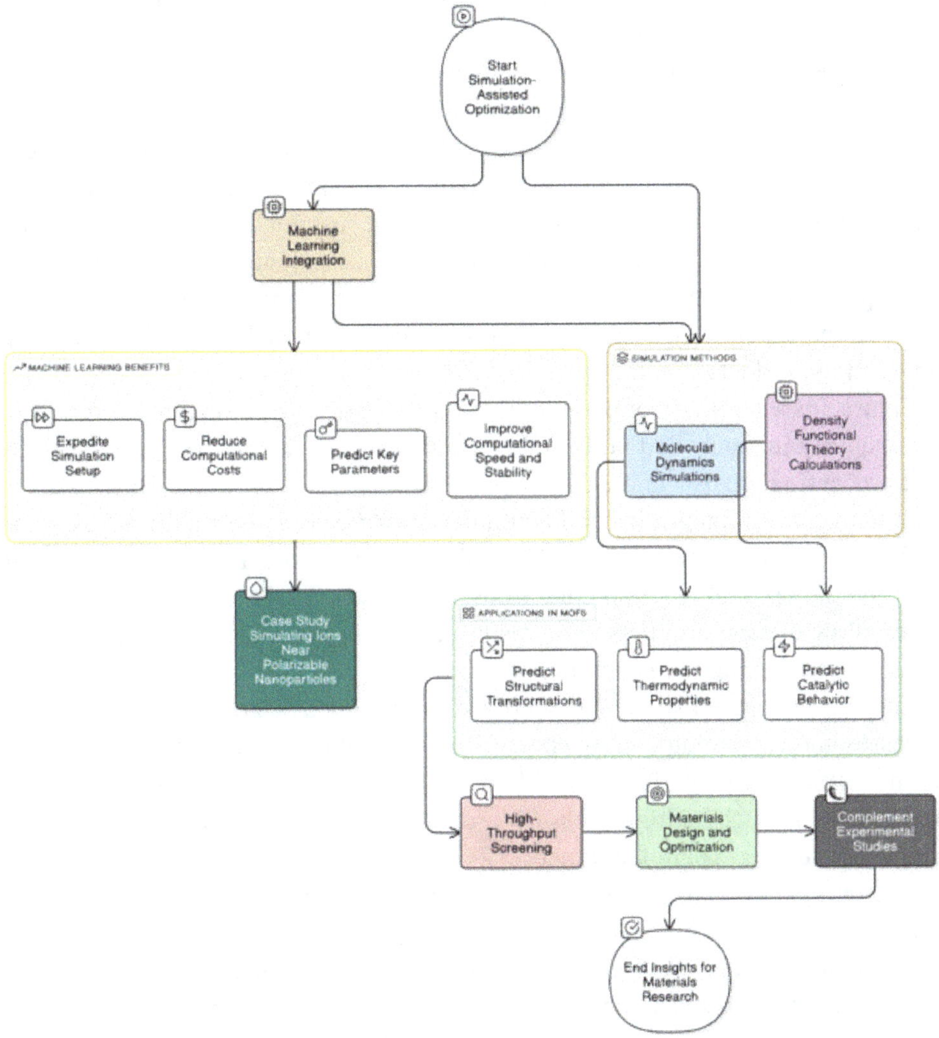

AI for Smart Process Control in Synthesis and Treatment

AI representing other industries is transforming the concept of smart process control in synthesis and treatment. Currently, chemical synthesis passes through the steps of optimization of reaction conditions, prediction of results and optimization of the quality of products using the machine learning algorithms and neural networks.

Production processes, quality control, waste and their management, and the process itself are being radically changed in the food industry with the implementation of AI. With the integration of AI with high pressure processing, UV and cold plasma processing, the strengthening of food safety and product shelf life has been achieved (Dhal and Kar, 2025). In the energy industry, AI is defining how to optimally use resources and more effectively apply them in smart buildings to achieve sustainability goals. In the same manner, AI vision technology is finding application in crop farms to score crops in real time, forecast analytics and resource optimization, and significantly enhance agriculture productivity and sustainability. This is how these innovations can transform the performance of process control in other areas, with the help of AI. AI is not only refining products to be more exact, responsive and efficient, the technology is assisting to create more sustainable practice in creating and treating products.

Figure 3 is a summary of the integration of AI offering intelligent, adaptive and sustainability in controlled processes within various fields of chemical synthesis, food processing, energy and agriculture. There is the AI flow that starts with the construction of AI integration and continues with the domain specific applications.

Within chemical synthesis, an AI is applied through machine learning algorithms to decide the optimal reaction conditions and the result of the synthesis. The processes allow the end products to be manufactured in a manner that will lead to the overall quality being good and precise.

The food industry also uses AI driven systems aimed at streamlining production processes and enhancing quality control and streamline waste management. In addition, more sophisticated methods of processing which assist in further improving the food safety such as high-pressure processing, set using ultraviolet light, cold plasma assisted by AI are facilitated.

The energy sector is one of the sectors in which AI technologies are used to optimize the process of resources utilization and enhance operational willingness. This involves the empowerment of smart building management that is of extremely great significance in sustainable energy practice.

Real time crop evaluation vision systems will allow the agriculture sector to take advantage of AI use; yield forecasting predictive analytics and resource management tools that control such factors as irrigation, fertilizers and pesticides. Technology enhances productivity and generates sustainability of farming activities in the long term.

In a broader sense, the four sector results are coming together to deliver the transformative nature of AI, automated and optimally responsive processes and enhance the quality and yields of the products. Collectively, such initiatives have been proceeding towards greener operations and AI is at the heart of transforming industrial/environmental operations.

Figure 3. System integration of AI across industries in smart process control and sustainable production

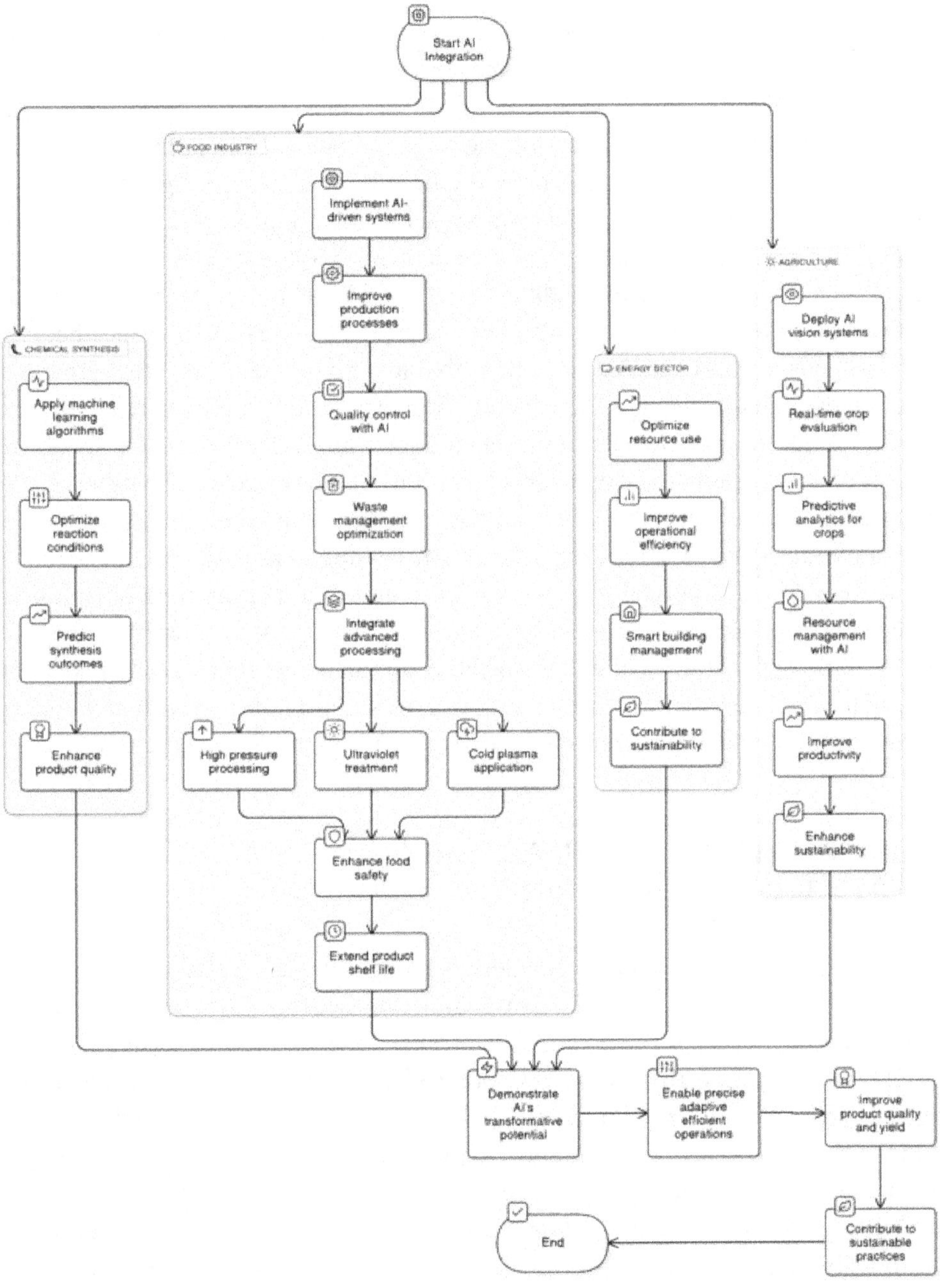

Real-Time Feedback and Learning from Experimental Outcomes

The actual feedback and experiment learning has contributed unimaginable importance in enhancing the engagement of both the students and the effectiveness of the learning of the students. The studies show that reasonable feedback systems that integrate advice messages and caution messages can be used to promote emotional and cognitive involvement of learners. Such systems can be designed to provide timely recommendations and interventions using learning analytics and machine learning algorithms that can boost the student participation rates along with the student outcome rates (Tirumanadham et al., 2025). It is also worth mentioning that the kind of feedback and the way it is provided may be the difference between effective feedback and not. An example is that just giving feedback after game play led to a great enhancement of the capacity of players to differentiate real news and misinformation in misinformation in inoculation games. However, one should remember that timing and form of feedback can have varied effects on learning results. Although instant feedback may be a valuable tool in both retention and transfer in the multimedia learning set up, mean timetables of augmented kinesthetic feedback have demonstrated superiority to every trial structure in the learning of motor capabilities. Therefore, real-time feedback and learning through experimental results present some opportunities of enhancing educational experiences. Nonetheless, the design and implementation of feedback systems should be done with serious care to ensure that they are maximally effective in different learning situations.

AI-DRIVEN SMART MANUFACTURING IN MATERIALS ENGINEERING

The defect identification of the materials engineering is experiencing the transformation with the help of predictive analytics based on AI. The ML and deep learning models compare real time sensor and imaging data with identifying an anomaly in the process that would have otherwise been overlooked by conventional means. Convolutional neural networks (CNNs) are employed in the additive manufacturing to detect defects and categorize the material properties, whereas support vector machines (SVMs) are employed in the additive manufacturing with high reliability (Mondal and Shubhra Goswami, 2024). However, and nevertheless, the methods significantly enhance the manufacturing as they lessen the amount of waste materials used in manufacturing, enhance manufacturing accuracy and increase product uniformity. Simultaneously, the AI agent paradigms enhance the material design by use of informative and economical modelling approaches (Wang et al.,

2024). The same AI solutions have enhanced the domain of quality control in the food industry to microbiologically safe food with the prolongation of the shelf life (Dhal and Kar, 2025).

The AI has found its way in manufacturing efficiency and intelligence through the integration of Industry 4.0 ecosystems. This proactive decision making, predictive maintenance, and blockchain secured data processing can be achieved with the use of the Internet of Things (IoT) and AI in combination with big data analytics (Soori et al., 2023). Smart factories are used today to apply AI intelligent digital twins and real time analytics to self-monitor and self-optimise, as well as self-heal. However, this development demands unified standards of operation, powerful governance and ethical data management.

Under the additive manufacturing, AI is used to automate 3D printing, real time monitoring and adjusting of the parameters. The AI2AM systems use the ML algorithms to determine the printability and aid slicing and defect detection (Sani et al., 2024). The bio composite development and scalable customization is further developed and supported by application.

The AI strategies of improving energy efficiency also create sustainability in manufacturing. The current industries use smart systems of MAM and ISO 50001 model and ESW policies, which assist the industries to control energy consumption and emissions. However, sludge-based biodiesel is a new project with the combination of AI and renewable energy generation (Aryanfar et al., 2024). The significant areas of AI technologies implementation in smart manufacturing, as shown in table 3, are primarily related to quality, sustainability, and automated work.

Table 3. AI-Driven Applications in Smart Manufacturing: Domains, Tools, and Industrial Benefits"

Manufacturing Domain	AI Application	Industry Benefits
Quality Control	Real-time anomaly detection using sensor data and CNNs for defect classification.	Minimizes waste, improves reliability, lowers inspection cost.
Additive Manufacturing	Intelligent parameter tuning, printability prediction, and closed-loop process control using AI2AM.	Reduces support structure waste, increases precision and designs freedom.
Energy Management	AI-based analytics for minimizing idle power, real-time energy optimization in processing plants.	Improves sustainability and cost savings in manufacturing pipelines.
Industry 4.0	AI integrates with IoT, blockchain, and digital twins to enable predictive maintenance and automated operations.	Supports autonomous factories with enhanced productivity and safety.

ADVANCED CHARACTERIZATION TECHNIQUES POWERED BY AI

AI has revolutionized microscopy and spectroscopy significantly and enhanced image capture, resolution and analysis. Microscopy algorithms based on AI are applied to focus the microscope automatically, enhance noise reduction and super resolution images to lift the image quality (Shin et al., 2021). Like the electron microscopy, ML methods accelerate the characterization of the materials by processing automatically the particles detection and classification (Sha et al., 2020). Spectroscopy AI can tremendously reduce the time spent on interpreting Raman and infrared spectra to determine the chemical structure in very high accuracy (Batra et al., 2020). The spectral unmixing is also improved with AI.

The analysis of a large-scale data encountered in materials and biological imaging is a challenging task which is dealt with through automated image analysis and feature extraction. AI can label high content screening (HCS) systems in a fast and efficient manner by classifying and segmenting images of cells (Kraus and Frey, 2016). Therefore, the accuracy of ML induced segmentation methods is as high as 95 percent when analysing microstructure and are applied to predicting the properties of porous and geological materials (Wijaya et al., 2024).

Moreover, AI accelerates the data analysis and enhances the degree of accuracy in materials strategies. Crystal system classification has been done through automated systems with convolutional neural networks with noisy conditions. SQL integrated ML implementations accelerate the process of database queries, facilitating experimental workflows (Islam, 2024). Today, models of statistical physics combine the modelling, imaging and experiment to discover data driven (Vasudevan et al., 2019).

Other important core characterization methods include SEM, XRD and Raman spectroscopy. SEM provides morphology analysis at high resolution (Akhtar et al., 2018), and crystal structures of XRD. Raman spectroscopy can be used to analyse graphene and other similar materials on a detailed basis (Hu et al., 2017). And it is these techniques that are frequently combined to demonstrate structural evolution within the process of reactions (Madadkhani et al., 2024). Table 4 is a summary of AI methods applied to materials characterization, where they have helped to improve images, spectral analysis and multimodal feature extraction.

Table 4. AI Techniques in Materials Characterization: Imaging, Spectral Analysis, and Feature Extraction

Technique	AI Capability	Benefit
Microscopy	AI enables autofocusing, defect recognition, and phase identification using CNNs and edge detection.	Reduces manual effort, increases reproducibility.
Spectroscopy	ML interprets Raman, FTIR, and XPS spectra to quantify compounds and analyze functional groups.	Accelerates chemical composition identification.
Image Segmentation & Feature Extraction	Automated microstructure analysis using deep learning to classify grains, pores, and defects in SEM/TEM images.	Enables high-throughput analysis of material phases.
Hybrid Modalities	Combines imaging and spectral data (e.g., SEM + Raman) using multimodal fusion models for enhanced material diagnostics.	Improves accuracy in complex material systems analysis.

CHALLENGES AND LIMITATIONS

Challenges related to data in materials engineering, such as sparsity, heterogeneity, and the absence of standards, hinder the successful use of AI. Sparseness of datasets and integration and analysis of heterogenous data limit model training: simulations to process logs. 99 percent of data sharing and reuse is impeded by a lack of standardised data formats across institutions (Al-Qaseemi et al. 2016). In these types of data, data efficiency in storage and interoperability are the most important aspect (Mateus et al., 2024). Following the example of such systems as Dataspace Management systems (DSMS) that implement the principles of FAIR to simplify the process of data linkage and exploration (Nahshon et al., 2024), this design will attempt to accomplish the objectives of consistent structured data.

Explainability of the model is another significant issue. Machine learning helps in design and material analysis through predictive insights, but black box models may reduce the level of trust to the users (Hu et al., 2024). Studies have shown that XAI interpretable model decisions, however, they also affect human decision-making in a mixed way (Alufaisan et al., 2021). SHAP tools provide clear results on the outputs of the models on material applications and build trust (Ponce-Bobadilla et al., 2024).

The hybrid modeling (combined physics and AI) is the new trend. It integrates first principles simulations with data-based learning to enhance the predictability as well as learn a functional that can be generalized in property forecasting or process optimization.

With regards to its use in critical areas, i.e., the issues of ethics also play a significant part. In both nanomedicine and gene vaccines, safety and regulatory compliance are essential due to the biological interactions (Souto et al., 2024). It is not only that the issue of food sustainability is subject to ethical challenges, (Varzakas and Antoniadou, 2024), but also the establishment of Responsible Leather (Navarro et al., 2020). They ought to bear moral responsibility towards the effects of the environment, just as the other sustainability policies and waste management policies.

CASE STUDIES AND REAL-WORLD APPLICATIONS

AI and ML have transformed materials engineering completely enabling it to be faster in discovering, processing and even making accurate predictions on material properties. The applications of AI in academic research have predicted polymorphism, crystal structure analysis, cocrystal property analysis, and lower academic research development costs and time (Heng et al., 2021). In the industrial setting, AI can be employed in predictive maintenance that allows minimizing the operational downtime of Industry 4.0 (Ahmed et al., 2022). Synergy between AI and 5G technology has also improved wireless communication reliability and management of resources (Rizvi, 2023). On the applications of AI in optimization of designs and preforming verification, Bappy (2024) expands the application of AI in the field of mechanical engineering, to enhance automation, performance and reliability of the system. However, even though the total count of Socialbots registered in the Socialbot Directory is rather large, there is still less experience related to applying AI technologies into the real world in the application domain of infectious disease management (Theodosiou and Read, 2023). This demonstrates the validity and domain specific measures.

In addition, energy, catalysis, and environmental application breakthroughs have also been made with a breakthrough in AI allowing things to become possible. AI accelerates the creation of solar cells, batteries and cycles of the hydrogen system; and materials in the general CO 2 capture. And these, though not all of them are AI, have also changed catalysis, including the ability of neural networks to predict the behavior of nanostructured catalysts to understand CO_2 reduction in the presence of bimetallic catalysts (Al-Akayleh et al., 2024). Krzywanski et al. (2024) have made it possible to integrate AI into real-world simulations and obtain greater accuracy and efficiency in the material modeling process. Then the "AI laddar" framework aids in cogitating this advance in simple anticipation models to divide hypothesis supported experimentation.

Without teamwork, it is not possible to create an innovation. Such disciplines as chemistry, physics, nanotechnology and computer science have been combined to endow material with biomaterials and smart materials with embedded intelligence and thus achieve materials engineering (Warr and West, 2020; Page et al., 2021).

One of the paths to sustained transforming results is the models such as Certus, which encourages the ongoing collaboration of interdisciplinary teams (Marijan and Gotlieb, 2020).

FUTURE DIRECTIONS AND RESEARCH OPPORTUNITIES

The nature of materials engineering is shifting as emerging technologies, such as quantum computing and edge AI, integrate into materials engineering. Among the most thrilling uses of quantum computing is the fact that it can solve problems of the form of quantum mechanics that is too hard to solve with classical computing. This potential may be utilized to make potential revolutionary capabilities of modelling materials and can predict new materials with predetermined properties. Meanwhile, edge AI allows decentralized computation by carried out computation at source, which would be useful in space-constrained manufacturing. In sensor networks, real time analytics can be utilized to make the utilization of dynamic optimization of process parameters dynamic and therefore, enhance the efficiency, quality control and predictive maintenance of the production lines. Incorporating the concept of IoT makes these systems smart and flexible in the form of intelligent and adaptive manufacturing processes. Most of this technological progress is due to the field of materials science itself: work in the quantum materials is still facilitating the production of hardware to run quantum computer; energy efficient materials are needed to optimize the behaviour of edge AI devices. Machine learning and quantum algorithms Quantum machine learning is the fusion of machine learning with quantum algorithms, accelerating the creation of materials in particular catalysts, battery material and smart manufacturing systems.

The need to improve materials engineering innovation is even greater with the help of open data repository and collaborative platform. These platforms make access and reuse of data, reproducibility of models and experimental outcomes and resource efficiency feasible as data can be shared in the form of datasets, computational models and experimental outcomes. The open repositories also reduce the redundant work, as well as enable the interdisciplinary research by means of the cross-domain data reuse. Despite the obstacles, such as the requirement of standardized metadata, extensive documentation, strict validation, the usage and application of it are more thoroughly justified than all the other economic information in existence. Though these barriers complicate the advancement of open data initiatives, nonetheless, the open data initiatives contribute to the development of the models, give the possibility to conduct resourceful simulation, and give room between the research and industry implementation.

Generative AI as a new type of materials design is engaged in the development of such a phenomenon. Synthetic materials that are related to an already formed latent image can be produced to give new composites, which have desirable characteristics. Nonetheless, these models have made the material processing easier and most proximity between structure performance and property. Through the support of transfer learning and graph neural networks, generative AI can now generalise to the large material systems. Even though increased design space of inorganic materials has become more challenging than organics, discoveries in structural representations have opened inorganic materials discoveries and next generation material design.

CONCLUSION

The application of AI and ML in advanced materials science has a transformative implication for research, discovery, and industrial application. This chapter has described how AI/ML methodologies – from supervised and unsupervised learning to reinforcement learning and deep generative models – are transforming the way material design, property prediction and process optimisation have been carried out. With the help of domain-specific tools like DeepChem, GT4SD, and Matminer, now researchers can accelerate the screening of materials, strengthen structure-property correlatives, and direct synthesis protocols with unprecedented accuracy. Moreover, the functionality of AI in smart manufacturing, real time feedback systems, and simulation-assisted optimization point to its versatility along the materials lifecycle. However, important challenges remain about standardization of data, generalizability of model and interpretability, and therefore there is need for strong hybrid frameworks that combine domain knowledge physics-based constraints. Interdisciplinarity, open data platforms, and explainable AI will prove essential to close the gap between theoretical models and experimental verification. With quantum computing, edge-AI, and generative models being on the growth curve, their intersection with materials science will unlock paths to intelligent adaptive and sustainable material ecosystem. In the end, this merger (AI, materials engineering) has the possibility to transform innovation in energy, healthcare, aerospace and environmental technologies.

REFERENCES

Ahmed, I., Jeon, G., & Piccialli, F. (2022). From Artificial Intelligence to Explainable Artificial Intelligence in Industry 4.0: A Survey on What, How, and Where. *IEEE Transactions on Industrial Informatics*, *18*(8), 5031–5042. DOI: 10.1109/TII.2022.3146552

Akhtar, K., Khan, S. B., Asiri, A. M., & Khan, S. A. (2018). Scanning Electron Microscopy: Principle and Applications in Nanomaterials Characterization (pp. 113–145). springer. DOI: 10.1007/978-3-319-92955-2_4

Al-Akayleh, F., Ali Agha, A. S. A., Al-Remawi, M., & Abdel Rahem, R. A. (2024). A mini review on the applications of artificial intelligence (AI) in surface chemistry and catalysis. *Tenside, Surfactants, Detergents*, *61*(4), 285–296. DOI: 10.1515/tsd-2024-2580

Al-Qaseemi, S. A., Almulhim, M. F., Chaudhry, S. R., & Almulhim, H. A. (2016, December 1). IoT architecture challenges and issues: Lack of standardization. DOI: 10.1109/FTC.2016.7821686

Ali, A., & Mashwani, W. (2023). A Supervised Machine Learning Algorithms: Applications, Challenges, and Recommendations. Proceedings of the Pakistan Academy of Sciences: A. Physical and Computational Sciences, 60(4). DOI: 10.53560/PPASA(60-4)831

Almansoori, A., Balázsi, K., & Balázsi, C. (2024). Advances, Challenges, and Applications of Graphene and Carbon Nanotube-Reinforced Engineering Ceramics. *Nanomaterials (Basel, Switzerland)*, *14*(23), 1881. Advance online publication. DOI: 10.3390/nano14231881 PMID: 39683269

Alnuaimi, A. F. A. H., & Albaldawi, T. H. K. (2024). An overview of machine learning classification techniques. BIO Web of Conferences, 97, 00133. DOI: 10.1051/bioconf/20249700133

Alufaisan, Y., Bakdash, J. Z., Zhou, Y., Marusich, L. R., & Kantarcioglu, M. (2021). Does Explainable Artificial Intelligence Improve Human Decision-Making? *Proceedings of the AAAI Conference on Artificial Intelligence*, *35*(8), 6618–6626. DOI: 10.1609/aaai.v35i8.16819

Aryanfar, Y., Wu, W., Alcaraz, J. L. G., Fernandez, J. B., Sadabad, A. N., & Keçebaş, A. (2024). Production of Biodiesel from Industrial Sludge: Recent Progress, Challenges, Perspective. *Processes (Basel, Switzerland)*, *12*(11), 2517. DOI: 10.3390/pr12112517

Badini, S., Regondi, S., & Pugliese, R. (2023). Unleashing the Power of Artificial Intelligence in Materials Design. *Materials (Basel), 16*(17), 5927. DOI: 10.3390/ma16175927 PMID: 37687620

Bai, J., Yang, Y., Wen, C., Chen, J., Zhou, G., Jiang, B., Peng, X., & Pan, F. (2023). Applications of magnesium alloys for aerospace: A review. *Journal of Magnesium and Alloys, 11*(10), 3609–3619. DOI: 10.1016/j.jma.2023.09.015

Bappy, M. A. (2024). Exploring the Integration of Informed Machine Learning in Engineering Applications: A Comprehensive Review. *American Journal of Science and Learning for Development, 3*(2), 11–21. DOI: 10.51699/ajsld.v3i2.3459

Barsce, J. C., Martinez, E. C., & Palombarini, J. A. (2017). Towards autonomous reinforcement learning: Automatic setting of hyper-parameters using Bayesian optimization. 36, 1–9. DOI: 10.1109/CLEI.2017.8226439

Batra, R., Song, L., & Ramprasad, R. (2020). Emerging materials intelligence ecosystems propelled by machine learning. *Nature Reviews. Materials, 6*(8), 655–678. DOI: 10.1038/s41578-020-00255-y

Brandić, I., Voća, N., Gunjača, J., Lončar, B., Bilandžija, N., Peter, A., Šurić, J., & Pezo, L. (2024). Biomass higher heating value prediction machine learning insights into ultimate, proximate, and structural analysis datasets. *Energy Sources. Part A, Recovery, Utilization, and Environmental Effects, 46*(1), 2842–2854. DOI: 10.1080/15567036.2024.2309303

Chan, Q.-H., Alias, S. A., Quek, S. W., Ng, C. Y., & Ku Marsilla, K. I. (2023). A review of the preparations, properties, and applications of smart biodegradable polymers. Polymer-Plastics Technology and Materials, ahead-of-print(ahead-of-print), 1273–1289. DOI: 10.1080/25740881.2023.2204954

Choy, L. H. T., & Ho, W. K. O. (2023). The Use of Machine Learning in Real Estate Research. *Land (Basel), 12*(4), 740. DOI: 10.3390/land12040740

Dhal, S. B., & Kar, D. (2025). Leveraging artificial intelligence and advanced food processing techniques for enhanced food safety, quality, and security: A comprehensive review. *Discover Applied Sciences, 7*(1), 75. Advance online publication. DOI: 10.1007/s42452-025-06472-w

Eckhardt, C. M., Ollivier, M., Madjarova, S. J., Williams, R. J., Pareek, A., Karlsson, J., & Nwachukwu, B. U. (2022). Unsupervised machine learning methods and emerging applications in healthcare. *Knee Surgery, Sports Traumatology, Arthroscopy : Official Journal of the ESSKA, 31*(2), 376–381. DOI: 10.1007/s00167-022-07233-7 PMID: 36378293

El Abbassi, M., Perrin, M. L., Calame, M., Braun, O., Overbeck, J., & van der Zant, H. S. J. (2021). Benchmark and application of unsupervised classification approaches for univariate data. *Communications on Physics*, *4*(1), 50. Advance online publication. DOI: 10.1038/s42005-021-00549-9

Epps, R. W., Volk, A. A., Reyes, K. G., & Abolhasani, M. (2021). Accelerated AI development for autonomous materials synthesis in flow. *Chemical Science (Cambridge)*, *12*(17), 6025–6036. DOI: 10.1039/D0SC06463G PMID: 34976336

Ferraz-Caetano, J., Teixeira, F., & Cordeiro, M. N. D. S. (2023). Explainable Supervised Machine Learning Model To Predict Solvation Gibbs Energy. *Journal of Chemical Information and Modeling*, *64*(7), 2250–2262. DOI: 10.1021/acs.jcim.3c00544 PMID: 37603608

Formalik, F., Shi, K., Snurr, R. Q., Wang, X., & Joodaki, F. (2023). Exploring the Structural, Dynamic, and Functional Properties of Metal-Organic Frameworks through Molecular Modeling. *Advanced Functional Materials*. Advance online publication. DOI: 10.1002/adfm.202308130

Geng, X., Wang, S., Mao, X., Wang, F., Wu, G., Zhao, H., Zhang, C., Gao, J., & Wu, H. (2023). Data-driven and artificial intelligence accelerated steel material research and intelligent manufacturing technology. *Materials Genome Engineering Advances*, *1*(1), e10. Advance online publication. DOI: 10.1002/mgea.10

Harren, T., Rarey, M., Hessler, G., Grebner, C., & Gutermuth, T. (2024). Modern machine-learning for binding affinity estimation of protein–ligand complexes: Progress, opportunities, and challenges. *Wiley Interdisciplinary Reviews. Computational Molecular Science*, *14*(3), e1716. Advance online publication. DOI: 10.1002/wcms.1716

Heidari, M., Zad, S., & Rafatirad, S. (2021)... *Ensemble of Supervised and Unsupervised Learning Models to Predict a Profitable Business Decision.*, *7*, 1–6. DOI: 10.1109/IEMTRONICS52119.2021.9422649

Heng, T., Zhang, L., Wang, R., Lu, Y., Yang, D., & Du, G. (2021). Progress in Research on Artificial Intelligence Applied to Polymorphism and Cocrystal Prediction. *ACS Omega*, *6*(24), 15543–15550. DOI: 10.1021/acsomega.1c01330 PMID: 34179597

Ho, W. K. O., Tang, B.-S., & Wong, S. W. (2020). Predicting property prices with machine learning algorithms. *Journal of Property Research*, *38*(1), 48–70. DOI: 10.1080/09599916.2020.1832558

Hu, X., Dai, Y., & Liu, A. (2024). Combining ChatGPT and knowledge graph for explainable machine learning-driven design: a case study. Journal of Engineering Design, ahead-of-print(ahead-of-print), 1–23. DOI: 10.1080/09544828.2024.2355758

Islam, S. (2024). FUTURE TRENDS IN SQL DATABASES AND BIG DATA ANALYTICS: IMPACT OF MACHINE LEARNING AND ARTIFICIAL INTELLIGENCE. *International Journal of Science and Engineering, 1*(4), 47–62. DOI: 10.62304/ijse.v1i04.188

Kadupitiya, J., Fox, G. C., & Jadhao, V. (2020). Machine learning for parameter auto-tuning in molecular dynamics simulations: Efficient dynamics of ions near polarizable nanoparticles. *International Journal of High Performance Computing Applications, 34*(3), 357–374. DOI: 10.1177/1094342019899457

Kim, H., Lee, K., Kim, C., Lim, J., & Kim, W. Y. (2023). DFRscore: Deep Learning-Based Scoring of Synthetic Complexity with Drug-Focused Retrosynthetic Analysis for High-Throughput Virtual Screening. *Journal of Chemical Information and Modeling, 64*(7), 2432–2444. DOI: 10.1021/acs.jcim.3c01134 PMID: 37651152

Kraus, O. Z., & Frey, B. J. (2016). Computer vision for high content screening. *Critical Reviews in Biochemistry and Molecular Biology, 51*(2), 102–109. DOI: 10.3109/10409238.2015.1135868 PMID: 26806341

Krzywanski, J., Sosnowski, M., Grabowska, K., Zylka, A., Lasek, L., & Kijo-Kleczkowska, A. (2024). Advanced Computational Methods for Modeling, Prediction and Optimization-A Review. *Materials (Basel), 17*(14), 3521. DOI: 10.3390/ma17143521 PMID: 39063813

Lau, M. L., Terry, J., Long, M., & Burleigh, A. (2023). Materials characterization: Can artificial intelligence be used to address reproducibility challenges? *Journal of Vacuum Science & Technology. A, Vacuum, Surfaces, and Films, 41*(6), 060801. Advance online publication. DOI: 10.1116/6.0002809

Liu, H., Liu, H., Liu, H., Chen, P., Cheng, M., Wang, H., Wang, X., Wang, H., Hu, B., Wang, X., Cheng, M., Wang, S., Chen, P., Chen, P., Wang, J., Wang, J., Wang, H., Lin, B., Wang, S., & Hu, B. (2024). Docking Score ML: Target-Specific Machine Learning Models Improving Docking-Based Virtual Screening in 155 Targets. *Journal of Chemical Information and Modeling, 64*(14), 5413–5426. DOI: 10.1021/acs.jcim.4c00072 PMID: 38958413

Liu, Y., Liu, D., Avdeev, M., Shi, S., Zou, X., Ma, S., & Yang, Z. (2023). Data quantity governance for machine learning in materials science. *National Science Review, 10*(7), nwad125. Advance online publication. DOI: 10.1093/nsr/nwad125 PMID: 37323811

Lu, D., Gangrade, S., Kao, S.-C., Konapala, G., & Painter, S. L. (2021). Streamflow simulation in data-scarce basins using Bayesian and physics-informed machine learning models. *Journal of Hydrometeorology, 1*(6). Advance online publication. DOI: 10.1175/JHM-D-20-0082.1

Ma, S., Wang, S., Du, Y., & Su, Y. (2023). Application of machine learning in material corrosion research. *Corrosion Reviews, 41*(4), 417–426. DOI: 10.1515/corrrev-2022-0089

Madadkhani, S., Nandy, S., Chae, K. H., Aleshkevych, P., & Najafpour, M. M. (2024). Advances in Understanding Tungsten Disulfide Dynamics during the Hydrogen-Evolution Reaction: An Initial Step in Elucidating the Mechanism. *The Journal of Physical Chemistry Letters, 15*(19), 5112–5119. DOI: 10.1021/acs.jpclett.4c00924 PMID: 38709179

Manica, M., Das, P., Wehden, K., Hoffman, S. C., Padhi, I., Mchugh, L., Chenthamarakshan, V., Born, J., Christofidellis, D., Dave, A., Cadow, J., Teukam, Y., Schilter, O., Kishimoto, A., Donovan, T., Khrabrov, A., Hamada, L., Zipoli, F., Giannone, G., & Smith, J. (2023). Accelerating material design with the generative toolkit for scientific discovery. *npj Computational Materials, 9*(1), 69. Advance online publication. DOI: 10.1038/s41524-023-01028-1

Marijan, D., & Gotlieb, A. (2020). Industry-Academia research collaboration in software engineering: The Certus model. *Information and Software Technology, 132*, 106473. DOI: 10.1016/j.infsof.2020.106473

Mateus, P., Moonen, J., Beran, M., Bron, E., Jansen, W., Heuvelink, J., Van Der Landen, S. M., Birhanu, M., Dekker, A., Oomens, J., Cats, D., Harms, A. G. J., Wolters, F. J., Bermejo, I., Mei, H., Schram, M. T., & Jaarsma, E. (2024). Data harmonization and federated learning for multi-cohort dementia research using the OMOP common data model: A Netherlands consortium of dementia cohorts case study. *Journal of Biomedical Informatics, 155*, 104661. DOI: 10.1016/j.jbi.2024.104661 PMID: 38806105

Mathotaarachchi, K. V., Mahmood, S., & Hasan, R. (2024). Advanced Machine Learning Techniques for Predictive Modeling of Property Prices. *Information (Basel), 15*(6), 295. DOI: 10.3390/info15060295

Mok, D. H., & Back, S. (2024). Generative Pretrained Transformer for Heterogeneous Catalysts. *Journal of the American Chemical Society, 146*(49), 33712–33722. DOI: 10.1021/jacs.4c11504 PMID: 39576215

Mondal, S., & Shubhra Goswami, S. (2024). Machine learning techniques for quality assurance in additive manufacturing processes. *International Journal of AI for Materials and Design*, *0*(0), 3455. DOI: 10.36922/ijamd.3455

Mounsaveng, S., Vazquez, D., Laradji, I., Ayed, I. B., & Pedersoli, M. (2021).. . *Learning Data Augmentation with Online Bilevel Optimization for Image Classification.*, *13*, 1690–1699. DOI: 10.1109/WACV48630.2021.00173

Muhammad, A., Ishaq, A. A., B, I. O., & Idris, M. B. (2023). Artificial Intelligence and Machine Learning for Real-time Energy Demand Response and Load Management. *Journal of Technology Innovations and Energy*, *2*(2), 20–29. DOI: 10.56556/jtie.v2i2.537

Nahshon, Y., Kumaraswamy, K., Zierep, P., Weber, M., Helm, D., Morand, L., De Andrés, P., & Büschelberger, M. (2024). Semantic Orchestration and Exploitation of Material Data: A Dataspace Solution Demonstrated on Steel and Copper Applications. *Advanced Engineering Materials*. Advance online publication. DOI: 10.1002/adem.202401448

Nakka, R., Harursampath, D., & Ponnusami, S. A. (2023). A generalised deep learning-based surrogate model for homogenisation utilising material property encoding and physics-based bounds. *Scientific Reports*, *13*(1), 9079. Advance online publication. DOI: 10.1038/s41598-023-34823-3 PMID: 37277405

Navarro, D., Fullana-I-Palmer, P., Lin, W., Wu, J., & Puig, R. (2020). Life cycle assessment and leather production. *Journal of Leather Science and Engineering*, *2*(1), 26. Advance online publication. DOI: 10.1186/s42825-020-00035-y

Ninduwezuor-Ehiobu, N., Tula, O., Daraojimba, C., Ogunjobi, O., Egbokhaebho, B., Ofonagoro, K., Gidiagba, J., & Banso, A. (2023). Tracing The Evolution Of Ai And Machine Learning Applications In Advancing Materials Discovery And Production Processes. *Engineering Science & Technology Journal*, *4*(3), 66–83. DOI: 10.51594/estj.v4i3.552

Onyiriuka, E. (2023). Predictive modelling of thermal conductivity in single-material nanofluids: A novel approach. *Bulletin of the National Research Center*, *47*(1), 140. Advance online publication. DOI: 10.1186/s42269-023-01115-9

Page, S. B., Bryson, J. M., Stone, M. M., Crosby, B. C., & Seo, D. (2021). Ambidexterity in Cross-Sector Collaborations Involving Public Organizations. *Public Performance & Management Review*, *44*(6), 1161–1190. DOI: 10.1080/15309576.2021.1937243

Park, D., Park, K., Ryu, S., & Lee, J. (2023). Hierarchical Generative Network: A Hierarchical Multitask Learning Approach for Accelerated Composite Material Design and Discovery. *Advanced Engineering Materials*, *25*(21), 2300867. Advance online publication. DOI: 10.1002/adem.202300867

Ponce-Bobadilla, A. V., Schmitt, V., Maier, C. S., Mensing, S., & Stodtmann, S. (2024). Practical guide to SHAP analysis: Explaining supervised machine learning model predictions in drug development. *Clinical and Translational Science*, *17*(11), e70056. Advance online publication. DOI: 10.1111/cts.70056 PMID: 39463176

Rizvi, S. (2023). Unveiling the Potential of Artificial Intelligence and Machine Learning in the 5G Network Landscape: A Comprehensive Review. *Asian Journal of Research in Computer Science*, *16*(4), 23–31. DOI: 10.9734/ajrcos/2023/v16i4367

Roggia, M., Natale, B., Amendola, G., Di Maro, S., & Cosconati, S. (2023). Streamlining Large Chemical Library Docking with Artificial Intelligence: The PyRMD2Dock Approach. *Journal of Chemical Information and Modeling*, *64*(7), 2143–2149. DOI: 10.1021/acs.jcim.3c00647 PMID: 37552222

Sani, A. R., Kouzani, A. Z., & Zolfagharian, A. (2024). Artificial Intelligence-Augmented Additive Manufacturing: Insights on Closed-Loop 3D Printing. *Advanced Intelligent Systems*, *6*(10), 2400102. Advance online publication. DOI: 10.1002/aisy.202400102

Schaarschmidt, J., Kahle, L., Wenzel, W., Hickel, T., Yuan, J., Bölle, F. T., Kondov, I., Huber, S. P., Castelli, I. E., Hanke, F., Neugebauer, J., Vegge, T., Pizzi, G., Rêgo, C. R. C., & Strunk, T. (2021). Workflow Engineering in Materials Design within the BATTERY 2030+ Project. *Advanced Energy Materials*, *12*(17), 2102638. DOI: 10.1002/aenm.202102638

Sengottaiyan, S., Shaisundaram, V. S., Shameer Basha, M., Deepanraj, B., & Senthilkumar, N. (2025). Effect of Duplex and Austenite Stainless Steel Fillers on Mechanical, Microstructural, and Corrosion Characteristics of Dissimilar Welds Using Cold Metal Transfer Welding. *Results in Engineering*, *26*, 104818. DOI: 10.1016/j.rineng.2025.104818

Senthilkumar, N., Deepanraj, B., Shaik, F., & Elsehly, E. M. "Artificial Intelligence Based Investigation on Drilling of MWCNT-Seashell-Nylon66 Hybrid Polymer Composites by Fuzzy Inference System," 2024 3rd International Conference for Innovation in Technology (INOCON), Bangalore, India, 2024, pp. 1-6, DOI: 10.1109/INOCON60754.2024.10511569

Senthilkumar, N., Ramu, S., & Deepanraj, B. (2024). Heat Conduction through Natural Fiber/Carbon Nanotubes Filler Matrix Polymer Composite Slabs: An Experimental and Analytical Comparisons (No. 2023-01-5107). SAE Technical Paper. DOI: 10.4271/2023-01-5107

Sha, W., Lu, S., Cao, Y.-C., Yuan, Q., Zhang, X., Guo, X., Guo, Y., Cheng, S., & Tang, S. (2020). Artificial Intelligence to Power the Future of Materials Science and Engineering. *Advanced Intelligent Systems*, *2*(4), 1900143. DOI: 10.1002/aisy.201900143

Shen, C., Jiang, S. B., Chen, L., Jia, X., & Gonzalez, Y. (2018). Intelligent Parameter Tuning in Optimization-Based Iterative CT Reconstruction via Deep Reinforcement Learning. *IEEE Transactions on Medical Imaging*, *37*(6), 1430–1439. DOI: 10.1109/TMI.2018.2823679 PMID: 29870371

Shen, C., Shen, C., Song, J., Hsieh, C.-Y., Hsieh, C.-Y., Cao, D., Kang, Y., Ye, W., Wu, Z., Wang, J., Zhang, O., Zhang, X., Zeng, H., Cai, H., Chen, Y., Chen, L., Luo, H., Zhao, X., Jian, T., & Hou, T. (2024). DrugFlow: An AI-Driven One-Stop Platform for Innovative Drug Discovery. *Journal of Chemical Information and Modeling*, *64*(14), 5381–5391. DOI: 10.1021/acs.jcim.4c00621 PMID: 38920405

Shen, T., Dong, H., & Wang, P. (2024). Research progress on nanocellulose and its composite materials as orthopedic implant biomaterials. *Alexandria Engineering Journal*, *87*, 575–590. DOI: 10.1016/j.aej.2024.01.003

Shin, Y., Kim, S., & Lee, Y. H. (2021). AI musculoskeletal clinical applications: How can AI increase my day-to-day efficiency? *Skeletal Radiology*, *51*(2), 293–304. DOI: 10.1007/s00256-021-03876-8 PMID: 34341865

Soori, M., Dastres, R., & Arezoo, B. (2023). AI-powered blockchain technology in industry 4.0, a review. *Journal of Economy and Technology*, *1*, 222–241. DOI: 10.1016/j.ject.2024.01.001

Souto, E. B., Blanco-Llamero, C., Blanco-Llamero, C., Krambeck, K., Kiran, N. S., Yashaswini, C., Postwala, H., Severino, P., Severino, P., Priefer, R., Prajapati, B. G., & Maheshwari, R. (2024). Regulatory insights into nanomedicine and gene vaccine innovation: Safety assessment, challenges, and regulatory perspectives. *Acta Biomaterialia*, *180*, 1–17. DOI: 10.1016/j.actbio.2024.04.010 PMID: 38604468

Su, J., Chen, L., Tan, C., Okulov, I., Zhu, H., Jiang, F., Zhang, L., Requena, G., Teng, J., Yan, M., & Wang, Y. M. (2024). Recent innovations in laser additive manufacturing of titanium alloys. *International Journal of Extreme Manufacturing*, *6*(3), 032001. DOI: 10.1088/2631-7990/ad2545

Theodosiou, A. A., & Read, R. C. (2023). Artificial intelligence, machine learning and deep learning: Potential resources for the infection clinician. *The Journal of Infection*, *87*(4), 287–294. DOI: 10.1016/j.jinf.2023.07.006 PMID: 37468046

Tirumanadham, N. S. K. M. K., Thaiyalnayaki, S., & Ganesan, V. (2025). Towards Smarter E-Learning: Real-Time Analytics and Machine Learning for Personalized Education. *International Journal of Computational and Experimental Science and Engineering*, *11*(1). Advance online publication. DOI: 10.22399/ijcesen.786

Varzakas, T., & Antoniadou, M. (2024). A Holistic Approach for Ethics and Sustainability in the Food Chain: The Gateway to Oral and Systemic Health. *Foods*, *13*(8), 1224. DOI: 10.3390/foods13081224 PMID: 38672896

Vasudevan, R. K., Smith, R., Mehta, A., Hattrick-Simpers, J., Vlcek, L., Kusne, G., Ziatdinov, M., Tavazza, F., Choudhary, K., & Kalinin, S. V. (2019). Materials Science in the AI age: High-throughput library generation, machine learning and a pathway from correlations to the underpinning physics. *MRS Communications*, *9*(3), 821–838. DOI: 10.1557/mrc.2019.95 PMID: 32166045

Wang, W. Y., Lu, J., Li, G., Wang, X., Su, Y., Li, J., Zhang, S., Ren, Y., Gao, X., & Song, H. (2024). Artificial intelligence enabled smart design and manufacturing of advanced materials: The endless Frontier in AI$^+$ era. *Materials Genome Engineering Advances*, *2*(3), e56. Advance online publication. DOI: 10.1002/mgea.56

Wang, Y., Zhang, Q., Wang, G.-G., & Cheng, H. (2024). The application of evolutionary computation in generative adversarial networks (GANs): A systematic literature survey. *Artificial Intelligence Review*, *57*(7), 182. Advance online publication. DOI: 10.1007/s10462-024-10818-y

Warr, M., & West, R. E. (2020). Bridging Academic Disciplines with Interdisciplinary Project-based Learning. *The Interdisciplinary Journal of Problem-Based Learning*, *14*(1). Advance online publication. DOI: 10.14434/ijpbl.v14i1.28590

Wijaya, A., Wagner, J., Sartory, B., & Brunner, R. (2024). Analyzing microstructure relationships in porous copper using a multi-method machine learning-based approach. *Communications Materials*, *5*(1), 59. Advance online publication. DOI: 10.1038/s43246-024-00493-5

Xu, C., Jiang, X., & Coen-Pirani, P. (2023). Empirical Study of Overfitting in Deep Learning for Predicting Breast Cancer Metastasis. *Cancers (Basel)*, *15*(7), 1969. DOI: 10.3390/cancers15071969 PMID: 37046630

Yadav, K., Bidnyk, S., & Balakrishnan, A. (2024). Artificial intelligence and machine learning in optics: Tutorial. *Journal of the Optical Society of America. B, Optical Physics*, *41*(8), 1739. DOI: 10.1364/JOSAB.525182

Yan, Y., Zhang, N., Wang, L., Qin, X., Fan, J., & Wu, Y. (2015). A restricted Boltzmann machine based two-lead electrocardiography classification. 1–9. DOI: 10.1109/BSN.2015.7299399

Younis, M., Ahmad, S., Amjad Farooq, M., Atiq, A., Huang, M., & Abbas, M. (2023). Recent Progress in Azobenzene-Based Supramolecular Materials and Applications. *Chemical Record (New York, N.Y.)*, *23*(11), e202300126. Advance online publication. DOI: 10.1002/tcr.202300126 PMID: 37435961

Zhu, L., Han, Y., Li, Q., Luo, Q., Zhang, Y., Zhang, L., Hu, B., & Chen, Q. (2024). Prediction of ultimate tensile strength of Al-Si alloys based on multimodal fusion learning. *Materials Genome Engineering Advances*, *2*(1), e26. Advance online publication. DOI: 10.1002/mgea.26

Chapter 2
Beyond Algorithms:
Real-World Constraints in Energy Materials Modeling and Design

Mona Albalawi
https://orcid.org/0009-0001-4594-8570
University of Tabuk, Saudi Arabia

ABSTRACT

The world stands at a critical energy crossroads. While renewable energy capacity grows exponentially, our ability to store, convert, and utilize this energy efficiently still depends on breakthroughs in materials science. This chapter isn't just about promising lab results – it's about real-world solutions that must balance scientific innovation with engineering pragmatism and industrial realities. Furthermore, the world is racing toward an energy revolution, but the path isn't as simple as swapping fossil fuels for solar panels and batteries. Behind every headline about a "breakthrough" material, there's a messy, often untold story of trial, error, and real-world constraints. This chapter isn't just about the science of energy materials—it's about the people behind them, the frustrations they face, and the hard-won lessons that separate lab-scale promise from industrial reality.

BEYOND ALGORITHMS: REAL-WORLD CONSTRAINTS IN ENERGY MATERIALS MODELING AND DESIGN

The realization of a clean energy future hinges on the successful translation of advanced energy materials from laboratory discovery to industrial-scale application. While research labs frequently yield promising breakthroughs, the scaling challenges inherent in transitioning these innovations to tangible products remain a significant impediment [Figure 1]. This energy materials dilemma underscores the

DOI: 10.4018/979-8-3373-3556-8.ch002

critical divergence between lab-scale feasibility and the complexities of industrial production (Ozin & Loh, 2022).

Figure 1. Bioreactor Scale Up Process

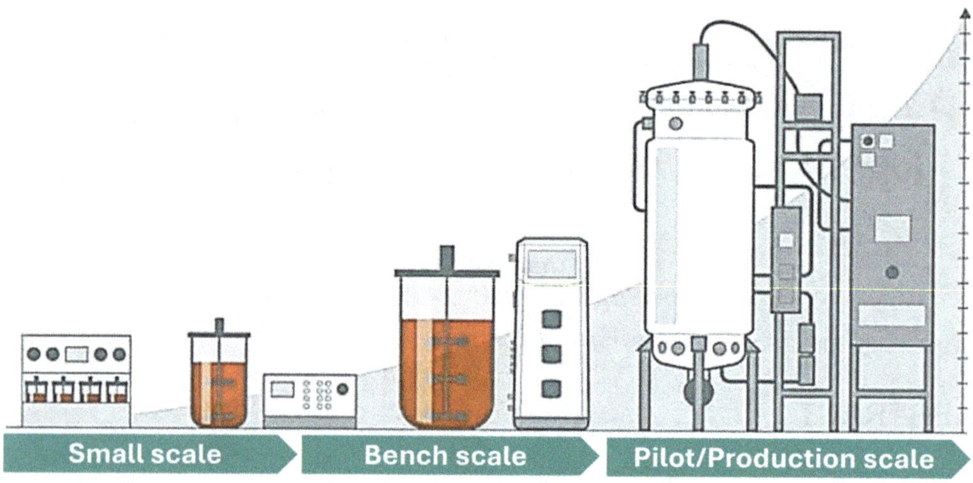

Synthesizing novel energy materials often involves intricate, multi-step processes that are difficult to replicate at industrial volumes. Longevity observed in accelerated aging tests may not translate to real-world performance under unpredictable stresses. Based on the available evidence, the environmental benefits of "green" materials can be negated by hidden costs within their production chains. A particularly significant hurdle involves the scaling of nanomaterials and thin films, where properties observed at gram-scale often deviate or fail entirely at industrial batch sizes (Mortazavi, 2025).

The commercialization landscape is replete with examples of ventures faltering due to these scaling issues. The high cost and scarcity of platinum-group metal catalysts exemplify this, hindering widespread adoption despite their superior performance. While earth-abundant alternatives like manganese, iron, cobalt, and nickel offer potential solutions, they often suffer from issues such as degradation or poor conductivity, necessitating complex strategies like carbon support or single-atom architecture, which themselves present scaling challenges (Mitchell et al., 2021; Ozin & Loh, 2022).

Polymers for solid-state electrolytes also illustrate the scaling paradox. Balancing ionic conductivity with mechanical strength requires meticulous optimization, and scaling production often involves environmentally problematic solvents and achieving uniform thin films in continuous manufacturing processes (Polizos et al., 2024). Similarly, nanomaterials like Metal–Organic Frameworks [MOFs] face challenges related to high production costs and sensitivity to environmental conditions (Liu et al., 2022; Mitchell et al., 2021; Mortazavi, 2025; Polizos et al., 2024).

Even with successful synthesis and scale-up, long-term performance in real-world conditions often deviates from accelerated testing results, highlighting the need for multi-stress aging protocols. Finally, current research suggests the pursuit of sustainable energy materials necessitates a comprehensive life-cycle analysis to ensure genuine environmental benefits.

Overcoming the scaling challenges in energy materials requires a holistic approach that integrates innovative scientific discovery with a deep understanding of engineering, economics, and environmental impact. Learning from past experiences and embracing comprehensive design and production strategies are crucial for realizing a future powered by sustainable energy solutions (Jafarizadeh et al., 2024).

LAB-TO-INDUSTRY TRANSITION ENERGY MATERIALS; CROSSING THE VALLEY OF DEATH

The "Valley of Death" represents the critical bottleneck between laboratory breakthroughs and industrial deployment in energy materials, where fewer than 5% of promising innovations achieve commercial viability beyond pilot-scale studies. This phenomenon encompasses the complex intersection of technical, economic, and engineering challenges that prevent advanced catalysts, solid-state electrolytes, and nanostructured systems from transitioning successfully to market applications (Fry & Cox, 2020).

Understanding the Valley of Death

The Valley of Death concept describes the gap between Technology Readiness Levels [TRL] 2-3 and TRL 4-5, where public research funding ends but private investment for commercial development has not yet materialized. For energy materials, this transition is particularly treacherous due to stringent real-world demands that innovations like high-entropy alloys and perovskite photovoltaics must satisfy. The commercialization challenge is fundamentally different from software technologies, as energy materials require substantial capital investment for scaling with marginal cost reductions that may not justify the investment (Fry & Cox, 2020).

Core Technical Barriers

Catalyst Degradation and Stability:

Heterogeneous catalysts face severe challenges when transitioning from laboratories to industrial environments. Surface reconstruction and poisoning by trace impurities present in industrial feedstocks represent critical issues rarely captured in idealized laboratory settings. These catalysts must maintain performance under harsh industrial conditions while handling complex feedstock compositions that differ significantly from pure laboratory reagents (García-Salaberri, 2024).

Solid-State Electrolyte Manufacturing:

The commercialization of solid-state electrolytes confronts three primary engineering challenges: precursors, processing, and pressure. Currently, solid electrolyte costs are two orders of magnitude higher than liquid alternatives, primarily due to immature supply chains and lack of scalable synthesis understanding. The persistent trade-off between mechanical properties and ionic conductivity becomes more pronounced under industrial processing conditions, where solvent evaporation kinetics and shear forces introduce additional complexity (García-Salaberri, 2024).

Nanostructured Material Scaling:

Despite their advantages in surface area and electrical conductivity, nanostructured materials suffer from agglomeration during processing and hidden costs of seemingly simple modifications. The challenge lies in maintaining uniformity and surface sensitivity while achieving cost-effective large-scale production. Traditional synthesis methods often fail to translate from laboratory milligram quantities to industrial kilogram requirements without significant performance degradation (Fry & Cox, 2020; García-Salaberri, 2024).

Economic and Process Engineering Challenges

The economic barriers extend beyond material costs to encompass entire process economics. Cost-performance imbalances frequently emerge where marginal performance gains require disproportionate investment. For high-entropy alloys, concerns about ease of manufacture, cost, and recyclability often outweigh potential advantages over conventional materials. The energy transition valley of death is characterized by largely opaque costs that are not immediately apparent during early-stage development.

Traditional accelerated aging tests fail to predict real-world degradation mechanisms, neglecting factors like calendar aging and microcurrent-induced phenomena that become critical during extended operation1. Understanding interfacial chemistry evolution under actual operating conditions represents a paramount challenge for long-term material performance prediction (Fry & Cox, 2020; García-Salaberri, 2024).

Scalable Synthesis Solutions

Flame Spray Pyrolysis:

This technique enables robust, scalable synthesis of nanomaterials through continuous-flow gas-phase processes. Production rates can be scaled from 100-500 g/h while maintaining consistent particle properties when high-temperature particle residence time is kept constant. The process allows controlled production with minimal environmental impact compared to traditional wet chemistry approaches (Kang & Wolverton, 2022).

Mechanochemical Synthesis:

Operating under environmental conditions without solvents or toxic components, mechanochemical synthesis offers very low process costs and high reaction yields typically above 90-95%. This solid-state reaction approach has been successfully applied to chalcogenides and nitrides for photovoltaic applications, enabling synthesis of materials with incompatible physical characteristics (Kang & Wolverton, 2022).

Continuous-Flow Synthesis:

The development of active heterogeneous catalysts for continuous flow systems has achieved higher turnover numbers compared to batch systems. Flow chemistry provides several advantages including improved mixing, short diffusion paths, and elimination of intermediate isolation requirements. This approach is particularly valuable for pharmaceutical and fine chemical production where consistent quality and reduced waste are critical (García-Salaberri, 2024).

Advanced Characterization and Predictive Approaches

Physics-Informed Machine Learning:

This emerging approach incorporates governing physical principles into machine learning algorithms, producing more physically realistic predictions while significantly reducing training data requirements. For energy materials, this methodology enables prediction of long-term behavior under various operating conditions by encoding fundamental physical laws into the learning process (Kang & Wolverton, 2022).

Operando Characterization:

Real-time monitoring of material behavior during operation provides critical insights into degradation mechanisms that cannot be captured through traditional pre- and post-mortem analyses. These techniques enable direct correlation between microstructure, chemical composition, and performance, offering unprecedented understanding of dynamic material behavior under actual operating conditions (Kang & Wolverton, 2022).

Pathways to Successful Translation

Bridging the Valley of Death requires early industrial collaboration to establish realistic performance targets and market alignment. Degradation-aware design principles must incorporate real-world stressors into early-stage testing protocols, moving beyond idealized laboratory conditions to simulate complex operational environments (Janasik et al., 2023).

Manufacturing integration from inception ensures that synthesis routes are inherently scalable and compatible with existing industrial infrastructure. Advanced validation protocols combining operando characterization with predictive modeling enable more accurate assessment of commercial viability.

Life cycle assessment becomes essential for developing truly sustainable energy materials, considering upstream energy costs, end-of-life recyclability, and solvent selection implications. Comprehensive sustainability evaluation prevents superficial environmental claims and guides material design toward genuinely sustainable solutions (Janasik et al., 2023).

Future Outlook

Successfully navigating the energy materials Valley of Death demands paradigm shifts in research approach, funding mechanisms, and industry-academia collaboration. Materials scientists must embrace process engineering considerations from the earliest stages of development, prioritizing scalable synthesis routes over singular performance metrics. The integration of advanced characterization techniques with predictive modeling approaches offers promising pathways for accelerating the translation of laboratory discoveries into transformative commercial technologies. Only through this comprehensive, interdisciplinary approach can the energy materials community deliver the advanced materials necessary for sustainable energy applications while overcoming the persistent challenges of the Valley of Death.

INDUSTRIAL CONSTRAINTS CATALYST COMMERCIALIZATION

The translation of innovative catalyst discoveries from the laboratory to impactful industrial application in energy materials science encounters significant obstacles. While advancements such as single-atom catalysts, non-precious metal alternatives, and bio-inspired catalytic systems demonstrate remarkable potential, their deployment in industrial chemical processes presents considerable challenges (Cong & Zhang, 2022; Janasik et al., 2023).

The attributes that define a promising catalyst in a controlled laboratory setting often diverge from the critical parameters for industrial viability. Beyond activity and selectivity, commercial adoption necessitates consideration of scalable and consistent synthesis, tolerance to impurities prevalent in industrial feedstocks, robustness under dynamic plant operating conditions, economically feasible production costs, and adherence to environmental sustainability regulations regarding usage and disposal (Cong & Zhang, 2022; Janasik et al., 2023).

This analysis examines these practical limitations through the perspectives of materials chemistry, reaction engineering, and relevant case studies. The objective is to advocate for a paradigm shift in early-stage catalyst design, integrating industrial realities as fundamental considerations.

SCALE-UP LIMITATIONS: BRIDGING THE GAP BETWEEN LABORATORY SYNTHESIS AND INDUSTRIAL PRODUCTION

Many novel catalysts with intricate architecture or specialized compositions rely on synthesis methodologies that are inherently challenging to translate to continuous, large-scale manufacturing. Examples include time-intensive, temperature-programmed synthesis within confined spaces and ALD, which exhibits limited throughput. Promising alternatives include continuous flow microwave synthesis, which enhances reaction rates and product uniformity [e.g., in cobalt-manganese oxide catalysts for air pollution control], and solvent-free mechanochemical synthesis, successfully scaled for molybdenum sulfide hydrodesulfurization catalysts (Cong & Zhang, 2022).

Furthermore, high-surface-area support materials crucial for catalyst dispersion, such as alumina and carbon nanotubes, can exhibit instability under industrial operating stresses. Carbon supports can undergo corrosion in fuel cells, leading to the degradation of platinum catalysts, while alumina can experience structural transformations at elevated temperatures in automotive exhaust catalysts (Khandan et al., 2024). The literature seems to converge on, emerging solutions involve the development of conductive ceramic supports with enhanced stability and the encapsulation of active nanoparticles [e.g., nickel in graphene for methane reforming] to prevent agglomeration and extend catalyst lifespan (Zhang, Xie, Tian et al, 2021).

Impact of Industrial Feedstock Impurities and Dynamic Operating Conditions

Industrial feedstocks invariably contain trace contaminants that can severely impede catalyst performance. Sulfur compounds can strongly chemisorb onto active sites of numerous catalysts, including nickel, cobalt, and precious metals, leading to deactivation. Early-stage experiments of Siloxanes present in biogas can irreversibly poison palladium hydrogenation catalysts. Mitigation strategies include the development of bifunctional materials capable of contaminant trapping and release [e.g., ceria-zirconia for sulfur] and the implementation of guard beds utilizing sacrificial materials like zinc oxide to selectively remove impurities upstream of the primary catalyst (De Blasio et al., 2021).

Industrial reactors also operate under dynamic conditions characterized by frequent startups, shutdowns, and fluctuations in process parameters. These variations can induce catalyst degradation, such as sintering of copper-zinc-alumina methanol synthesis catalysts due to thermal excursions during plant cycling, and necessitate catalysts capable of accommodating transient feed compositions, as exemplified by CO_2 methanation catalysts requiring resilience to intermittent hydrogen supply from renewable sources. Materials design strategies to address these challenges include core-shell catalyst architectures that provide thermal protection to active components and the incorporation of promoters like cerium and manganese that can buffer changes in reactant concentrations through oxygen storage and release mechanisms (Cong & Zhang, 2022; Giglio et al., 2021).

Economic and Environmental Considerations in Catalyst Commercialization

The economic viability of industrial catalysts is significantly influenced by the cost of constituent materials, particularly precious metals. Consequently, the efficient recovery of valuable components from spent catalysts is paramount. Automotive catalysts are increasingly designed with palladium-rhodium alloys to facilitate recycling, and alloying platinum with refractory metals like tantalum can minimize material losses during the recovery process (Van Allsburg et al., 2022).

Environmental regulations are also driving innovation in catalyst synthesis methodologies. The phasing out of traditional organic solvents necessitates the development of sustainable alternatives, such as water-based synthesis routes, which have been successfully employed for the production of platinum-nickel octahedra. Supercritical carbon dioxide is also emerging as a solvent-free medium for nanoparticle synthesis, with commercial applications in fuel cell catalyst manufacturing (Osman et al., 2024).

Accelerated and Realistic Catalyst Evaluation Methodologies

Conventional laboratory-scale catalyst testing often fails to accurately predict performance and durability under real-world industrial conditions. These tests may not capture phenomena such as chemical condensation leading to catalyst poisoning during cold starts or physical attrition due to mechanical forces within industrial reactors. Advanced evaluation techniques are therefore crucial, including multi-parameter testing rigs capable of simulating simultaneous vibrational and thermal stresses, and *in situ* characterization methods that provide real-time insights into catalyst behavior under realistic operating conditions. Furthermore, data-driven approaches, such as training computational models on historical refinery catalyst performance data and leveraging materials degradation mechanisms to predict catalyst lifetimes, are gaining importance (Flaherty & Bhan, 2024).

Towards Industrially Relevant Catalyst Design

Bridging the gap between laboratory innovation and industrial implementation in catalysis requires a fundamental shift towards a design philosophy that prioritizes real-world constraints from the initial stages of development. This necessitates proactive engagement with process engineers to define relevant operating parameters, the adoption of catalyst evaluation protocols that incorporate industrial impurities and dynamic conditions, and a focus on the end-of-life management of catalytic materials, including the recovery of valuable components and the regeneration of catalyst supports. Readers interested in deeper analysis may refer to the next generation of catalytic materials must exhibit robust performance and longevity within the complexities of industrial reactors to realize their full potential in advancing energy materials science (Cong & Zhang, 2022; De Blasio et al., 2021; Giglio et al., 2021; Khandan et al., 2024; Osman et al., 2024; Van Allsburg et al., 2022; Zhang, Xie, Tian et al, 2021).

INDUSTRIAL CONSTRAINTS IN POLYMER COMMERCIALIZATION

The Polymer Commercialization Conundrum

The translation of novel polymers from academic laboratories to widespread industrial application represents a significant challenge in materials science. Despite scholarly reports highlighting polymers with exceptional ionic conductivity, specific strength, or self-repair capabilities, a limited fraction achieves commercial viability. This divergence stems from fundamental differences in evaluation criteria between academic research and the essential demands of industrial implementation. Successful industrial polymer commercialization necessitates a balanced consideration of processing feasibility, environmental resilience, material sourcing, regulatory adherence, and economic practicality, aspects often receiving varying attention in academic settings (Thomas et al., 2023).

Challenges in Manufacturing Scale-Up and Durability

Laboratory polymer synthesis frequently prioritizes ultimate material characteristics, potentially compromising industrial processability. For instance, conductive polymer composites with high filler content, optimized for conductivity in research, often exhibit viscosities incompatible with standard industrial processing techniques (Herdiana et al., 2022). While advancements in shear-thinning nanocomposites offer potential solutions, the interplay between material properties and processing remains critical. Furthermore, the shift to industrial production introduces complexities in solvent usage regarding cost, recovery, and emission control, impacting the commercialization of materials like Poly[3,4-ethylenedioxythiophene]: Poly[styrenesulfonate] [PEDOT:PSS]. Standard accelerated aging protocols often fail to predict real-world performance due to the exclusion of synergistic degradation mechanisms encountered in operational environments, as exemplified by the premature failure of Polyvinylidene fluoride [PVDF] membranes. Interfacial degradation in polymer composites, particularly in solid-state batteries, also presents a significant challenge (Bhat et al., 2021).

Navigating Economic and Regulatory Obstacles

The commercialization of advanced polymers is further constrained by the availability and sourcing of specialized monomers, often characterized by limited global production, single suppliers, and complex synthesis routes. Regulatory changes, such as restrictions on perfluorinated monomers, can also significantly impact development timelines. Moreover, increasingly stringent recycling regulations pose challenges for high-performance polymers, many of which exhibit property degradation upon reprocessing. While chemical recycling and compatibilizing additives offer partial solutions, the development of inherently recyclable polymers, such as thermally reversible polyurethane networks, represents a promising direction (Powell & Housz, 2023).

The Way Forward: Designing for Industrial Needs

Successfully bridging the gap between laboratory innovation and commercial realization requires an early and thorough assessment of processing characteristics using industrially relevant metrics, formulation strategies prioritizing readily available components, proactive consideration of evolving regulatory landscapes, and the incorporation of design principles facilitating efficient recycling. The polymers that will shape our technological future will be those that achieve an optimal balance between their inherent properties and their feasibility for manufacturing, long-term reliability, and environmental sustainability throughout their entire lifecycle. This necessitates enhanced collaboration between synthetic chemists, process engineers, and end-users from the initial stages of research and development (Powell & Housz, 2023).

ENVIRONMENTAL DEGRADATION OF ENERGY MATERIALS

The Challenge of Real-World Stability

A significant disparity exists between the performance of energy materials in controlled laboratory settings and their operational longevity under real-world conditions. Unanticipated environmental degradation mechanisms, arising from complex, coupled stresses, frequently lead to premature failures, accounting for a substantial proportion of field deployment issues. Addressing these multidimensional stability challenges, encompassing climatic stresses, chemical interactions, mechanical wear, and biological fouling, is crucial for advancing energy technologies (Senthilkumar, Ramu, & Deepanraj, 2024).

Atmospheric and Temperature-Induced Degradation

Water vapor penetration poses a considerable threat to solid-state energy devices, accelerating detrimental phase formation in Solid Oxide Fuel Cells [SOFCs] and inducing mechanical stresses in Proton Exchange Membrane Fuel Cells [PEMFCs]. Industrial and marine atmospheric contaminants, such as chloride and sulfur dioxide, exhibit corrosive effects on various energy material components, including perovskite solar cells and metallic heat exchangers. Extreme temperatures and thermal cycling further exacerbate degradation. Cryogenic conditions promote lithium dendrite formation in batteries, while phase change materials suffer microstructural damage. High temperatures lead to sintering and chemical transformation in catalysts and defect formation in refractory ceramics (Zhang, 2022).

Synergistic Degradation and Emerging Solutions

The combined effects of photochemical-thermal aging and mechanical-environmental stress coupling create particularly aggressive degradation scenarios in outdoor energy materials and structural components. Addressing these complex interactions necessitates the development of advanced protective coatings, more representative accelerated aging test protocols, and self-healing material systems (Blaiszik et al., 2015; Zhang, 2022).

Toward Environmentally Resilient Materials

Developing durable energy materials requires a fundamental shift towards designing for environmental resilience. This necessitates interdisciplinary collaboration and the utilization of advanced characterization techniques to understand and mitigate complex degradation mechanisms. The ultimate objective is to create material systems capable of sustaining high performance across diverse and challenging real-world environments, thereby enabling truly sustainable energy solutions (Blaiszik et al., 2015; Zhang, Li, Luo et al, 2021).

SINGLE-ATOM CATALYST STABILITY CHALLENGES

Single-atom catalysts [SACs] represent a significant advancement in heterogeneous catalysis, offering near-theoretical atomic efficiency and unique electronic structures. However, their commercial viability is hampered by fundamental stability challenges, revealing a disconnect between exceptional lab-scale performance and real-world applicability. The very atomic dispersion that endows SACs with remarkable catalytic properties also renders them thermodynamically susceptible to aggregation, leaching, and poisoning under practical conditions (Giulimondi et al., 2023).

The high surface free energy of isolated metal atoms inherently drives them towards aggregation, even with minor thermal fluctuations, following an Ostwald ripening mechanism at lower temperatures than nanoparticle sintering. This instability is exacerbated by the low coordination numbers of single atoms on supports [typically 2-4 ligands], contrasting with the 10-12 coordinated neighbors stabilizing nanoparticle surfaces. Support material selection critically influences migration energy barriers, a factor often underestimated in initial studies (Liu et al., 2024).

Beyond aggregation, SACs face multifaceted chemical degradation pathways. In PEMFCs, the combination of low pH, high potential, and oxygen creates conditions conducive to electrochemical dissolution, with free energy calculations indicating lower dissolution potentials for SACs compared to nanoparticles due to weakened metal-support bonds. Furthermore, ubiquitous impurities in industrial feedstocks, such as sulfur compounds and chlorides, can permanently deactivate SACs through strong chemical bonding or the formation of volatile metal complexes, respectively (Giulimondi et al., 2023; Liu et al., 2024; Zhang, Li, Luo et al, 2021).

Stabilization strategies are evolving from empirical approaches to rational design. Support engineering beyond high surface area, such as introducing defect anchoring sites or utilizing zeolite confinement, can enhance metal-support interactions and restrict aggregation. Dynamic restructuring, where SACs undergo reversible aggregation under reaction conditions followed by redispersion, presents another avenue for stabilization [Liu, 2024].

Traditional accelerated stress tests [ASTs] designed for nanoparticles are often inadequate for evaluating SAC stability, as they may not capture SAC-specific failure modes like metal-support decohesion or low-temperature degradation pathways [Moguchikh, 2023]. Emerging testing protocols, including coupled electrochemical-microcalorimetry and multi-modal *operando* platforms, offer more insightful real-time analysis of SAC degradation mechanisms [Li, Donghao, et al., 2023].

The future of SACs necessitates a shift in focus from maximizing initial activity to designing systems with intrinsic stability under realistic operating conditions. This includes adopting industrial testing conditions early in the development process, establishing SAC-specific characterization standards beyond initial dispersion, and embracing dynamic material design principles (Bae et al., 2023). The most impactful SACs will be those that maintain atomic dispersion over extended periods of real-world operation, demanding a convergence of materials innovation and engineering pragmatism [Song, Wanqing, et al. 2024].

COMPUTATIONAL MODELING LIMITATIONS IN CATALYST DESIGN

The Promise and Peril of *In Silico* Discovery

Computational approaches, including density functional theory [DFT], molecular dynamics [MD], and machine learning [ML], have shown immense promise in revolutionizing catalyst design (Kee, 2023). These methods have successfully predicted active sites, optimized material compositions, and screened vast libraries of hypothetical materials. However, a significant disconnect persists, with many computationally designed catalysts failing experimental validation. This discrepancy arises from fundamental limitations in current models to fully capture the complexities of catalytic systems operating under realistic industrial conditions [Dell'Angelo, D. 2022].

The Electronic Structure Dilemma: When DFT Encounters Disorder

Standard DFT calculations often rely on idealized, defect-free surfaces at 0K, conditions rarely encountered in operational catalysts. Real catalytic materials exhibit dynamic surface reconstructions, coverage-dependent active sites, and kinetics often dominated by defects. Emerging solutions such as grand canonical DFT, cluster expansion methods, and active learning MD are attempting to address these limitations by incorporating potential and pH effects, modeling disordered surfaces, and identifying metastable intermediates [Sobrino, N., Jacob, D., & Kurth, S. 2023].

The Time-Scale Challenge

The vast disparity between catalyst lifetimes [years] and the timescales accessible by *ab initio* MD [nanoseconds] obscures crucial phenomena like sintering pathways, slow poisoning mechanisms, and creep-induced structural changes (Zhu & Gao, 2023). Bridging strategies involve kinetic Monte Carlo simulations, accelerated MD techniques, and the assimilation of *operando* data to constrain simulations [Xing, Yunhua, et al. 2022]

Complexity Chasm: From Model Systems to Real Catalysts

Simulations often focus on idealized single-crystal facets, while real catalysts comprise multi-faceted particles with complex support interactions and bifunctional mechanisms. Multiscale modeling approaches, including Quantum Mechanics / Molecular Mechanics [QM/MM], phase-field models, and graph neural networks, are being developed to incorporate these complexities. Furthermore, the typical neglect of ppm-level impurities in computational screening, which can significantly impact catalyst deactivation, necessitates the development of poisoning-aware descriptor spaces and high-throughput microkinetic studies with impurity inclusion [Napchi, O. 2021].

The Data Desert: When Machine Learning Lacks Ground Truth

The reliance of many ML models on DFT-generated data, overly clean experimental data, and small datasets with limited chemical space coverage leads to issues such as false positives, failure to predict pressure gaps, and missed azeotrope effects. Addressing this requires federated learning, automated literature mining, and active experimentation to generate more robust datasets. The interpretability crisis of black-box ML models also hinders understanding and necessitates the advancement of explainable Artificial Intelligence [AI] techniques [Exarchou, Dimitrios-Marios, et al. 2021].

The Interface Challenge: Bridging Simulation and Experiment

A mismatch exists between common experimental characterization techniques, which often provide ensemble averages and *ex situ* information, and the atomistic details provided by simulations [An, Rong, et al. 2024]. *Operando* modeling synergies, such as DFT-assisted Extended X-ray Absorption Fine Structure [EXAFS] fitting and microkinetic-guided Temperature-Programmed Desorption [TPD], along with neural network potentials for realistic microscopy image simulation, are crucial for bridging this gap. The traditional linear workflow between simulation and experiment also needs to evolve into cybernetic approaches, such as self-driving laboratories and Bayesian optimization, to accelerate the feedback loop and optimize catalyst design [Wang, Yalan, et al. 2021]

Toward Computationally Aware Catalyst Engineering

The future of catalyst design lies in recognizing computational modeling as a powerful yet limited tool that complements experimental efforts. Key priorities include developing "dirty DFT" methods, creating multiscale workflows, building open experimental databases, and fostering closer collaborations between theoretical and industrial researchers. The integration of advanced computational power with a deeper understanding of the complexities inherent in real catalytic systems will be paramount for future breakthroughs.

DEGRADATION MECHANISMS IN ZINC-AIR BATTERY CATALYSTS

The Zinc-Air Catalyst Paradox

Zinc-air batteries [ZABs] present a promising avenue for large-scale energy storage, boasting theoretical energy densities significantly exceeding lithium-ion systems. However, their widespread commercialization is impeded by the rapid degradation of air-electrode catalysts, crucial for the oxygen reduction [ORR] and evolution [OER] reactions. While academic research often highlights catalysts with exceptional initial performance, the complex and interconnected degradation pathways encountered under practical operating conditions remain a significant challenge. The catalyst's triple-phase boundary, where solid, liquid, and gas interact, subjects it to electrochemical corrosion, mechanical stress from zinc dendrites, and chemical poisoning from ambient air components. Understanding these degradation mechanisms necessitates employing *operando* characterization techniques to elucidate time-dependent failure modes, moving beyond initial activity assessments (Pattanayak et al., 2025).

Atomic-Scale Degradation

Transition metal [Fe, Co, Mn] Nx-C active sites, common in non-precious ZAB catalysts, undergo cation leaching in alkaline electrolytes and phase segregation in bimetallic systems during cycling. Mitigation strategies involve enhancing metal-support interactions through nitrogen-rich carbon shells and employing potential-cycling protocols that avoid prolonged high potentials. Carbon matrix corrosion, particularly edge-site oxidation and microporous collapse during OER, can be mitigated by utilizing graphitized carbon nanowires or ceramic-carbon hybrids (Nematollahi & Neyts, 2024).

System-Level Failure Modes

Electrolyte-driven degradation includes carbonate precipitation from CO_2 ingress and zincate deposition, both of which can block active sites. Solid-state electrolytes and anion-exchange membranes offer potential solutions. Ambient air impurities, such as CO_2 and SO_2, can also poison catalysts. Air scrubbing layers and redox mediators represent engineering adaptations to address this (Shabeer et al., 2025).

Characterization Gaps and Emerging Solutions

Standard rotating disk electrode [RDE] tests often underpredict degradation by neglecting factors like flooding-drying cycles and local pH gradients. Advanced diagnostics, such as neutron imaging and scanning electrochemical microscopy [SECM], provide more realistic insights. Common accelerated stress tests [ASTs] also fail to replicate real-world failure modes like startup/shutdown corrosion and thermal cycling. Proposed protocols include combined electrochemical-thermal cycling and humidity swing tests. Emerging material solutions encompass self-healing catalysts with dynamic bonds or liquid metal matrices and asymmetric electrode designs that decouple ORR and OER catalysts or employ graded porosity (Tarasova et al., 2024).

A Multiscale Approach to Durability

Enhancing ZAB catalyst durability requires a multiscale approach, optimizing atomic-level active sites, designing corrosion-resistant carbon architectures, and integrating system-level solutions for air purification and thermal management. The focus must shift from solely pursuing high initial activity to engineering inherently degradation-resistant systems through interdisciplinary collaboration (Choi et al., 2024).

NON-PRECIOUS METAL CATALYSTS

The Promise and Perils of Precious Metal Alternatives

Replacing precious metal catalysts with earth-abundant alternatives is a critical pursuit in sustainable energy. While platinum group metals dominate key applications, their high costs and supply chain vulnerabilities necessitate the development of non-precious metal catalysts [NPMCs]. However, the transition from promising academic findings to robust industrial implementation has proven challenging.

Recent advancements in coordination chemistry and materials engineering have yielded NPMCs with activities approaching or rivaling precious metals in specific conditions [e.g., Fe-N-C for oxygen reduction, cobalt phosphides for hydroprocessing, Ni-Fe layered double hydroxides for water electrolysis]. Yet, successful commercial translation remains limited (Singh et al., 2024).

The Stability Conundrum: Operational Lifespan Beyond Initial Activity

Academic research often emphasizes initial catalytic activity under idealized conditions, overlooking the critical aspect of operational stability in real-world industrial environments. NPMCs face significant deactivation pathways rarely observed in controlled laboratory settings. For instance, Fe-N-C catalysts in fuel cells can undergo demetalation during start-stop cycles (Yang et al., 2024), while transition metal phosphides in hydroprocessing can lose phosphorus in the presence of trace water. Such degradation significantly impacts long-term performance, a factor often underestimated by short-term activity tests.

Synthesis Scalability: Overcoming Manufacturing Bottlenecks

Scaling NPMC synthesis from laboratory milligrams to industrial kilograms presents a major obstacle. Many high-performing materials rely on complex, costly, and difficult-to-scale preparation methods, such as those involving metal-organic frameworks. Pyrolysis, common in NPMC synthesis, can lead to batch-to-batch variability and material loss. Continuous flow synthesis offers potential solutions for more consistent production, but energy efficiency and capital costs at industrial scales remain key considerations (Fernandes et al., 2023; Yang et al., 2024).

The Support Stability: Ensuring Catalyst Bed Integrity

Catalyst supports are crucial for NPMC longevity, often dictating practical lifetime. Conventional carbon blacks can corrode under oxidizing conditions, destabilizing the catalyst bed and potentially contaminating downstream components (Rana et al., 2021). Alternative supports like conductive metal oxides or nitrogen-doped carbon nanotubes face challenges related to mass transport limitations or cost. Hierarchical carbon architectures show promise in enhancing both durability and active site accessibility (Fernandes et al., 2023; Rana et al., 2021; Xu et al., 2021).

Economic Realities: A Holistic Cost Assessment

The economic viability of NPMCs extends beyond raw material costs. Lower intrinsic activity may necessitate higher catalyst loadings and larger reactor volumes. Shorter operational lifetimes can increase maintenance costs and production downtime. A comprehensive technoeconomic analysis must also consider the availability and cost of specialized precursors and the underdeveloped infrastructure for NPMC recovery and recycling (Xie et al., 2022).

Pathways Forward: Integrated Design for Industrial Relevance

Addressing these challenges requires a shift towards designing NPMCs with industrial viability as a primary objective. Key principles include employing accelerated stability testing that mimics real-world conditions, developing simplified and scalable synthesis routes using earth-abundant precursors, creating self-protecting catalyst architectures, and evaluating performance based on total lifetime productivity rather than just initial activity. While significant hurdles remain, aligning fundamental research with industrial realities is crucial for realizing the potential of NPMCs in achieving a sustainable energy future. The widespread adoption of these materials will be pivotal in scaling and Affordable clean energy technologies for global impact.

THE FUNDAMENTAL TRADE-OFF IN POLYMER ELECTROLYTES: RECONCILING MECHANICAL INTEGRITY WITH IONIC CONDUCTIVITY

The Central Dilemma of Solid Polymer Electrolytes

The development of high-performance solid polymer electrolytes [SPEs] constitutes a significant challenge in contemporary energy materials science. These materials are critical for next-generation energy storage, offering safer alternatives to liquid electrolytes and enabling flexible device designs. However, a fundamental inverse relationship exists between a polymer's mechanical strength and its ionic conductivity. This inherent trade-off has impeded the commercialization of SPE-based technologies for decades, necessitating innovative solutions at molecular, mesoscopic, and device levels (Du et al., 2024).

The core scientific issue arises from the conflicting structural requirements for these two essential properties. High ionic conductivity necessitates low crystallinity, segmental mobility of polymer chains, and minimal crosslinking to facilitate ion transport. Conversely, mechanical robustness requires high crystallinity, extensive crosslinking, or rigid backbone structures to resist deformation. The following discussion examines recent strategies to address this dichotomy through advanced polymer architectures, nanocomposite engineering, and interfacial stabilization, moving beyond conventional poly[ethylene oxide] [PEO]-based systems (Lu et al., 2023).

Molecular Design Strategies for Balanced Properties

Recent advances in alternative polymer chemistries offer promising pathways beyond the limitations of PEO. Rod-coil block copolymers, incorporating rigid aromatic polyimides and oligoether segments, demonstrate a decoupling of mechanical and conductive functions, achieving high storage moduli and acceptable conductivity. Ladder-type polymers with fused ring structures exhibit dimensional stability with strategically placed ionic side groups (Fernandes et al., 2023; Reinoso & Frechero, 2022). Dynamic covalent chemistry has enabled vitrimer-type networks with self-healing capabilities while maintaining ionic transport through temperature-dependent exchange kinetics, as seen in disulfide-containing poly[ionic liquids] (Irzhak et al., 2022).

Nanocomposite Approaches to Property Decoupling

The incorporation of high-aspect ratio nanomaterials offers a route to enhance mechanical properties without severely compromising ionic transport. Aligned nanofillers, such as cellulose nanocrystals with grafted polymer brushes, provide mechanical reinforcement and continuous ion channels. Percolating ionic networks, utilizing conductive nanofillers like Lithium Aluminum Titanium Phosphate [LATP] nanowires and modified graphene oxide, create continuous mechanical and ionic pathways through optimized interfacial chemistry, often employing zwitterionic modifications (Ahmad et al., 2023; Du et al., 2024; Irzhak et al., 2022; Lu et al., 2023; Reinoso & Frechero, 2022; Xie et al., 2022).

Device-Level Integration Strategies

Moving beyond homogeneous materials, asymmetric electrolyte architectures, such as stiff interfacial layers and soft conductive central regions, allow for localized property optimization. Ultraviolet [UV]-patterned electrolytes with spatially varying crosslink density, achieved through 3D printing, demonstrate the potential for stable cycling at improved current densities (Ahmad et al., 2023; Qadir et al., 2021; Xu et al., 2021). Self-regulating electrolyte systems, incorporating thermo-responsive polymers or electroresponsive composites, offer feedback mechanisms to automatically adjust properties under operational stresses, enhancing safety and performance (Qadir et al., 2021).

Pathways Toward Practical Implementation

The field has progressed significantly through approaches that decouple functional properties via block copolymers, create hierarchical structures through nanoscale engineering, and incorporate adaptive responses. However, challenges remain in scalable manufacturing, long-term stability, and cost-effectiveness. Future progress will likely emerge from interdisciplinary efforts in advanced polymer synthesis, precision processing, and innovative device engineering, ultimately overcoming the traditional conductivity-mechanical property trade-off (Nepal et al., 2023).

POLYETHYLENE OXIDE MODIFICATIONS FOR ADVANCED ENERGY STORAGE: NAVIGATING THE TRADE-OFFS BETWEEN IONIC TRANSPORT AND MECHANICAL STABILITY

Enduring Relevance of Polyethylene Oxide

Polyethylene oxide [PEO] remains a prominent polymer matrix for solid-state electrolytes, attributed to its lithium salt solvation capability, non-toxicity, and processability. However, achieving simultaneous fast ion transport and robust mechanical integrity at operational temperatures presents a fundamental challenge for PEO-based electrolytes and the broader field of polymer science in energy storage. Recent advancements strategically modify PEO to address crystallinity suppression, interfacial engineering, and hybrid architectures, aiming to decouple mechanical and transport properties (Crespilho et al., 2025). Here we are critically examines these modification strategies, considering both polymer physics and industrial processing requirements, highlighting successful transitions to pilot-scale production.

Addressing the Crystallinity and Interfacial Challenges

PEO's semi-crystalline morphology inherently limits ionic conductivity within crystalline domains while amorphous regions, enabling ion transport, lack mechanical stability above its glass transition temperature. Breakthroughs include star-branched PEO architectures and random copolymerization with propylene oxide to reduce crystallinity and enhance conductivity. Plasticization strategies, employing oligomeric ionic liquids or zwitterionic copolymers, offer alternatives to traditional plasticizers that compromise mechanical properties (Wang et al., 2024). Interfacial instability at

Li/PEO interfaces, leading to uneven lithium deposition and chemical degradation, is being tackled through *in situ* interphase engineering using fluorinated additives or Janus-functionalized PEO, and through the integration of 3D scaffolds like electrospun PEO/PAN blends or vertically aligned Carbon nanotube [CNT] arrays.

Hybrid Architectures and Processing Innovations

Hybrid systems incorporating inorganic fillers such as LATP nanowires or MOF-incorporated PEO aim to create percolated ion transport pathways and coordinate polymer chains with anions. Block copolymer architectures, like PS-PEO-PS triblocks or PEO-b-PVDF systems, are being explored to combine high modulus and conductivity. Manufacturing innovations are crucial for scale-up, moving from solution casting to reactive extrusion and UV-curable PEO formulations. Controlling film thickness through multilayer coextrusion and electrospray deposition remains a key challenge for commercial viability (Sikdar et al., 2021).

The Future of Modified PEO

Commercially promising modified PEO systems often exhibit asymmetric design, multi-scale ordering, and process-aware chemistry. Ongoing research suggests that optimized PEO variants hold the potential to achieve high energy densities in solid-state lithium metal batteries, provided that fundamental scientific advancements are effectively translated into engineering solutions (Espasa-Valdepeñas et al., 2021; Sikdar et al., 2021; Wang et al., 2024).

SCALING CHALLENGES IN POLYMER NANOCOMPOSITES: BRIDGING THE DIVIDE BETWEEN LABORATORY INNOVATION AND INDUSTRIAL PRODUCTION

The Nanocomposite Dilemma

Polymer nanocomposites exhibit transformative potential across various applications; however, their translation from laboratory-scale breakthroughs to commercially viable products present significant complexities. While academic research often highlights exceptional property enhancements achieved under controlled conditions, industrial adoption necessitates reproducibility, cost-effectiveness, and stability in real-world operating environments—criteria that many promising nanocomposites struggle to meet. This paper examines the fundamental differences in priorities between academic research, focused on idealized performance metrics, and industrial applications, which demand process scalability, material stability, economic viability, and regulatory compliance (Das et al., 2025). These challenges are analyzed through materials science, processing engineering, and industrial case studies to provide a roadmap for designing nanocomposites suitable for both high performance and scalable production (Deepanraj et al., 2025).

Homogeneity and Processing Challenges

Achieving uniform nanofiller dispersion at scale is a critical bottleneck. While lab-scale methods like sonication can yield excellent results, they often fail in large-scale processing due to viscosity gradients and re-agglomeration. In-situ polymerization and the use of compatibilizers represent emerging solutions. Furthermore, batch-to-batch variability due to nanofiller polydispersity and processing parameter sensitivity necessitates advanced inline monitoring and machine learning-assisted formulation optimization for robust production (Hiremath et al., 2021).

Processing limitations also hinder translation. The reliance on toxic solvents in solution casting and the energy intensity of solvent evaporation pose environmental and economic concerns. Melt compounding and supercritical CO_2-assisted dispersion offer alternatives but present their own challenges, particularly with high filler loadings. Additionally, thermal and shear degradation of polymers and nanofillers during industrial extrusion necessitate innovations like reactive extrusion and low-shear compounding techniques (Nanta et al., 2025).

Performance-Property Trade-offs and Economic/Regulatory Hurdles

Real-world applications often involve performance-property trade-offs. Enhancing electrical conductivity with conductive fillers can compromise mechanical strength and processability. Hybrid filler systems and segregated network architectures offer potential balanced solutions. Environmental stability gaps, such as hydrolytic and UV degradation, require stabilization strategies like multi-layer encapsulation and the use of inherently stable matrices (Crespilho et al., 2025; Espasa-Valdepeñas et al., 2021; Giglio et al., 2021; Nepal et al., 2023; Qadir et al., 2021; Sikdar et al., 2021; Wang et al., 2024).

Economic and regulatory hurdles further complicate commercialization. The high cost of certain nanofillers and functionalization processes necessitates exploring cost-reduction pathways such as waste-derived fillers and in-situ synthesis. Finally, navigating evolving nanomaterial regulations requires proactive compliance measures, including strategies to minimize nanoparticle release and early engagement with regulatory bodies (Kim et al., 2024).

A Systems Approach to Scalable Nanocomposites

Bridging the lab-to-industry gap for polymer nanocomposites requires a holistic approach emphasizing co-development of materials, processes, and applications. Prioritizing design for manufacturability, implementing standardized aging protocols encompassing relevant stress factors, and integrating circular economy principles are crucial. The next generation of nanocomposites must be engineered for reliable long-term performance under real-world conditions, guided by a systems-level perspective that considers both performance metrics and the practicalities of industrial production (Khatoon & Velidandi, 2025; Kim et al., 2024).

INDUSTRIAL PRODUCTION OF POLYMER ELECTROLYTES: NAVIGATING THE MINEFIELD BETWEEN LABORATORY INNOVATION AND MANUFACTURING REALITY

The Polymer Electrolyte Paradox in Energy Storage

Polymer electrolytes hold significant promises for safer, higher-energy-density energy storage devices. However, the high ionic conductivities reported in academic literature [e.g., 10–3 S/cm at room temperature] often contrast with challenges encountered in pilot production, such as film brittleness, dendrite formation, and inconsistent batch performance (Amici et al., 2022). This discrepancy arises from differing evaluation criteria between academia, focused on maximizing conductivity, and industry, which prioritizes processability, mechanical robustness, environmental stability, and cost scalability. The transition from laboratory breakthroughs to industrial production necessitates navigating the complex interplay between molecular design and manufacturing constraints. The subsequent sections examine these challenges through the perspectives of materials science, polymer physics, and manufacturing engineering, proposing a framework for developing polymer electrolytes suitable for large-scale deployment (Senthilkumar, Deepanraj, Shaik et al, 2024; Xie et al., 2022).

Scalability Challenges: From Laboratory to Continuous Production

Solvent selection presents a significant hurdle. While academic research often employs toxic aprotic solvents [e.g., NMP, DMF] for polymer dissolution, industrial processes face increasing regulatory pressure (Jordan et al., 2022). Aqueous processing and solvent-free techniques using supercritical CO_2 offer alternatives but introduce their own complexities in terms of processing control and cost. For instance, residual moisture in $scCO_2$-processed electrolytes can severely impact conductivity (Yang et al., 2022).

Achieving thin, uniform films required for high energy density in industrial battery production [<20 μm with <5% variation across meter-wide rolls] exposes fundamental limitations in polymer physics, such as stress cracking in high-molecular-weight systems and plasticizer migration. Emerging solutions include reactive coating of oligomeric precursors and layer-by-layer electrostatic deposition for precise thickness control (Yang et al., 2021).

The Interplay of Mechanical Stability and Ionic Transport

Enhancing mechanical properties through crosslinking often compromises ionic conductivity, creating a fundamental trade-off. Recent advancements explore topologically constrained networks and nanofiber-reinforced composites to decouple these properties. Interfacial instabilities, such as electrode delamination during calendaring and lithium dendrite penetration in large-format cells, are also exacerbated during industrial production and require innovative material designs like gradient modulus electrolytes and glass-ceramic-polymer hybrids for mitigation (Bostwick et al., 2022; Yang et al., 2021).

Quality Control in High-Volume Manufacturing

Maintaining consistent quality in large-scale polymer electrolyte production presents considerable challenges. Batch-to-batch variability in the crystallinity of semi-crystalline polymers like PEO, which directly affects conductivity, necessitates in-line monitoring techniques such as Raman spectroscopy and the use of nucleating agents. Managing moisture ingress during production, which can lead to hydrolysis of lithium salts, requires advanced drying protocols and moisture-scavenging separators (Bostwick et al., 2022; Werker et al., 2023; Yang et al., 2021)

Cost Engineering: A Critical Consideration

The economic viability of polymer electrolytes for large-scale energy storage is heavily influenced by raw material costs and manufacturing efficiency. Supply chain vulnerabilities for key components like lithium salts and complex block copolymers necessitate exploration of alternative materials, such as sodium- or potassium-based electrolytes and biopolymer hybrids. Achieving high web speeds in continuous production demands rapid curing chemistries and self-healing formulations to minimize defects (Maham et al., 2025).

A Systems-Level Approach to Industrialization

Successful industrial production of polymer electrolytes requires a holistic, systems-level approach that co-optimizes materials design, equipment integration, and quality control systems. The next generation of these materials must be engineered not solely for optimal ionic conductivity but also for robust manufacturability, where performance is ultimately validated by high production yields and minimal defect rates in large-scale manufacturing environments (Maham et al., 2025; Yang et al., 2022).

ENZYMATIC FUEL CELLS: PRACTICAL LIMITATIONS AND PATHWAYS TO COMMERCIAL VIABILITY

The Promise and Peril of Bioelectrocatalysis

Enzymatic fuel cells [EFCs] represent a compelling convergence of biotechnology and energy science, offering the potential for environmentally benign power generation through the oxidation of renewable fuels. Utilizing enzymes as substrate-specific catalysts under mild conditions theoretically enables operation in physiological environments, suggesting applications in implantable medical devices and self-powered biosensors. However, despite three decades of research, EFCs remain largely confined to academic laboratories due to fundamental challenges encountered under real-world operating conditions. These limitations encompass electron transfer kinetics, operational stability, mass transport constraints, and system integration complexities (Barelli et al., 2021). We are trying critically to examine these limitations through bioelectrochemistry, materials science, and device engineering, highlighting emerging strategies for commercial viability.

The Electron Transfer Dilemma: Bridging Proteins and Electrodes

The insulating protein shells surrounding redox enzyme catalytic centers impede efficient charge collection, necessitating either direct electron transfer [DET] or mediated electron transfer [MET], both with inherent limitations. DET suffers from the exponential decay of electron tunneling probability and strict orientation requirements. MET encounters issues with mediator depletion, enzyme inactivation, and potential toxicity. Recent advances in nanoscale electrode engineering, such as quantum tunneling wires, conductive protein binders, and plasmonic enhancement, offer partial solutions. Furthermore, the potential mismatch between enzymatic systems and practical electrode materials necessitates overpotentials that reduce cell voltage (Barelli et al., 2021; Yuwen et al., 2023).

Stability Constraints: Biological Meets Engineering

Enzymes exhibit catastrophic failure outside narrow pH and temperature windows, unlike the gradual degradation of inorganic catalysts. Temperature sensitivity, interfacial denaturation upon electrode adsorption, and oxidative damage from reactive oxygen species significantly limit operational lifetime. Accelerated aging tests often underestimate these effects. Modern stabilization strategies combine physical encapsulation [e.g., silica sol-gel matrices], genetic engineering [directed evolution for thermostability and reactive oxygen species [ROS] resistance], and biohybrid systems [enzyme conjugation to redox polymers or MOF encapsulation] (Akbarian & Chen, 2022; Barelli et al., 2021; Yuwen et al., 2023).

Mass Transport and System-Level Challenges

Even with ideal catalysts, EFCs face intrinsic current limitations due to substrate diffusion, leading to concentration polarization at both anode and cathode. Microfluidic architectures employing laminar flow and electroosmotic pumping offer partial mitigation (Yuwen et al., 2023). Stacking EFCs to increase voltage introduces challenges such as enzyme inhibition by product accumulation, pH gradients destabilizing enzymes, and parasitic losses from redox mediator crossover. Biocompatible ionic liquids show promise in buffering pH, reducing mediator solubility, and enhancing enzyme stability (Akbarian & Chen, 2022; Yuwen et al., 2023).

Commercialization Pathways: Learning from Failure

Analysis of abandoned EFC projects highlights recurring pitfalls, including fibrous encapsulation hindering glucose diffusion in medical implants and significant power loss due to temperature fluctuations and microbial fouling in self-powered sensors. Emerging solutions systematically address these issues through subdermal designs with vasculature interfaces, enzyme-free backup systems, and antifouling coatings (Barelli et al., 2021).

The Road Ahead

While EFCs may not compete with batteries in raw power density, their biocompatibility and substrate specificity define niche applications. The path forward requires transparent performance metrics reflecting real-world conditions, hybrid enzyme-inorganic catalyst approaches, and standardized testing protocols. Near-term applications likely lie in transient medical devices, environmental sensors, and educational tools, demanding a focus on understanding biological complexity and learning from past setbacks to realize this technology's potential (Akbarian & Chen, 2022).

PHOTOSYNTHESIS-INSPIRED ENERGY MATERIALS: TRANSLATING NATURE'S BLUEPRINT INTO SCALABLE TECHNOLOGIES

Three Billion Years of Inspiration

Natural photosynthesis, an exceptionally efficient and self-sustaining energy conversion system, offers profound inspiration for addressing limitations in conventional photovoltaic and electrochemical technologies. This overview examines the application of photosynthetic principles to the development of advanced energy materials, focusing on bioinspired light harvesting architectures, water oxidation catalysts, and energy transduction networks (Kathpalia & Verma, 2021).

Bioinspired Light Harvesting

Nature's light-harvesting complexes achieve remarkable energy transfer efficiency. Synthetic efforts are mimicking these capabilities through porphyrin-peptide nanotubes, DNA-templated dye arrays, and quantum dot-phycobiliprotein hybrids. A key challenge remains in maintaining structural integrity outside biological environments while mitigating quenching effects. Integrating photonic structures inspired by Morpho butterflies with photosynthetic principles, such as in chiral nematic cellulose films and diatom frustule replicas, also demonstrate enhanced light absorption and ion transport (Kathpalia & Verma, 2021; Mu et al., 2024).

Water Oxidation Catalysts: Lessons from the Mn_4CaO_5 Cluster

The oxygen-evolving complex [OEC] in photosynthesis operates with high efficiency. Biomimetic systems, including Mn-terpyridine frameworks and Co-polyphenol catalysts, are showing promise by mimicking the OEC's structure and redox mechanisms. Replicating the protein matrix's role in stabilizing catalytic intermediates through metal-organic frameworks and conducting polymers is a critical focus. Furthermore, self-healing mechanisms observed in the OEC are inspiring materials strategies like light-driven metal reinsertion and redox-active ligand systems (Capone et al., 2021).

Energy Transduction Networks and CO_2 Fixation

Artificial Z-scheme heterostructures, inspired by photosynthesis's two-photon system, aim to overcome single-bandgap limitations in solar energy conversion. Examples include $BiVO_4/Fe_2O_3$ tandem cells and organic-inorganic hybrids. The Calvin cycle's enzymatic cascade is also informing new catalytic materials for CO_2 fixation, such as MOF-encapsulated formate dehydrogenase and Cu-polymer composites mimicking rubisco's CO_2 concentration mechanisms (Liu et al., 2021).

Towards Hybrid Bio-Artificial Systems

The most promising advancements lie in extracting fundamental design principles from photosynthesis – hierarchical organization, dynamic self-repair, proton-coupled electron transfer, and multi-functional integration – for synthetic materials. Recent progress in *operando* characterization and machine learning-assisted design is facilitating this translation. The coming decade may witness the emergence of commercial systems leveraging these bioinspired principles in solar fuels production, photoelectrochemical water treatment, and self-powered carbon capture (Liu et al., 2021; Wang et al., 2022).

BIOLOGICAL SELF-HEALING MATERIALS FOR NEXT-GENERATION ENERGY STORAGE: HARNESSING NATURE'S BLUEPRINTS FOR RESILIENT ELECTROCHEMICAL SYSTEMS

The Need for Autonomous Repair in Energy Systems

Modern energy storage technologies inherently face material degradation from electrochemical processes, leading to mechanical stress, chemical decomposition, and interfacial instabilities. Nature offers inspiration for autonomous repair mechanisms, observed across scales from molecular bond reformation to tissue-level vascular networks. Translating these biological principles to energy materials necessitates interdisciplinary research, integrating supramolecular chemistry, polymer physics, and electrochemical engineering (Fang et al., 2021). The biological self-repair mechanisms relevant to energy storage are being discussed, material platforms enabling biomimetic healing in electrochemical devices, performance trade-offs, and scalability challenges, focusing on systems demonstrating enhanced electrochemical cycling stability.

Biological Blueprints for Autonomous Repair

Biological tissues utilize reversible bonding strategies analogous to those in energy materials. Metal-ligand coordination, as seen in marine organisms, inspires self-healing hydrogels and polymer electrolytes with reversible crosslinks that can reform after damage. π–π stacking in conductive polymers, mimicking fungal melanins, informs self-repairing electrode coatings where disrupted structures can reconfigure during cycling. Microvascular networks in plants and animals provide a blueprint for mobilizing healing agents to damage sites through microencapsulated redox mediators and electrochemically responsive hydrogels that release repair agents upon triggering (Tagliaferri et al., 2024).

Material Realizations: Bio-Inspiration in Electrochemical Function

Recent advances in battery electrolytes challenge the perceived trade-off between mechanical robustness and ionic conductivity. Phase-separated block copolymers, mimicking cartilage, exhibit both high mechanical strength and ionic conductivity. Metallosupramolecular electrolytes with photo-triggered reconfiguration capabilities enable localized repair. In electrode design, lessons from biomineralization, such as bone remodeling, inspire organic-inorganic interpenetration strategies and bio-molecular templating to create electrodes with enhanced stability and self-repairing conductive pathways (Capone et al., 2021; Liu et al., 2021; Mu et al., 2024; Wang et al., 2022).

The future of Bio-Inspired Energy Materials

Future progress requires mechanistic studies linking molecular-scale healing to macroscopic performance, the design of multifunctional materials where healing enhances energy storage, and strong industry-academic collaborations to facilitate the transition from fundamental research to industrial applications (Fang et al., 2021; Wang et al., 2022).

THE RISE AND FALL OF MOF-BASED HYDROGEN STORAGE: A POST-MORTEM ANALYSIS OF COST-DRIVEN ABANDONMENT

The Promise and Peril of MOFs for Energy Storage

Metal-organic frameworks [MOFs] garnered significant attention in the early 2000s as potential candidates for hydrogen storage, exhibiting exceptional surface areas and theoretically tunable pore structures. Academic research highlighted cryogenic H_2 uptake approaching 10 wt% in specific materials. However, nearly two decades later, MOF-based hydrogen storage has not achieved commercial viability in energy applications. This divergence between academic promise and industrial reality stems from a fundamental mismatch between material properties and economic constraints, rather than inherent scientific limitations.

This analysis deconstructs the key technical and financial barriers contributing to the abandonment of MOF hydrogen storage initiatives, examining process economics, engineering realities, competition from established technologies, and offering lessons for future porous material development (Hangarter et al., 2024).

The Cost Culprits: Economic Failures in MOF Synthesis

The synthesis of MOFs relies on multidentate organic linkers, often derived from petrochemicals. Even for purportedly "green" alternatives like 2,5-furandicarboxylic acid [FDCA], costs remain prohibitively high for large-scale production. Noble-metal-containing MOFs further escalate costs due to expensive metal precursors. Scale-up attempts have revealed significant yield reductions compared to laboratory synthesis, as demonstrated by BASF's pilot production of Basolite Z377. Solvent-free mechanochemical synthesis, while reducing costs, introduced performance-degrading byproducts. Furthermore, the high-temperature vacuum activation required to remove solvent from MOF pores represents an energy-intensive step incompatible with continuous production and can lead to reduced hydrogen capacity due to incomplete activation (Hangarter et al., 2024; Manyepedza & Makuve, 2024).

System-Level Challenges: Overlooked Engineering Hurdles

The exothermic adsorption and endothermic release of hydrogen by MOFs necessitate complex thermal management systems, adding significant cost and weight, as seen in prototype tanks. Slow hydrogen adsorption kinetics at ambient temperatures often necessitate cryogenic operation, negating potential energy density advantages. Repeated hydrogen adsorption and desorption cycles can also lead to mechanical instability, including linker hydrolysis in the presence of water and particle pulverization, significantly increasing bed resistance over time (Hangarter et al., 2024; Semelsberger et al., 2021).

The Cost Competition: Outperformance by Conventional Methods

Compressed hydrogen storage in Type IV carbon fiber tanks has achieved lower projected system costs and significantly faster refueling times compared to MOF-based systems. Liquid organic hydrogen carriers [LOHCs] offer ambient-condition stability and compatibility with existing petroleum infrastructure, presenting another cost-competitive alternative (Ali & Shaikh, 2024).

Lessons for Next-Generation Materials

The focus on gravimetric capacity in MOF research often overshadowed more critical industrial metrics such as volumetric capacity and cycle lifetime. Future porous material development should prioritize water-stable linkers, cost-effective one-pot syntheses, and consider hybridization strategies to enhance properties like thermal conductivity. A crucial pivot in porous materials development is necessary, emphasizing the co-development of synthesis and engineering, targeting cost-driven benchmarks alongside performance, and recognizing that manufacturability at scale is paramount for successful commercialization. The experience with MOF-based hydrogen storage provides invaluable lessons for ensuring the next generation of energy storage materials avoids similar pitfalls (Ali & Shaikh, 2024; Semelsberger et al., 2021; Ye et al., 2021).

FUTURE PATHWAYS IN MODELING: CYBER-PHYSICAL INTEGRATION AND DATA CO-GENERATION

The evolution of modeling techniques—spanning density functional theory [DFT], molecular dynamics [MD], and machine learning [ML]—demands transformative pathways to bridge computational accuracy and real-world applicability. Future advancements must prioritize cyber-physical integration and experimental co-generation to overcome existing limitations in energy materials design.

Cyber-Physical Systems for Real-Time Validation

Cyber-physical frameworks merge computational models with physical hardware, enabling dynamic feedback loops between simulation and experimentation. For instance, autonomous laboratories equipped with ML-driven synthesis robots can iteratively refine catalyst designs by correlating DFT predictions with real-time performance data under industrial conditions. Such systems accelerate the discovery of stable single-atom catalysts [SACs] by simulating degradation pathways [e.g., aggregation, leaching] while validating mitigation strategies [e.g., zeolite confinement, defect anchoring]. This closed-loop approach reduces reliance on idealized DFT assumptions and incorporates real-world stressors like impurity poisoning and thermal fluctuations (Fiedler et al., 2022).

Data Co-Generation with Operando Experiments

Integrating operando characterization [e.g., in-situ X-ray absorption spectroscopy, neutron imaging] with ML models addressing critical gaps in computational accuracy. For example:

- Dynamic Data Assimilation: Combining ML interatomic potentials [e.g., Allegro, FLARE] with operando electrochemical data enables real-time prediction of zinc-air battery degradation mechanisms, such as carbonate precipitation or dendrite formation.
- Physics-Informed ML: Hybrid models trained on both DFT-derived descriptors and experimental datasets [e.g., solvent viscosity, ionic conductivity] improve predictions for polymer electrolyte stability beyond traditional force fields. Schrödinger's AI-driven workflows demonstrate this by optimizing battery electrolytes using MD-informed ML models, achieving 90% accuracy in viscosity prediction (Shi et al., 2021).

Unified Data Infrastructures

Future workflows require standardized data ecosystems to facilitate collaboration:

- Federated Learning Platforms: Secure, decentralized databases aggregating experimental and computational data [e.g., degradation rates, synthesis parameters] enable cross-institutional model training without compromising proprietary information.

- Digital Twins: Multi-scale digital replicas of energy systems [e.g., fuel cells, solid-state batteries] simulate aging under dynamic stressors, guiding materials redesign for industrial scalability (Fiedler et al., 2022; Shi et al., 2021).

CONCLUSION

The synergy of cyber-physical integration and experimental co-generation represents a paradigm shift toward predictive, resilient energy materials. By embedding ML within autonomous experimentation and unifying data architectures, researchers can transcend traditional trade-offs [e.g., conductivity vs. mechanical strength in polymers] and accelerate the lab-to-industry transition.

REFERENCES

Ahmad, Z., Kumar, S., Trinh, C. K., Shim, J. J., & Lee, J. S. (2023). Decoupling electrochemical parameters of molecular-level-controlled polypyrrole and graphene oxide nanocomposite. *Applied Surface Science*, *610*, 155464.

Akbarian, M., & Chen, S. H. (2022). Instability challenges and stabilization strategies of pharmaceutical proteins. *Pharmaceutics*, *14*(11), 2533.

Ali, A., & Shaikh, M. N. (2024). Recent developments in catalyst design for liquid organic hydrogen carriers: Bridging the gap to affordable hydrogen storage. *International Journal of Hydrogen Energy*, *78*, 1–21.

Amici, J., Asinari, P., Ayerbe, E., Barboux, P., Bayle-Guillemaud, P., Behm, R. J., & Edström, K. (2022). A roadmap for transforming research to invent the batteries of the future designed within the European large scale research initiative battery. *Advanced Energy Materials*, *12*(17), 2102785.

Bae, G., Han, S., Oh, H. S., & Choi, C. H. (2023). Operando stability of single-atom electrocatalysts. *Angewandte Chemie*, *135*(19), e202219227.

Barelli, L., Bidini, G., Pelosi, D., & Sisani, E. (2021). Enzymatic biofuel cells: A review on flow designs. *Energies*, *14*(4), 910.

Bhat, M. A., Rather, R. A., & Shalla, A. H. (2021). PEDOT and PEDOT: PSS conducting polymeric hydrogels: A report on their emerging applications. *Synthetic Metals*, *273*, 116709.

Blaiszik, B., Kramer, S. L. B., Grady, M. E., McIlroy, D. A., Moore, J. S., White, S. R., & Sottos, N. R. (2015). Autonomic restoration of electrical conductivity. *Advanced Materials*, *27*(40), 5962–5967.

Bostwick, J. E., Zanelotti, C. J., Yu, D., Pietra, N. F., Williams, T. A., Madsen, L. A., & Colby, R. H. (2022). Ionic interactions control the modulus and mechanical properties of molecular ionic composite electrolytes. *Journal of Materials Chemistry. C, Materials for Optical and Electronic Devices*, *10*(3), 947–957.

Capone, M., Narzi, D., & Guidoni, L. (2021). Mechanism of oxygen evolution and Mn_4CaO_5 cluster restoration in the natural water-oxidizing catalyst. *Biochemistry*, *60*(30), 2341–2348.

Choi, J. S., Fortunato, G. V., Jung, D. C., Lourenço, J. C., Lanza, M. R., & Ledendecker, M. (2024). Catalyst durability in electrocatalytic H_2O_2 production: Key factors and challenges. *Nanoscale Horizons*, *9*(8), 1250–1261.

Cong, H., & Zhang, N. (2022). Perspectives in translating microfluidic devices from laboratory prototyping into scale-up production. *Biomicrofluidics*, *16*(2).

Crespilho, F. N., Costa, C. M., & Lanceros-Méndez, S. (2025). Sustainable Battery Biomaterials. *ChemElectroChem*, *12*(3), e202400530.

Das, N., Panda, S., Das, D. K., Nayak, S. K., Parthasarathy, S., Kirgiz, M. S., & Patnaik, P. (2025). Polymer Nanocomposites: Innovations in Material Design and Applications. In *Exploring Nanomaterial Synthesis, Characterization, and Applications* (pp. 169–196). IGI Global.

De Blasio, C., Salierno, G., & Magnano, A. (2021). Implications on feedstock processing and safety issues for semi-batch operations in supercritical water gasification of biomass. *Energies*, *14*(10), 2863.

Deepanraj, B., Saravanan, A. M., Senthilkumar, N., Afzal, A. A., & Afzal, A. R. (2025). Drilling Characteristics Optimization of Polymer Composite Fortified with Eggshells using Box-Behnken Design and Zebra Optimization Algorithm. *Results in Engineering*, •••, 104102.

Du, A., Lu, H., Liu, S., Chen, S., Chen, Z., Li, W., & Yang, C. (2024). Breaking the Trade-Off between Ionic Conductivity and Mechanical Strength in Solid Polymer Electrolytes for High-Performance Solid Lithium Batteries. *Advanced Energy Materials*, *14*(31), 2400808.

Espasa-Valdepeñas, A., Vega, J. F., Cruz, V., Ramos, J., Müller, A. J., & Martínez-Salazar, J. (2021). Revisiting polymer–particle interaction in PEO solutions. *Langmuir*, *37*(13), 3808–3816.

Fang, Y., Yang, X., Lin, Y., Shi, J., Prominski, A., Clayton, C., & Tian, B. (2021). Dissecting biological and synthetic soft–hard interfaces for tissue-like systems. *Chemical Reviews*, *122*(5), 5233–5276.

Fernandes, C., Jathar, M., Sawant, B. K. S., & Warde, T. (2023). Scale-up of nanoparticle manufacturing process. In *Pharmaceutical Process Engineering and Scale-up Principles* (pp. 173–203). Springer Nature Switzerland.

Fiedler, L., Shah, K., Bussmann, M., & Cangi, A. (2022). Deep dive into machine learning density functional theory for materials science and chemistry. *Physical Review Materials*, *6*(4), 040301.

Flaherty, D. W., & Bhan, A. (2024). Improving the rigor and reproducibility of catalyst testing and evaluation in the laboratory. *Journal of Catalysis*, *431*, 115408.

Fry, C., & Cox, D. (2020). Crossing the Valley of Death: Solutions to the Challenges of Translating Academic Research into Industrial Innovation. *Nature Reviews. Materials*, *5*(9), 637–643.

García-Salaberri, P. A. (2024) Battery manufacturing—from laboratory to industry challenges, Elsevier, 465-489.

Giglio, E., Pirone, R., & Bensaid, S. (2021). Dynamic modelling of methanation reactors during start-up and regulation in intermittent power-to-gas applications. *Renewable Energy*, *170*, 1040–1051.

Giulimondi, V., Mitchell, S., & Pérez-Ramírez, J. (2023). Challenges and opportunities in engineering the electronic structure of single-atom catalysts. *ACS Catalysis*, *13*(5), 2981–2997.

Hangarter, C. M., Dyatkin, B., Laskoski, M., Palenik, M. C., Miller, J. B., Tyagi, M., & Klug, C. A. (2024). A cobalt-based layered MOF material for supercapacitor applications. *Journal of Energy Storage*, *89*, 111476.

Herdiana, Y., Wathoni, N., Shamsuddin, S., & Muchtaridi, M. (2022). Scale-up polymeric-based nanoparticles drug delivery systems: Development and challenges. *OpenNano*, *7*, 100048.

Hiremath, A., Murthy, A. A., & Thipperudrappa, S., & KN, B. (2021). Nanoparticles filled polymer nanocomposites: A technological review. *Cogent Engineering*, *8*(1), 1991229.

Irzhak, V. I., Uflyand, I. E., & Dzhardimalieva, G. I. (2022). Self-healing of polymers and polymer composites. *Polymers*, *14*(24), 5404.

Jafarizadeh, H., Yamini, E., Zolfaghari, S. M., Esmaeilion, F., Assad, M. E. H., & Soltani, M. (2024). Navigating challenges in large-scale renewable energy storage: Barriers, solutions, and innovations. *Energy Reports*, *12*, 2179–2192.

Janasik, M., Jagusiak-Kocik, M., & Ulewicz, R. (2023). Barriers to technology transfer and commercialization of research findings: Case study. *Organization and Management Series*, (179), 164–180.

Jordan, A., Hall, C. G., Thorp, L. R., & Sneddon, H. F. (2022). Replacement of less-preferred dipolar aprotic and ethereal solvents in synthetic organic chemistry with more sustainable alternatives. *Chemical Reviews*, *122*(6), 6749–6794.

Kang, D., & Wolverton, C. (2022). Data-driven materials discovery and manufacturing: Challenges and future directions. *MRS Bulletin*, *47*(3), 227–233.

Kathpalia, R., & Verma, A. K. (2021). Bio-inspired nanoparticles for artificial photosynthesis. *Materials Today: Proceedings*, *45*, 3825–3832.

Kee, C. W. (2023). Molecular Understanding and Practical In Silico Catalyst Design in Computational Organocatalysis and Phase Transfer Catalysis—Challenges and Opportunities. *Molecules (Basel, Switzerland)*, *28*(4), 1715.

Khandan, A., Khosravi, M., Roustazadeh, D., & Aghadavoudi, F. (2024). Impact of alumina and carbon nanotubes on mechanical properties of a composite: Molecular dynamic (MD) simulation. *Iranian Journal of Chemistry and Chemical Engineering*, *43*(8), 2866–2877.

Khatoon, U. T., & Velidandi, A. (2025). An Overview on the Role of Government Initiatives in Nanotechnology Innovation for Sustainable Economic Development and Research Progress. *Sustainability*, *17*(3), 1250.

Kim, S., Kim, J., Lee, Y., Lee, W. B., Lee, C. W., Jin, K., & Na, J. (2024). Economic and environmental insights into the hybrid ethylene oxide production processes. *Chemical Engineering Journal*, *494*, 152879.

Liu, D., Chen, S., Li, R., & Peng, T. (2021). Review of Z-scheme heterojunctions for photocatalytic energy conversion. *Wuli Huaxue Xuebao*, *37*(6).

Liu, D., Gu, W., Zhou, L., Wang, L., Zhang, J., Liu, Y., & Lei, J. (2022). Recent advances in MOF-derived carbon-based nanomaterials for environmental applications in adsorption and catalytic degradation. *Chemical Engineering Journal*, *427*, 131503.

Liu, L., Chen, T., & Chen, Z. (2024). Understanding the Dynamic Aggregation in Single-Atom Catalysis. *Advancement of Science*, *11*(13), 2308046.

Lu, G., Zhang, Y., Zhang, J., Du, X., Lv, Z., Du, J., & Cui, G. (2023). Trade-offs between ion-conducting and mechanical properties: The case of polyacrylate electrolytes. *Carbon Energy*, *5*(2), e287.

Maham, M., Rabia, C., Anum, I., & Ahad, I. U. (2025). Critical and Strategic Raw Materials for Energy Storage Devices. *Batteries*, *11*(4), 163.

Manyepedza, T., & Makuve, N. (2024). Catalysts for biomass upgrading: Transforming 5-hydroxymethyl furfural into 2, 5-furan dicarboxylic acid for fine chemicals, featuring mil-100 (Fe) Mof as a novel catalyst. *ChemistrySelect*, *9*(28), e202401412.

Mitchell, S., Qin, R., Zheng, N., & Pérez-Ramírez, J. (2021). Nanoscale engineering of catalytic materials for sustainable technologies. *Nature Nanotechnology*, *16*(2), 129–139.

Mortazavi, B. (2025). Recent Advances in Machine Learning-Assisted Multiscale Design of Energy Materials. *Advanced Energy Materials*, *15*(9), 2403876.

Mu, B., Hao, X., Luo, X., Yang, Z., Lu, H., & Tian, W. (2024). Bioinspired polymeric supramolecular columns as efficient yet controllable artificial light-harvesting platform. *Nature Communications*, *15*(1), 903.

Nanta, P., Kasemwong, K., & Grisdanurak, N. (2025). Production and characterization of starch microspheres using supercritical carbon dioxide assisted spray drying. *International Journal of Biological Macromolecules*, *305*, 141197.

Nematollahi, P., & Neyts, E. C. (2024). Enhancement of Mn-N-C single atom catalysts via sulfur and/or oxygen co-doping for oxygen reduction in acidic conditions: Unveiling the catalyst durability in fuel cells. *Molecular Catalysis*, *553*, 113745.

Nepal, D., Kang, S., Adstedt, K. M., Kanhaiya, K., Bockstaller, M. R., Brinson, L. C., & Heinz, H. (2023). Hierarchically structured bioinspired nanocomposites. *Nature Materials*, *22*(1), 18–35.

Osman, A. I., Ayati, A., Krivoshapkin, P., Tanhaei, B., Farghali, M., Yap, P. S., & Abdelhaleem, A. (2024). Coordination-driven innovations in low-energy catalytic processes: Advancing sustainability in chemical production. *Coordination Chemistry Reviews*, *514*, 215900.

Ozin, G. A., & Loh, J. Y. (2022). *Energy Materials Discovery: Enabling a Sustainable Future*. Royal Society of Chemistry.

Pattanayak, S. R., Thakur, P., & Narayanan, T. N. (2025). The oxygen paradox in alkaline zinc-air batteries creates radical-induced anodic corrosion. *Cell Reports. Physical Science*.

Polizos, G., Goswami, M., Keum, J. K., He, L., Jafta, C. J., Sharma, J., & Li, J. (2024). Nanoscale ion transport enhances conductivity in solid polymer-ceramic lithium electrolytes. *ACS Nano*, *18*(4), 2750–2762.

Powell, P. C., & Housz, A. I. (2023). *Engineering with polymers*. CRC Press.

Qadir, A., Le, T. K., Malik, M., Min-Dianey, K. A. A., Saeed, I., Yu, Y., & Pham, P. V. (2021). Representative 2D-material-based nanocomposites and their emerging applications: A review. *RSC Advances*, *11*(39), 23860–23880.

Rana, M. S., AlHumaidan, F. S., Bouresli, R., & Navvamani, R. (2021). Guard-bed catalyst: Impact of textural properties on catalyst stability and deactivation rate. *Molecular Catalysis*, *502*, 111375.

Reinoso, D. M., & Frechero, M. A. (2022). Strategies for rational design of polymer-based solid electrolytes for advanced lithium energy storage applications. *Energy Storage Materials*, *52*, 430–464.

Semelsberger, T., Graetz, J., Sutton, A., & Rönnebro, E. C. (2021). Engineering challenges of solution and slurry-phase chemical hydrogen storage materials for automotive fuel cell applications. *Molecules (Basel, Switzerland)*, *26*(6), 1722.

Senthilkumar, N., Deepanraj, B., Shaik, F., & Elsehly, E. M. (2024, March). Artificial Intelligence Based Investigation on Drilling of MWCNT-Seashell-Nylon66 Hybrid Polymer Composites by Fuzzy Inference System. In *2024 3rd International Conference for Innovation in Technology (INOCON)* (pp. 1-6). IEEE.

Senthilkumar, N., Ramu, S., & Deepanraj, B. "Heat Conduction through Natural Fiber/Carbon Nanotubes Filler Matrix Polymer Composite Slabs: An Experimental and Analytical Comparisons," *SAE Technical Paper 2023-01-5107*, 2024, https://doi.org/DOI: 10.4271/2023-01-5107

Shabeer, Y., Madani, S. S., Panchal, S., Mousavi, M., & Fowler, M. (2025). Different Metal–Air Batteries as Range Extenders for the Electric Vehicle Market: A Comparative Study. *Batteries*, *11*(1), 35.

Shi, X., Lin, X., Luo, R., Wu, S., Li, L., Zhao, Z. J., & Gong, J. (2021). Dynamics of heterogeneous catalytic processes at operando conditions. *JACS*, *1*(12), 2100–2120.

Sikdar, S., Menezes, P. V., Maccione, R., Jacob, T., & Menezes, P. L. (2021). Plasma electrolytic oxidation (PEO) process—Processing, properties, and applications. *Nanomaterials (Basel, Switzerland)*, *11*(6), 1375.

Singh, M., Sharma, H. M., Gupta, R. K., & Kumar, A. (2024). Recent advancements and prospects in noble and non-noble electrocatalysts for materials methanol oxidation reactions. *Discover Nano*, *19*(1), 128.

Tagliaferri, S., Gaspard, L., Au, H., Mattevi, C., Titirici, M. M., & Crespo-Ribadeneyra, M. (2024). Nature-inspired batteries: From biomaterials to biomimetic design strategies. *Green Chemistry*, *26*(12), 6944–6958.

Tarasova, L., Gnann, S., Yang, S., Hartmann, A., & Wagener, T. (2024). Catchment characterization: Current descriptors, knowledge gaps and future opportunities. *Earth-Science Reviews*, •••, 104739.

Thomas, J., Patil, R. S., Patil, M., & John, J. (2023). Addressing the sustainability conundrums and challenges within the polymer value chain. *Sustainability*, *15*(22), 15758.

Van Allsburg, K. M., Tan, E. C., Super, J. D., Schaidle, J. A., & Baddour, F. G. (2022). Early-stage evaluation of catalyst manufacturing cost and environmental impact using CatCost. *Nature Catalysis*, *5*(4), 342–353.

Wang, C., O'Hagan, M. P., Willner, B., & Willner, I. (2022). Bioinspired artificial photosynthetic systems. *Chemistry (Weinheim an der Bergstrasse, Germany)*, *28*(9), e202103595.

Wang, M., Li, Y., Yang, H., Ding, Z., & Liu, R. (2024). Enhancing ion conductivity in polyethylene oxide-based polymers through semi-interpenetrating networks with liquid crystalline polymer for lithium metal batteries. *Journal of Energy Storage*, *100*, 113781.

Werker, A., Pei, R., Kim, K., Moretto, G., Estevez-Alonso, A., Vermeer, C., & de Vries, E. (2023). Thermal pre-processing before extraction of polyhydroxyalkanoates for molecular weight quality control. *Polymer Degradation & Stability*, *209*, 110277.

Xie, X., Du, L., Yan, L., Park, S., Qiu, Y., Sokolowski, J., & Shao, Y. (2022). Oxygen evolution reaction in alkaline environment: Material challenges and solutions. *Advanced Functional Materials*, *32*(21), 2110036.

Xu, Z., Deng, W., & Wang, X. (2021). 3D hierarchical carbon-rich micro-/nanomaterials for energy storage and catalysis. *Electrochemical Energy Reviews*, *4*, 269–335.

Yang, X., Adair, K. R., Gao, X., & Sun, X. (2021). Recent advances and perspectives on thin electrolytes for high-energy-density solid-state lithium batteries. *Energy & Environmental Science*, *14*(2), 643–671.

Yang, X., Huang, Z., Du, L., Li, Q., & Ye, S. (2024). Active Sites and Stability Study of Fe/N/C Catalyst in PEMFCs: A Decade of Stunning Progress and the Remaining Challenges. *ACS Catalysis*, *14*(20), 15471–15488.

Yang, Y., Yuan, W., Hou, J., & You, Z. (2022). Review on physical and chemical factors affecting fines migration in porous media. *Water Research*, *214*, 118172.

Ye, Z., Jiang, Y., Li, L., Wu, F., & Chen, R. (2021). Rational design of MOF-based materials for next-generation rechargeable batteries. *Nano-Micro Letters*, *13*, 1–37.

Yuwen, T., Shu, D., Zou, H., Yang, X., Wang, S., Zhang, S., & Zang, G. (2023). Carbon nanotubes: A powerful bridge for conductivity and flexibility in electrochemical glucose sensors. *Journal of Nanobiotechnology*, *21*(1), 320.

Zhang, H., (2022). Microstructural degradation of phase change materials under thermal cycling. *Renewable & Sustainable Energy Reviews*, *156*, 111987.

Zhang, L., Xie, Y., Tian, Z., Liu, Y., Geng, C., & Xu, S. (2021). Thermal conductive encapsulation enables stable high-power perovskite-converted light-emitting diodes. *ACS Applied Materials & Interfaces, 13*(25), 30076–30085.

Zhang, Y., Li, X., Luo, X., & Huang, Y. (2021). Multifactor accelerated aging of photovoltaic materials: Protocols and mechanisms. *Progress in Photovoltaics: Research and Applications, 29*(4), 379–397.

Zhu, G., & Gao, Y. (2023). Multi-objective optimal scheduling of an integrated energy system under the multi-time scale ladder-type carbon trading mechanism. *Journal of Cleaner Production, 417*, 137922.

Chapter 3
Machine Learning as a Catalyst in Advanced Materials Research and Development

Muhammad Akram
https://orcid.org/0000-0002-0079-7270
Abertay University, UK

Waqas Uddin Choudhry
Wien University of Applied Sciences, Hochschule, Austria

ABSTRACT

Machine learning is changing how materials are designed, tested, and improved. It does this by learning from data to guess properties, suggest new recipes, and pick better experiments. There are three main types of machine learning: supervised learning, which learns from labelled examples; unsupervised learning, which finds hidden groups in unlabeled data; and reinforcement learning, which learns by trying steps and getting feedback. Machine learning works on data from lab experiments, computer simulations, and past papers, but this data needs to be cleaned, features picked, and dimensions reduced, or models can get confused. With good data, ML can predict things like bandgap, hardness, stability, or lifetime. It can also do inverse design, which means starting from a target and working backward to a material or recipe, and optimize processes to get better results with fewer trials across polymers, alloys, nanomaterials, and composites. Some new directions are autonomous labs that loop measure-learn-decide, AI-guided design that ties simulations and experiments together more closely, and work that tries to make materials greener and less wasteful. There are still issues, such as a lack of high-quality data, labels that are

DOI: 10.4018/979-8-3373-3556-8.ch003

noisy, and models that don't work with new chemistries or conditions. Because of this, it's important to combine physical rules and domain knowledge with machine learning and to use explainable methods so that decisions make sense. It's also important to keep an eye on ethical issues like bias, safety, and reproducibility. Overall, this review tries to give a clear and simple look at where machine learning helps and where it struggles in advanced materials and sustainability. It also tries to share ideas that will help future studies and encourage teams from different fields to work together to build and test new ideas.

1. INTRODUCTION

Machine learning is becoming a big help in materials science. It makes it easier as well as faster to solve tough engineering problems in the real world. The old trial-and-error method is too slow and not accurate enough anymore. It just doesn't keep up with what people need today. ML can look at a lot of data, find patterns, and suggest better options without having to do so many tests. People who work in aerospace, energy, electronics, and biomedicine need materials with very specific properties, such as parts that are lighter but stronger, batteries that are safer, catalysts that are cleaner, and implants that are biocompatible. (Lu et al. 2024). We need systems that can quickly screen and predict that can handle large compositional and processing spaces to make these materials. In this case, ML gives us a flexible, data-centered framework that can learn from data that has been tested or simulated. This lets researcher do predictive modelling, inverse design, and find materials more quickly. ML doesn't replace the methods we already have; it just makes them better and adds to them. This helps researchers get useful information from big datasets and make their guesses more accurate (Yue Liu et al. 2017).

ML is important in materials science because it can work with data that is high-dimensional, nonlinear, and noisy. This is because of different experimental settings and changes in composition. For example, predicting properties like thermal conductivity, dielectric strength, mechanical hardness, or corrosion resistance may necessitate the integration of various factors, including microstructural attributes, synthesis conditions, and processing histories (P. Sharma et al. 2022; Prateek et al. 2024). When trained on reliable data, ML models can discover hidden relationships between these variables and the material's large-scale behaviour. This lets them make quick guesses that would normally take a lot of lab work or very accurate simulations (Natrayan et al. 2023). By moving from a purely empirical or computational approach to one that uses AI, we can cut down on development time, resources, and unnecessary experiments. It also makes it possible to make materials that can do many things, such as battery electrodes, polymer nanocomposites, and shape-memory alloys (X. Liu, Zhang, and Pei 2023). Figure 1 (G. Chen and Tang 2024) shows the different parts of ML-based materials research.

Figure 1. Various aspects of ML-based materials research (G. Chen and Tang 2024)

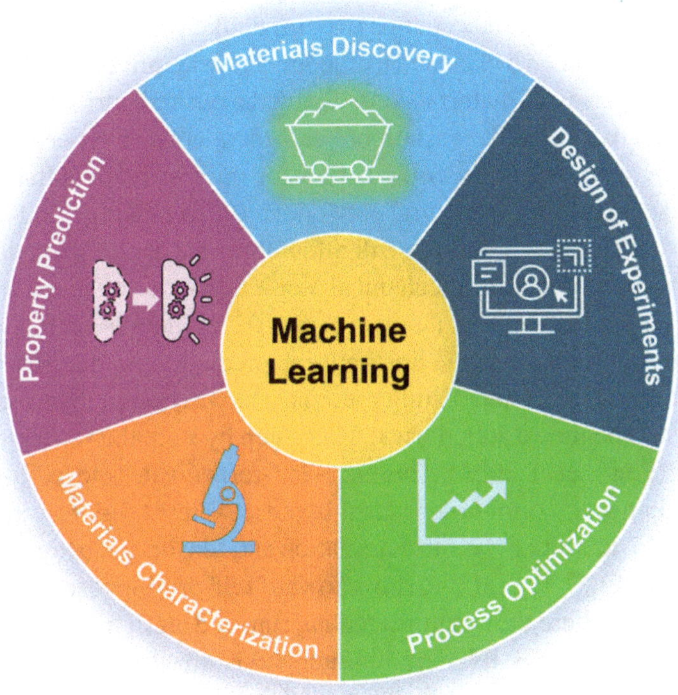

It has grown over time as experimental materials characterization, physics-informed modelling, and data science have become more alike. The Materials Genome Initiative (MGI) and other high-throughput computing projects were the main drivers of materials informatics at first. These projects aimed to create databases of computed attributes accessible to all (Yingli Liu et al. 2020). Some of the repositories that have made ML possible by giving researchers organized and vetted datasets are the Materials Project, AFLOWLIB, and OQMD (Shen et al. 2022; Calderon et al. 2015). Instrumentation and characterization have also improved a lot, which has added to the data that can be used to train models. This includes measurements taken in situ, electron microscopy, and spectroscopy. When researchers got their hands on these large, diverse datasets, they began using supervised learning techniques to guess the properties of materials and make the conditions for synthesis better (Kanti, Sharma, et al. 2023). Support vector machines and random forests are two examples of kernel-based methods that were among the first to show that they could be used to group crystal structures, find phase transitions, and estimate bandgaps (Zhuo et al., 2018; Mattur et al. 2022).

Deep learning has become more important in recent years, especially for problems that need to extract features in a hierarchy. For instance, convolutional neural networks are used to look at pictures of microstructure, and recurrent neural networks are used to describe how materials change over time, like when they get tired or creep. These methods have been broadened to encompass generative models that facilitate the reverse engineering of materials possessing the desired attributes (White 2022). Researchers can use tools like variational autoencoders and generative adversarial networks to come up with new compositions or microstructures that meet certain performance standards. Graph neural networks are also looking good for modelling structures on the atomic scale. This will allow for more accurate modelling of molecular dynamics and quantum-mechanical properties without excessive dependence on manually crafted descriptors (Yaman et al. 2023; Choubisa et al. 2020).

In modern materials work, the link between predictions and experiments is very important. Models choose interesting materials, experimental teams test them, and the new measurements are added back. Then, step by step, the models are retrained and get better (Pogue et al. 2023). This back-and-forth creates closed-loop discovery systems, also known as autonomous labs. In these systems, ML tells robots what to make and how to measure it. As lab automation and sensors get better, these systems become more useful. The goal is to move toward self-driving research and development in materials, where the system can plan, run, and learn from experiments with as few manual steps as possible (Abolhasani and Kumacheva 2023).

ML still needs domain experts, even though it has come a long way. We can get wrong or not very useful answers if we only look at data and not at physics and the real world (Pogue et al. 2023). So the field is moving toward physics-informed ML, which means putting known laws, limits, and symmetries into the models. For example, neural nets should follow the law of conservation of energy, and features should use crystal symmetry. This makes models clearer and more reliable instead of just a black box (Pogue et al. 2023; Abolhasani and Kumacheva 2023).

In summary, machine learning is changing the way we study, design, and make materials better. It combines data science, physical modelling, and experimentation into one process that can quickly predict properties, find flaws, do inverse design, and run self-optimizing trials (Pogue et al. 2023). As computers get faster, data access gets better, and algorithms get better, machine learning will be easier to use in everyday tasks. This will change how things are done and open up new ways for innovation in many fields.

2. CORE ML TECHNIQUES AND ALGORITHMS

ML has become an important tool in modern scientific research, especially in materials science. It helps scientists understand complex data, guess what new materials will be like, and make experiments better. There are three main types of ML: supervised, unsupervised, and reinforcement learning (P. Sharma and Bora 2022). Depending on the problem's structure and nature, each one gives researcher a different way to look at and understand data. The most common way is supervised learning. It learns a mapping function that takes input variables and gives the desired output by using labelled training data. A lot of people have used this method to predict the mechanical, thermal, and electrical properties of materials, like hardness, band gap, or thermal conductivity, when they have big datasets with known compositions and qualities (Fuhr and Sumpter 2022). Linear regression, decision trees, support vector machines, and neural networks are all common models in this situation. In materials science, decision trees clarify the decision-making process, aiding scientists in understanding how particular properties, such as atomic radius or bond energy, influence material behaviour (Kanti, Shrivastav, et al. 2023). They can deal with nonlinear relationships, which makes them useful for figuring out what kinds of alloys are in a sample or predicting phase changes. But they might become unstable with small changes in the data and are likely to overfit, which is a problem that ensemble methods can help with. Figure 2 (Talaei Khoei and Kaabouch 2023) shows a list of different types of ML approaches.

Figure 2. Classification of ML approaches (Talaei Khoei and Kaabouch 2023)

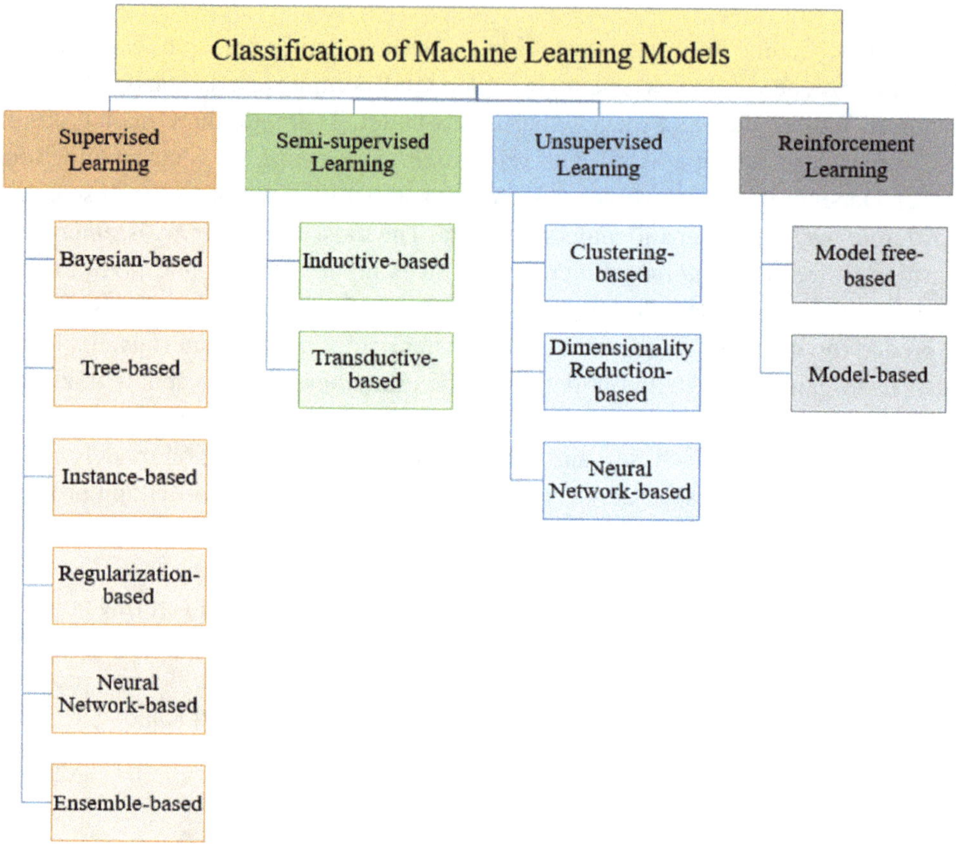

On the other hand, support vector machines (SVM) make hyperplanes in high-dimensional space to separate different classes or make a regression model. They are useful for tasks like sorting crystal structures or finding flaws in scanning electron microscope pictures because they can handle data with many dimensions and are not affected by outliers (Le et al. 2023). The kernel method in SVM makes it useful for nonlinear problems by changing the input features into higher-dimensional spaces where a linear decision boundary can be drawn. Even though SVM models have these features, they are hard to understand and cost a lot more to run when there are a lot of data points. Neural networks are made up of layers of connected nodes (neurons) that work like the brain does. These layers make it possible to model interactions that are very nonlinear and complex. Deep learning, which is a part of neural networks, has been very helpful for imaging materials, finding microstructures, and making sense of spectroscopic data. People often use convolutional neural networks (CNNs) to work with images, like finding grain boundaries or separating phases in micrographs. Recurrent neural networks (RNNs) are particularly adept at handling time-series data, exemplified by the analysis of system degradation or behaviour during thermal cycling (Alharkan, Habib, and Islam 2023; X. Gao et al., 2024; Jung, Han, and Choi 2021). Neural networks are very good at making predictions, especially when they have a lot of data to work with. However, they also need a lot of computing power and are usually thought of as "black box" models because they aren't very clear.

Unsupervised learning uses data that doesn't have labels on it to try to find patterns or structures that aren't obvious. We use clustering methods like k-means, hierarchical clustering, or density-based spatial clustering (DBSCAN) to find similarities and natural groups in how materials are made or how they come together. This is used in high-throughput screening, which tests thousands of possible materials at once (Q. Liu et al. 2012). Clustering can put things together that are similar in structure or performance. That makes it easier to find people who would be good to follow up with. PCA and t-SNE are two common tools for reducing the number of dimensions in this area (Spiwok and Kříž 2020; Jolliffe, 2022). They reduce complex, high-dimensional data to fewer numbers while preserving significant patterns or clusters. Think of spectral fingerprints or big simulation outputs. These tools help researcher see and understand them better. PCA is often used to reduce the number of variables in big screening datasets while keeping most of the variance. This makes later analysis easier.

Reinforcement learning (RL) is getting more and more popular. It teaches an agent how to choose actions in an environment that will give them the most reward over time. People use RL in materials labs to test it for autonomous experiments, process tuning, and robot control (Rajak, Wang, et al. 2021). The agent tries out different combinations of materials or synthesis settings. It changes its policy based on what it learns. This ongoing feedback loop is useful for finding the best processing parameters in additive manufacturing or for quickly switching from one reaction route to another. RL frequently employs model-free methodologies such as Q-learning and policy gradient algorithms (Rajak, Krishnamoorthy, et al. 2021; Abel et al. 2018). These are especially helpful when it's hard to understand or doesn't know how the environment works. For example, these strategies have been used to control the deposition of thin films or to guide the search for high-performance polymers in a way that changes. RL applications in materials science are still limited because they are costly to implement and developing reward functions that align with scientific objectives is challenging (Sui et al. 2021).

Ensemble methods like random forests, gradient boosting, and stacking are very popular because they can combine the predictions of many base models to make the whole model more accurate and stable. Groups of decision trees called random forests are very good at classifying and regressing tasks that involve finding new materials and predicting defects (Said et al. 2023; V. G. Nguyen et al. 2024). They help with the problem of overfitting that happens when one only use one tree. They also tell us how important each characteristic is, which helps us understand how different descriptors change the way materials work. XGBoost and LightGBM are examples of gradient boosting models that work well with structured data (Yu xiao Liu et al. 2023; Ke et al. 2017). They have been utilized to predict material properties, such as glass transition temperature, dielectric strength, and fatigue life. These models make decision trees one after the other. Each new tree tries to fix the problems that the trees that came before it made. The end result is a perfect model that can handle nonlinear relationships and interactions between variables. SHAP (SHapley Additive exPlanations) and other similar methods are more useful in scientific settings because they make it easier to understand how they work. This helps scientists figure out which parts of the data have the biggest effect on predictions (V. N. Nguyen et al. 2024; K. V. Sharma et al. 2023).

Using these important ML methods in materials research has made it much easier to understand complicated relationships, cut down on the work needed for experiments, and speed up the discovery process. These methods are expected to become more important not only for predicting outcomes but also for making hypotheses, running experiments, and finding the physical laws that govern how materials behave as data collection and computing power continue to get better. The next step is to make sure that these algorithmic tools work well with domain knowledge so that they can be used to make predictions that are both scientifically sound and useful in real life.

3. DATA SOURCES AND PREPROCESSING IN MATERIALS SCIENCE

In the constantly evolving discipline of materials science, data is the base layer that modern computational models and ML algorithms are built on. Materials systems are very different and complicated, so they need to get information from a lot of different places. Polymers, alloys, ceramics, and composites are all parts of these systems. These sources are different in more than just where they come from. They also have different resolutions, sizes, and how they fit into the bigger picture (Chong et al. 2024; Han et al. 2021). It is very important to collect, preprocess, and curate this kind of data correctly so that ML algorithms can give us strong, clear, and scientifically useful insights. Experimental data remains one of the most direct and widely utilized sources in materials research. This has information from lab tests like spectroscopy, microscopy, mechanical testing, thermal analysis, and more. This kind of data is usually very accurate and comes from controlled physical situations.

But experimental datasets are usually small because it costs a lot of time and money to figure out what materials are like. This makes it hard to train ML models that need a lot of data, like deep neural networks that need a lot of input to learn how to generalize. This is a problem, but experimental data can help us test models and find real-world connections between composition, processing, and performance (Fuhr and Sumpter 2022; Elton et al. 2018). Figure 3 (Frydrych et al. 2021) shows how ML can help with investigations that have to do with materials.

Figure 3. ML-based workflow for materials investigation (Frydrych et al. 2021)

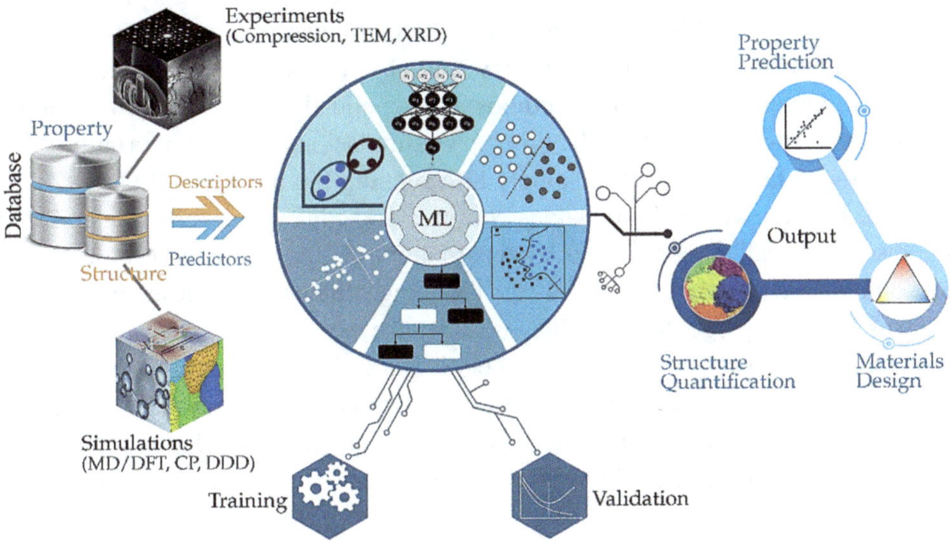

Alternatively, simulation data can be utilized in lieu of actual experiments in a scalable manner. Density functional theory (DFT), molecular dynamics (MD), and finite element analysis (FEA) are examples of computational methods that are now very useful for making a lot of high-resolution data over a wide range of lengths and times (Deepthi Jayan and Ragin Ramdas 2022; Hospital et al., 2015; Hinton & Owen, 1986). These simulations can look at parameter spaces that are hard to get to in real life and help us understand how materials act at the atomic or meso-scale level. There are, however, problems with simulation data, like assumptions, approximations, and model dependencies that may not fully show how things work in real life. It is often a good idea to use both simulation and experimental data to make ML models that are more complete and can be used in other situations. The scientific literature is becoming a big and new source of information. Text mining and natural language processing have been employed to extract structured data from unstructured texts, such as journal articles, patents, and technical reports. These data usually contain implicit information, like patterns, heuristics, and qualitative insights that aren't always easy to find in structured databases. Data from the literature can greatly improve training datasets by filling in holes in experimental or simulation data. But to make sure it is reliable, it usually needs to be carefully validated and standardized (Prateek et al. 2024; Shen et al. 2022; Paunovska et al. 2019).

Preprocessing is an important step after getting data because it cleans it up, makes it standard, and gets it ready for ML. Feature engineering is the most important part of this method. This means picking and putting together the right factors that best explain the materials phenomenon. In materials science, features may encompass elemental characteristics, structural descriptors, phase compositions, or process parameters (Xu et al. 2023). To do good feature engineering, you need to know a lot about the materials system, chemistry, and physics. When there is a lot of data, teams sometimes use automated methods like genetic algorithms as well as mutual information criteria. Reducing the number of dimensions is also important; it tries to make the data easier to understand while keeping the most useful parts (Z. Liu et al. 2021). High-dimensional data is common and can cause overfitting and slow models. This is called the "curse of dimensionality." PCA, t-SNE, and autoencoders are some of the most common tools for reducing dimensions. These help you see patterns, find groups, get rid of duplicates, and make models easier for comprehension and work better (Tariq et al. 2022; Zou et al. 2022). The quality of the data is still a big problem in materials ML. Missing values, noisy measurements, mismatched units, and label errors in datasets all make predictions less reliable. Imputation is a common way to fill in gaps. For simple cases, it can be as simple as replacing the mean, and for more complicated cases, it can be k-nearest neighbours or regression-based imputations (Cichy and Rass 2019).

To make sure that learning algorithms treat features fairly and to reduce variance, it is also important to make sure that units are consistent, that scaling is done right, and that normalisation is done right. It is important to find and remove outliers in some situations so that strange data points don't throw off model predictions. In supervised learning, labelled datasets are essential for instructing models on how to predict specific material characteristics or classifications (Nandy, Duan, and Kulik 2022; León et al., 2020). In materials science, obtaining labelled data can be challenging; consequently, researchers must utilize semi-supervised or weakly supervised learning techniques that incorporate both labelled and unlabeled data. This problem has made people interested in transfer learning, which lets you use models that you learned on one task or dataset on another task or dataset with less data. This makes it easier to sort through a lot of data.

One also considers about how easy it is to reproduce and understand data flows. Standardized data formats, like the ones suggested by the Materials Genome Initiative or the Citrine Informatics platform, have made it easier for scientists to share data and test their models. Open repositories like the Materials Project, OQMD (Open Quantum Materials Database), and AFLOWLIB have made it easier to find high-quality datasets of materials. This has made it easier for people to work together on research and made discoveries happen faster (Saal et al. 2013; Kirklin et al. 2015). Another area that is growing is the integration of multi-modal data, which means putting experimental, simulation, and literature-based data into a single framework. This type of integration allows for the creation of more detailed information representations and their comparison, which makes ML models more reliable and useful. However, when you combine data from different places, it's very important to make sure that the formats, ontologies, and semantics of the data match up. Because of this, ontological frameworks and data fusion methods have been made just for use in materials science.

Also, more and more people are using active learning techniques to make data more useful. In active learning, the model finds areas of uncertainty and gives you more data points to get, either through experiments or simulations. This iterative loop helps make models that are more accurate with fewer data points, which is good for the practical limits of experimental research. To get the most out of data in materials research, experimentalists, computational scientists, and data scientists need to work together closely. When we know the limits and assumptions of the different data sources we use, models stay based in the real world.

Simultaneously, data preparation techniques must be tailored to the requirements of materials datasets, striking a balance between increasing complexity and enhancing utility. As ML becomes more common in materials research, the need for high-quality, diverse, and well-curated data will only grow. This shows how important it is to come up with new ideas in the field.

4. APPLICATIONS IN MATERIALS DISCOVERY AND DESIGN

ML has quickly gone from being a helpful analytical tool to being a key part of workflows for finding and designing materials. By learning patterns from large datasets, ML models help us look into complicated material systems faster and for less money. This could change how we find, test, and improve new materials. This change is most obvious in applications like predicting property values, inverse design, and multi-objective optimization (Deng et al. 2020). In combination, they make a clear plan for how to make advanced materials that are great for some purposes. You can use these methods on more than one type of material. They are becoming more common in polymers, alloys, nanomaterials, and composites. Each of these has its own set of problems and possibilities. One of the most advanced and widely used applications of ML in materials research is property prediction (Rehman, Shatanawi, and Çolak 2023; Ahmadi et al. 2019). When trying to figure out things like thermal conductivity, bandgap, hardness, or corrosion resistance, traditional experimental or computational methods often take a lot of time and effort. Data-driven models, on the other hand, may be able to quickly predict these traits if they are trained on representative datasets. You can use ML techniques like random forests, support vector regression, and deep neural networks to guess a wide range of material properties based on simple descriptors like elemental composition, crystal structure, or processing parameters. These predictive models not only speed up screening, but they also find nonlinear relationships that might be hard to see with other methods (Singh, Tiwari, and Paliwal 2023; Maleki, Haghighi, and Mahariq 2021). Figure 4 shows the diagrams of ML-based material investigation (Konstantopoulos, Koumoulos, and Charitidis 2022).

Figure 4. AI-Enabled Closed-Loop Framework for Experiment Optimization in Materials Research (Konstantopoulos, Koumoulos, and Charitidis 2022)

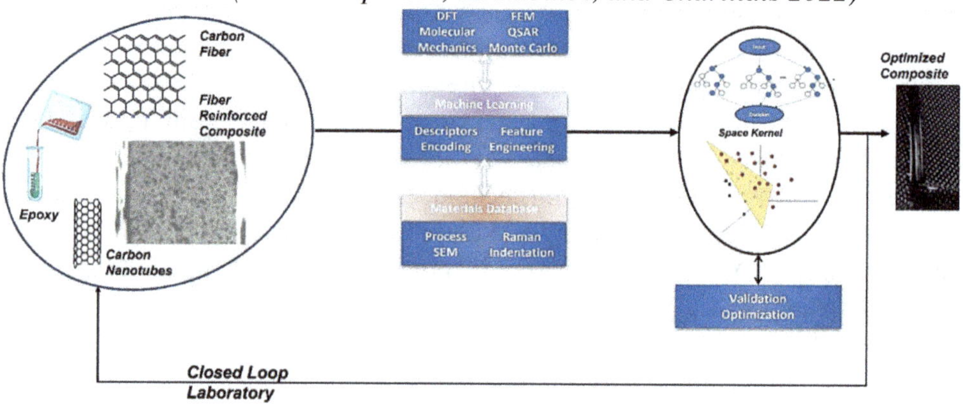

Predicting properties is not the same as designing things backwards. In inverse design, we begin with a desired property and seek structures or formulas that can achieve it. This is difficult because the search space is very large and often split into many islands (Kadulkar et al. 2022). Generative models are useful here. VAEs, GANs, and reinforcement learning can figure out how to go from property space to material space. They can suggest new candidates that are appropriate for a given use. For instance, VAEs have been used to make organic molecules with specific bandgaps and strong oscillators in optoelectronics (C. T. Chen and Gu 2020; Noh et al. 2019). We still do optimization after generation. One goal at a time, or sometimes many goals at once. Real materials need to be strong, flexible, and stable at high temperatures all at once. Genetic algorithms, Bayesian optimization, and particle swarm optimization are all multi-objective tools that now use ML to find trade-offs and choose the best sets within limits (Oliva et al. 2022; P. Zhang et al. 2021; Nagarajan et al. 2022). For example, Bayesian optimization has adjusted the settings for additive manufacturing to get good microstructure and mechanics while keeping porosity low (Said et al. 2022).

ML has worked well with polymers. It can predict with trained model the glass transition temperature, tensile strength, and thermal conductivity, which are important for everything from packaging to aerospace. There are a lot of monomers, chain lengths, and architectures in polymer space. Classic trials have a hard time scanning it quickly. High-throughput screens and machine learning can quickly connect structure to property. Recent research indicates that graph neural networks can capture the structure of polymer chains and predict their elasticity or permeability based on the sequence and mixture of monomers. Reinforcement learning has been employed to propose polymer backbones that optimize ease of processing and thermal stability (Q. Gao et al. 2024; Savari et al. 2017). In alloys, property prediction has helped create new mixes for superalloys, high-entropy alloys, and light metals. Small changes in chemistry, grain size, and phases can change hardness, yield strength, and fracture toughness. ML trained on metallurgy data can make these predictions and suggest new composition areas that should be tested. Researchers have used decision-tree methods to guess whether certain combinations of alloys will make single-phase solid solutions or multiphase mixes, which cuts down on the number of experiments (Q. Nguyen et al. 2020; Tian et al. 2021). Ensemble models also help choose the right heat treatment schedules and thermo-mechanical conditions so that advanced steels and titanium alloys behave the way they should (Q. Chen et al. 2024; Zhu et al. 2021).

ML might also be useful for nanomaterials because their properties change with size and their shapes are hard to understand. We need to use a multi-scale modelling method to predict the properties of nanoparticles, quantum dots, and carbon-based nanostructures because both atomic-scale chemistry and mesoscale form affect them (Kurzydlowski 2010). Researchers have employed ML models, particularly deep neural networks and kernel-based methodologies, to forecast characteristics such as specific surface area, optical absorbance, and catalytic activity, utilizing descriptors including atomic coordination, defect density, and synthesis conditions. Convolutional neural networks trained on microscope images have been employed to categorize and quantify the shapes and distributions of nanoparticles in catalytic materials (Timoshenko et al. 2020). Also, generative models have been used to suggest new nanoparticle shapes that work best as catalysts. This is a more logical way to design catalysts. Composites are hard to work with because they have a hierarchical structure and the different phases interact with each other in different ways. They usually mix several phases to make mechanical or thermal performance better. ML offers a method for simulating these interactions without the expense of finite element simulations. Scientists have used supervised learning models to guess things like thermal conductivity, elastic modulus, and fracture toughness based on how much filler there is, how it is oriented, and how well it sticks to other things. Support vector machines and gradient boosting have been used to find out how the way carbon nanotubes are lined up in a polymer matrix affects the flow of heat and electricity in different directions. Researchers have used Bayesian optimization frameworks in a new way with experimental platforms to find the best fiber-matrix combinations for aerospace composites that can take the most damage (Wang, Wang, and Zhang 2024).

The performance of these applications is often determined by the quality and detail of the input descriptors used to describe materials. In polymers, constitutional descriptors might be atomic ratios, while higher-order descriptors might be chain length and branching index (Himanen et al. 2020). Alloys can be described by differences in atomic size, electronegativity, and thermodynamic stability. Nanomaterials often use geometric properties like surface curvature, aspect ratio, and pore connectivity. The properties of the interface, the shape of the filler, and the quality of the dispersion are the most important things for composites. It is still very hard to choose or design these traits, especially when researcher want to make a model that is both easy to understand and makes accurate predictions. People are also starting to look for new content in different ways because of active learning and data-driven experimentation (Trinh et al. 2023). In this loop, ML models look for areas of uncertainty or high potential in the design space and suggest more tests to run. Then, these are put back into the model to make the predictions more accurate. This method has worked well for finding perovskite materials that work very well with solar energy. With each new set of data, the model gets better at understanding how composition affects properties.

Even though things have gotten a lot better, there are still problems to solve. One of these problems is how well models can make predictions, especially when it comes to inverse design. Many models can interpolate between data sets that already exist, but they usually get the predictions wrong when they try to extrapolate to completely new compositions or structures. To address this issue, ML processes are progressively adopting uncertainty quantification techniques such as Bayesian neural networks and dropout methods (Sulzer et al. 2021; Temraz and Keane 2022). How to use domain information, like thermodynamic rules or symmetry constraints, in learning algorithms is another problem that keeps coming up. This mixed method, which is often called physics-informed ML, makes sure that the model's outputs are not only correct but also realistic in the real world. Transfer learning is becoming more common in materials science, especially when there isn't a lot of data available. This is another important change. Researchers can modify models trained on extensive datasets from analogous domains to forecast characteristics in a distinct yet related material system using a limited amount of new data. This has been especially useful in new areas like biodegradable polymers and sustainable materials, where there isn't a lot of experimental data yet (Verma et al. 2024; Moshood et al. 2022).

In conclusion, ML has made great strides in the last few years in finding and making new materials. It is a helpful addition to the theoretical and experimental methods that are already there. ML is helping researchers look at huge design areas faster and more accurately than ever before, whether they are predicting attributes, doing inverse design, or optimizing for multiple goals. There are case studies on polymers, alloys, nanomaterials, and composites, which shows how versatile and useful these methods can be. As materials science continues to combine computing power, algorithmic complexity, and data accessibility, the synergy between ML and materials engineering will become more and more important for solving scientific problems now and in the future.

5. CHALLENGES AND LIMITATIONS

Combining ML with materials science has opened up new ways to speed up discovery and improve properties. But, like any new field that crosses borders, there are a few big problems and limitations that need to be fixed before it can reach its full potential. Some of the biggest problems are not having enough data, models that can't be used in other situations, algorithms that are hard to understand, and the need for easy integration with basic laws of physics. All of these things could change how reliable, transferable, and useful ML applications are in materials research.

5.1 Lack of Data and Unbalanced Datasets

But, like every new field that crosses traditional ones, there are a lot of big issues and limitations that need to be worked out before it can reach its full potential. Some of the biggest problems are not having enough data, having models that don't work in many situations, having algorithms that are hard to understand, and having to follow basic laws of physics. All of these traits could affect how useful, reliable, and portable ML apps are for materials research.

One of the biggest problems with using ML in materials science is that there isn't enough good data that is well-organized. For example, image recognition and natural language processing have huge labelled datasets. On the other hand, materials databases are usually small, broken up, and only useful in certain areas. This is mostly because it costs a lot of time and money to run simulations like molecular dynamics or density functional theory (Noor et al., 2024; Alzubaidi et al. 2023).

5.2 Limitations of Model Generalization and Extrapolation

Also, current databases are often unbalanced because they have too many well-studied material classes (like metals and oxides) and not enough new or specialized elements. This imbalance could make model predictions less accurate and make it harder to use them on groups that aren't as well represented. For instance, a model that was mostly trained on crystal structures with good symmetry would not be able to predict the properties of systems that are amorphous or have bad symmetry very well. To get around these problems, people are working on things like data augmentation, transfer learning, and active learning. But the main problem with data sparsity is still a big one. Generalization, or how well a model can work with new data, is very important for any real-world use. However, in materials science, many ML models work well while they're being trained but not as well when they're used in other areas. This problem happens when models find patterns that only work with the training data and not with new data (Li et al. 2023; Fuhg et al. 2023).

It is becoming more and more important to be able to talk about and measure how uncertain a model is. Some of the methods that are becoming more popular for giving predictions and confidence intervals are Bayesian neural networks, ensemble learning, and Monte Carlo dropout. Researchers can utilize these methods to distinguish between high-confidence and low-confidence predictions. This is very important when using ML for making decisions in experiments or for autonomous synthesis.

5.3 Understanding Models and Being Honest in Science

Some ML models, such as linear regression and decision trees, are simple to understand. People usually think of other models, especially deep learning models, as "black boxes." This lack of openness is a problem in materials science, where it's just as important to know why a prediction is made as it is to make it. It is also important for domain experts to be able to understand the results so that they can trust them, since they might not believe in conclusions that are based only on statistics. SHAP (Shapley Additive Explanations) and LIME (Local Interpretable Model-Agnostic Explanations) are two new ways to make AI explainable (XAI) that look like they could help get feature significance and decision logic out of complicated models. These methods may help determine whether atomic, structural, or processing characteristics are responsible for a specific material property prediction. It is still hard to find the right balance between how complex a model is and how easy it is to understand (Wojtuch, Jankowski, and Podlewska 2021; Sadeghi et al. 2022; Man and Chan 2021). Figure 5 (Abdullah, Zahid, and Ali 2021) shows a trade-off between accuracy and how easy it is to understand.

Figure 5. Trade-off between accuracy and interpretability (Abdullah, Zahid, and Ali 2021)

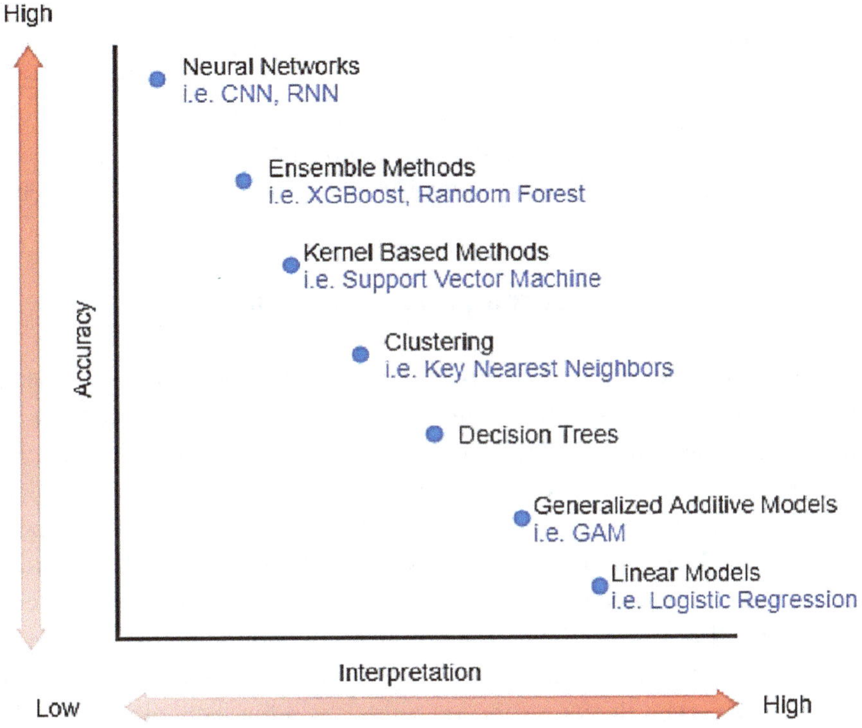

5.4 Integration with Domain Knowledge and Physical Constraints

One of the fundamental problems with scientific ML, as opposed to commercial applications, is that it has to include physical principles and domain knowledge in the learning process. A lot of traditional ML algorithms don't care about basic rules like symmetry, thermodynamic stability, or energy conservation. Because of this, they could make predictions that are mathematically correct yet break established physical rules or create things that can't exist chemically. This problem is not only an academic one in materials science; it may also lead to false predictions and wasted experimental expenditures (B. Liu et al. 2024; Yin et al. 2021). For instance, a model that recommends a novel molecule with great mechanical characteristics could not take into account how unstable it is at room temperature or how it doesn't work with certain synthesis methods. As a result, one of the most important areas of study is putting physics into ML models. Adding restrictions to the loss function, adding symmetry-aware descriptors, or utilizing hybrid models that mix empirical equations with data-driven parts are all approaches to achieve this (Lee and Popovics 2022; E. Zhang et al. 2022).

Physics-informed neural networks (PINNs) are one way to close this gap by explicitly putting governing equations into the training process. Graph neural networks (GNNs) are also good at following chemical and crystallographic rules because they can easily include bonding and structural information. These methods are still being worked on, but they might help make ML models in materials research more accurate and trustworthy.

5.5 Computational Resources and Scalability

As models become increasingly complicated, particularly when deep learning and high-dimensional datasets are included, they need a lot more processing power. Investigators may need access to high-performance computer clusters or specialized hardware like GPUs or TPUs to train a deep neural network to forecast the properties of materials or do inverse design. This makes things more difficult and expensive, especially for smaller research groups or universities that don't have a lot of resources. Also, scalability becomes a problem when trying to connect prediction, synthesis, and characterisation in real time. Systems like autonomous labs or closed-loop materials discovery platforms need to be able to make quick inferences and learn all the time, even when they are short on time. It is a big technological challenge to get this level of performance while keeping the model's correctness and physical validity (Budiutama et al. 2023; Kim et al. 2024; Teferra and Rowenhorst 2021).

6. FUTURE DIRECTIONS AND OPPORTUNITIES

Combining ML with materials science has demonstrated a lot of promise. But the area keeps changing. New trends suggest ways to study, analyze, and create materials that are more seamless, automated, and long-lasting. There are a few things that stand out as likely to affect the next round of study.

6.1 Autonomous Laboratories

The increasing number of autonomous laboratories is a big change that is coming. These are research systems that can do synthesis, characterization, and decision-making on their own, so machines don't need any help from people. ML algorithms help plan the trials, learn from the results, and change their future decisions based on what they learn (Xie et al. 2023; Jeffrey A. Bennett and Abolhasani 2022). This cycle, also known as the closed-loop or self-driving lab, could lead to discoveries much faster than current human-centered projects. In an autonomous lab, robots do chemical synthesis or process materials. Instruments collect data in real time and send it to a ML model. The model looks at the results and suggests what to do next. This process keeps going in a circle, which cuts down on trial and error and makes research more useful. The main benefit is that it is fast. It could take months to get anything done in a regular lab, but it could only take days or hours in a new lab. Autonomous laboratories are also free of biases, so they can explore areas of materials space that people wouldn't normally think to look into (J. A. Bennett et al. 2024; Butakova et al. 2022).

This method has proven effective well in a few test projects. For example, some labs use robotic arms to mix chemicals, start reactions, and then use spectroscopy to look at the results. Some people use automated microscopes to look at structures and sort them. Researchers can go through hundreds or thousands of samples with hardly any work thanks to these methods. But there are still problems. For autonomous laboratories to work, their hardware, software, and data infrastructure need to be closely linked. Robots can't currently conventionally handle many materials synthesis processes. The models also require trustworthy input in real-time to learn well. Investing in automation, data collection tools, and strong ML frameworks will be very important for making these laboratories function on a large scale.

6.2 AI-Materials Integration

Beyond automation, another future trend is the deeper integration of artificial intelligence with materials science. It's not simply about applying ML to look at data. It's about putting AI into every part of the materials cycle, from theory to use. For example, AI models may be added to simulation software to make it less expensive to run. In quantum chemistry or atomistic modelling, deep learning models may take the role of costly simulations (Badini, Regondi, and Pugliese 2023). Variational autoencoders (VAEs) and generative adversarial networks (GANs) are examples of generative models that may propose new chemical compounds with certain attributes. Researchers may be able to use large language models to better search across materials databases or scientific papers (Asperti, Evangelista, and Loli Piccolomini 2021; S. Zhang et al. 2023).

AI may also be used for design work. For instance, inverse design frameworks might start with the desired qualities of a material and then work backward to suggest a structure or composition. Agents that use reinforcement learning can find their way around complicated design areas and work toward several goals at once. As data and algorithms become better, these technologies will become more and more ubiquitous. Integration also implies that people and AI work together, which is important. AI won't replace scientists; instead, it will work with them.

6.3 Toward Sustainable Development

Machine intelligence also opens up new approaches to making materials that last. Because of climate change, resource scarcity, and environmental deterioration, this is becoming more and more important. Research on materials may be very helpful in making energy systems cleaner, manufacturing greener, and economies that are more circular. One way that ML may help is by finding materials that have less of an effect on the environment. For instance, computers may look for catalysts that don't need as many rare earth metals or polymers that break down more quickly. ML may also make industrial processes better so that they use less energy, make less waste, or have less harmful byproducts. ML can assist find electrodes or electrolytes that last longer and are better for the environment when designing batteries (Moses et al. 2021; Peng and Sadaghiani, 2024).

Life cycle thinking is very important for reaching sustainability objectives. This means looking at not just how well a material works, but also how much damage it does from the time it is mined until it is thrown away. Combining ML with life cycle assessment (LCA) techniques may offer a whole picture. For innovation to survive, open-source data sharing and collaboration platforms are also very important.

7. CONCLUSION

ML has quickly gone from being a helpful analytical tool to being a key part of modern materials science. It can find patterns, predict properties, and make things better, which has led to new ways to find and make materials with certain functions. This study has demonstrated the application of ML methodologies, encompassing traditional algorithms to deep learning models, to expedite innovation across various materials. We have talked about how important it is to have good data, how hard it is to understand models, and how important it is to combine domain knowledge with algorithmic efficiency. These methods will only be effective if there are numerous, well-curated datasets, models carefully designed to adhere to physical laws, and teams from diverse fields collaborating. Several things are about to happen that will change the way research is done. AI-driven decision-making and real-time data loops will make autonomous labs more likely to accelerate and reduce the cost of discovering new materials. Adding sustainability indicators to ML pipelines will also become increasingly relevant for driving responsible innovation. Even if things have improved, significant challenges remain, particularly in areas where data is scarce, such as when AI is applied to high-impact applications or generalized to new materials systems.

In conclusion, ML is a handy tool for solving hard issues in materials science. But its success does not simply depend on algorithms; it also depends on careful, open, and context-aware implementation. As the discipline moves forward, it will be important to continue investing in data infrastructure, educating people from other fields, and using open scientific methods. The future of new materials will probably be more intelligent, more automated, and more in line with the world's aims for sustainability.

REFERENCES

Abdullah, T. A. A., Mohd, S. M. Z., & Ali, W. (2021). A Review of Interpretable Ml in Healthcare: Taxonomy, Applications, Challenges, and Future Directions. *Symmetry*, *13*(12), 2439. Advance online publication. DOI: 10.3390/sym13122439

Abel, D., Arumugam, D., Lehnert, L., & Littman, M. (2018, July). State abstractions for lifelong reinforcement learning. In *International Conference on Machine Learning* (pp. 10-19). PMLR.

Abolhasani, M., & Kumacheva, E. (2023). The Rise of Self-Driving Labs in Chemical and Materials Sciences. *Nature Synthesis*, *2*(6), 483–492. Advance online publication. DOI: 10.1038/s44160-022-00231-0

Ahmadi, M. H., Ahmadi, M. A., Nazari, M. A., Mahian, O., & Ghasempour, R. (2019). A Proposed Model to Predict Thermal Conductivity Ratio of Al2O3/EG Nanofluid by Applying Least Squares Support Vector Machine (LSSVM) and Genetic Algorithm as a Connectionist Approach. *Journal of Thermal Analysis and Calorimetry*, *135*(1), 271–281. DOI: 10.1007/s10973-018-7035-z

Alharkan, H., Habib, S., & Islam, M. (2023). Solar Power Prediction Using Dual Stream CNN-LSTM Architecture. *Sensors (Basel)*, *23*(2), 945. Advance online publication. DOI: 10.3390/s23020945

Alzubaidi, L., Bai, J., Al-Sabaawi, A., Santamaría, J., Albahri, A. S., Al-dabbagh, B. S. N., Fadhel, M. A., Manoufali, M., Zhang, J., Al-Timemy, A. H., Duan, Y., Abdullah, A., Farhan, L., Lu, Y., Gupta, A., Albu, F., Abbosh, A., & Gu, Y. (2023). A Survey on Deep Learning Tools Dealing with Data Scarcity: Definitions, Challenges, Solutions, Tips, and Applications. *Journal of Big Data*, *10*(1), 46. Advance online publication. DOI: 10.1186/s40537-023-00727-2

Asperti, A., Evangelista, D., & Piccolomini, E. L. (2021). A Survey on Variational Autoencoders from a Green AI Perspective. *SN Computer Science*, *2*(4), 301. Advance online publication. DOI: 10.1007/s42979-021-00702-9

Badini, S., Regondi, S., & Pugliese, R. (2023). Unleashing the Power of Artificial Intelligence in Materials Design. *Materials (Basel)*, *16*(17), 5927. Advance online publication. DOI: 10.3390/ma16175927

Bennett, J. A., & Abolhasani, M. (2022). Autonomous Chemical Science and Engineering Enabled by Self-Driving Laboratories. *Current Opinion in Chemical Engineering*, *36*, 100831. Advance online publication. DOI: 10.1016/j.coche.2022.100831

Bennett, J. A., Orouji, N., Khan, M., Sadeghi, S., Rodgers, J., & Abolhasani, M. (2024). Autonomous Reaction Pareto-Front Mapping with a Self-Driving Catalysis Laboratory. *Nature Chemical Engineering*, *1*(3), 240–250. Advance online publication. DOI: 10.1038/s44286-024-00033-5

Budiutama, G., Li, R., Manzhos, S., & Ihara, M. (2023). Hybrid Density Functional Tight Binding (DFTB)—Molecular Mechanics Approach for a Low-Cost Expansion of DFTB Applicability. *Journal of Chemical Theory and Computation*, *19*(15), 5189–5198. Advance online publication. DOI: 10.1021/acs.jctc.3c00310

Butakova, M. A., Chernov, A. V., Kartashov, O. O., & Soldatov, A. V. (2022). Data-Centric Architecture for Self-Driving Laboratories with Autonomous Discovery of New Nanomaterials. *Nanomaterials (Basel, Switzerland)*, *12*(1), 12. Advance online publication. DOI: 10.3390/nano12010012

Calderon, C. E., Plata, J. J., Toher, C., Oses, C., Levy, O., Fornari, M., Natan, A., Mehl, M. J., Hart, G., Buongiorno Nardelli, M., & Curtarolo, S. (2015). The AFLOW Standard for High-Throughput Materials Science Calculations. *Computational Materials Science*, *108*, 233–238. Advance online publication. DOI: 10.1016/j.commatsci.2015.07.019

Chen, C. T., & Gu, G. X. (2020). Generative Deep Neural Networks for Inverse Materials Design Using Backpropagation and Active Learning. *Advanced Science (Weinheim, Baden-Wurttemberg, Germany)*, *7*(5), 1902607. Advance online publication. DOI: 10.1002/advs.201902607

Chen, G., & Tang, D.-M. (2024). Machine Learning as a 'Catalyst' for Advancements in Carbon Nanotube Research. *Nanomaterials (Basel, Switzerland)*, *14*(21), 1688. DOI: 10.3390/nano14211688

Chen, Q., He, Z., Zhao, Y., Liu, X., Wang, D., Zhong, Y., Hu, C., Hao, C., Lu, K., & Wang, Z. (2024). Stacking Ensemble Learning Assisted Design of Al-Nb-Ti-V-Zr Lightweight High-Entropy Alloys with High Hardness. *Materials & Design*, *246*(October), 113363. DOI: 10.1016/j.matdes.2024.113363

Chong, S. S., Ng, Y. S., Hui, Q. W., & Jin, C. Z. (2024). Advances of Machine Learning in Materials Science: Ideas and Techniques. *Frontiers of Physics*, *19*(1), 13501. Advance online publication. DOI: 10.1007/s11467-023-1325-z

Choubisa, H., Askerka, M., Ryczko, K., Voznyy, O., Mills, K., Tamblyn, I., & Sargent, E. H. (2020). Crystal Site Feature Embedding Enables Exploration of Large Chemical Spaces. *Matter*, *3*(2), 433–448. Advance online publication. DOI: 10.1016/j.matt.2020.04.016

Cichy, C., & Rass, S. (2019). An Overview of Data Quality Frameworks. *IEEE Access : Practical Innovations, Open Solutions, 7*, 24634–24648. Advance online publication. DOI: 10.1109/ACCESS.2019.2899751

Deepthi Jayan, K., & Ragin Ramdas, M. (2022). Exploring the Photovoltaic Capabilities of Sc4C3 MXene Using Density Functional Theory. *Materials Letters, 318*, 132217. Advance online publication. DOI: 10.1016/j.matlet.2022.132217

Deng, T., Ran, Y., Yin, Y., & Liu, P. (2020). Multi-Objective Optimization Design of Thermal Management System for Lithium-Ion Battery Pack Based on Non-Dominated Sorting Genetic Algorithm II. *Applied Thermal Engineering, 164*, 114394. Advance online publication. DOI: 10.1016/j.applthermaleng.2019.114394

Elton, D. C., Boukouvalas, Z., Butrico, M. S., Fuge, M. D., & Chung, P. W. (2018). Applying Machine Learning Techniques to Predict the Properties of Energetic Materials. *Scientific Reports, 8*(1), 9059. Advance online publication. DOI: 10.1038/s41598-018-27344-x

Frydrych, K., Karimi, K., Pecelerowicz, M., Alvarez, R., Dominguez-Gutiérrez, F. J., Rovaris, F., & Papanikolaou, S. (2021). Materials Informatics for Mechanical Deformation: A Review of Applications and Challenges. *Materials (Basel), 14*(19), 5764. Advance online publication. DOI: 10.3390/ma14195764

Fuhg, J. N., Hamel, C. M., Johnson, K., Jones, R., & Bouklas, N. (2023). Modular Machine Learning-Based Elastoplasticity: Generalization in the Context of Limited Data. *Computer Methods in Applied Mechanics and Engineering, 407*, 115930. Advance online publication. DOI: 10.1016/j.cma.2023.115930

Fuhr, A. S., & Sumpter, B. G. (2022). Deep Generative Models for Materials Discovery and Machine Learning-Accelerated Innovation. *Frontiers in Materials, 9*, 865270. Advance online publication. DOI: 10.3389/fmats.2022.865270

Gao, Q., Dukker, T., Schweidtmann, A. M., & Weber, J. M. (2024). Self-Supervised Graph Neural Networks for Polymer Property Prediction. *Molecular Systems Design & Engineering, 9*(11), 1130–1143. DOI: 10.1039/D4ME00088A

Gao, X., Jamil, N., Ramli, M. I., & Syed, M. Z. S. Z. A. (2024). A Comparative Analysis of Combination of CNN-Based Models with Ensemble Learning on Imbalanced Data. *JOIV: International Journal on Informatics Visualization, 8*(1), 456. Advance online publication. DOI: 10.62527/joiv.8.1.2194

Han, G., Sun, Y., Feng, Y., Lin, G., & Lu, N. (2021). *Machine Learning Regression Guided Thermoelectric Materials Discovery – A Review*. ES Materials & Manufacturing., DOI: 10.30919/esmm5f451

Himanen, L., Jäger, M. O. J., Morooka, E. V., Canova, F. F., Ranawat, Y. S., Gao, D. Z., Rinke, P., & Foster, A. S. (2020). DScribe: Library of Descriptors for Machine Learning in Materials Science. *Computer Physics Communications*, *247*, 106949. Advance online publication. DOI: 10.1016/j.cpc.2019.106949

Hinton, E., & Owen, D. R. J. (1986). *Finite elements in plasticity: Theory and practice*. Pineridge Press Limited.

Hospital, A., Goñi, J. R., Orozco, M., & Gelpí, J. L. (2015). Molecular Dynamics Simulations: Advances and Applications. *Advances and Applications in Bioinformatics and Chemistry : AABC*, 37. Advance online publication. DOI: 10.2147/AABC.S70333

Jolliffe, I. (2022). A 50-Year Personal Journey through Time with Principal Component Analysis. *Journal of Multivariate Analysis*, *188*, 104820. Advance online publication. DOI: 10.1016/j.jmva.2021.104820

Jung, Y. J., Han, S. H., & Choi, H. J. (2021). Explaining CNN and RNN Using Selective Layer-Wise Relevance Propagation. *IEEE Access : Practical Innovations, Open Solutions*, *9*, 18670–18681. Advance online publication. DOI: 10.1109/ACCESS.2021.3051171

Kadulkar, S., Sherman, Z. M., Ganesan, V., & Truskett, T. M. (2022). Machine Learning-Assisted Design of Material Properties. *Annual Review of Chemical and Biomolecular Engineering*, *13*(1), 235–254. Advance online publication. DOI: 10.1146/annurev-chembioeng-092220-024340

Kanti, P. K., Sharma, P., Maiya, M. P., & Sharma, K. V. (2023). The stability and thermophysical properties of Al2O3-graphene oxide hybrid nanofluids for solar energy applications: Application of robust autoregressive modern machine learning technique. *Solar Energy Materials and Solar Cells*, *253*, 112207.

Kanti, P. K., Shrivastav, A. P., Sharma, P., & Maiya, M. P. (2024). Thermal performance enhancement of metal hydride reactor for hydrogen storage with graphene oxide nanofluid: Model prediction with machine learning. *International Journal of Hydrogen Energy*, *52*, 470–484.

Ke, G., Meng, Q., Finley, T., Wang, T., Chen, W., Ma, W., Ye, Q., & Tie, Y. L. 2017. "LightGBM: A Highly Efficient Gradient Boosting Decision Tree." In *Advances in Neural Information Processing Systems*. Vol. 2017-December.

Kim, S., Jeon, H., Koo, J. M., Oh, D. X., & Park, J. (2024). Practical Applications of Self-Healing Polymers Beyond Mechanical and Electrical Recovery. *Advanced Science (Weinheim, Baden-Wurttemberg, Germany)*, *11*(16), 2302463. Advance online publication. DOI: 10.1002/advs.202302463

Kirklin, S., Saal, J. E., Meredig, B., Thompson, A., Doak, J. W., Aykol, M., Rühl, S., & Wolverton, C. (2015). The Open Quantum Materials Database (OQMD): Assessing the Accuracy of DFT Formation Energies. *npj Computational Materials*, *1*(1), 15010. Advance online publication. DOI: 10.1038/npjcompumats.2015.10

Konstantopoulos, G., Koumoulos, E. P., & Charitidis, C. A. (2022). Digital Innovation Enabled Nanomaterial Manufacturing; Machine Learning Strategies and Green Perspectives. *Nanomaterials (Basel, Switzerland)*, *12*(15), 2646. Advance online publication. DOI: 10.3390/nano12152646

Kurzydlowski, K. J. (2010). Modelling of the Microstructure and Properties in the Length Scales Varying from Nanoto Macroscopic. *Bulletin of the Polish Academy of Sciences. Technical Sciences*, *58*(2). Advance online publication. DOI: 10.2478/v10175-010-0021-7

Le, A. T., Pandey, A., Sirohi, R., Sharma, P., Chen, W.-H., Nguyen, D. K. P., Tran, V. D., Nguyen, X. P., & Hoang, A. T. (2023). Precise Prediction of Biochar Yield and Proximate Analysis by Modern Machine Learning and SHapley Additive ExPlanations. *Energy & Fuels*, *37*(22), 17310–17327. DOI: 10.1021/acs.energyfuels.3c02868

Lee, S., & Popovics, J. S. (2022). Applications of Physics-informed Neural Networks for Property Characterization of Complex Materials. *RILEM Technical Letters*, *7*, 178–188. Advance online publication. DOI: 10.21809/rilemtechlett.2022.174

León, G., Fernanda, M., & Dewulf, J. (2020). Data Quality Assessment Framework for Critical Raw Materials. The Case of Cobalt. *Resources, Conservation and Recycling*, *157*, 104564. Advance online publication. DOI: 10.1016/j.resconrec.2019.104564

Li, K., DeCost, B., Choudhary, K., Greenwood, M., & Hattrick-Simpers, J. (2023). A Critical Examination of Robustness and Generalizability of Machine Learning Prediction of Materials Properties. *npj Computational Materials*, *9*(1), 55. Advance online publication. DOI: 10.1038/s41524-023-01012-9

Liu, B., Wang, Y., Rabczuk, T., Olofsson, T., & Lu, W. (2024). Multi-Scale Modeling in Thermal Conductivity of Polyurethane Incorporated with Phase Change Materials Using Physics-Informed Neural Networks. *Renewable Energy*, *220*, 119565. Advance online publication. DOI: 10.1016/j.renene.2023.119565

Liu, Q., Deng, M., Shi, Y., & Wang, J. (2012). A Density-Based Spatial Clustering Algorithm Considering Both Spatial Proximity and Attribute Similarity. *Computers & Geosciences*, *46*, 296–309. Advance online publication. DOI: 10.1016/j.cageo.2011.12.017

Liu, X., Zhang, J., & Pei, Z. (2023). Machine Learning for High-Entropy Alloys: Progress, Challenges and Opportunities. *Progress in Materials Science*, *131*, 101018. Advance online publication. DOI: 10.1016/j.pmatsci.2022.101018

Liu, Y., Niu, C., Wang, Z., Gan, Y., Zhu, Y., Sun, S., & Shen, T. (2020). Machine Learning in Materials Genome Initiative: A Review. *Journal of Materials Science and Technology*, *57*, 113–122. Advance online publication. DOI: 10.1016/j.jmst.2020.01.067

Liu, Y., Zhao, T., Ju, W., & Shi, S. (2017). Materials Discovery and Design Using Machine Learning. *Journal of Materiomics*, *3*(3), 159–177. Advance online publication. DOI: 10.1016/j.jmat.2017.08.002

Liu, Z., Shi, Y., Chen, H., Qin, T., Zhou, X., Huo, J., Dong, H., Yang, X., Zhu, X., Chen, X., Zhang, L., Yang, M., Gao, Y., & Ma, J. (2021). Machine Learning on Properties of Multiscale Multisource Hydroxyapatite Nanoparticles Datasets with Different Morphologies and Sizes. *npj Computational Materials*, *7*(1), 142. Advance online publication. DOI: 10.1038/s41524-021-00618-1

Liu, Y. X., Dong, Y. W., Jiang, Z. H., Li, Y. S., Zha, W., Du, Y. X., & Du, S. Y. (2023). XGBoost-based model for predicting hydrogen content in electroslag remelting. *Journal of Iron and Steel Research International*, *30*(5), 887–896.

Lu, B., Xia, Y., Ren, Y., Xie, M., Zhou, L., Vinai, G., Morton, S. A., Wee, A. T. S., van der Wiel, W. G., Zhang, W., & Wong, P. K. J. (2024). When Machine Learning Meets 2D Materials: A Review. *Advanced Science (Weinheim, Baden-Wurttemberg, Germany)*, *11*(13), 2305277. Advance online publication. DOI: 10.1002/advs.202305277

Maleki, A., Haghighi, A., & Mahariq, I. (2021). Machine Learning-Based Approaches for Modeling Thermophysical Properties of Hybrid Nanofluids: A Comprehensive Review. *Journal of Molecular Liquids*, *322*(January), 114843. Advance online publication. DOI: 10.1016/j.molliq.2020.114843

Man, X., & Chan, E. P. (2021). The Best Way to Select Features? Comparing MDA, LIME, and SHAP. *The Journal of Financial Data Science*, *3*(1), 127–139. Advance online publication. DOI: 10.3905/jfds.2020.1.047

Mattur, M. N., Nagappan, N., Rath, S., & Thomas, T. (2022). Prediction of nature of band gap of perovskite oxides (ABO3) using a machine learning approach. *Journal of Materiomics*, *8*(5), 937–948.

Moses, O. A., Chen, W., Adam, M. L., Wang, Z., Liu, K., Shao, J., & Yu, X. (2021). Integration of data-intensive, machine learning and robotic experimental approaches for accelerated discovery of catalysts in renewable energy-related reactions. *Materials Reports: Energy*, *1*(3), 100049.

Moshood, T. D., Nawanir, G., Mahmud, F., Mohamad, F., & Ahmad, M. H., & AbdulGhani, A. (2022). Sustainability of biodegradable plastics: New problem or solution to solve the global plastic pollution? *Current Research in Green and Sustainable Chemistry*, *5*, 100273.

Nagarajan, V., Solaiyappan, A., Mahalingam, S. K., Nagarajan, L., Salunkhe, S., Nasr, E. A., Shanmugam, R., & Hussein, H. M. A. M. (2022). Meta-Heuristic Technique-Based Parametric Optimization for Electrochemical Machining of Monel 400 Alloys to Investigate the Material Removal Rate and the Sludge. *Applied Sciences (Basel, Switzerland)*, *12*(6), 2793. Advance online publication. DOI: 10.3390/app12062793

Nandy, A., Duan, C., & Kulik, H. J. (2022). Audacity of Huge: Overcoming Challenges of Data Scarcity and Data Quality for Machine Learning in Computational Materials Discovery. *Current Opinion in Chemical Engineering*, *36*, 100778. Advance online publication. DOI: 10.1016/j.coche.2021.100778

Natrayan, L., Janardhan, G., Paramasivam, P., & Dhanasekaran, S. (2023). Enhancing Mechanical Performance of TiO2 Filler with Kevlar/Epoxy-Based Hybrid Composites in a Cryogenic Environment: A Statistical Optimization Study Using RSM and ANN Methods. *Frontiers in Materials*, *10*, 1267514. Advance online publication. DOI: 10.3389/fmats.2023.1267514

Nguyen, Q., Rizvandi, R., Karimipour, A., Malekahmadi, O., & Bach, Q. V. (2020). A Novel Correlation to Calculate Thermal Conductivity of Aqueous Hybrid Graphene Oxide/Silicon Dioxide Nanofluid: Synthesis, Characterizations, Preparation, and Artificial Neural Network Modeling. *Arabian Journal for Science and Engineering*, *45*(11), 9747–9758. DOI: 10.1007/s13369-020-04885-w

Nguyen, V. G., Sharma, P., Ağbulut, Ü., Le, H. S., Truong, T. H., Dzida, M., Tran, M. H., Le, H. C., & Tran, V. D. (2024). Machine Learning for the Management of Biochar Yield and Properties of Biomass Sources for Sustainable Energy. *Biofuels, Bioproducts & Biorefining*, *18*(February), 567–593. Advance online publication. DOI: 10.1002/bbb.2596

Nguyen, V. N., Tarełko, W., Sharma, P., El-Shafay, A. S., Chen, W.-H., Phuoc, Q. P. N., Nguyen, X. P., & Hoang, A. T. (2024). Potential of Explainable Artificial Intelligence in Advancing Renewable Energy: Challenges and Prospects. *Energy & Fuels*, *38*(3), 1692–1712. DOI: 10.1021/acs.energyfuels.3c04343

Noh, J., Kim, J., Stein, H. S., Sanchez-Lengeling, B., Gregoire, J. M., Aspuru-Guzik, A., & Jung, Y. (2019). Inverse Design of Solid-State Materials via a Continuous Representation. *Matter*, *1*(5), 1370–1384. Advance online publication. DOI: 10.1016/j.matt.2019.08.017

Noor, B., Fahel, M., Yasmin, N., & Besara, T. (2024). Machine Learning in High-Entropy Alloys: Phase Formation Predictions with Artificial Neural Networks. *Future Sustainability*, *2*(1), 47–58. Advance online publication. DOI: 10.55670/fpll.fusus.2.1.5

Oliva, D., Houssein, E. H., Abualigah, L., & Elaziz, M. A. (2022). *Integrating Meta-Heuristics and Machine Learning for Real-World Optimization Problems*. Springer International Publishing.

Paunovska, K., Loughrey, D., Sago, C. D., Langer, R., & Dahlman, J. E. (2019). Using Large Datasets to Understand Nanotechnology. *Advanced Materials*, *31*(43), 1902798. DOI: 10.1002/adma.201902798

Peng, W., & Sadaghiani, O. K. (2024). Machine Learning for Sustainable Reutilization of Waste Materials as Energy Sources–a Comprehensive Review. *International Journal of Green Energy*, *21*(7), 1641–1666. Advance online publication. DOI: 10.1080/15435075.2023.2255647

Pogue, E. A., New, A., McElroy, K., Le, N. Q., Pekala, M. J., McCue, I., Gienger, E., Domenico, J., Hedrick, E., McQueen, T. M., Wilfong, B., Piatko, C. D., Ratto, C. R., Lennon, A., Chung, C., Montalbano, T., Bassen, G., & Stiles, C. D. (2023). Closed-Loop Superconducting Materials Discovery. *npj Computational Materials*, *9*(1), 181. Advance online publication. DOI: 10.1038/s41524-023-01131-3

Prateek, S., Garg, R., & Saxena, K. K. (2024). Data-Driven Materials Science: Application of ML for Predicting Band Gap. *Advances in Materials and Processing Technologies*, *10*(2), 708–717. Advance online publication. DOI: 10.1080/2374068X.2023.2171666

Rajak, P., Krishnamoorthy, A., Mishra, A., Kalia, R., Nakano, A., & Vashishta, P. (2021). Autonomous Reinforcement Learning Agent for Chemical Vapor Deposition Synthesis of Quantum Materials. *npj Computational Materials*, *7*(1), 108. Advance online publication. DOI: 10.1038/s41524-021-00535-3

Rajak, P., Wang, B., Nomura, K., Luo, Y., Nakano, A., Kalia, R., & Vashishta, P. (2021). Autonomous Reinforcement Learning Agent for Stretchable Kirigami Design of 2D Materials. *npj Computational Materials*, *7*(1), 102. Advance online publication. DOI: 10.1038/s41524-021-00572-y

Rehman, K. U., Shatanawi, W., & Çolak, A. B. (2023). Levenberg–Marquardt Training Technique Analysis of Thermally Radiative and Chemically Reactive Stagnation Point Flow of Non-Newtonian Fluid with Temperature Dependent Thermal Conductivity. *Mathematics*, *11*(3), 753. Advance online publication. DOI: 10.3390/math11030753

Saal, J. E., Kirklin, S., Aykol, M., Meredig, B., & Wolverton, C. (2013). Materials Design and Discovery with High-Throughput Density Functional Theory: The Open Quantum Materials Database (OQMD). *Journal of the Minerals Metals & Materials Society*, *65*(11), 1501–1509. Advance online publication. DOI: 10.1007/s11837-013-0755-4

Sadeghi, G., Mehrali, M., Shahi, M., & Brem, G. (2022). Experimental analysis of Shape-Stabilized PCM applied to a Direct-Absorption evacuated tube solar collector exploiting sodium acetate trihydrate and graphite. *Energy Conversion and Management*, *269*, 116176.

Said, Z., Sharma, P., Bora, B. J., & Pandey, A. K. (2023). Sonication impact on thermal conductivity of f-MWCNT nanofluids using XGBoost and Gaussian process regression. *Journal of the Taiwan Institute of Chemical Engineers*, *145*, 104818.

Said, Z., Sharma, P., Sundar, L. S., Nguyen, V. G., Tran, V. D., & Le, V. V. (2022). Using Bayesian Optimization and Ensemble Boosted Regression Trees for Optimizing Thermal Performance of Solar Flat Plate Collector under Thermosyphon Condition Employing MWCNT-Fe3O4/Water Hybrid Nanofluids. *Sustainable Energy Technologies and Assessments*, *53*(October), 102708. DOI: 10.1016/j.seta.2022.102708

Savari, M., Moghaddam, A. H., Amiri, A., Shanbedi, M., & Ayub, M. N. B. (2017). Comprehensive heat transfer correlation for water/ethylene glycol-based graphene (nitrogen-doped graphene) nanofluids derived by artificial neural network (ANN) and adaptive neuro-fuzzy inference system (ANFIS). *Heat and Mass Transfer*, *53*(10), 3073–3083.

Sharma, K. V., & Sesha Talpa Sai, P. H. V. (2023). Prognostic Modeling of Polydisperse SiO2/Aqueous Glycerol Nanofluids' Thermophysical Profile Using an Explainable Artificial Intelligence (XAI) Approach. *Engineering Applications of Artificial Intelligence*, *126*(November), 106967. DOI: 10.1016/j.engappai.2023.106967

Sharma, P., & Bora, B. J. (2022). A Review of Modern Machine Learning Techniques in the Prediction of Remaining Useful Life of Lithium-Ion Batteries. *Batteries*, *9*(1), 13. DOI: 10.3390/batteries9010013

Sharma, P., Said, Z., Kumar, A., Nižetić, S., Pandey, A., Hoang, A. T., Huang, Z., Afzal, A., Li, C., Le, A. T., Nguyen, X. P., & Tran, V. D. (2022). Recent Advances in Machine Learning Research for Nanofluid-Based Heat Transfer in Renewable Energy System. *Energy & Fuels*, *36*(13), 6626–6658. DOI: 10.1021/acs.energyfuels.2c01006

Shen, J., Griesemer, S. D., Gopakumar, A., Baldassarri, B., Saal, J. E., Aykol, M., Hegde, V. I., & Wolverton, C. (2022). Reflections on One Million Compounds in the Open Quantum Materials Database (OQMD). *JPhys Materials*, *5*(3), 031001. Advance online publication. DOI: 10.1088/2515-7639/ac7ba9

Singh, S. K., Tiwari, A. K., & Paliwal, H. K. (2023). A state-of-the-art review on the utilization of machine learning in nanofluids, solar energy generation, and the prognosis of solar power. *Engineering Analysis with Boundary Elements*, *155*, 62–86.

Spiwok, V., & Kříž, P. (2020). Time-Lagged t-Distributed Stochastic Neighbor Embedding (t-SNE) of Molecular Simulation Trajectories. *Frontiers in Molecular Biosciences*, *7*, 132. Advance online publication. DOI: 10.3389/fmolb.2020.00132

Sui, F., Guo, R., Zhang, Z., Gu, G. X., & Lin, L. (2021). Deep Reinforcement Learning for Digital Materials Design. *ACS Materials Letters*, *3*(10), 1433–1439. Advance online publication. DOI: 10.1021/acsmaterialslett.1c00390

Sulzer, V., Mohtat, P., Aitio, A., Lee, S., Yeh, Y. T., Steinbacher, F., Khan, M. U., Lee, J. W., Siegel, J. B., Stefanopoulou, A. G., & Howey, D. A. (2021). The Challenge and Opportunity of Battery Lifetime Prediction from Field Data. *Joule*, *5*(8), 1934–1955. Advance online publication. DOI: 10.1016/j.joule.2021.06.005

Talaei Khoei, T., & Kaabouch, N. (2023). Machine Learning: Models, Challenges, and Research Directions. *Future Internet*, *15*(10), 332. Advance online publication. DOI: 10.3390/fi15100332

Tariq, R., Torres-Aguilar, C. E., Sheikh, N. A., Ahmad, T., Xamán, J., & Bassam, A. (2022). Data Engineering for Digital Twining and Optimization of Naturally Ventilated Solar Façade with Phase Changing Material under Global Projection Scenarios. *Renewable Energy*, *187*, 1184–1203. Advance online publication. DOI: 10.1016/j.renene.2022.01.044

Teferra, K., & Rowenhorst, D. J. (2021). Optimizing the Cellular Automata Finite Element Model for Additive Manufacturing to Simulate Large Microstructures. *Acta Materialia*, *213*, 116930. Advance online publication. DOI: 10.1016/j.actamat.2021.116930

Temraz, M., & Keane, M. T. (2022). Solving the Class Imbalance Problem Using a Counterfactual Method for Data Augmentation. *Machine Learning with Applications*, *9*(September), 100375. DOI: 10.1016/j.mlwa.2022.100375

Tian, S., Arshad, N. I., Toghraie, D., Eftekhari, S. A., & Hekmatifar, M. (2021). Using Perceptron Feed-Forward Artificial Neural Network (ANN) for Predicting the Thermal Conductivity of Graphene Oxide-Al2O3/Water-Ethylene Glycol Hybrid Nanofluid. *Case Studies in Thermal Engineering*, *26*(August), 101055. DOI: 10.1016/j.csite.2021.101055

Timoshenko, J., Jeon, H. S., Sinev, I., Haase, F. T., Herzog, A., & Cuenya, B. R. (2020). Linking the Evolution of Catalytic Properties and Structural Changes in Copper-Zinc Nanocatalysts Using: Operando EXAFS and Neural-Networks. *Chemical Science (Cambridge)*, *11*(14), 3727–3736. Advance online publication. DOI: 10.1039/D0SC00382D

Trinh, C., Tbatou, Y., Lasala, S., Herbinet, O., & Meimaroglou, D. (2023). On the Development of Descriptor-Based Machine Learning Models for Thermodynamic Properties: Part 1—From Data Collection to Model Construction: Understanding of the Methods and Their Effects. *Processes (Basel, Switzerland)*, *11*(12), 3325. DOI: 10.3390/pr11123325

Verma, S. K., Arbind Prasad, S., & Katiyar, V. (2024). State of Art Review on Sustainable Biodegradable Polymers with a Market Overview for Sustainability Packaging. *Materials Today Sustainability*, *26*(June), 100776. DOI: 10.1016/j.mtsust.2024.100776

Wang, Y., Wang, K., & Zhang, C. (2024). Applications of Artificial Intelligence/Machine Learning to High-Performance Composites. *Composites. Part B, Engineering*, *285*(October), 111740. DOI: 10.1016/j.compositesb.2024.111740

White, A. D. (2022). Deep Learning for Molecules and Materials. *Living Journal of Computational Molecular Science*, *3*(1), 1499. Advance online publication. DOI: 10.33011/livecoms.3.1.1499

Wojtuch, A., Jankowski, R., & Podlewska, S. (2021). How Can SHAP Values Help to Shape Metabolic Stability of Chemical Compounds? *Journal of Cheminformatics*, *13*(1), 74. Advance online publication. DOI: 10.1186/s13321-021-00542-y

Xie, Y., Sattari, K., Zhang, C., & Lin, J. (2023). Toward Autonomous Laboratories: Convergence of Artificial Intelligence and Experimental Automation. *Progress in Materials Science*, *132*, 101043. Advance online publication. DOI: 10.1016/j.pmatsci.2022.101043

Xu, P., Ji, X., Li, M., & Lu, W. (2023). Small Data Machine Learning in Materials Science. *npj Computational Materials*, *9*(1), 42. Advance online publication. DOI: 10.1038/s41524-023-01000-z

Yaman, M. Y., Kalinin, S. V., Guye, K. N., Ginger, D. S., & Ziatdinov, M. (2023). Learning and Predicting Photonic Responses of Plasmonic Nanoparticle Assemblies via Dual Variational Autoencoders. *Small*, *19*(25), 2205893. Advance online publication. DOI: 10.1002/smll.202205893

Yin, M., Zheng, X., Humphrey, J. D., & Karniadakis, G. E. (2021). Non-Invasive Inference of Thrombus Material Properties with Physics-Informed Neural Networks. *Computer Methods in Applied Mechanics and Engineering*, *375*, 113603. Advance online publication. DOI: 10.1016/j.cma.2020.113603

Zhang, E., Dao, M., Karniadakis, G. E., & Suresh, S. (2022). Analyses of Internal Structures and Defects in Materials Using Physics-Informed Neural Networks. *Science Advances*, *8*(7), eabk0644. Advance online publication. DOI: 10.1126/sciadv.abk0644

Zhang, P., Zhen, Y. Y., Yin, F. J., Chan, T. H. T., & Fu, P. G. (2021). Intelligent Modelling of Clay Compressibility Using Hybrid Meta-Heuristic and Machine Learning Algorithms. *Geoscience Frontiers*, *12*(1), 441–452. DOI: 10.1016/j.gsf.2020.02.014

Zhang, S., Qian, Z., Huang, K., Zhang, R., Xiao, J., He, Y., & Lu, C. (2023). Robust Generative Adversarial Network. *Machine Learning*, *112*(12), 5135–5161. Advance online publication. DOI: 10.1007/s10994-023-06367-0

Zhu, C., Li, C., Di Wu, W. Y., Shi, S., Ming, H., Zhang, X., & Zhou, K. (2021). A Titanium Alloys Design Method Based on High-Throughput Experiments and Machine Learning. *Journal of Materials Research and Technology*, *11*(March), 2336–2353. DOI: 10.1016/j.jmrt.2021.02.055

Zhuo, Y., Tehrani, A. M., & Brgoch, J. (2018). Predicting the Band Gaps of Inorganic Solids by Machine Learning. *The Journal of Physical Chemistry Letters*, *9*(7), 1668–1673. Advance online publication. DOI: 10.1021/acs.jpclett.8b00124

Zou, M., Jiang, W.-G., Qin, Q.-H., Liu, Y.-C., & Li, M.-L. (2022). Optimized XGBoost Model with Small Dataset for Predicting Relative Density of Ti-6Al-4V Parts Manufactured by Selective Laser Melting. *Materials (Basel)*, *15*(15), 5298. DOI: 10.3390/ma15155298

Chapter 4
Nanomaterials and Nanotechnology Applications

Patrick Ehi Imoisili
https://orcid.org/0000-0002-5435-9794
University of Johannesburg, South Africa

ABSTRACT

Nano-materials (NMs) and nano-technology (NT) are some of the most amazing things that have happened in science, technology, and business in the last few years. They let us make materials that can do many things. Nanotechnology has also been materially advanced by various industrial sectors, encompassing quantum confinement, enhanced surface area-to-volume ratios, and properties that are tunable optically, mechanically, and electronically. This chapter provides a comprehensive history of the growth of nanoscience and nanotechnology, with a focus on the types and properties of (NMs). The possible uses of NMs were discussed, with a focus on how nanotechnology can benefit the economy and society in the construction and manufacturing industries. Information and communication technologies, sustainable energy, ballasting technology, environmental sustainability, health, and medical uses. The problems and chances that NMs and NT present were also discussed.

1.0 INTRODUCTION

Nanotechnology (NT) and nanomaterials (NMs) are two of the most sophisticated domains in contemporary engineering and science. They use biology, chemistry, physics, and materials science to build usable systems at the nanoscale. The Greek word "nanos," which means "dwarf," is where the term "nano" derives from. It

DOI: 10.4018/979-8-3373-3556-8.ch004

discusses structures that are typically between one and one hundred nanometers (nm) in length. These sizes and materials possess biological, chemical, and physical properties that distinguish them from their bulk equivalents. This is due to factors such as surface quantum confinement, surface plasmon resonance (SPR), and a significantly higher interface-to-volume ratio (Gupta, 2023; Pisano & Durlo, 2025).

Particles that range from 1 to 100 nanometers are known as nanoparticles (NPs). Their chemical and physical properties are barely visible to the human eyes, and differ significantly from their larger counterparts. Only a few hundred atoms make up most of the nanoparticles. Particles exceeding 500 nm in size or fibers and tubes measuring less than 100 nm in only two dimensions can occasionally be referred to as NPs (Gupta, 2023; Norton, 2023; Pisano & Durlo, 2025).

The prevailing notion is that nanotechnology as a scientific field originated with Richard Feynman's pivotal 1959 lecture, "There's Plenty of Room at the Bottom," in which he examined the potential for influencing matter at the atomic and molecular scale (Norton, 2023). The invention of the Atomic Force Microscope (AFM) and Scanning Tunneling Microscope (STM) in the 1980s marked the first time that direct imaging and manipulation of individual atoms and nanostructures became possible. Since then, the field has evolved into a multidisciplinary industry that combines top-down approaches, such as lithography and etching, with bottom-up techniques, including chemical vapor deposition (CVD) and sol-gel processing, to create customized nanostructured materials (NSMs) (Rodríguez-Galván & Contreras-Torres, 2022).

There are four types of NMs according on their size: zero-dimensional (0D), one-dimensional (1D), two-dimensional (2D), and three-dimensional (3D). Other types of nanocomposites are metal-metal oxide, polymer nanocomposites, hybrid nanocomposites, and carbon-based (Xu et al., 2013). These materials have become part of NMs in various ways due to their diverse structures and architectures. For example, they are used in energy storage devices, new coatings, medication delivery systems, biosensors, and technologies for cleaning up the environment (Lin et al., 2022).

The utilization of NT has become a critical tool for addressing pressing global challenges and enhancing existing technologies. Scientists are currently developing NMs to enhance the efficacy of renewable energy technologies, including solar power generation and hydrogen fuel cell production (Li et al., 2020; Imoisili et al., 2020a). NPs have been developed from NT for medical applications to enhance the effectiveness of therapies and reduce adverse effects by enabling tailored medication distribution and improved imaging diagnostics (Khan et al., 2021). NPs are also utilized for long-lasting water treatment, new farming practices, building and construction, as they assist in efficient utilization of resources and environmental protection (Ahmed et al., 2022). This integration of NMs underscores the substantial and commonly underestimated impact of NT on everyday life. The application of NT has raised questions regarding its societal, environmental implications as well as the necessity for risk assessment and informed regulations.

1.1 Significance of Nanotechnology in Materials Science and Engineering

The significance of NT lies in its capacity to establish structure-property relationships at dimensions comparable to fundamental length scales in nature, thereby transcending the limitations of conventional bulk materials. Bandgap tunability and quantum confinement are two unique characteristics of NSMs, which are missing in bulk materials and are intricately linked to their electrical, optical, and catalytic properties (Park et al., 2021). These unique properties are also exhibited by NMs, such as quantum dots, which enable size-tunable photoluminescence for application in optoelectronic and biomedical imaging through the formation of distinct electronic states as their size decreases below the exciton Bohr radius (Rekhviashvili, 2023).

Likewise, NPs like gold (Au) and silver (Ag) demonstrate localized SPR, which enhances their interaction between matter and light, making it suitable for application in sensors, photovoltaics, and photothermal therapy (Khurana & Jaggi, 2021). The ability of NMs to connect diverse scientific disciplines, including biology, physics, chemistry, and engineering, makes them particularly useful for global technology and innovation compared to traditional scientific fields. The advancement of functional NMs with adaptable attributes for applications, including graphene, metal-organic frameworks (MOFs), and carbon nanotubes (CNTs), has accelerated due to the interdisciplinary nature of NMs and NT (Gupta et al., 2022). For example, the 2D honeycomb structure of graphene demonstrates remarkable attributes in terms of thermal conductivity, mechanical strength, and electron mobility. Consequently, 2D nanomaterials are an excellent option for use in sensors, nanoelectronics, and energy storage. These innovative applications of NMs illustrate how NT integrates technological innovation with basic sciences (Ma et al., 2022).

NT is widely recognized as a critical component of industry and national research initiatives. Because of the enormous potential for NT production and utilization, global economies have made significant investments in research initiatives. In 2000, the US government initiated the National Nanotechnology Initiative (NNI), with federal funding totaling approximately $29 billion by 2020. The European Union's Horizon 2020 initiative, as well as robust national projects in China, South Korea, and Japan, have underscored the geopolitical importance of nanoscale science and engineering (Roco 2023).

Despite significant advancements in technology, there are still serious concerns regarding the life-cycle effects and safe deployment of engineered NMs, owing to their potential toxicity and environmental persistence. NPs can interact with tissues, cells, and ecosystems in unpredictable ways due to their high surface reactivity and capability to navigate across biological boundaries. This condition, characterized by the unclear risk and transformational potential, often defined the recent debate over the significance of NT. In integrating technological development with the sustainability principle, research is increasingly engrossed in designing NMs that balance functionality and ecological safety (Wadhawan et al., 2024).

1.2 Historical Evolution of Nanoscience and Nanotechnology

The historical development of NT showcases the gradual alignment of scientific innovation, technological progress, and interdisciplinary collaboration. Nonetheless, the theoretical foundations of NT have been present for centuries; however, the formal establishment of the field as a discipline was ignited during the second half of the twentieth century, especially following Richard Feynman's influential 1959 speech “There's Plenty of Room at the Bottom” (Norton, 2023).

In this revolutionary speech, Richard Feynman articulated the potential for exploiting atomic and molecular-scale matter, forewarning of the ability to develop unique materials with specific characteristics through atomistic control. Although speculative at the time, Richard Feynman's theories laid the theoretical groundwork for nanoscale science and inspired researchers from future cohorts to investigate bottom-up material design for nanomaterials (Behera et al., 2025).

Until the advancements in technology in the 1980s, which allowed for the direct manipulation of individual atoms, Richard Feynman's conceptual vision remained purely aspirational. Researchers can now study surfaces at atomic resolutions, with the revolutionary discovery of the scanning tunneling microscope (STM) by Gerd Binning and Heinrich Rohrer in 1981 (Yaminsky, 2024).

AFM was developed shortly after by Gerd Binnig, Calvin Quate, and Christoph Gerber, thus increasing the capability and ease of researchers to characterize both conductive and insulating surfaces. These also enable access to the atomic world, as these tools not only confirm Richard Feynman's vision but also pave the way for nanoscience evolution (Behera et al., 2025; Yaminsky, 2024; Joshi et al., 2022)

NT were made tangible at the same time by developments in material synthesis. The 1996 Nobel Prize in Chemistry was awarded to Kroto, Curl, and Smalley, who discovered buckminsterfullerene, also known as the buckyball (C_{60}). Key scientific facts that helped scientists discover buckyballs include the understanding that all matter is composed of atoms, the way atoms combine to form molecules, and the precise mass of carbon atoms (Hargittai, 2022). Helical microtubules were discovered in carbonaceous soot from arc discharge, as reported by Dr. Iijima in 1991. This led to a novel allotrope of carbon, called CNTs, taking center stage. CNTs display outstanding properties owing to their unique structure, which has generated considerable interest and expectations for their prospective applications (Li, 2021). In 2004, the successful isolation of graphene was accomplished by Professors Andre Geim and Kostya Novoselov. This led to their discovery of 2D-NMs and earned them the 2010 Nobel Peace Prize in physics. These momentous discoveries have revolutionized our fundamental understanding of quantum electron-phonon interactions, transport, and dimensionality-dependent features, in addition to expanding the classification of NMs (Al Ghamdi & Al-Ghamdi, 2023).

A variety of approaches, including top-down and bottom-up techniques, have also marked the development of nanotechnology NT. The miniaturization trend promised by Moore's Law in microelectronics was initially driven by top-down approaches, such as photolithography, focused ion beam etching, and electron-beam lithography, which enable nanoscale patterning (Bespalov et al., 2022). The combination of computer modeling and simulation marked another turning point in NT development. Computational simulations, including density functional theory (DFT), and molecular dynamics (MD), were utilized to complement experimental methodologies, thus providing an atomistic insight into the structure-property interaction of NMs

The amalgamation of simulation and computer modeling marked yet another turning point in the development of NT.

To enhance experimental methodologies and accelerate the logical design of nanosystems, computational simulations, including Monte Carlo techniques, DFT, and MD, offer atomistic discernments into the structure-property interactions of nanomaterials (Dai & Chen, 2022).

The computational revolution has highlighted the transition of NT from empirical research to a predictive and design-oriented field. Three interconnected stages can be use to view the development stages of NT (Figure 1). These three stages are;

i. **The discovery phase (1950s–1980s):** The discovery stage was the phase of conception of NT. This was the phase of conceptual groundwork and development of tools for NMs characterizations.
ii. **Materials Phase (1980s–2000s):** The materials phase marks the advancement of new functional NMs such as graphene, fullerenes, and CNTs. This phase was also accompanied by new advancements in methodology.
iii. **Integration Phase (2000s–present):** The integration phase is the policy initiatives, industrial applications, and integration into interdisciplinary fields like nano-medicine, nano-electronics, nano-catalysis etc, have all contributed to the institution of NT

This timeline has demonstrated how NT has evolved from a theoretical field to an applied science that supports the development of new technologies. NT is still evolving as new ideals are pushing the limits in fields like quantum NT, nano-robotics, and nano-informatics (Yao et al., 2024; Dai & Chen, 2022).

Figure 1. Schematic of The Different Phases of Nanotechnology Evolution

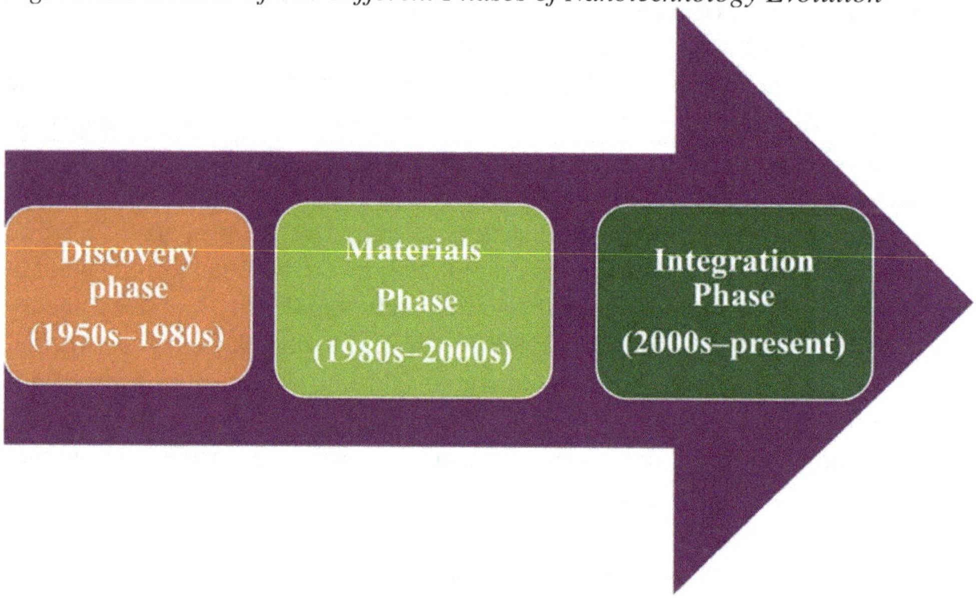

1.3 Classification and Characteristics of Nanomaterials

NMs can be classified (Figure 2) according to chemical compositions, structural morphology and dimensionality. These are essential for both tailoring NMs for specific technological purposes and comprehending the basic mechanisms driving their physicochemical characteristics. Beyond classification, the special characteristics of NMs stem from their quantum phenomena, interfacial dominance, and confinement effects, which make them unique from their bulk colleagues (Mekuye & Abera 2023).

1.3.1 Nanomaterial Classification by Origin

NMs can be categorized by their origin (Figure 2), such as natural, incidental, and engineered.

i. **Natural Nanomaterials:** Mineral colloids, volcanic ash, and biological nanostructures, such as bacterial magnetosomes, are examples of natural nanomaterials that don't require human intervention. Their research gives templates for bio-inspired design as well as insights on evolutionary nanofabrication (Mekuye & Abera, 2023; Zhao et al., 2021)
ii. **Incidental Nanomaterials:** Byproducts of human activities, such as combustion (soot, fly ash), industrial effluents, or wear and tear from tires and machinery, are known as incidental nanomaterials. Environmental nanoscience studies are necessary because, although not being purposefully created, their presence in the environment has caused ecological and health issues (Barhoum et al., 2022; Zhao et al., 2021).
iii. **Engineered Nanomaterials:** These are purposefully created and produced with precise regulation of chemical makeup, surface chemistry, dimensions, and morphology. These consist of functionalized 2D materials, metal nanoparticles, quantum dots, and nanofibers. Because of their specialized qualities for electronics, energy, healthcare, and environmental cleanup, engineered nanomaterials are the subject of the most study (Albalawi et al, 2021; Barhoum et al., 2022)

Figure 2. Schematic of the Classification of Nanomaterials by Origin

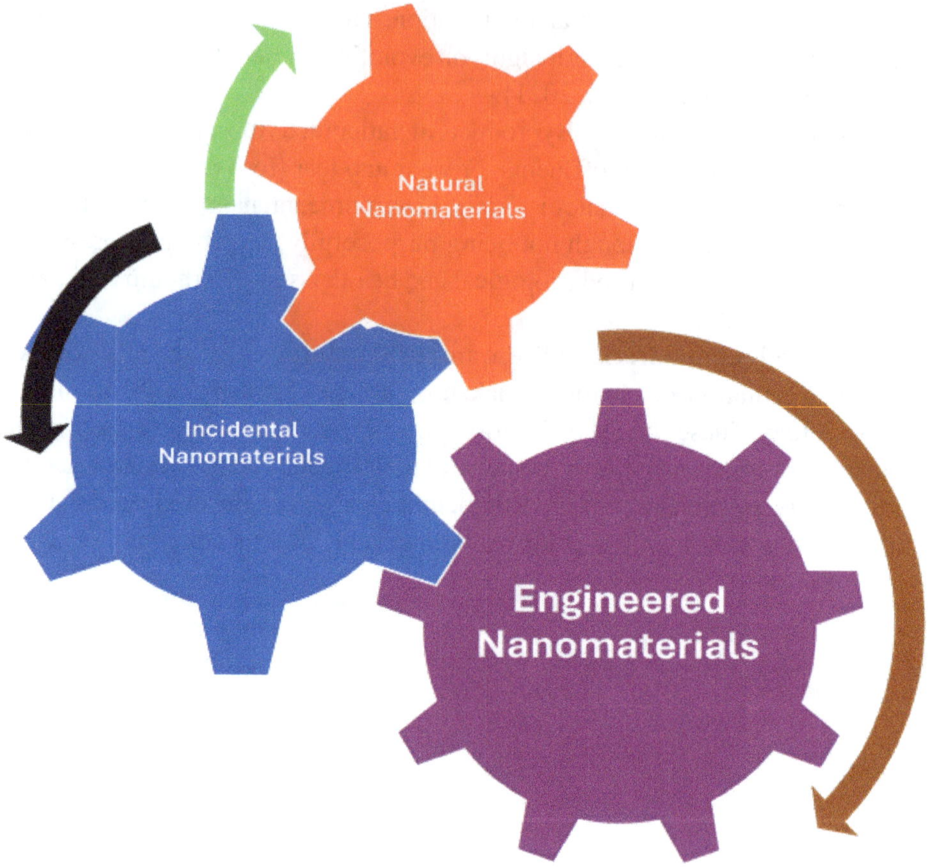

1.3.2 Nanomaterial Classification by Dimension

NMs are classified using a commonly utilized framework according to the quantity of dimensions confined to the nanoscale domain, as illustrated in Figure 3.

Figure 3. Schematic of the Classification of Nanomaterials by Dimensions

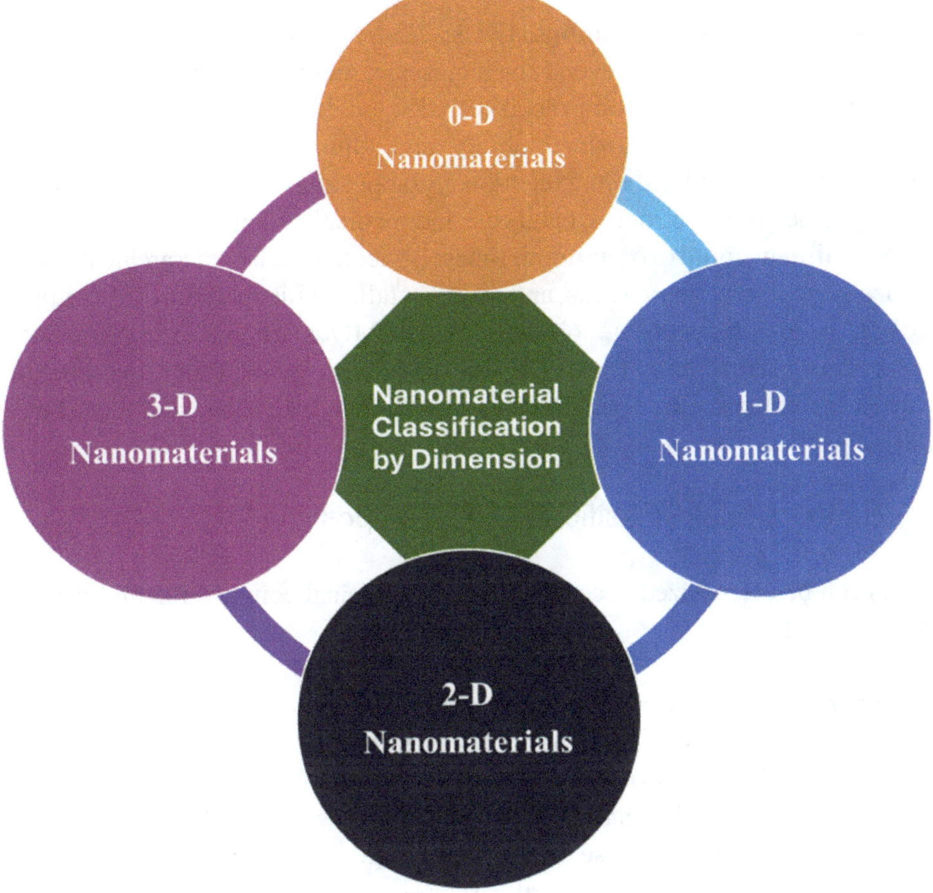

i. **Zero-dimensional (0D) nanomaterials:** Zero-dimensional (0D) NMs refer to structures, such as NPs, nanoclusters, and quantum dots, that have all three spatial dimensions less than 100 nm. Quantum effects predominate because of complete confinement, resulting in size-tunable band gaps, discrete energy levels, and photoluminescence. These phenomena are demonstrated by colloidal quantum dots (such as CdSe and PbS), which allow for tunable optoelectronic applications (Boretti, 2025; Gui & Jin, 2025).
ii. **One-dimensional (1D) nanoparticles:** One-dimensional (1D) nanoparticles, such as nanotubes, nanorods, and nanowires, are confined in two dimensions and extend in three dimensions. They are especially well-suited for use in mechanical reinforcement, photonics, and nanoelectronics due to their high aspect ratios and anisotropic characteristics. For example, CNTs exhibit tensile strengths an

order of magnitude greater than steel and ballistic electron transport (Machín et al., 2021)

iii. **Two-dimensional (2D) nanoparticles:** 2D-NMs are atomically thin entities with lateral dimensions beyond the nanoscale, including graphene, hexagonal boron nitride, and transition metal dichalcogenides (TMDs). Their extensive surface area, elevated carrier mobility, and adjustable electrical and optical characteristics render them exceptionally appealing for utilization in energy storage, flexible electronics, catalysis, and medical systems.

iv. **Three-dimensional (3D) nanoparticles:** Structures such as hierarchical assemblies, nanoporous frameworks, and nanocrystalline bulk materials are examples of 3D nanoparticles. Nanoporous silica and MOFs are two examples. These structures offer high surface area, adjustable porosity, and multifunctionality by fusing macroscopic form parameters with nanoscale characteristics (Lee et al., 2010; Van Aert et al., 2011)

1.3.3 Nanomaterial Classification by Composition

NMs can be categorized according to their chemical composition and the type of bond formed.

i. **Carbon-Based Nanomaterials:** As shown in Figure 4, examples of carbon-based NMs include carbon nanofibers, graphene, fullerenes, carbon dots, nanodiamonds, nanohorn, etc. They possess remarkable electrical, mechanical, and thermal properties that their sp^2-hybridized networks offer, making them promising materials for use in energy storage, electronics, and composites (Ayanda et al., 2024; Imoisili et al., 2020[b]).

Figure 4. Schematic of the type of Carbon-Based Nanomaterials

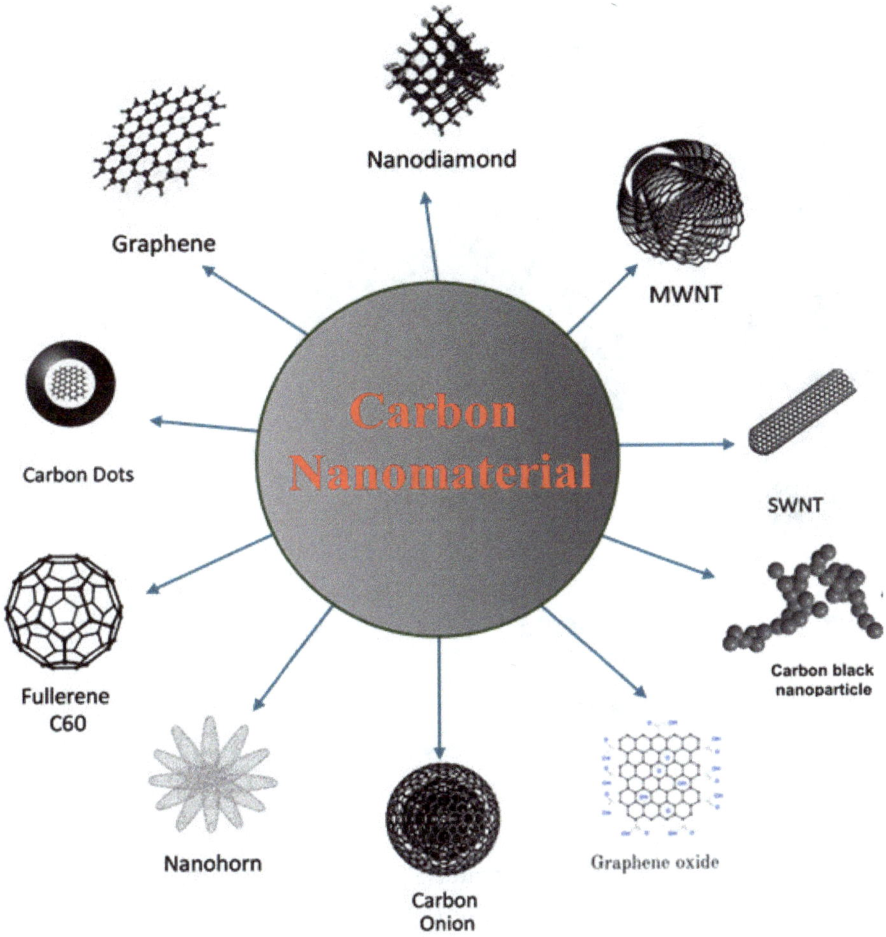

ii. **Metal-Oxide Nanomaterials:** Examples of Metal-oxide NMs are TiO_2, ZnO, Fe_2O_3, CuO, MgO, V_2O_5, CaO, ZnO, K_2CrO_2, and CeO_2 (Figure 5). These NMs exhibit distinctive catalytic, electrical, magnetic, and optical properties, which are attributed to their elevated surface area and defect-rich architectures. They are extensively applied in photocatalysis, gas sensing, energy storage, and biomedical fields. Their tunable band gaps and surface chemistry make them dynamic in environmental remediation and sustainable technologies. While TiO_2, ZnO, and FeO_4 offer photocatalytic, magnetic, and semiconducting properties, noble metals like Pt, Ag, and Au exhibit localized SPR (Kamyab et al., 2023).

Figure 5. Different Types of Metal Oxide-Based Nanomaterials

iii. **Polymer Nanomaterials:** These materials are developed from organic polymers, which are utilized in fields such as drug delivery, biosensing, and nanocomposites, owing to their high surface area, biodegradability, controllable physicochemical characteristics, and capability to encapsulate active compounds. Polymeric NPs such as micelles, liposomes, and nano-capsules. Polymer Nanocomposites include CNTs in epoxy, polycarbonate, polyethylene, and block copolymer-based NMs, which are a few examples (Figure 5) (Yudaev et al., 2022).

Figure 6. Different Types of Polymer Nanomaterials

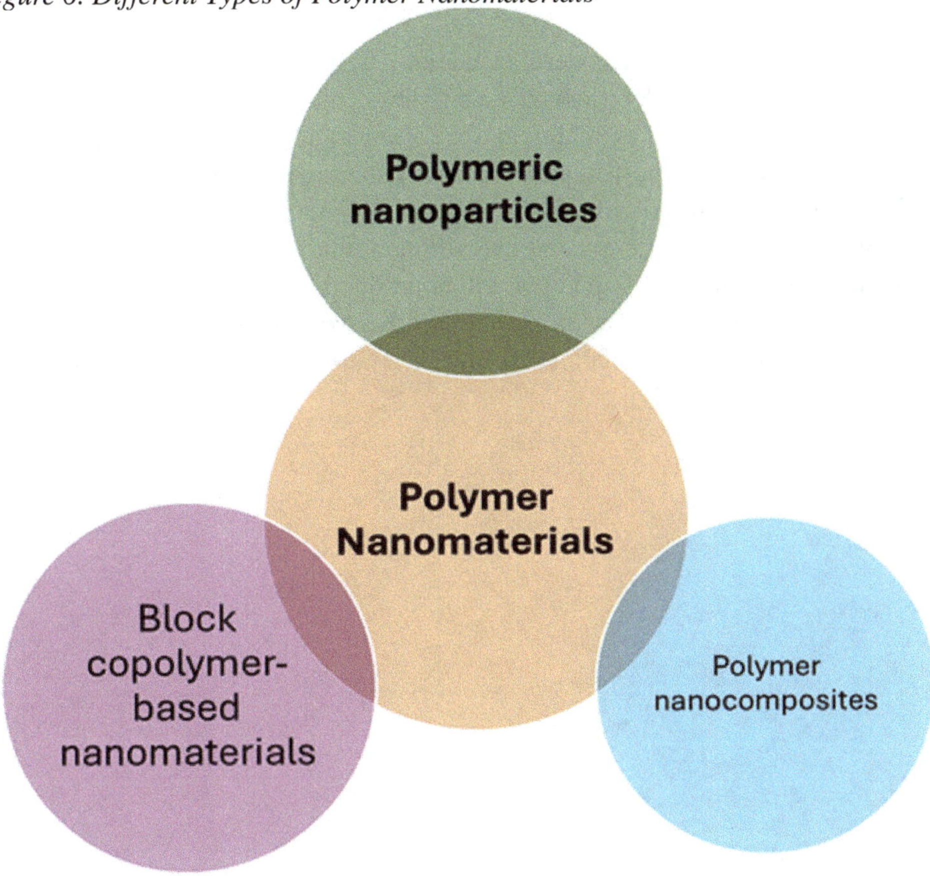

iv. **Composite and Hybrid Nanomaterials:** Composite refers to a material that is composed of two or more component substances with distinct qualities. When two or more nanoscale materials are combined to enhance their properties or provide them with multiple functions. To make composite materials robust, more conductive, or stable, they are often filled with NPs or nanotubes in a matrix. While hybrid NPs are materials that mix organic and inorganic phases to alter their functionality, such as catalysts, optics, or biometrics. They are utilized in various domains, notably energy, environment, and healthcare (David et al., 2021; Godja & Munteanu, 2024).

1.4 Methods of Nanomaterial Synthesis and Fabrication

NMs have revolutionized modern science and engineering due to their exceptional physical, chemical, and electrical characteristics. The synthesis and fabrication of NMs have been developed as vital fields for research and development, with diverse purposes (Al-Harbi & Abd-Elrahman, 2024; Alshammari et al., 2023).

The preferred synthesis methods influence the morphology, size distribution, crystallinity, surface chemistry, and overall functionality of the NMs. Hence, it is essential to select the precise and controlled production methods. As illustrated in Figure 7, Top-down and bottom-up methods are the two general categories into which NMs production procedures fall, each with its own set of applications (Hachem et al., 2022; Baig et al., 2021)

Figure 7. Methods of Nanomaterial Synthesis and Fabrication

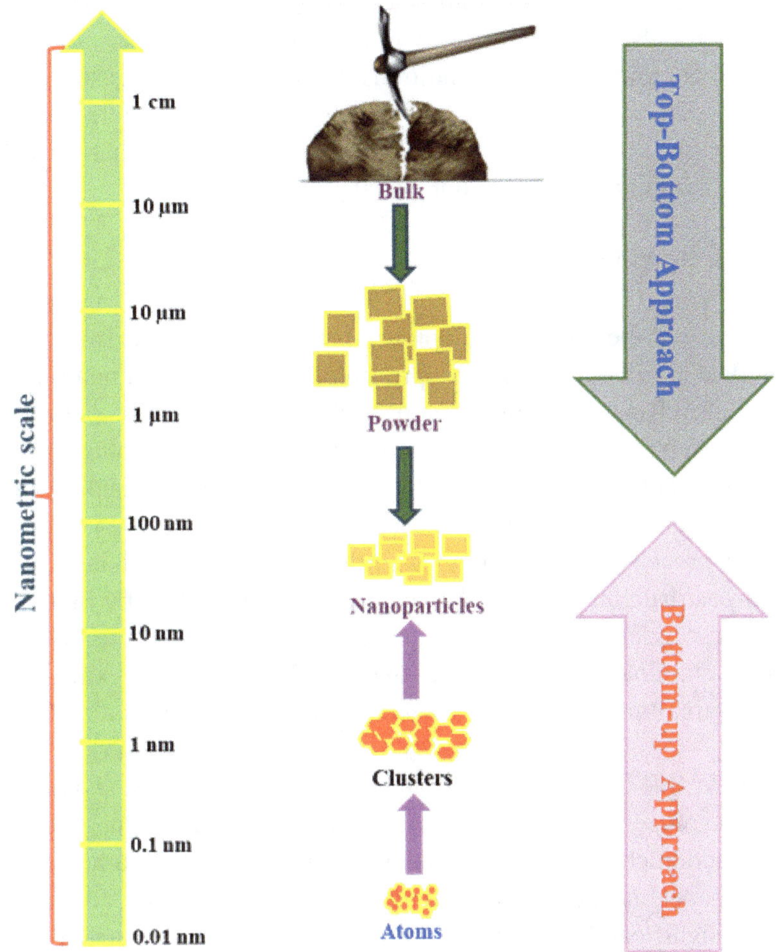

1.4.1 The Top-Bottom Approach

The top-bottom approach for NMs production involves mechanical crushing of the base materials. Ball milling is a commonly employed top-down mechanical synthesis technique for fabricating NMs. In this method, bulk powders are placed in a rotating cylindrical chamber with hardened steel or ceramic balls. These steel or ceramic balls, when they collide with the powders at high energy, result in a reduction of the particle size to the nanoscale. The ball milling procedure is a resourceful and cost-effective method frequently employed in the preparation of NPs, nanocomposites, and nanostructured alloys. Structural flaws and mill contaminations are some of the challenges associated with the method (Karnwal et al., 2024; Verma et al., 2022).

Laser ablation also offers versatility in regulating particle size and shape, as it can be carried out under various conditions, including liquids, gases, and vacuum (Khairani et al., 2023; Jiang et al., 2022). Nanoscale surface patterns of metals and semiconductors are also being produced by chemical techniques that utilize selective dissolution (Khairani et al., 2023; Jiang et al., 2022).

1.4.2 The Bottom-Up Approach

The objective of the bottom-up approach is to assemble NMs from atomic, molecular, or cluster progenitors. This approach closely resembles natural processes, where materials self-assemble into organized structures. Examples of the bottom-up method include biomineralization, sol-gel procedures, CVD, hydrothermal synthesis, self-assembling, and electrospinning. The most commonly used method for fabricating NMs is the sol-gel process (Abid et al., 2022; Zain et al., 2024; Nwanna et al., 2022). In this process, a colloidal gel is further processed into nanoparticles, thin films, or porous structures, or metal alkoxide or salt in solution by first undergoing hydrolysis and condensation. The sol-gel method is commonly used to syntheses silica, titania, alumina, and other metal oxide NMs owing to its exact control over particle structure, homogeneity, and size distribution (Haider et al., 2022; Khutso et al., 2020).

CVD is a flexible method for preparing thin films, nanotubes, and nanowires. In this fabrication process, a layer of NMs is deposited by introducing volatile precursors into a reaction chamber, where they react chemically with a heated substrate. Processed parameters, such as pressure, gas flow rate, precursor composition, and temperature, significantly influenced the shape, crystallinity, and purity of the deposited NMs. Additional control is achieved through processes such as Atomic Layer Deposition (ALD) and plasma-enhanced CVD (Nwanna et al., 2020; Zhao et al., 2024). PVD is another process in which NMs can be deposited onto surfaces through a physical process, rather than through chemical reactions. PVD typically requires high-vacuum conditions and meticulous process control; it is wisely utilized in thin-film fabrication for electronics, optics, and protective coatings (Ichou et al., 2023; Minello et al., 2025).

An array of NMs possessing distinct chemical and physical properties, with prospective uses across multiple industries, can be manufactured via the solvothermal process. The solvothermal process often occurs in the supercritical phase at temperatures exceeding 300 °C. Over the past decade, several investigations have explored this approach to nanocrystal formation under comparatively mild conditions (i.e., at temperatures below 300 °C). A range of NMs, encompassing metals, metal oxides, semiconductors, alloys, composites, hydroxides, zeolites, and ceramics in diverse morphologies such as NPs, nanospheres, nanosheets, nanotubes, and nanorods, have been manufactured using the solvothermal method to date (Imoisili & Jen, 2021; Peixoto et al., 2023; Srivastava et al., 2020).

Electrochemical deposition is an effective technique for synthesizing metal NPs nonetheless, it is typically less employed than wet-chemical approaches. This is a technique wherein metal ions in a solution are conveyed by an electric field and are deposited onto the surface of a substrate. The deposition process may involve either a cathodic or anodic reaction, contingent upon the workpiece to be coated (cathode or anode), whereby metallic ions are drawn to the cathode and reduced to their metallic state. The diagram below illustrates the process, where (1) represents the material to be deposited (anode), (2) denotes the deposited metal (NPs) on the cathode, and (3) signifies the metallic ions. Nanostructures, such as nanocrystalline, nanocomposite, amorphous films, or layered materials, can be achieved by regulating electrolysis parameters during electrodeposition. The predominant approaches employed are (a) pulse current deposition to control deposit growth, (b) the use of additives and surfactants to modify deposit grain size, and (c) the incorporation of nanoparticles into deposits to create nanocomposites. This approach may exhibit certain limitations regarding the dimensions and permissible morphologies of nanomaterials; however, it offers numerous benefits, particularly in terms of quick synthesis, the elimination of chemical reductants or oxidants, and the absence of unwanted by-products (Liu et al., 2025; Zou et al., 2025; Yin et al., 2025; Tonelli et al., 2019).

Self-assembly denotes the autonomous arrangement of elements into broader periodic arrangements without external mechanical interference. Self-organization is a common occurrence in nature across diverse length scales. In NT, self-assembling is an advantageous technique for fabricating functional structures and interfaces utilizing NMs and molecules. The increasing fascination with self-assembly arises from its capacity for self-organization at the nanoscale, where mechanical intervention is infeasible. Consequently, in addition to the viability of distinct, spontaneously developing structures that are unattainable by other synthesis or nanofabrication methods, self-assembly also offers a cost-effective alternative for nanofabrication technologies (Liu et al., 2025; Zou et al., 2025). Although thin-film techniques represent the most advanced methods for fabricating nanostructures and nanostructured surfaces, they are accompanied by numerous hurdles and inherent limitations. For example, geometries produced during atomic or molecular deposition and subsequent etching procedures are often confined to basic planar shapes of a single element. As a result, applications requiring complex geometries, acet-dependent chemistry, hybrid metal/metal and metal/metal oxide contacts, along with metal/metal oxide and organic couplings, are not easily attainable using those methods. In certain domains, such as plasmonics, the synthesis of intricate geometries, including 3D nanostars, helical nanostructures, and nanocages, poses challenges when utilizing traditional thin-film methodologies. Furthermore, the production of sub-30-nanometer structures using lithographic methods remains a significant difficulty. Conversely, the assembly process is inherently devoid of these disadvantages. The swift progress in wet-chemical synthesis techniques enables the facile and economical production of nanostructures with intricate geometries and compositions, achieving sizes as small as <5 nm, such as quantum dots. The self-assembly of nanostructures as component parts for functional interfaces offers enhanced flexibility in achieving complex designs that lithographic approaches cannot access (Liu et al., 2025; Zou et al., 2025; Yin et al., 2025; Borah et al., 2023).

Electrospinning has been understood for about a century and can be considered a variation of the electrospray process. During the electro spraying process, the liquid droplet elongates as the electric field strengthens. When the repulsive force produced by the charge distribution on the droplet's surface balances with the liquid's surface tension, the fluid droplet becomes a conical shape. When the repulsive force exceeds the surface tension, a liquid jet is ejected from the apex of the cone. Minute droplets arise from the varicose breakdown of the jet in low-viscosity liquids (Gao et al., 2025; Su et al., 2025; Megahed et al., 2025). When this phenomenon is added to polymer solutions, a solid fiber is produced rather than fragmenting into discrete droplets due to the electrostatic repulsions between surface charges and the solvent's vaporization. The initial depiction of a device for this electrospinning method was presented by Formhals in 1934 as a patent; nonetheless, it garnered little interest until the early 1990s (Lu et al., 2009). Subsequently, some researchers have made significant contributions with this technique. They investigated electrospinning technology for several polymers. Furthermore, they elucidated the mechanism underlying the creation of polymer fibers. Electro spun fibers are utilized in various applications due to their small dimensions and extensive surface areas, including templates for nanostructure fabrication, filtration, chemical, catalysis, and biological sensing, nanofiber reinforcement, wound healing, tissue engineering, drug delivery, and tumor treatment (Lu et al., 2009; Yarin et al., 2001; Gao et al., 2025; Su et al., 2025; Megahed et al. 2025).

Electrospinning was primarily utilized to produce elastomeric nanofibers. As of now, more than one hundred types of natural and synthetic polymers have been successfully electrospun into nanofibers. including polyacrylamide (PAAm), polyethylene-co-vinyl acetate (PEVA), polyvinyl alcohol (PVA), poly(vinylidene fluoride) (PVDF), polyethylene oxide (PEO), polybenzimidazole (PBI), Nylon-6,6 (PA-6,6), polyether imide (PEI), polyamide (PA), polyacrylonitrile (PAN), poly(vinyl pyrrolidone) (PVP), polycarbonate (PC), polystyrene (PS), polyurethane (PU), polyvinyl carbazole (PVK), polymethacrylate (PMMA), polyethylene glycol (PEG), polycaprolactone (PCL), among others (Lu et al., 2009; Huang et al., 2003; Xie et al., 2008)

1.5 Applications of Nanotechnology

NTs are promising candidates for various applications across multiple domains due to their unique dimensions (size, shape, and structure) and exceptional biological, physical, and chemical properties. Figure 8 illustrates the applications of NT throughout several domains (Megahed et al., 2025; Su et al., 2025; Ojodun et al., 2025[a]; Meng et al., 2025).

Figure 8. Applications of Nanotechnology

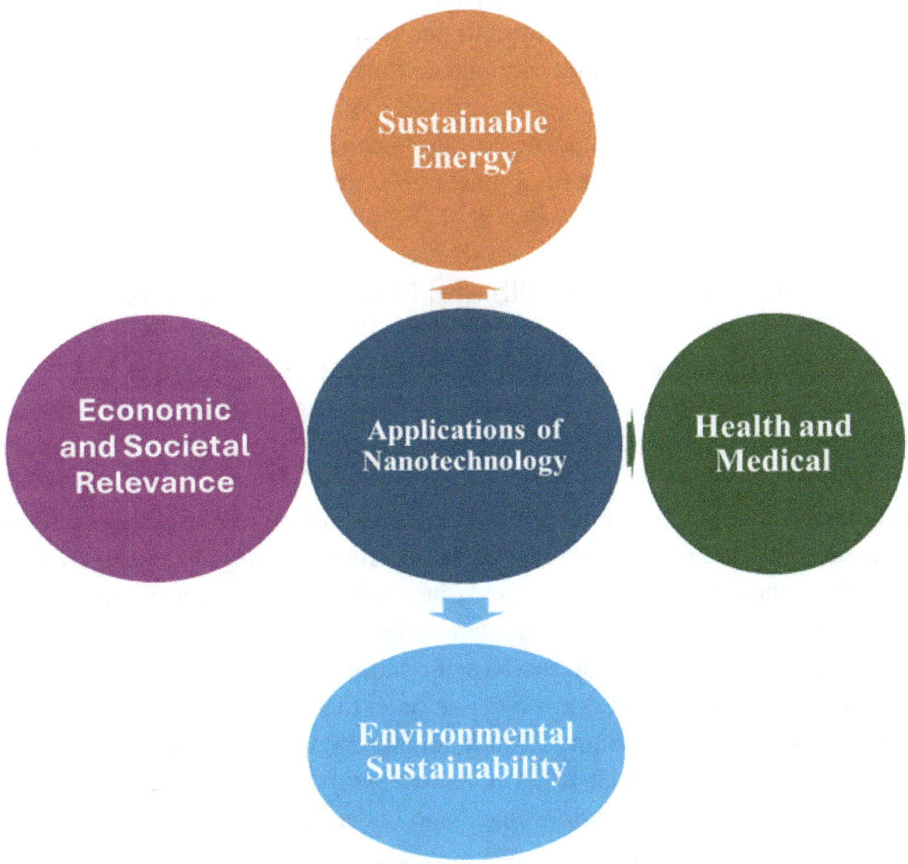

1.5.1 Nanotechnology in Sustainable Energy Application

A major technological problem of the 21st century is the advancement of renewable energy technology, driven by critical issues associated with energy generation and consumption. A rapidly expanding area of research known as nanotechnology is currently considered one of the most viable solutions to this issue. NT provides, for the first time, instruments to establish new enterprises founded on economic and efficient practices, significantly aiding sustainable economic development. In the energy sector, NT has the capacity to markedly diminish the effects of energy generation, storage, and consumption. Although we remain distant from achieving a truly sustainable energy system, the scientific community is exploring advancements in energy nanotechnologies. The development of NT-based products for renewable energy offers numerous advantages. Moreover, nanotechnology can enhance renewable energy sources; for instance, the efficiency of wind energy can be augmented by employing lightweight, high-strength nanomaterials for rotor blades. Biomass energy employs nano-based precision agriculture to enhance crop production for biofuel generation. Nano-coatings are utilized to prevent corrosion in tidal energy apparatuses, whereas nanocomposites are applied to improve the resilience to fatigue of drilling machinery in geothermal power contexts (Ojodun et al., 2025b; Mei et al., 2020; Polman et al., 2016; Saeloo et al. 2024).

1.5.2 Nanotechnology in Health and Medical Applications

NT can significantly impact the identification and therapy strategies for a disease. The unparalleled sensitivity and functionality, augmented reliability and versatility, as well as distinctive physicochemical characteristics of NMs have been employed for healthcare diagnosis for the prompt identification of illnesses, in targeted clinical therapies, and in regenerative medicine for the restoration of damaged tissues. Nanomedicine (NMD) is a comprehensive discipline that integrates various medical applications, including medical care and diagnostics, disease prevention, analgesic technologies, health promotion for humans, nanoscale therapies for injuries caused by trauma, and therapeutic alternatives for ailments. Thus, an interdisciplinary approach is being employed to leverage the outcomes of nanomaterials, biological technology, biomedical technology, and genetic engineering, commonly referred to as NMd (Patel & Nanda, 2015). The nanoscaling of medical supplies promotes effectiveness, accelerates reaction times, and enhances functional performance in numerous chemical and biological processes related to pharmaceutical manufacturing. Recognizing the advantageous features of NMd, researchers are endeavoring to utilize NT for disease diagnosis, treatment, and prevention. There are also prospective applications in the development of medical instruments and methodologies for the emerging cohort of health scientists (Malik et al., 2023; Patel & Nanda, 2015).

A multidisciplinary strategy is being employed to implement the consequences of biotechnological NMs, and genetic engineering, collectively referred to as nanomedicine. The nanoscaling of medical technologies enhances productivity, accelerates response rates, and improves functional efficacy in many biological and chemical compounds utilized in the assembly of medicinal products. Consequently, research consistently offers optimism for the forthcoming utilization of nanomedicine (Bansal et al., 2020; Malik et al., 2023).

Au-NPs and magnetic nanostructures have become very sensitive tools for molecular imaging and biosensing. This has made it possible to diagnose diseases with never-before-seen accuracy (Bansal et al., 2020). The lipid nanoparticle platform in mRNA vaccines against COVID-19 is a new example of NT's importance in keeping track of global health problems. In addition to pharmaceuticals and diagnostics, nanostructure scaffolds are also important in tissue engineering because they help cells stick together, grow, and change in regenerative applications (Chung et al., 2020).

1.5.3 Nanotechnology in Environmental Sustainability

NT in environmental sustainability is seen as pivotal in the evolution of contemporary environmental engineering and science. Traditional environmental remediation methods appear to be largely ineffective against the significantly increasing burden of contaminants that infiltrate the air, water, and soil (Jabeen et al., 2024; Biswal, 2025). Nanotechnology can facilitate the purification of air and water resources by applying nanoparticles as catalysts and/or sensor devices. The application of nanomaterials, such as synthetic mordenite nanocrystals, in environmental conservation and sustainability has been investigated for removing contaminants from water. It has evolved into a versatile framework, proficient in delivering methodical, economical, and environmentally friendly solutions. NT is crucial for the development of smart materials that effectively detect and eliminate hazardous chemical pollutants in the environment. Certain NPs, such as semiconductive nanomaterials such as titanium dioxide (TiO_2) and tungsten trioxide (WO_3), zero-valent metals such as copper (Cu), iron (Fe), and zinc(Zn), and metallic nanocomposites such as iron/palladium (Fe/Pd) and Au/Ag, play a vital role in the sustainability of the environment due to their exceptional reactivity with pollutants, including aromatic dyes, toxic heavy metals, and pesticides (Jabeen et al., 2023).

NT has been employed for the removal of Co (III) from wastewater, and it may also eliminate colors, pharmaceuticals, phenols, organic waste, and pesticides from wastewater using NMs. In the manufacturing sector, the dissemination of toxic metals such as cadmium, arsenic, and lead, as well as chlorinated compounds, due to landfills and leaks from subterranean storage, presents a significant concern. The primary operational processes of NMs are nano-adsorption and photocatalytic degradation (Jabeen et al., 2023; Imoisili & Jen, 2024; Opetubo et al., 2023).

1.5.4 Nanotechnology in Economic and Societal Relevance

NT's role in societal development is a crucial examination of the complex relationship between nanotechnology and societal progress. In an era marked by rapid technological progress, nanotechnology emerges as a revolutionary force with diverse implications across multiple domains. A sustainable global economy relies on industrialization. It is the principal catalyst for innovation and the generation of high-value labor worldwide. NT is set to revolutionize global economies and businesses by providing unparalleled precision, control, and production capabilities at the atomic level. It guarantees superior materials, advanced technology, enhanced industrial processes, and novel opportunities for sector-wide expansion. Nonetheless, environmental, legal, health, ethical, and social implications must be reconciled with financial benefits to realize its complete potential. The world's NT sector is projected to attain $74.1 billion by 2032, exhibiting an overall growth rate of 34.5%, up from approximately $3.78 billion in 2020 (Tawiah et al., 2024). The National Science Foundation forecasts that by 2025, the worldwide NT sector will require approximately 6 million jobs. Moreover, substantial growth is anticipated in semiconductor production, energy, pharmaceutical delivery, potable water accessibility, life sciences, textiles, culinary technology, and consumer products, all of which are enhanced by NPs (Madkour 2019; Tawiah et al., 2024). The most commercially significant applications of NMs currently and expected soon are in magnetic and optical components, tissue engineering, quantum computing, mechanical and electrical devices. Alongside enhancing battery catalysts, energy efficiency in solar cells, and filtration systems, along with sustainable manufacturing methods, diminish gasoline expenses in the transportation sector. The shelf life and security of food are enhanced through antimicrobial nano-packaging, while medical nanomedicine seeks to reduce healthcare expenses by facilitating early diagnosis, tailored treatment, and expedited healing applications. The anticipated increase in nanoscale product sales, contingent upon upholding scientific integrity and ethical considerations, yields both direct and indirect economic benefits (Tawiah et al., 2024; Amutha et al., 2024; Kumar et al., 2024; Selvarani et al., 2025)

1.6 Challenges and Future Prospects of Nanotechnology

The progress in the technological industry is escalating daily due to advancements in NT. The utilization of nanostructures and nanomaterials is anticipated to increase rapidly in the impending years. NT may not have any adverse consequences or effects on humanity. However, NPs, which are the primary focus of this technology, may have repercussions for the health of living organisms, including humans, animals, and insects, as well as for the environment. Furthermore, nanotechnology may have significant societal implications in the future (Tahir et al., 2021).

NMs, which comprise nanoparticles, may pose health risks to humans and the living world due to their heightened reactivity and mobility. The behavior of NPs depends on their size, shape, and surface reactivity with adjacent tissue. In addition to the implications of non-degradable or partially degradable NPs accumulating in organs, there is concern regarding their potential interactions with biological processes within the body. Due to their extensive surface area, NPs will promptly adsorb macromolecules from tissues and fluids upon exposure. Moreover, nanodevices may exhibit side effects, particularly those employed in the treatment of surgical wounds. Certain nano liquids can be highly detrimental when they interact with the human body (Tahir et al., 2021; Egwu et al., 2022).

The production of nanodevices entails environmental consequences due to the production of waste. These are termed nano waste, which may be hazardous due to their minuscule dimensions. These nano-wastes can be emitted into the atmosphere and readily infiltrate plants, resulting in their deoxygenation upon reaction to carbon dioxide. The impacts of nano-waste materials vary at different phases of manufacturing; at times, their adverse consequences result in fatalities of cats and other small animals due to the inhalation of toxic gases or consumption of polluted water or liquids discharged by manufacturing facilities. To accurately assess the health risks, each phase during production must be thoroughly examined individually, followed by the implementation of measures to mitigate waste. The harmful impacts on human health also encompass societal ramifications. Scientists propose that the societal implications of NT are complex and not immediately comprehensible to the general populace. Certain applications, such as NT in the filtering processes of Au and diamonds, are unsuitable for the global market due to increased costs, which adversely impact consumers' purchasing power (Tahir et al., 2021; Rana & Rathore, 2023; Singh et al., 2023).

Future screens, utilizing CNTs and field-effect displays, are anticipated to be slimmer and of superior quality compared to conventional televisions. Furthermore, NT is anticipated to enhance storage size and processing speed in electronic and communication equipment by applying nanolayers and quantum dots. This application will transform communication methods. NPs are anticipated to enhance environmental conditions by increasing the efficiency of combustion in power plants and cars as catalysts. Innovative smart materials will be employed for vacuum applications, as well as for lubrication and sealing purposes. NT will transform the medical business by implementing novel drug delivery mechanisms. The phenomena depicted in films will subsequently be realized by the application of NT. The engineering of materials will undergo a transformation due to innovative nano-hybrid methodologies and instruments. The energy crisis is expected to be addressed through renewable energy sources. Outer space can be better comprehended using lightweight vehicles and miniature robotic systems. NT can also enhance safety standards by integrating novel materials and safety devices (Tahir et al., 2021; Rana & Rathore, 2023; Singh et al., 2023).

1.7 CONCLUSION

NT has redefined new possibilities for materials design, performance, and application. NT finds applications across diverse disciplines owing to its distinctive characteristics and enhanced chemical reactivity. NT facilitates efficient solar energy collection, sophisticated energy storage, catalysis, and water purification in various energy and environmental applications. To fully optimize NT's potential, we must address sustainability, economic, and technological challenges, ensure safety, and include ethical considerations in its design, production, and application. NT is revolutionizing the fields of materials science and engineering. Its capacity to create novel, adaptable materials is essential for addressing global challenges in environment, technology, energy, and health. NT can foster responsible, ethical, and sustainable innovation to harmonize technological growth with human and environmental well-being. These will aid NT in developing new solutions for present and future global demands.

REFERENCES

Abid, N., Khan, A. M., Shujait, S., Chaudhary, K., Ikram, M., Imran, M., Haider, J., Khan, M., Khan, Q., & Maqbool, M. (2022). Synthesis of nanomaterials using various top-down and bottom-up approaches, influencing factors, advantages, and disadvantages: A review. *Advances in Colloid and Interface Science*, *300*, 102597. DOI: 10.1016/j.cis.2021.102597

Ahmed, S. F., Mofijur, M., Ahmed, B., Mehnaz, T., Mehejabin, F., Maliat, D., Hoang, A. T., & Shafiullah, G. M. (2022). Nanomaterials as a sustainable choice for treating wastewater. *Environmental Research*, *214*, 113807. DOI: 10.1016/j.envres.2022.113807

Al Ghamdi, H. S., & Al-Ghamdi, A. A. (2023). Opening band gap of multi-color graphene quantum dots using D-fructose as a green precursor. *Alexandria Engineering Journal*, *79*, 155–163. DOI: 10.1016/j.aej.2023.08.019

Al-Harbi, N., & Abd-Elrahman, N. K. (2024). Physical methods for preparation of nanomaterials, their characterization and applications: A review. *Journal of Umm Al-Qura University for Applied Sciences*, *11*(2), 356–377. DOI: 10.1007/s43994-024-00165-7

Albalawi, F., Hussein, M. Z., Fakurazi, S., & Masarudin, M. J. (2021). Engineered Nanomaterials: The Challenges and Opportunities for Nanomedicines. *International Journal of Nanomedicine*, *16*, 161–184. DOI: 10.2147/IJN.S288236

Alshammari, B. H., Lashin, M. M. A., Mahmood, M. A., Al-Mubaddel, F. S., Ilyas, N., Rahman, N., Sohail, M., Khan, A., Abdullaev, S. S., & Khan, R. (2023). Organic and inorganic nanomaterials: Fabrication, properties and applications. *RSC Advances*, *13*(20), 13735–13785. DOI: 10.1039/D3RA01421E

Amutha, C., Gopan, A., Pushbalatatha, I., Ragavi, M., & Reneese, J. A. (2024). Nanotechnology and Governance: Regulatory Framework for Responsible Innovation. Nanotechnology in Societal Development, 481–503. DOI: 10.1007/978-981-97-6184-5_14

Ayanda, O. S., Mmuoegbulam, A. O., Okezie, O., Durumin Iya, N. I., Mohammed, S. E., James, P. H., Muhammad, A. B., Unimke, A. A., Alim, S. A., Yahaya, S. M., Ojo, A., Adaramoye, T. O., Ekundayo, S. K., Abdullahi, A., & Badamasi, H. (2024). Recent progress in carbon-based nanomaterials: Critical review. *Journal of Nanoparticle Research*, *26*(5), 106. Advance online publication. DOI: 10.1007/s11051-024-06006-2

Baig, N., Kammakakam, I., & Falath, W. (2021). Nanomaterials: A review of synthesis methods, properties, recent progress, and challenges. *Materials Advances, 2*(6), 1821–1871. DOI: 10.1039/D0MA00807A

Bansal, S. A., Kumar, V., Karimi, J., Singh, A. P., & Kumar, S. (2020). Role of gold nanoparticles in advanced biomedical applications. *Nanoscale Advances, 2*(9), 3764–3787. DOI: 10.1039/D0NA00472C

Barhoum, A., García-Betancourt, M. L., Jeevanandam, J., Hussien, E. A., Mekkawy, S. A., Mostafa, M., Omran, M. M. S., Abdalla, M., & Bechelany, M. (2022). Review on Natural, Incidental, Bioinspired, and Engineered Nanomaterials: History, Definitions, Classifications, Synthesis, Properties, Market, Toxicities, Risks, and Regulations. *Nanomaterials (Basel, Switzerland), 12*(2), 177. DOI: 10.3390/nano12020177

Behera, M., Philip, A., & Kumar, A. R. (2025). Optimized Hydroxyapatite–Carbon Nanotube Composite Electrodes for Enhanced Supercapacitor Performance. ChemistrySelect, 10(29). Portico. DOI: 10.1002/slct.202501845

Bespalov, V. A., Dyuzhev, N. A., & Kireev, V. Yu. (2022). Possibilities and Limitations of CMOS Technology for the Production of Various Microelectronic Systems and Devices. *Nanobiotechnology Reports, 17*(1), 24–38. DOI: 10.1134/S2635167622010037

Biswal, M. (2025). Advances in nanotechnology for heavy metal removal from contaminated water sources. Journal of Nanomaterials and Devices. DOI: 10.61577/jnd.2025.100003

Borah, R., AG, K. R., Minja, A. C., & Verbruggen, S. W. (2023). A Review on Self-Assembly of Colloidal Nanoparticles into Clusters, Patterns, and Films: Emerging Synthesis Techniques and Applications. Small Methods, 7(6). Portico. DOI: 10.1002/smtd.202201536

Boretti, A. (2025). A narrative review of breakthroughs and future horizons in zero-dimensional nanomaterials. *Optical and Quantum Electronics, 57*(6), 370. Advance online publication. DOI: 10.1007/s11082-025-08288-9

Chung, Y. H., Beiss, V., Fiering, S. N., & Steinmetz, N. F. (2020). COVID-19 Vaccine Frontrunners and Their Nanotechnology Design. *ACS Nano, 14*(10), 12522–12537. DOI: 10.1021/acsnano.0c07197

Dai, X., & Chen, Y. (2022). Computational Biomaterials: Computational Simulations for Biomedicine. Advanced Materials, 35(7). Portico. DOI: 10.1002/adma.202204798

David, M. E., Ion, R.-M., Grigorescu, R. M., Iancu, L., Holban, A. M., Nicoara, A. I., Alexandrescu, E., Somoghi, R., Ganciarov, M., Vasilievici, G., & Gheboianu, A. I. (2021). Hybrid Materials Based on Multi-Walled Carbon Nanotubes and Nanoparticles with Antimicrobial Properties. *Nanomaterials (Basel, Switzerland)*, *11*(6), 1415. DOI: 10.3390/nano11061415

Egwu, C. N., Babalola, R., Udoh, T. H., & Esio, O. O. (2022). Nanotechnology: Applications, Challenges, and Prospects. Advanced Manufacturing in Biological, Petroleum, and Nanotechnology Processing, 3–15. DOI: 10.1007/978-3-030-95820-6_1

Gao, Y., Li, Y., Niu, Q., & Zhang, P. (2025). Transforming petroleum asphalt into carbon fibers and related metal/oxide composites by electrospinning synthesis. *Green Chemistry*, *27*(26), 8007–8017. DOI: 10.1039/D5GC01775K

Godja, N.-C., & Munteanu, F.-D. (2024). Hybrid Nanomaterials: A Brief Overview of Versatile Solutions for Sensor Technology in Healthcare and Environmental Applications. *Biosensors (Basel)*, *14*(2), 67. DOI: 10.3390/bios14020067

Gui, R., & Jin, H. (2025). Synthesis, characterizations, properties, modifications and applications of zero-dimensional monoelemental selenium nanomaterials. *Coordination Chemistry Reviews*, *535*, 216626. DOI: 10.1016/j.ccr.2025.216626

Gupta, D., Varghese, B. S., Suresh, M., Panwar, C., & Gupta, T. K. (2022). Nanoarchitectonics: Functional nanomaterials and nanostructures—A review. *Journal of Nanoparticle Research*, *24*(10), 196. DOI: 10.1007/s11051-022-05577-2

Gupta, P. K. (2023). Introduction and historical background. In *Nanotoxicology in Nanobiomedicine* (pp. 1–22). Springer International Publishing., DOI: 10.1007/978-3-031-24287-8_1

Hachem, K., Ansari, M. J., Saleh, R. O., Kzar, H. H., Al-Gazally, M. E., Altimari, U. S., Hussein, S. A., Mohammed, H. T., Hammid, A. T., & Kianfar, E. (2022). Methods of Chemical Synthesis in the Synthesis of Nanomaterial and Nanoparticles by the Chemical Deposition Method: A Review. *BioNanoScience*, *12*(3), 1032–1057. DOI: 10.1007/s12668-022-00996-w

Haider, J., Shahzadi, A., Akbar, M. U., Hafeez, I., Shahzadi, I., Khalid, A., Ashfaq, A., Ahmad, S. O. A., Dilpazir, S., Imran, M., Ikram, M., Ali, G., Khan, M., Khan, Q., & Maqbool, M. (2022). A review of synthesis, fabrication, and emerging biomedical applications of metal-organic frameworks. *Biomaterials Advances, 140*, 213049. DOI: 10.1016/j.bioadv.2022.213049

Hargittai, I. (2022). Robert F. Curl, Jr. (1933–2022) structural chemist, Nobel laureate co-discoverer of buckminsterfullerene. *Structural Chemistry*, *33*(6), 2265–2266. DOI: 10.1007/s11224-022-02041-8

Huang, Z.-M., Zhang, Y.-Z., Kotaki, M., & Ramakrishna, S. (2003). A review on polymer nanofibers by electrospinning and their applications in nanocomposites. *Composites Science and Technology*, *63*(15), 2223–2253. DOI: 10.1016/S0266-3538(03)00178-7

Ichou, H., Arrousse, N., Berdimurodov, E., & Aliev, N. (2023). Exploring the Advancements in Physical Vapor Deposition Coating: A Review. *Journal of Bio- and Tribo-Corrosion*, *10*(1), 3. Advance online publication. DOI: 10.1007/s40735-023-00806-0

Imoisili, P. E., & Jen, T.-C. (2021). Facile Extraction and Characterization of Silica Nanoparticles from Corn Stalk by Sol-Gel Hydrothermal Methods. *IOP Conference Series. Materials Science and Engineering*, *1107*(1), 012030. DOI: 10.1088/1757-899X/1107/1/012030

Imoisili, P. E., & Jen, T. C. (2024). [a]). Microwave-assisted sol–gel synthesis of V-TiO2 nanocatalyst for wastewater treatment. *Materials Today: Proceedings*, *105*, 247–252. Advance online publication. DOI: 10.1016/j.matpr.2023.09.021

Imoisili, P. E., & Jen, T.-C. (2024). [b]). Synthesis and characterization of amorphous nano silica from South African coal fly ash. *Materials Today: Proceedings*, *105*, 21–26. DOI: 10.1016/j.matpr.2023.06.077

Imoisili, P. E., Makhatha, M. E., & Jen, T.-C. (2024). Artificial Intelligence prediction and optimization of the mechanical strength of modified Natural Fibre/MWCNT polymer nanocomposite. *Journal of Science. Advanced Materials and Devices*, *9*(2), 100705. DOI: 10.1016/j.jsamd.2024.100705

Imoisili, P. E., Ukoba, K., & Jen, T.-C. (2020b). Physical, mechanical and thermal properties of high frequency microwave treated plantain (Musa Paradisiaca) fibre/MWCNT hybrid epoxy nanocomposites. *Journal of Materials Research and Technology*, *9*(3), 4933–4939. DOI: 10.1016/j.jmrt.2020.03.012

Imoisili, P. E., Ukoba, K. O., & Jen, T. C. (2020a). Synthesis and characterization of amorphous mesoporous silica from palm kernel shell ash. boletín de la sociedad española de cerámica y vidrio, 59(4), 159-164. DOI: 10.1016/j.bsecv.2019.09.006

Jabeen, S., Khan, T., Jaiswal, A., & Bala, S. (2024). Green Nanotechnology for Clean Energy and Environmental Sustainability. Sustainable Nanomaterials, 1–20. DOI: 10.1007/978-981-97-2761-2_1

Jabeen, S., Khan, T., Jaiswal, A., & Bala, S. (2024). Green Nanotechnology for Clean Energy and Environmental Sustainability. Sustainable Nanomaterials, 1–20. DOI: 10.1007/978-981-97-2761-2_1

Jiang, Z., Li, L., Huang, H., He, W., & Ming, W. (2022). Progress in Laser Ablation and Biological Synthesis Processes: "Top-Down" and "Bottom-Up" Approaches for the Green Synthesis of Au/Ag Nanoparticles. *International Journal of Molecular Sciences*, *23*(23), 14658. DOI: 10.3390/ijms232314658

Joshi, J., Homburg, S. V., & Ehrmann, A. (2022). Atomic Force Microscopy (AFM) on Biopolymers and Hydrogels for Biotechnological Applications—Possibilities and Limits. *Polymers*, *14*(6), 1267. DOI: 10.3390/polym14061267

Kamyab, H., Chelliapan, S., Hayder, G., Yusuf, M., Taheri, M. M., Rezania, S., Hasan, M., Yadav, K. K., Khorami, M., Farajnezhad, M., & Nouri, J. (2023). Exploring the potential of metal and metal oxide nanomaterials for sustainable water and wastewater treatment: A review of their antimicrobial properties. *Chemosphere*, *335*, 139103. DOI: 10.1016/j.chemosphere.2023.139103

Karnwal, A., Kumar Sachan, R. S., Devgon, I., Devgon, J., Pant, G., Panchpuri, M., Ahmad, A., Alshammari, M. B., Hossain, K., & Kumar, G. (2024). Gold Nanoparticles in Nanobiotechnology: From Synthesis to Biosensing Applications. *ACS Omega*, *9*(28), 29966–29982. DOI: 10.1021/acsomega.3c10352

Khairani, I. Y., Mínguez-Vega, G., Doñate-Buendía, C., & Gökce, B. (2023). Green nanoparticle synthesis at scale: A perspective on overcoming the limits of pulsed laser ablation in liquids for high-throughput production. *Physical Chemistry Chemical Physics*, *25*(29), 19380–19408. DOI: 10.1039/D3CP01214J

Khan, S., Naushad, M., Al-Gheethi, A., & Iqbal, J. (2021). Engineered nanoparticles for removal of pollutants from wastewater: Current status and future prospects of nanotechnology for remediation strategies. *Journal of Environmental Chemical Engineering*, *9*(5), 106160. DOI: 10.1016/j.jece.2021.106160

Khurana, K., & Jaggi, N. (2021). Localized surface plasmonic properties of Au and Ag nanoparticles for sensors: A review. *Plasmonics*, *16*(4), 981–999. DOI: 10.1007/s11468-021-01381-1

Khutso Ntobeng, M., Ehi Imoisili, P., & Jen, T.-C. (2020). Synthesis and photocatalytic performance of Ag-TiVOX nanocomposite. *Journal of King Saud University. Science*, *32*(7), 3103–3110. DOI: 10.1016/j.jksus.2020.08.021

Kumar, A., Tyagi, P. K., Tyagi, S., & Ghorbanpour, M. (2024). Integrating green nanotechnology with sustainable development goals: A pathway to sustainable innovation. *Discover Sustainability*, *5*(1), 364. Advance online publication. DOI: 10.1007/s43621-024-00610-x

Lee, H., You, S., Pikhitsa, P. V., Kim, J., Kwon, S., Woo, C. G., & Choi, M. (2010). Three-Dimensional Assembly of Nanoparticles from Charged Aerosols. *Nano Letters*, *11*(1), 119–124. DOI: 10.1021/nl103787k

Li, R., Zhou, Y., Li, W., Zhu, J., & Huang, W. (2020). Structure Engineering in Biomass-Derived Carbon Materials for Electrochemical Energy Storage. *Research*, *2020*, 2020/8685436. Advance online publication. DOI: 10.34133/2020/8685436

Li, Y. (2021). Carbon Nanotube Research in Its 30th Year. *ACS Nano*, *15*(6), 9197–9200. DOI: 10.1021/acsnano.1c04972

Lin, S. P., Khumsupan, D., Chou, Y. J., Hsieh, K. C., Hsu, H. Y., Ting, Y., & Cheng, K. C. (2022). Applications of atmospheric cold plasma in agricultural, medical, and bioprocessing industries. *Applied Microbiology and Biotechnology*, *106*(23), 7737–7750. DOI: 10.1007/s00253-022-12252-y

Liu, C., Ma, H., Yuan, S., Jin, Y., & Tian, W. (2025). Living Cell-Mediated Self-Assembly: From Monomer Design and Morphology Regulation to Biomedical Applications. *ACS Nano*, *19*(2), 2047–2069. DOI: 10.1021/acsnano.4c16669

Lu, X., Wang, C., & Wei, Y. (2009). One-Dimensional Composite Nanomaterials: Synthesis by Electrospinning and Their Applications. Small, 5(21), 2349–2370. Portico. DOI: 10.1002/smll.200900445

Ma, M.-Y., Han, D., Chen, N.-K., Wang, D., & Li, X.-B. (2022). Recent Progress in Double-Layer Honeycomb Structure: A New Type of Two-Dimensional Material. *Materials (Basel)*, *15*(21), 7715. DOI: 10.3390/ma15217715

Machín, A., Fontánez, K., Arango, J. C., Ortiz, D., De León, J., Pinilla, S., Nicolosi, V., Petrescu, F. I., Morant, C., & Márquez, F. (2021). One-Dimensional (1D) Nanostructured Materials for Energy Applications. *Materials (Basel)*, *14*(10), 2609. DOI: 10.3390/ma14102609

Madkour, L. H. (2019). Introduction to Nanotechnology (NT) and Nanomaterials (NMs). Nanoelectronic Materials, 1–47. DOI: 10.1007/978-3-030-21621-4_1

Malik, S., Muhammad, K., & Waheed, Y. (2023). Emerging Applications of Nanotechnology in Healthcare and Medicine. *Molecules (Basel, Switzerland)*, *28*(18), 6624. DOI: 10.3390/molecules28186624

Malik, S., Muhammad, K., & Waheed, Y. (2023). Emerging Applications of Nanotechnology in Healthcare and Medicine. *Molecules (Basel, Switzerland)*, *28*(18), 6624. DOI: 10.3390/molecules28186624

Megahed, S. H., Abdel-Halim, M., El-shabrawy, Y. I., Saad, E. M., Hefnawy, A., Handoussa, H., Mizaikoff, B., & El Gohary, N. A. (2025). Design, synthesis and medical prospects of electrospun molecularly imprinted fibers. *Scientific Reports*, *15*(1), 26082. Advance online publication. DOI: 10.1038/s41598-025-11114-7

Mei, J., Wang, J., Gu, H., Du, Y., Wang, H., Yamauchi, Y., Liao, T., Sun, Z., & Yin, Z. (2020). Nano Polymorphism-Enabled Redox Electrodes for Rechargeable Batteries. Advanced Materials, 33(8). Portico. DOI: 10.1002/adma.202004920

Mekuye, B., & Abera, B. (2023). Nanomaterials: An overview of synthesis, classification, characterization, and applications. Nano Select, 4(8), 486–501. Portico. DOI: 10.1002/nano.202300038

Meng, Y., Feng, Y., Bai, X., Yu, Q., Zhou, J., & Wang, J. (2025). Application of nanotechnology in agricultural sustainability: Absorption, translocation, and challenges of nanoparticles. *Current Plant Biology*, *42*, 100492. DOI: 10.1016/j.cpb.2025.100492

Minello, L. V. P., Ahmad, I., Aguzzoli, C., & Sperotto, R. A. (2025). Surface engineering of nanoparticles via physical vapor deposition (PVD): Opportunities and challenges for sustainable agriculture. *Plant Nano Biology*, *13*, 100187. DOI: 10.1016/j.plana.2025.100187

Norton, M. G. (2023). There's Still Plenty of Room at the Bottom. In *A Modern History of Materials: From Stability to Sustainability* (pp. 81–100). Springer International Publishing., DOI: 10.1007/978-3-031-23990-8_5

Nwanna, E. C., Imoisili, P. E., & Jen, T.-C. (2020). Fabrication and synthesis of SnOX thin films: A review. *International Journal of Advanced Manufacturing Technology*, *111*(9–10), 2809–2831. DOI: 10.1007/s00170-020-06223-8

Nwanna, E. C., Imoisili, P. E., & Jen, T.-C. (2022). Synthesis and characterization of SnO2 thin films using metalorganic precursors. *Journal of King Saud University. Science*, *34*(5), 102123. DOI: 10.1016/j.jksus.2022.102123

Ojodun, O. E., Imoisili, P. E., & Jen, T.-C. (2025). [a]). Thermal annealing-induced structural modifications and electrochemical enhancement of NiO/CNT electrodes synthesized by spray pyrolysis for high-performance supercapacitors. *Electrochemistry Communications*, *178*, 108010. DOI: 10.1016/j.elecom.2025.108010

Ojodun, O. E., Imoisili, P. E., & Jen, T.-C. (2025[b]). Chemical Solution–Based Deposition Techniques for Asymmetric Supercapacitors: Progress in Electrode Materials for High-Efficiency Energy Storage. Journal of Nanotechnology, 2025(1). Portico. DOI: 10.1155/jnt/5551493

Opetubo, O. R., Kitalu, R., Oviroh, P. O., Oyinbo, S. T., Imoisili, P. E., & Jen, T. C. (2023). A mini-review on MoS2 membrane for water desalination: Recent development and challenges. *Nanotechnology Reviews*, *12*(1), 20220563. DOI: 10.1515/ntrev-2022-0563

Park, H., Shin, D. J., & Yu, J. (2021). Categorization of quantum dots, clusters, nanoclusters, and nanodots. *Journal of Chemical Education*, *98*(3), 703–709. DOI: 10.1021/acs.jchemed.0c01403

Patel, S., & Nanda, R. (2015). Nanotechnology in Healthcare: Applications and Challenges. *Medicinal Chemistry (Sunnyvale, Calif.)*, *05*(12). Advance online publication. DOI: 10.4172/2161-0444.1000312

Peixoto, D. A., Silva, S. C., Borges, P. H. S., Lima, R. C., & Nossol, E. (2023). Hydrothermal synthesis as a versatile tool for the preparation of metal hexacyanoferrates: A review. *Journal of Materials Science*, *58*(7), 2993–3024. DOI: 10.1007/s10853-023-08190-3

Pisano, R., & Durlo, A. (2025). Nanosciences and Nanotechnologies: A Scientific–Historical Introductory Review. Nanoscience & Nanotechnologies: Critical Problems, Science in Society, Historical Perspectives, 3-90. DOI: 10.1007/978-3-031-85122-3_1

Polman, A., Knight, M., Garnett, E. C., Ehrler, B., & Sinke, W. C. (2016). Photovoltaic materials: Present efficiencies and future challenges. *Science*, *352*(6283), aad4424. Advance online publication. DOI: 10.1126/science.aad4424

Rana, B., & Rathore, S. S. (2023). Industry 4.0 – Applications, challenges and opportunities in industries and academia: A review. *Materials Today: Proceedings*, *79*, 389–394. DOI: 10.1016/j.matpr.2022.12.162

Rekhviashvili, S. S. (2023). Effect of Polarization on Exciton Properties in a Semiconductor Quantum Dot. *Russian Physics Journal*, *65*(12), 2058–2067. DOI: 10.1007/s11182-023-02871-x

Roco, M. C. (2023). National Nanotechnology Initiative at 20 years: Enabling new horizons. *Journal of Nanoparticle Research*, *25*(10), 197. Advance online publication. DOI: 10.1007/s11051-023-05829-9

Rodríguez-Galván, A., & Contreras-Torres, F. F. (2022). Scanning tunneling microscopy of biological structures: An elusive goal for many years. *Nanomaterials (Basel, Switzerland)*, *12*(17), 3013. DOI: 10.3390/nano12173013

Saeloo, B., Saisopa, T., Chavalekvirat, P., Iamprasertkun, P., Jitapunkul, K., Sirisaksoontorn, W., Lee, T. R., & Hirunpinyopas, W. (2024). Role of Transition Metal Dichalcogenides as a Catalyst Support for Decorating Gold Nanoparticles for Enhanced Hydrogen Evolution Reaction. *Inorganic Chemistry*, *63*(40), 18750–18762. DOI: 10.1021/acs.inorgchem.4c02668

Selvarani, K., Theivasanthi, T., & Ajitha, M. (2025). Nanotechnology in Agriculture: Enhancing Seed Quality, Yield and Contribution to Sustainable Development Goals (SDGs). Journal of Advances in Biology &amp. *Journal of Advances in Biology & Biotechnology*, *28*(1), 22–30. DOI: 10.9734/jabb/2025/v28i11857

Singh, R., Dutt, S., Sharma, P., Sundramoorthy, A. K., Dubey, A., Singh, A., & Arya, S. (2023). Future of Nanotechnology in Food Industry: Challenges in Processing, Packaging, and Food Safety. Global Challenges, 7(4). Portico. DOI: 10.1002/gch2.202200209

Srivastava, N., Srivastava, M., Mishra, P. K., & Gupta, V. K. (Eds.). (2020). *Green Synthesis of Nanomaterials for Bioenergy Applications*., DOI: 10.1002/9781119576785

Su, H., Zheng, J., Tang, L., Liu, F., Lu, M., Wang, F., Wang, L., Zheng, H., Cai, B., & Yu, W. W. (2025). One-step electrospinning synthesis of lead-free perovskite nanofiber-based triboelectric nanogenerator for wearable intelligent fall alarm. *Journal of Colloid and Interface Science*, *696*, 137836. DOI: 10.1016/j.jcis.2025.137836

Tahir, M. B., Sagir, M., & Rafique, M. (2021). Challenges and Future of Nanotechnology. *Nanotechnology*, *141–142*, 141–142. Advance online publication. DOI: 10.1007/978-981-15-9437-3_6

Tahir, M. B., Sagir, M., & Rafique, M. (2021). Challenges and Future of Nanotechnology. *Nanotechnology*, *141–142*, 141–142. Advance online publication. DOI: 10.1007/978-981-15-9437-3_6

Tawiah, B., Ofori, E. A., & George, S. C. (2024). Nanotechnology in Societal Development. Nanotechnology in Societal Development, 1–64. DOI: 10.1007/978-981-97-6184-5_1

Tonelli, D., Scavetta, E., & Gualandi, I. (2019). Electrochemical Deposition of Nanomaterials for Electrochemical Sensing. *Sensors (Basel)*, *19*(5), 1186. DOI: 10.3390/s19051186

Van Aert, S., Batenburg, K. J., Rossell, M. D., Erni, R., & Van Tendeloo, G. (2011). Three-dimensional atomic imaging of crystalline nanoparticles. *Nature*, *470*(7334), 374–377. DOI: 10.1038/nature09741

Verma, R., Jha, S., Harimurugan, D., Kasthurirengan, S., Shivaprakash, N. C., & Behera, U. (2022). Nanomaterials' Synthesis Approaches for Energy Storage and Electronics Applications. Applications of Nanomaterials in Energy Storage and Electronics, 240–257. DOI: 10.2174/9789815050714122030017

Wadhawan, S., Wadhawan, D., Jain, A., & Mehta, S. K. (2024). Toxic implication of nanoparticles: A review of factors, mechanism, exposure and control strategies. *International Journal of Environmental Science and Technology*, *22*(2), 1203–1224. DOI: 10.1007/s13762-024-05810-6

Xie, J., Li, X., & Xia, Y. (2008). Putting electrospun nanofibers to work for biomedical research. *Macromolecular Rapid Communications*, *29*(22), 1775–1792. DOI: 10.1002/marc.200800381

Xu, M., Liang, T., Shi, M., & Chen, H. (2013). Graphene-like two-dimensional materials. *Chemical Reviews*, *113*(5), 3766–3798. DOI: 10.1021/cr300263a

Yaminsky, I. V. (2024). Scanning Probe Bionanoscopy: Methods and Equipment, Results and Achievements. *Moscow University Physics Bulletin*, *79*(6), 839–855. DOI: 10.3103/S0027134924702473

Yao, R., Sun, K., Zhang, K., Wu, Y., Du, Y., Zhao, Q., Liu, G., Chen, C., Sun, Y., & Li, J. (2024). Stable hydrogen evolution reaction at high current densities via designing the Ni single atoms and Ru nanoparticles linked by carbon bridges. *Nature Communications*, *15*(1), 2218. Advance online publication. DOI: 10.1038/s41467-024-46553-9

Yarin, A. L., Koombhongse, S., & Reneker, D. H. (2001). Taylor cone and jetting from liquid droplets in electrospinning of nanofibers. *Journal of Applied Physics*, *90*(9), 4836–4846. DOI: 10.1063/1.1408260

Yin, J., Wang, H., Pyle, D., Choi, S., Liu, Y., Wen, J., Guest, J. R., Lyding, J. W., & Dong, G. (2025). Synthesis and Self-Assembly of Monodisperse Graphene Nanoribbons: Access to Submicron Architectures with Long-Range Order and Uniform Orientation. *ACS Nano*, *19*(4), 4366–4376. DOI: 10.1021/acsnano.4c12313

Yudaev, P., Chuev, V., Klyukin, B., Kuskov, A., Mezhuev, Y., & Chistyakov, E. (2022). Polymeric Dental Nanomaterials: Antimicrobial Action. *Polymers*, *14*(5), 864. DOI: 10.3390/polym14050864

Zain Ul Abidin, M., Ikram, M., Moeen, S., Nazir, G., Kanoun, M. B., & Goumri-Said, S. (2024). A comprehensive review on the synthesis of ferrite nanomaterials via bottom-up and top-down approaches advantages, disadvantages, characterizations and computational insights. *Coordination Chemistry Reviews*, *520*, 216158. DOI: 10.1016/j.ccr.2024.216158

Zhao, R., Pandyaswargo, A. H., & Onoda, H. (2024). A bottom-up approach for greenhouse gas emission estimation at the community level: A case study in Japan. *Energy*, *307*, 132530. DOI: 10.1016/j.energy.2024.132530

Zhao, Y., Zhang, Z., Pan, Z., & Liu, Y. (2021). Advanced bioactive nanomaterials for biomedical applications. Exploration, 1(3). Portico. DOI: 10.1002/EXP.20210089

Zou, X., Gu, F., Zhang, X., & Zhang, X. (2025). Self-assembly synthesis of FeOOH/Cu2O heterostructure porous microspheres with enhanced photocatalytic activity. *Journal of Alloys and Compounds*, *1010*, 178017. DOI: 10.1016/j.jallcom.2024.178017

Chapter 5 Addressing Energy, Environment, and Housing Challenges Through Nanotechnology: Materials of the Future

Shehzad Ahmad
University of the Punjab, Lahore, Pakistan

Zainab Maqsood
University of the Punjab, Lahore, Pakistan

Ali Hasan
University of Agriculture, Faisalabad, Pakistan

Shahid Mahmood
University of the Punjab, Lahore, Pakistan

Ali Hussain
https://orcid.org/0000-0002-8105-3029
University of the Punjab, Lahore, Pakistan

ABSTRACT

The worldwide demand of energy and change of the climate highpoint the advanced need of sustainable energy storage. During on wind reliance solar is requires for energy storage. The major source for emissions and energy use is a building sectors.

DOI: 10.4018/979-8-3373-3556-8.ch005

Advances nanostructured materials, alloys, ceramics and polymers are enable such next generation solar cells, flexible electronics and effective devices. In this chapter, explores economic aspects, application of advance nanostructured material in energy storage and in environment. Furthermore, explores challenges in building a sustainable future and fabrication strategies.

1.INTRODUCTION

As the sustained technological progress and the increase in the requirements to sophisticated materials have become key vehicles to innovation, they include nanomaterials (e.g., graphene, carbon nanotubes), high-performance polymers, ceramics, and special alloys (Eyube et al., 2025). These materials are distinguished by their amazing physical, chemical, and mechanical properties and are used in transformative applications in energy, electronics, aerospace, transportation, and healthcare (Skiba, 2025). They have facilitated inventions, including semiconducting polymers, which may be used in flexible, lightweight, energy-efficient circuits, superalloys, which may be used in power-generation turbines and aircraft engines, and which can withstand extreme pressures and temperatures (Holloway, 2025; Skiba, 2025).

Nanotechnology has been instrumental in facilitating the development of energy storage schemes by improving some of the most important parameters of energy storage like transport charge, surface area, and electrode stability as shown in Table 1. Nanostructured materials such as nanoparticles, nanotubes and nanowires, have very high surface-to-volume ratios than their bulk forms. There in it offers a large number of active sites in electrochemical reactions. The structural advantage is transferred in the form of increased energy densities and accelerated charge-discharge kinetics. Which are necessary to the enhanced performance of current batteries and supercapacitors as indicated in Table 1. The SDGs can also be achieved through nanotechnology because it facilitates domestic and more sustainable energy storage systems. Thus, nanostructured material can be beneficial when it comes to the global efforts around sustainable energy system shows in Table 1.

Advanced biomaterials increase the range of applications in healthcare in tissue engineering, drug delivery, and prosthetics, which will help to improve patient outcomes and quality of life (Erzen et al., 2025). Likewise, the innovations of renewable energy like perovskite solar cells will aid the process of transition to maintainable energy by providing competitive and cost-effective options to the usual photovoltaic technologies (Wheatley, 2024). The use and the prospects of advanced materials are one of the current fields of research in the twenty-first century, and their application to environmental issues, reducing the energy demand, and enhancing the human condition is of a special interest (Bai and Zhang. 2025; Hariram et al., 2023). Nevertheless, even with such opportunities, there are aspects of scalable manufacturing and environmental management that should be carefully considered (Simoes, 2024). The given book chapter is intended to present a general overview of the advanced materials, their various uses, recent advances in the sphere of these materials, and their contribution to the progressive and sustainable future, backed up with specific examples and the new changes.

The world today, as a whole, is witnessing an alarming increase in the rate of energy use coupled with the rapidly increasing climate change, which underscores the imperative need to shift towards sustainable and energy-efficient housing. The construction industry contributes 30 to 40 percent of the total energy consumption and 40 to 50 percent of CO_2 emissions worldwide (Chen et al., 2024). Based on these thoughts, residential sector provided 21 percent of energy requirement and 17 percent of emissions with construction, bricks, and glass industries following with another 10 percent, that points to a significant environmental load. These sectors are the ones that cause global warming and climate change, which, in turn, leads to extreme weather patterns and negative health effects, which is what makes the implementation of greenhouse gas mitigation and green practices particularly necessary (Asdrubali et al., 2025). Despite the significant scale of these effects, the world is expected to keep increasing its energy demand, and the construction industry is going to play a significant role (Tootkaboni et al., 2025). The construction industry has a significant energy-saving potential because a significant portion of the existing buildings is characterized by the lack of efficiency caused by the fact that they were constructed before the emergence of modern regulatory frameworks (Penazzato et al., 2024; Tian et al., 2022).

As the world population increases and the economy gains, there has been a demand to enhance sustainability and efficiency in energy storage. The growing development of renewable resources like wind and solar which by nature is subject to changes is a huge burden to reliability and grid stability. This means that more efforts should be put on developing sophisticated energy storage technologies that can be used in the integration of the renewable sources on a large scale and also in giving security to the energy in the long term. The past four decadses have seen increasing demand of energy with close correlation to the GDP growth, and historical evidence has shown that GDP in global growth 1 percent has seen a 0.6 percent increase in primary energy use, which has translated into a projected annual growth of approximately 2 percent (Bogmans et al., 2025; Abbasi et al., 2021; Semieniuk et al., 2021; Scheffran et al., 2020). The energy storage demands in the world are expected to grow four-fold by 2030 and advanced systems and equipment have to be implemented to ensure supply is secure and reliable (Islam et al., 2024). Such an increased demand has increased the research on energy storage technologies, which are essential to address the issue of renewable energy variability and the optimization of RES integration, as well as to limit the emission of carbon (Shafie et al., 2025; Adeyink et al., 2024).

In addition, manufacture of innovative energy storage devices that make use of nanostructured materials is also discussed in this chapter. The main strategies, such as top-down, bottom-up, and hybrid strategies are presented based on their scalability, accuracy and ability to utilize them in large-scale production. There are a few challenges related to nanomaterials as the part of energy storage systems, specifically stability, safety, and long-term operation, which should be properly organized to guarantee the reliability of the functioning (Islam et al., 2024). There is also the economic approach which covers the cost aspect of developing and commercialization of these emerging technologies. Together, these observations highlight the potential change and the practical constraints of the nanotechnology-based energy storage solutions and reiterate the need to conduct further research and innovation to ensure that the laboratory-level discoveries are applied in the industry.

2. ADVANCED MATERIALS FOR ENERGY APPLICATIONS

2.1 Nanostructured Materials for Energy Storage

Their nanoscale dimensions provide shortened ion diffusion pathways, improved electrode electrolyte contact, and enhanced mechanical tolerance to volume changes during charge discharge cycles. As a result, these materials contribute to faster charging rate, longer life cycle, and high energy densities as compared conventional electrode materials show in Table 1. The growing dependence on technology, from portable to hybrid devices and electronic vehicle has intensified demand for advanced storage of energy solutions. Lithium-ion batteries (LIBs) and available rechargeable system are regarded as the most mature and efficient technology, primarily due to their reliance on solid-state intercalation compounds capable of reversibly hosting lithium ions within electrode structures show in Table 1. Such nanostructured approaches enhance ion transport, increase surface area for electrochemical reactions, and improve structural stability, thereby providing pathways to overcome existing limitations and enabling lithium-ion batteries to next generation of high performance show in Table 1.

According to Mohammed et al. (2025) globally increasing the demand of energy highlights the need for sustainability, efficiently high-performance energy storage system. By enabling nanotechnology, the manipulation of material at the nanoscale provides remarkable opportunities. To enhance the functionality of storage devices owing to attributes such as superior electrical conductivity, and tunable structural properties. Similarly, in supercapacitors, corbon-based nanostructured material comprising carbon nanotubes, activated carbon and graphene offer large surface areas for charge accumulation and rapid ion transport, enabling enhanced capacitance and ultrafast charge–discharge rates. These properties make them particularly suitable for applications requiring frequent and high-power energy delivery, such as electric vehicles and grid stabilization show in Table 1.

According to Roy and Das (2025) recent years have witnessed a substantial increase in study on nanostructured material driving in all the discipline such biomedical, chemistry, physics, medicine, industrial chemistry. Nanotechnology as a highly interdisciplinary field with broad scientific and technological significance. Furthermore, study of metal oxide nanoparticles highlights a convergence of fundamental research and applied innovation, with potential to deliver breakthroughs in areas spanning environmental remediation, targeted drug delivery, advanced diagnostics, and therapeutic interventions show in Table 1.

Table 1. Nanostructured material area and types in energy applications

Application Area	Material Type	Key Properties	Benefits	References
Lithium-ion Batteries	Silicon nanowires, graphene sheets, transition metal oxides	Nanoscale dimensions, shortened ion diffusion pathways	Faster charging rates, longer life cycles, high energy densities	Khalid et al., 2025
Supercapacitors	Carbon nanotubes, activated carbon, graphene	Large surface areas, rapid ion transport	Enhanced capacitance, ultrafast charge-discharge rates	Priyantha and Karunathilake, 2025
Capacitors	Nanostructured dielectrics	Higher dielectric constants, greater energy storage capacity	Improved breakdown resistance, extended operational lifespans	Phogat et al., 2025
Fuel Cells	Carbon nanotubes, graphene-supported nanoparticles, alloy oxides	Enhanced catalytic activity, faster electron transfer	Improved efficiency and durability, reduced catalyst degradation	Chen et al., 2024
Solar Cells	Quantum dots, perovskite nanocrystals, nanoscale semiconductors	Superior light absorption, tunable bandgaps	Higher power conversion efficiencies, improved charge separation	Mangaiyarkkarasi, 2025
Metallic Nanoparticles	Iron, aluminum, silver, copper, zinc nanoparticles	Superparamagnetic behavior, biocompatibility, low cytotoxicity	Biomedical applications, unique physicochemical properties	Roy and Das, 2025

continued on following page

Table 1. Continued

Application Area	Material Type	Key Properties	Benefits	References
Metal Oxide Nanoparticles	Various metal oxide forms	Convergence of fundamental research and applied innovation	Environmental remediation, targeted drug delivery, advanced diagnostics	Roy and Das, 2025
LIB Electrode Materials	Solid-state intercalation compounds with nanostructured design	Reversible lithium ion hosting, enhanced ion transport	Improved voltage and capacity, structural stability	Yao et al., 2025; Khan et al., 2025
Energy Storage Systems	Nanoparticles, nanotubes, nanowires	High surface-to-volume ratios, superior electrical conductivity	Higher energy densities, faster charge-discharge kinetics	Mohammed et al., 2025
Electrode Design	Nanostructured electrode materials	Tunable structural properties, enhanced functionality	Improved electrochemical performance, abundant active sites	Pomerantseva et al., 2019; Asamoah et al., 2024
Electrolyte Modification	Novel nanostructured materials	Interfacial stability enhancement	Mitigation of parasitic reactions, improved performance	Mohammed et al., 2025
Separator Optimization	Nanostructured separator materials	Enhanced transport properties	Optimized electrochemical performance	Mohammed et al., 2025
Battery Performance	Nanostructured materials in battery systems	High surface area, improved charge transport	Enhanced energy densities, faster kinetics	Dutta et al., 2023; Hepel, 2023
Sustainable Energy	Nanostructured materials for sustainability	Domestic energy storage capabilities	Support for global sustainable energy efforts	Roy and Das, 2025

2.2 Photocatalytic and Photovoltaic Materials

At present, more than 80% of the global energy requirements are fulfilled by fossil fuel combustion. A practice that not only accelerates global warming but also poses severe environmental consequences (Kalair et al., 2021; Saini et al., 2025). Furthermore, due to limited fossil fuel will finally be end. Due to their high rate of consumption, thereby necessitating the search for clean, renewable, and cost-effective alternatives. Perovskite materials, with the general formula ABX_3, have arisen as extremely adaptable applicants for these applications (Mahmoud et al., 2024). Offering improved light absorption, charge separation, and carrier transport for next-generation solar energy technologies (Kiani, 2025; Ratnesh et al., 2025; Adeoye et al., 2023).

Several environmental and depletion of fossil fuel their impacts of combustion have accelerated the global pursuit of renewable and environmentally sustainable energy alternatives. Under such circumstances, solar energy stands out as a particularly promising option due to its inexhaustible supply and minimal ecological footprint. However, depends on achieving high-efficiency energy conversion technologies, positioning solar energy as a viable and transformative substitute for fossil fuels in the near future (Chen et al., 2023; Ojo et al., 2023). Plasmonic nanostructured materials are generally defined as matels or metal like materials. Which are characterized by high density of free electron and a negative real permittivity. Among them, noble metals including silver and such as gold (Au) and silver (Ag) have been widely explored for enhancing the performance of photovoltaic and photocatalytic systems.

2.2.1 Challenges of perovskites toxicity and stability

In recent years, perovskites material has attracted attentions due to their high efficiency solar cell and optoelectronics properties (Znidi et al., 2024). The perovskites layer can be cause of chemical degradation due to mild exposure of temperature or humidity and formation into lead iodide and other by-products (Niu et al., 2015). Scientist address these problems through cation substitution, encapsulation technologies and material engineering methods (Ma et al., 2022) Several alternative systems, including tin-based and bismuth-based perovskites, have been proposed to overcome this challenge. However, these lead-free materials often suffer from low efficiency and rapid oxidation, which further compromise their practical utility (Ke and Kanatzidis, 2019). Therefore, achieving a balance between stability, non-toxicity, and high performance remains a critical research direction in the ongoing development of sustainable perovskite solar technologies.

2.3 Fuel cells and Hydrogen Storage materials

The explosion of population globally caused an increase use of energy sources, so natural energy sources are decreasing day by day. The need for study is to develop sustainable and eco-friendly energy sources, so hydrogen is considered a renewable resources and environmental friendly. In electrochemical devices fuel cell in the form of hydrogen used as a fuel. As it has near-zero CO_2 emissions, emitting only warm air and water, and it is eco-friendly in contrast to fossil fuels show in Table 2.

According to Manzo et al (2025), major applications of fuel cells include public transportation, specialty vehicles, and electric vehicles. They also have stationary, portable, and grid-related applications. PEMFC (fuel cell) is the best choice for electric vehicles, whereas SOFC and MCFC are best for grid and related to utility applications. In future, the fuel cell will become the main source of sustainable energy. There is evolution in the transportation sector due to the development of fuel cell-based electric vehicles, and for fuel cell economy and energy efficiency V_2G technology is best. In contrast to BEVs, V_2G using FCEV is best, as the cost of the five fuel cell types varies. These findings are helpful for academics and engineering specialists who want to adopt low-emission energy solutions show in Table 2.

Our future economic fuel is hydrogen, as it is a cost-effective and green fuel. The most abundant, lightest, nontoxic, and environmental friendly element in nature is hydrogen. Hydrogen does not produce detrimental exhaust like natural gas and coal show in Table 2. Moreover, renewable and non-renewable energy sources are utilized for hydrogen production. Hydrogen is utilized as fuel due to its high energy density (120MJ/kg) that is more than conventional fuel. The transportation and storage of hydrogen faces main issues due to its pressure and very low energy density at ambient temperature show in Table 2. To store hydrogen fuel is required large volume, it is necessary to decrease volume by increasing its density. Solid storage is most significant approach for storage of hydrogen and in metal hydride exothermic/endothermic reactions are fruitful for designing various energy systems like fuel cells, heat transformers, heat pumps, cooling system, and energy storage system (Mekonnin et al., 2025).

For hydrogen compression and purification process the useful system is metal hydride-based system. The chemisorption and physisorption properties of nanostructured materials like CNTs and MOFs make them stable medium for hydrogen storage. In nanostructured materials hydrogen is adsorbed in a physical form show in Table 2. CNTs store hydrogen and are made of graphite sheet rolled into crystalline materials, have large surface area and adsorb hydrogen at 298K. The more efficient and effective system required reduced storage pressure when adsorption of hydrogen occurs. Nanostructured materials have low density and store huge hydrogen as compared to conventional hydrogen storage methods show in Table 2.

Table 2. Hydrogen fuel cells and storage technologies

Technologies	Key Characteristics	Advantages	References
Hydrogen as Renewable Fuel	Near-zero CO_2 emissions, emits only warm air and water	Eco-friendly alternative to fossil fuels, sustainable energy source	Hassan et al., 2023
Fuel Cell Applications	Public transportation, specialty vehicles, electric vehicles, stationary and grid applications	PEMFC best for electric vehicles, SOFC and MCFC for utility applications	Manzo et al., 2025
V_2G Technology with FCEV	Vehicle-to-grid technology using fuel cell electric vehicles	Better fuel cell economy and energy efficiency compared to BEVs	Manzo et al., 2025
Hydrogen Properties	Most abundant, lightest, nontoxic, environmentally friendly element	Cost-effective green fuel, no detrimental exhaust like natural gas and coal	MacCarley, 2025; Rolo et al., 2023
Hydrogen Energy Density	High energy density (120MJ/kg)	More energy dense than conventional fuels	Chakraborty et al., 2023; Xu, 2023; Zhang and Sun, 2022
Hydrogen Storage Challenge	Low energy density at ambient temperature, requires high pressure	Transportation and storage issues, requires large volume storage	Chakraborty et al., 2023; Xu, 2023; Zhang and Sun, 2022
Metal Hydride Storage	Solid storage approach using exothermic/endothermic reactions	Useful for fuel cells, heat transformers, heat pumps, cooling systems	Mekonnin et al., 2025
Hydrogen Storage Methods	Metal/chemical hydrides, sorbents, hydrogen tanks	Development efforts over last fifteen years for fuel cell vehicles	Agrawa et al., 2025; Mendrela et al., 2024; Amin et al., 2023
Metal Hydride Systems	Compression and purification of hydrogen	Useful system for hydrogen processing and storage	Abdulkadir et al., 2024; Dewangan et al., 2022
CNTs for Hydrogen Storage	Carbon nanotubes made from rolled graphite sheets	Large surface area, hydrogen adsorption at 298K, crystalline structure	Assyl et al., 2025; Dewangan et al., 2022
Nanostructured Materials	CNTs and MOFs with chemisorption and physisorption properties	Stable medium for hydrogen storage, physical hydrogen adsorption	Abdulkadir et al., 2024; Dewangan et al., 2022
Nanostructured Storage Benefits	Low density materials with huge hydrogen storage capacity	More efficient than conventional methods, reduced storage pressure required	Assyl et al., 2025; Dewangan et al., 2022

2.4. Thermoelectric and Piezoelectric Materials

In the last few years, research on piezoelectric and thermoelectric materials has occurred to utilize renewable energy in transportation and also to generate electricity. Piezoelectric transducers (PZTs) do not produce instant electricity output, whereas thermoelectric generators (TEGs) produce more electricity. There is a need for supercapacitors and rechargeable batteries to store electricity produced from these materials (Rausa et al., 2025; Hasan et al., 2021).

Thermoelectric materials can convert solar heat into electricity. In this study, we will analyze thermoelectricity principles, most advance development in thermoelectric material like lead telluride, bismuth telluride, organic and inorganic compounds (Baghel and Kumar, 2025). In order to mitigate environmental issues, we have developed certain applications of thermoelectric materials such as thermoelectric generators (TEG), thermoelectric coolers (TEC), thermoelectric dehumidifiers (TED) and thermoelectric heaters (TEH). Thermoelectric devices produce low voltages and when metal work together having varied electrical properties. Over the past 60 year this technology gained development (Baghel and Kumar, 2025; Che et al., 2025; Rausa et al., 2025).

Thermoelectric materials utilized the Seebeck effect to convert heat into energy. As these material have high stability, sustainability and thermoelectric efficiency to be utilized in the applications of solar energy. Major components of old thermoelectric materials are inorganic compounds like alloys of bismuth telluride (Bi_2Te_3), silicon-germanium (SiGe) alloy, lead telluride (PbTe), tin selenide (SnSe), and antimony telluride (Sb_2Te_3). In initial research era (1950-1960), the inorganic thermoelectric materials such as PbTe, Bi_2Te_3, and SiGe have been discovered (Che et al., 2025; Wang et al., 2025; Neeleshwar et al., 2022; Recatala-Gomez et al., 2020).

Their real-world application is based on material design innovation like electrical characteristics and thermal conductivity that provides better performance within wide temperature range. High precision, fast response and speed are properties of piezoelectric materials. These materials have wide range of applications like micro electromechanical systems (MEMS), nanoelectromechanical systems (NEMS), sensors, 3D-printing and actuators (Tsai et al., 2025; Sekhar et al., 2023; Wei et al., 2021). According to Aabid et al (2021) in the last thirty years, use of smart materials become popular. Piezoelectric materials have high voltage coefficient, piezoelectric charge, marvelous mechanical-to-electrical energy conversion properties and are utilized in industries as renewable energy source. Net electric charge is developed in piezoelectric materials when exposed to mechanical vibration that induces displaced ions in materials.

Piezoelectric material is used for structural actuation due their low cost, simple, light weight, and easy to control smart material. These material can be easily converted into different shapes like cylinders, fibers, thin films and patches. Piezoelectric materials that are mostly utilized in aerospace and automobile engineering are: Lead Magnesium Niobate (LMN), Lead Meta Niobate (LMN), Lead Zitconate Titanate (PZT), Lead Titanate (LT), and Sodium Potassium Niobate (SPN) (Khorsand et al., 2024; Panda and Sahoo, 2020). The most widely studied ferroelectric material is PZT, which has application as a pyroelectric material, mostly used to repair structures (Marino et al., 2017). PZT are utilized for shape control, repair of cracks, structural health monitoring and vibration control (Gao et al., 2018).

Nanotechnology has emerged as a key tool for improving the performance of existing technologies in the energy and environmental sectors. Conventional materials often show limitations in stability and efficiency. While nanostructured materials, because of their large surface area and tunable properties, offer far better functional behavior (Malik et al., 2023) In energy applications, they enhance solar and fuel cell efficiency and improve energy storage capacity by accelerating charge transfer. In environmental systems, nanostructured materials such as nanosorbents and photocatalysts are being used for wastewater treatment, pollutant removal, and carbon capture (Saleh, 2022). The integration of nanotechnology with current systems not only boosts overall efficiency but also supports the global move toward cleaner and more sustainable technologies.

3. ADVANCED MATERIALS FOR ENVIRONMENTAL APPLICATIONS

3.1 Adsorbent materials for Water and Wastewater Treatment

In recent years, there is massive growth of industries and urban development that creates a lot of wastewater. Different types of methods were applied to treat wastewater, but their use was hazardous due to chemical by product formation show in table 3 (Tariq and Mushtaq, 2023). Sorption is environment friendly method. Nanotechnology has also utilized for wastewater treatment. Nano-adsorbents are more beneficial than traditional adsorbent because they have ability to remove harmful substances from wastewater show in Table 3.

Nano-adsorbents are graphite, metal oxide-based nanoparticles (NPs), carbon nanotubes (CNTs), and plant nanocomposites (NCs) show in Table 3. Researchers are focused towards natural resources for production of adsorbents and these natural resources includes rice husks, coconut, fruit peel, peat, mud, sugarcane, shells, wheat straw, domestic and agricultural waste. Different categories of forest waste (pine needles, bark, sawdust, chestnut borer, etc) are used to remove PTEs as they have adsorbents properties as adsorption values for chestnut bur and its bark are

16.18 mg g^{-1} and 9.31 mg $g-^{1}$ respectively. Biochar is made from forestry waste, as it has highest ability to remove PTEs from wastewater. For wastewater treatment, agricultural waste-derived biochar is cheap and effective adsorbent. As they have stable, porous structure under oxygen-free environment and formed through pyrolysis. Carbon nanotubes (CNTs) by adsorption mechanism can remediate PTEs from wastewater. Scientists modify CNTs with multi-walled CNTs because tiny particles of CNTs as adsorbent creates problems show in Table 3.

A natural and cost effective cation exchange is algae. Cell all of brown algae is composed of cellulose, sulfated polysaccharides and alginic acid. Algal blooms mostly contain sulfates and carboxyl group. Red algae also play its role in biosorption process. PTEs are adsorbed by green algae as their cell wall contains heteropolysaccharides that play role in adsorption. Cell wall of *Chlorella* contains cellulose and proteins that forms glycoproteins show in Table 3.

NPs have been developing for treating various kinds of watewwater. NPs are divided into three types: nanofilms, nanoadsorbents and nanocatalysts. Nano-adsorbents have high adsorbing capacity and are very effective. We can remove various contaminants from wastewater by using nano-adsorbents. As small size particles have high adsorption capacity and chemical activity. Nano-adsorbents are divided into various groups that include metal oxide NPs, metal nanoparticles (NPs), magnetic NPs and nanostructured mixed oxides show in Table 3.

Table 3. Wastewater treatment methods, adsorbents and applications

Treatment Method	Category	Key Properties	Applications	References
Biosorption Process	Wastewater treatment method	Economical, produces less secondary pollutants	Best process to remove toxic pollutants from wastewater	Tariq and Mushtaq, 2023
Traditional Treatment Methods	Chemical methods	Adsorption, solvent extraction, reverse osmosis, oxidation, chemical precipitation, membrane filtration	Hazardous due to chemical by-product formation	Shabbirahmed et al., 2025
Nano-adsorbents	Nanotechnology-based adsorbents	Graphite, metal oxide-based NPs, CNTs, plant nanocomposites	More beneficial than traditional adsorbents, remove harmful substances	Pandey et al., 2022; Rezania et al., 2024
Natural Resource Adsorbents	Organic waste materials	Rice husks, coconut, fruit peel, peat, mud, sugarcane, shells, wheat straw	Domestic and agricultural waste utilization for adsorbent production	Ungureanu et al., 2023

continued on following page

Table 3. Continued

Treatment Method	Category	Key Properties	Applications	References
Forestry Waste Adsorbents	Natural adsorbents	Pine needles, bark, sawdust, chestnut borer, polyphenol and polysaccharides	Remove PTEs, adsorption values: chestnut bur (16.18 mg g^{-1}), bark (9.31 mg g^{-1})	Ungureanu et al., 2023; Younas et al., 2021
Biochar from Forest Waste	Processed forestry waste	Highest ability to remove PTEs	Efficacy greater than 69% for PTE removal	Younas et al., 2021
Peat Adsorbents	Natural peat materials	Sediment peat, moor peat, herbaceous peat, wood peat	Cheap, versatile, excellent cation exchange capacity	Auteri, 2023; Younas et al., 2021
Peat Composition	Organic compounds	Cellulose, fulvic and humic acid, lignin, polar functional groups	High affinity to adsorb Congo red dye from wastewater	Auteri, 2023; Younas et al., 2021
Agricultural Waste Adsorbents	Processed agricultural waste	Cellulose, lignin, -OH and -COOH groups	Remove PTEs: Ni, Pb, As, Hg, Cd(II), Cr(IV)	Nieder and Benbi, 2024; Mehmood et al., 2021; Younas et al., 2021
Agricultural Waste Biochar	Pyrolyzed agricultural waste	Stable, porous structure formed in oxygen-free environment	Cheap and effective adsorbent for wastewater treatment	Multiple references cited above
Carbon Nanotubes (CNTs)	Nanostructured materials	Multi-walled CNTs modifications	Remediate PTEs from wastewater through adsorption mechanism	Chung et al., 2022; Alayan et al., 2021
Algae-based Adsorbents	Natural biological materials	Brown algae: cellulose, sulfated polysaccharides, alginic acid	Natural and cost-effective cation exchange for PTE adsorption	Aslam et al., 2025; Yang et al., 2025
Green Algae Adsorbents	Biological adsorbents	Cell wall with heteropolysaccharides, Chlorella with cellulose and proteins	Forms glycoproteins, effective for PTE biosorption	Aslam et al., 2025; Yang et al., 2025
Nanoparticle Categories	Advanced nanomaterials	Metal oxide NPs, metal NPs, magnetic NPs, nanostructured mixed oxides	High adsorbing capacity, remove various contaminants effectively	Tripathy et al., 2024; Sheerazi and Ahmed, 2022

3.2 Photocatalysts for Pollutant Degradation

One possible approach is photocatalysis, which converts contaminants into nontoxic chemicals by channeling the energy of light (Rani et al., 2024; Bagheri et al., 2017). As one of the most popular materials for photocatalysis, titanium dioxide (TiO_2) is gaining a lot of interest due to its strong photocatalytic activity, chemical stability, nontoxicity, and affordability. Numerous TiO_2-based composites have been thoroughly investigated for environmental applications in recent years, such as TiO_2/graphene, TiO_2/carbon nanotubes (Purabgola et al., 2022; Bhanvase et al., 2017), and TiO_2/metal oxides (Rzaij and Abass, 2020). Industrial wave of globalization has led to environmental degradation specifically with water pollution that has develop a barrier to sustainability development. Despite Rhodamine B (RhB) having numerous applications in various other industries, such as printing, textile, and cosmetics. textiles, printing, and cosmetics (Chiou et al., 2018), the remarkable stability and aqueous solubility have enabled it to persistently accumulate in the environment, which may negatively affect water quality and create immense aesthetic contamination. A search of efficient, ecological, and organic dye degradation method (Al Miad et al., 2024; Sharma et al., 2024), especially in the slow degradation contaminant.

With its distinct benefits such as the elimination of secondary pollutants and the requirement for external energy photocatalytic technology has steadily gained traction as a capable technology, particularly in the treatment of pollution, and management (Noureen et al., 2023; Zhu et al., 2024; He et al., 2025). The unique 2D layered structure, effective photovoltaic conversion, exceptional thermal stability (600°C in air), emerging semiconducting catalyst, and exceptional sensitivity to visible light of g-C3N4 (Umeka et al., 2023), have all demonstrated great potential for environmental remediation. The scientific research community has quickly taken a keen interest in this photocatalyst (HU et al., 2024), g-C3N4 is a good photocatalytic material because it can efficiently use sunlight in photocatalytic reactions due to its band gap width, which is close to 2.7 eV (Fang et al., 2024).

3.3 Membrane Materials for Separation and Purification

Membrane separation technology has attracted a lot of attention lately, particularly because to the quick advancements in the desalination, biomedical, and ESS industries show in Figure 2. By doing away with the need for expensive and toxic solvents, the membrane separation process has advantages over conventional techniques. This procedure can greatly reduce the need for expensive post-separation processing by achieving the enhanced separation efficiency under simple, continuous operating conditions show in Figure 1 (Xu et al.,2016; Li et al., 2019).

The most recent membrane-based methods for lithium extraction. Generally speaking, membranes used in the recovery process can be separated into two categories: nonselective membranes that act as support and selective membranes that use adsorption and penetration techniques. Nonselective membranes are typically made to meet the needs of particular applications show in Figure 1Due to their energy efficiency and adaptability in a continuous process with possible commercial viability, membrane processes are among the lithium-extraction methods that show the most promise show in Figure 1 (Tian et al., 2025).

The increase in the demand to have sustainable solutions to energy and environmental problems has led to the development of innovative materials. They have gained popularity due to the porosities of nanofibrous membranes, which have large surface areas, adjustable porosity and can be used as adaptable mechanical properties (Surat'man et al., 2025). Due to their outstanding mechanical strength, controllable pore size, and large surface area-volume ratio, tunable pore size, and high surface area, nanofibrous membranes have widespread applications in energy storage, air filtration, and water purification show in Figure 2 (Cui et al., 2020; Lim, 2017; Yanet al., 2021). Synthetic polymers such as polystyrene (PS), polyacrylonitrile (PAN), and polyvinylidene fluoride (PVDF) were nanofibrous membranes (Cheng et al., 2020; Wu et al., 2018; Lu et al., 2018). These materials have been chosen due to their ease of processing, chemical resistance, and thermal stability, which have various industrial applications. However, it is impossible to disregard the impact of these traditional materials on the environment (Surat'man et al., 2025).

The benefits of the various procedures, including, scable design, and their molecular. Which permits varied solicitation across a wide range of surroundings, industrial and municipal systems as well as decentralized unit treatment (Nunes, 2021).

Because of their remarkable qualities, such as high great adsorption capacity, catalytic activity, surface area, antibacterial effects, and mechanical resilience, NMs such ZnO, TiO_2, Fe_2O_3, CuO, SiO_2, GO, and MOFs are chosen show in Figure 1 (Ben Dassi and Chamam, 2025).

Figure 1. Modern Membrane Technology for Environmental and Industrial Applications

Figure 2. Industrials applications

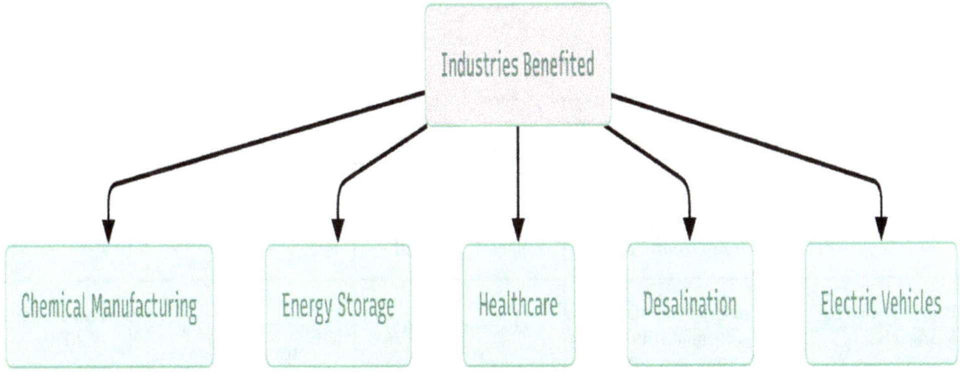

3.4 Smart Materials for Environmental Monitoring

Smart materials are also referred to as advanced material and innovative substances. No single definition can be used that will be sufficient to define them. They may be said to be high- tech materials that may react intelligently to variations in their environment, or they may be said to be materials that are able to revert to their initial shape in response to specific stimuli show in Table 4. The former, e.g. photochromatic glasses, can only change color in the presence of sunlight; they are unable to change their properties. A shape memory alloy (SMA) is one of the most prevalent kinds of smart materials. Among the properties of these materials which make them different than other materials, is the fact that they are able to maintain their original shape upon exposure to external forces, e.g. stress. With the variation of temperature these material can change shape. Moreover, also known as intelligent materials show in Table 4.

Magnetic-shaped memory materials have attracted a lot of curiosity both among researchers and technologists. They have been well researched since they have several applications in different spheres. These materials can be converted to useful electricity via thermoelectric modules and generators, which might complement renewable energy sources by using the waste heat show in Table 4.

Al Alim et al. (2025) assert that the invention of smart polymeric materials was the start of a new era in engineering and construction because rapid development of the material science occurred. This paper aims to provide an in-depth knowledge of their nature, applications and the possible impacts on sustainable development in the future show in Table 4. Smart materials have reshaped the water filtration industry by providing solutions that are very efficient and greenThey will provide flexible, effective, and sustainable solutions to satisfy the increasing demand of clean and safe water across the world show in Table 4.

Table 4. Smart materials: types, properties, Mechanisms and applications

Material Types	Key Properties	Mechanisms	Applications	References
Smart Materials (General)	High-tech materials, react intelligently to environment	Revert to initial shape in response to specific stimuli	Advanced materials for various technological applications	Zhang et al., 2019
Passive Smart Materials	Energy transmission capability	Transmit electromagnetic waves	Optical fiber applications for energy transmission	Bah et al., 2020
Active Smart Materials (Type 1)	Color-changing properties	Change color in presence of sunlight	Photochromatic glasses, light-responsive applications	Bah et al., 2020
Shape Memory Alloys (SMA)	Maintain original shape under external forces	Temperature-induced shape changes	Stress resistance, temperature-responsive applications	Bah et al., 2020
Magnetic Shape Memory Materials	Magnetic field sensitivity	Reversible shape change in magnetic fields	Robotics, aircraft, actuation technologies, precise sensors	Dadda et al., 2025
Magnetic Refrigeration Materials	Heat absorption and release capability	Respond to external magnetic field influence	Sustainable alternative to usual refrigerants	Dadda et al., 2025
Thermoelectric Smart Materials	Heat-to-electricity conversion	Convert waste heat to useful electricity	Thermoelectric modules, generators for renewable energy	Dadda et al., 2025
Smart Polymeric Materials	Multifunctional, environmentally sustainable	Better performance than traditional materials	Engineering, construction, building applications	Al Alim et al., 2025
Composite Smart Materials	Plant fiber, graphite fiber, polymeric fiber	Enhanced structural properties	Carbon nanotube-based composites, liquid crystal elastomers	Al Alim et al., 2025
Stimuli-Responsive Polymers	Environmental responsiveness	React to environmental changes	Water filtration, pollutant detection, adaptive systems	Namdeti et al., 2025
Smart Biomaterials	Biocompatible, multifunctional	Biological compatibility with smart responses	Water treatment, biomedical applications	Namdeti et al., 2025
Smart Nanocomposites	High removal efficiency, pollutant detection	Adaptation to environmental situations	Water purification, heavy metal removal, microbial treatment	Namdeti et al., 2025

3.5 Carbon-Based Materials for Air Pollution Control and CO_2 Capture

Today, carbon capture and storage seems to be one of the greatest tools in upcoming worldwide efforts to mitigate emission of greenhouse gases and their accumulation in the atmosphere. There are alternatives, and we are strongly of the opinion that measures such as energy efficient technology, renewable energy sources, and more nuclear power are also vital, but carbon capture and storage is a mandatory requirement when we continue to utilize fossil fuels in the predictable future show in figure 4. The primary aim of this study is to carefully examine carbon-based materials in optional carbon dioxide separation and capture processes show in Table 5.

Carbon-based sorbents or catalysts are used to capture CO_2 produced during the thermochemical conversion of biomass into biofuels or other useful products during biomass conversion processes show in Figure 4. Photoconversion methods of carbon dioxide conversion employ carbon-based photocatalysts, such as graphene or carbon nanotubes. These materials may even enhance the efficiency of light-dependent reactions when under the impact of light, and it becomes simpler to re-purpose collected CO_2 into useful fuels or chemicals show in Table 5.

Reduction of CO_2 emissions is necessary to avoid the climate change and protecting the ecosystem. Carbon dioxide emissions generated by power plants and industrial sources are by means of CCS technology show in Figure 4 which captures the emissions, transports them underground or finds numerous applications. In this way, CO_2 does not reach the atmosphere. Ravi et al. examined the application of palm shells as a biomass source of activated carbon to remove CO_2 in exhausts of diesel engines. The results indicated that as diesel was mixed with biodiesel ratio 80:20, it was able to collect 37 percent of CO_2, whereas diesel alone was in a position to collect a good percentage of 30 show in Table 5.

One of the innovative methods, liquid amine capture, which has been in use over a few generations has reached high maturity due to the numerous research undertakings which seek to advance CO_2 capture technology show in figure 4. Carbon-based aerogels are making large strides as carbon capture materials due to their high surface area, low density, three-dimensional, multiporous structure, and adjustable properties, all of which significantly influence CO_2 capture efficiency. The chemical and thermal stability of carbon aerogels enables high-temperature performance on a long-term basis show in Table 5 and Figure 4.

The most efficient approach is transportation, Carbon dioxide capture, transportation, usage, and storage (CCTUS). Energy and industrial sources of CO_2 emissions have to be eliminated, CO_2 needs to be transported to the storage facilities or deposits where they are used, CO_2 needs to be converted into useful resources. The nature of the combustion process, the quantity of CO_2, and whether the separation, transportation, or storage of that CO_2 is to be used or stored is among the variables that influence the performance of CCTUS show in Table 5 and Figure 3.

It has been estimated that microalgae (MA) has the ability to take up 183 tons of CO_2 per 100 tons of MA. With the help of Fig. 8, it is stated that algal growth is conditional upon favorable conditions including water, pH, light, nutrient, carbon dioxide, and light (Anwar et al., 2018). MA are simpler, more efficient in CO_2 fixing and develop faster than the terrestrial plants. The wastewater constitutes the raw material that can be used, thus, making it a more cost-effective and efficient process (Ghosh and Kiran, 2017). Some of the species that are capable of metabolizing carbon dioxide and yielding the best lipids needed in the manufacture of bio-diesel include *Chlorella sp., Botryococcus b., Chlorella vulgaris, Scenedesmus obliquus and Scenedesmus sp.* (Hariram et al., 2022). Nanomaterial adsorption capicity and nanomaterials is degrded by temperature rises that exceed a certain degree, as several studies have highlighted show in Table 5.

Biochar is one of the most promising materials for CO_2 collecting. Compared to other adsorbent materials, it provides a wider range of raw materials and has a lower environmental impact (Guo et al., 2022). There is a great deal of potential for the use of biochar in a variety of settings to cut greenhouse gas emissions by up to 2.56* 10^9t CO_2 per year, or 5.0% of global GHG emissions show in Table 5.

Carbon capture processes such as oxy-fuel combustion, post-combustion and pre-combustion are used in capturing CO_2 sources or in the atmosphere. Pre-combustion is incredibly efficient but it requires changing infrastructures. Post-combustion is adjustable though it consumes a large amount of energy. Combustion by use of oxygen-fuel achieves high purity, but energy is required. Industrial integration, cost-cutting, and research are the keys to future success show in Table 5.

Figure 3. CCTUS process

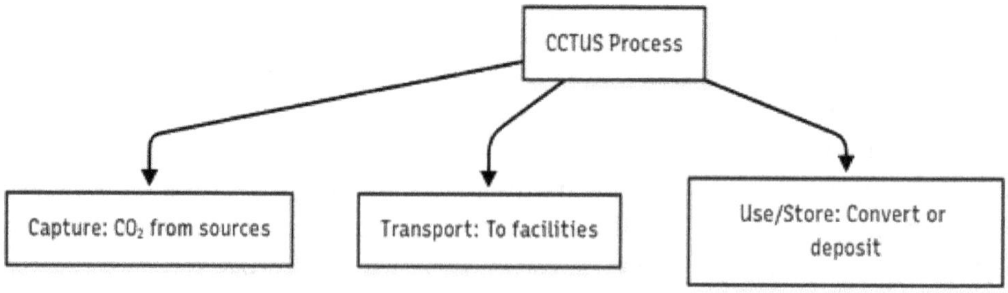

Table 5. Carbon capture and storage technologies, Mechanisms and applications

Technologies	Types	Mechanisms	Applications	References
Carbon Capture and Storage (CCS)	Greenhouse gas mitigation tool	Capture CO_2 emissions, transport underground	Mandatory requirement for continued fossil fuel use	Singh et al., 2025
Carbon-based Sorbents/Catalysts	Biomass conversion materials	Capture CO_2 during thermochemical conversion	Used in biofuel production, sustainable closed-loop systems	Shahab et al., 2024; Zheng et al., 2025
Carbon-based Char/Residue	Thermal conversion byproduct	High temperature pyrolysis/gasification with little oxygen	CO2 sorbents, increase carbon usage efficiency	Guo et al., 2025
Carbon-based Photocatalysts	Photoconversion materials	Graphene, carbon nanotubes for light-dependent reactions	Convert captured CO2 into useful fuels or chemicals	Zheng et al., 2025; Jia and Sundar, 2024
Palm Shell Activated Carbon	Biomass-derived adsorbent	Activated carbon from palm shells	37% CO2 capture with biodiesel (80:20), 30% with diesel alone	Hariram et al., 2025
Liquid Amine Capture	Mature CO_2 capture technology	Chemical absorption using amine solutions	High maturity, extensive research for advancement	Asrafali et al., 2025
Carbon-based Aerogels	Advanced capture materials	High surface area, low density, 3D multiporous structure	High CO_2 capture efficiency, chemical and thermal stability	Asrafali et al., 2025
CCTUS Technology	Comprehensive CO_2 management	Carbon dioxide capture, transportation, usage, and storage	Most efficient approach for industrial CO_2 management	Esiri et al., 2024
Microalgae (MA) CO_2 Fixation	Biological carbon capture	183 tons CO_2 uptake per 100 tonnes of MA	More efficient than terrestrial plants, uses wastewater	Anwar et al., 2018; Ghosh and Kiran, 2017
Microalgae Species for Biodiesel	Specific algal strains	*Chlorella sp., Botryococcus b., Chlorella vulgaris, Scenedesmus*	CO_2 metabolism with high-quality lipid production	Hariram et al., 2022
Nanomaterials for CO_2 Capture	Advanced nanomaterials	Temperature-sensitive adsorption capacity	Degraded performance at elevated temperatures	Al Atrach et al., 2024
Biochar CO_2 Adsorption	Promising adsorbent material	Wide range of raw materials, low environmental impact	Potential to reduce 2.56×10^9t CO_2/year (5.0% global GHG)	Guo et al., 2022; Ofuani, 2024
CO_2 Capture Processes	Industrial capture methods	Oxy-fuel, post-combustion, pre-combustion technologies	Pre: efficient but needs infrastructure; Post: adaptable, energy-intensive; Oxy: high purity, energy required	Joarder et al., 2025

Figure 4. Carbon capturing technologies

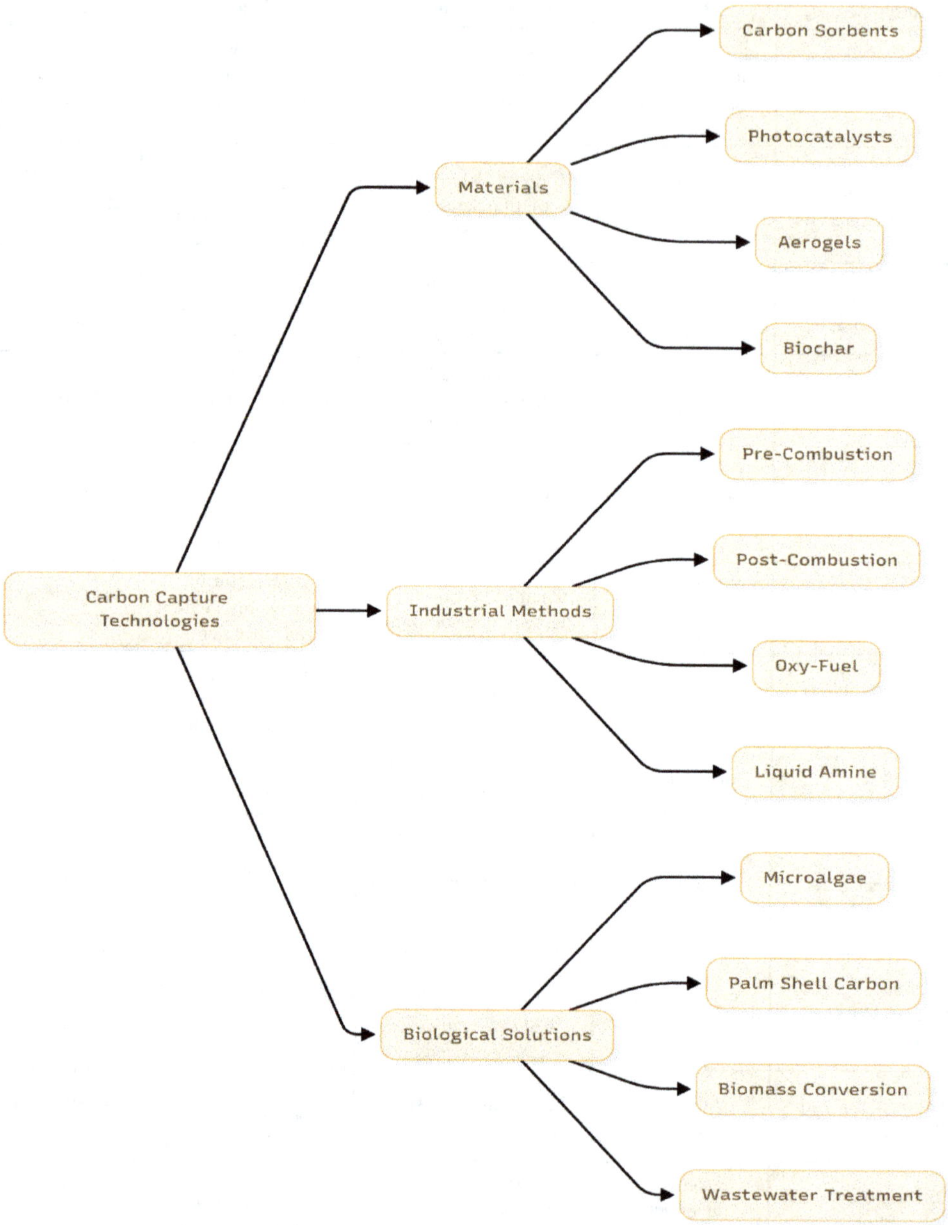

4. CHALLENGES AND FUTURE PERSPECTIVES

In Future Perspectives and Challenges, the evolving industry of electrochemical devices is discussed in connection with new advances, current issues, and opportunities of collaboration and innovation. Though there are several shortcomings, the future of the development of novel materials to be used in energy and environmental applications is promising. The requirement of the improved material features, industrial activities, and the cooperation of the stakeholders are significant issues. Sometimes, depending on under what conditions, a detailed knowledge of the properties of some complex materials, including 2D nanomaterials, is required to avoid early failures (Podgornik, 2023). Synthesis procedures should be innovated since the currently used manufacturing techniques may not be suitable to produce the complex substances needed (Kirubakaran et al., 2025). It is important to examine the lifecycle of new materials created to make sure that they do not cause pollution or drain resources (Othman et al., 2024; Umar and Muthusamy, 2023).

Advanced materials that can be utilized to store energy, remediate the environment, and generate renewable energy may include carbon-based materials and 2D nanomaterials (Talreja et al., 2024; Khanam et al., 2022). The industry and academia should work closer to develop viable resources that can facilitate the sustainability goals (Podgornik, 2023). On-going material design and surface modification can result in a high level of environmental benefits and performance (Panghal and Kumar, 2025; Nolan et al., 2016).

Environmental issues such as pollution, greenhouse gas emissions, sustainability, global warming, and climate change are the most serious issues of the present day. To this end, the more cost-effective use of energy and materials and technological innovation and optimization significantly contribute to the pursuit of a green society and green industry. Additionally, when component design is pushed to its limits, inaccurate and antiquated measurement techniques, measurement uncertainty, and divergence from specified material qualities can result in unforeseen early component failure and environmental issues (Miah et al., 2024; Ye et al., 2024).

Researchers and engineers tend to focus on creative ways to improve the performance, sustainability, and applicability of these composites as the need for materials that can do several tasks at once grows. The mechanical, thermal, and electrical properties of composites can be greatly improved by adding metal–organic frameworks (MOFs), graphene, and carbon nanotubes (CNTs). In recent studies, such as one, graphene-based natural fiber composites were found to be good in terms of mechanical and thermal properties, and this means that the product can be used in numerous applications with multipurpose applications, such as EMI shielding and structural health measurements. The properties of these composites can be tailored with the reinforcement of nanomaterials, which opens new potentials of application in the high-performance applications (Simoes, 2024).

Advanced materials have many advantages over traditional materials due to their unique properties; however, details about their availability of resources, energy and carbon footprint-intensive production, degradability and recyclability, biocompatibility, etc., may raise sustainability issues and present some significant environmental risks (Podgornik, 2023). Reducing energy use, limiting environmental impact, and cutting water production costs are the main obstacles facing the development of sustainable desalination technologies in this century (Roy & Ragunath, 2018). The increasing understanding of hybrid methods helps to reduce the amount of environmental resources used in energy production. The development of sustainable energy and water demands has recently seen a significant contribution from membrane-based technologies (Roy & Ragunath, 2018).

One major environmental protection challenge is how to reduce direct emissions and the CO_2 footprint during material manufacturing. The three largest sources of CO_2 emissions are the manufacture of copper, steel, and aluminum. Currently, the most practical and long-term way to achieve carbon-neutral production is to combine recycling, electrification, and the usage of green hydrogen. The main obstacle, though, will be generating enough green electricity from renewable sources (Panagiotopoulou et al., 2022). Modern design aiming for a green future depends on the advanced and trustworthy determination of material qualities and their correlations assisted by computational methods, machine learning, and artificial intelligence (Badini et al., 2023).

More effort is also required to create economical techniques for the synthesis of nanomaterials and to test efficiencies on a large scale for practical field applications. Future studies must address the following research gaps. Prior to conducting pilot-scale trials, cost-benefit analysis must be performed. It is necessary to investigate the effectiveness of the adsorbents under discussion in conjunction with alternative therapy modalities, particularly those based on carbon (Gautam et al., 2024; Younas et al., 2021). The future development relies on the research, cost minimization, regulatory provisions and the acceptance by people. Climate change cannot be solved with any of these without collaboration and cross-sector innovation (Pedersen et al., 2021; Compagnucci and Spigarelli, 2018).

The advancement of science will lead to the development of new materials with even more remarkable properties (Zhi et al., 2023). With responsible development and an ongoing dedication to problem-solving, advanced materials could help us move toward a future that is more resilient, sustainable, and technologically advanced (Murtagh et al.,2020). In the future, innovative molecules for personalized healthcare will be swiftly discovered and optimized by machine learning algorithms applied to innovative materials, revolutionizing medication development and treatment approaches (Verma and Verma, 2022). The explores of nanostructured material to addressing imporatance of environment and issues of energy and more used in future sustainable. Furthermore, advanced technology and facilities for research, universities, closely collaboration with industries can efficiently help for greener future of our society.

REFERENCES

Aabid, A., Raheman, M. A., Ibrahim, Y. E., Anjum, A., Hrairi, M., Parveez, B., & Mohammed Zayan, J. (2021). A systematic review of piezoelectric materials and energy harvesters for industrial applications. *Sensors (Basel), 21*(12), 414. DOI: 10.3390/s21124145

Abbasi, K. R., Abbas, J., & Tufail, M. (2021). Revisiting electricity consumption, price, and real GDP: A modified sectoral level analysis from Pakistan. *Energy Policy, 149*, 112087. DOI: 10.1016/j.enpol.2020.112087

Abdulkadir, B. A., Teh, L. P., Rahman Khan, M. M., & Setiabudi, H. D. (2024). Potential Applications of Metal-Doped Nanomaterials for Enhancing Solid-State Hydrogen Storage. *ChemistrySelect, 9*(14), e202400051. DOI: 10.1002/slct.202400051

Adeoye, A. E., Adeaga, O. A., Oloore, L. E., & Ogunmola, E. D. (2023). Review of Peculiarities of Multifunctional Nano Materials for Harvesting and Storing Light Energy. *Adeleke University Journal of Engineering and Technology, 6*(2), 329–349.

Al Alim, M., Hore, S., Zohora, F., Ghosh, D., & Hore, R. (2025). Advances in Smart Polymeric Materials for Sustainable and Next-generation Applications. *Civil Engineering & Building Science, 2*(1).

Al Atrach, J., Aitblal, A., Amedlous, A., Xiong, Y., Desmurs, M., Ruaux, V., Guillet-Nicolas, R., & Valtchev, V. (2024). Nanosized zeolite P for enhanced CO2 adsorption kinetics. *ACS Applied Materials & Interfaces, 16*(29), 38006–38016. DOI: 10.1021/acsami.4c05988

Al Miad, A., Saikat, S. P., Alam, M. K., Hossain, M. S., Bahadur, N. M., & Ahmed, S. (2024). Metal oxide-based photocatalysts for the efficient degradation of organic pollutants for a sustainable environment: A review. *Nanoscale Advances, 6*(19), 4781–4803. DOI: 10.1039/D4NA00517A

Alayan, H. M., Aljumaily, M. M., Alsaadi, M. A., Mjalli, F. S., & Hashim, M. A. (2021). A review exploring the adsorptive removal of organic micropollutants on tailored hierarchical carbon nanotubes. *Toxicological and Environmental Chemistry, 103*(4), 282–325. DOI: 10.1080/02772248.2021.1932888

Amin, M., Shah, H. H., Bashir, B., Iqbal, M. A., Shah, U. H., & Ali, M. U. (2023). Environmental assessment of hydrogen utilization in various applications and alternative renewable sources for hydrogen production: A review. *Energies, 16*(11), 4348. DOI: 10.3390/en16114348

Anwar, A., Liu, Y., Dong, R., Bai, L., Yu, X., & Li, Y. (2018). The physiological and molecular mechanism of brassinosteroid in response to stress: A review. *Biological Research, 51*(1), 51. DOI: 10.1186/s40659-018-0195-2

Asamoah, G. A., Korsah, M., Jeyasundar, P. G. S. A., Ahmed, M., Lau, S. Y., & Danquah, M. K. (2024). Nanotechnology-based lithium-ion battery energy storage systems. *Sustainability (Basel), 16*(21), 9231. DOI: 10.3390/su16219231

Asdrubali, F., Berardi, U., & Stasi, R. (2025). The impact of the building sector. In *Sustainability Certifications, Labels and Tools in the Built Environment* (pp. 4-21). Routledge. DOI: 10.1201/9781032705149-2

Aslam, A., Kanwal, F., Javied, S., Nisar, N., & Torriero, A. A. J. (2025). Microbial biosorption: A sustainable approach for metal removal and environmental remediation. *International Journal of Environmental Science and Technology, 22*(13), 1–32. DOI: 10.1007/s13762-025-06611-1

Asrafali, S. P., Periyasamy, T., & Bari, G. A. R. (2025). Advances in Carbon-Based Aerogels for CO2 Capture: Fundamental Design Strategies and Technological Progress. *Gels (Basel, Switzerland), 11*(5), 361. DOI: 10.3390/gels11050361

Assyl, S., Botakoz, S., & Saule, Z. (2025). Dimensions, structure, and morphology variations of carbon-based materials for hydrogen storage: A review. *Discover Nano, 20*(1), 1–41. DOI: 10.1186/s11671-025-04229-3

Auteri, N. (2023). Characterization of modified cactus pear (Opuntia ficus indica (L.) Mill) pruning waste as a sustainable and innovative solution to recycle phosphorus from agricultural streams.

Badini, S., Regondi, S., & Pugliese, R. (2023). Unleashing the power of artificial intelligence in materials design. *Materials (Basel), 16*(17), 5927. DOI: 10.3390/ma16175927

Baghel, N., & Kumar, A. (2025). Sustainable thermoelectric materials for solar energy applications: A review. *Solid State Sciences, 160*, 107784. DOI: 10.1016/j.solidstatesciences.2024.107784

Bagheri, S., TermehYousefi, A., & Do, T.-O. (2017). Photocatalytic pathway toward degradation of environmental pharmaceutical pollutants: Structure, kinetics and mechanism approach. *Catalysis Science & Technology, 7*(20), 4548–4569. DOI: 10.1039/C7CY00468K

Bai, X., & Zhang, X. (2025). Artificial intelligence-powered materials science. *Nano-Micro Letters, 17*(1), 1–30.

Ben Dassi, R., & Chamam, B. (2025). Nanomaterial-enhanced membranes for advanced water and wastewater treatment: A comprehensive review. *Reviews in Environmental Science and Biotechnology*, *24*(2), 1–35. DOI: 10.1007/s11157-025-09723-9

Bhanvase, B. A., Shende, T. P., & Sonawane, S. H. (2017). A review on graphene–TiO2 and doped graphene–TiO2 nanocomposite photocatalyst for water and wastewater treatment. *Environmental Technology Reviews*, *6*(1), 1–14. DOI: 10.1080/21622515.2016.1264489

Bogmans, C., Gomez-Gonzalez, P., Ganpurev, G., Melina, G., Pescatori, A., & Thube, S. (2025). Power hungry: How ai will drive energy demand. *IMF Working Paper*, *81*(4), 2025.

Chakraborty, A., Islam, M., Shahriyar, F., Uddin, M. M., Uddin, M. H., & Minar, M. R. (2023, July). Economic and Environmental Feasibility Study of Fossil Fuel, Electric and Hydrogen Fuel Cell Vehicles in Perspective of Bangladesh. In *2023 14th International Conference on Computing Communication and Networking Technologies (ICCCNT)* (pp. 1-6). IEEE.

Che, L., Li, N., Wei, W., Li, J., Ji, J., Zhao, X., Yu, B., & Wang, Q. (2025). Day-Night energy harvesting: Photovoltaics-driven moisture evaporation and absorption for simultaneous 24-hour power and dehumidification. *The Innovation Energy*, *2*(2), 100078–1. DOI: 10.59717/j.xinn-energy.2025.100078

Chen, F. D., Xie, Z. Y., Li, M. T., Chen, S. G., Ding, W., Li, L., Li, J., & Wei, Z. D. (2024). Series Reports from Professor Wei's Group of Chongqing University: Advancements in Electrochemical Energy Conversions (1/4): Report 1: High-Performance Oxygen Reduction Catalysts for Fuel Cells. *Dian Hua Xue*, *30*(7), 1. DOI: 10.61558/2993-074X.3459

Chen, S., Zhang, C., & Lu, X. (2023). Energy conversion from Fossil fuel to renewable energy. In *Handbook of Air Quality and Climate Change* (pp. 1–44). Springer Nature Singapore.

Cheng, X., Sun, Z., Yang, X., Li, Z., Zhang, Y., Wang, P., & Shao, L. (2020). Construction of superhydrophilic hierarchical polyacrylonitrile nanofiber membranes by in situ asymmetry engineering for unprecedently ultrafast oil–water emulsion separation. *Journal of Materials Chemistry. A, Materials for Energy and Sustainability*, *8*(33), 16933–16942. DOI: 10.1039/D0TA03011B

Chiou, C. S., Chen, H. W., & Chen, Z. T. (2018). Application of magnetic adsorbent with silicate and phenyl polymers to adsorb Rhodamine B. *International Journal of Environmental Science and Technology*, *15*(9), 1879–1886. DOI: 10.1007/s13762-017-1553-2

Chung, J. H., Hasyimah, N., & Hussein, N. (2022). Application of carbon nanotubes (CNTs) for remediation of emerging pollutants-a review. *Tropical Aquatic and Soil Pollution*, *2*(1), 13–26. DOI: 10.53623/tasp.v2i1.27

Compagnucci, L., & Spigarelli, F. (2018). Fostering cross-sector collaboration to promote innovation in the water sector. *Sustainability (Basel)*, *10*(11), 4154. DOI: 10.3390/su10114154

Cui, J., Li, F., Wang, Y., Zhang, Q., Ma, W., & Huang, C. (2020). Electrospun nanofiber membranes for wastewater treatment applications. *Separation and Purification Technology*, *250*, 117116. DOI: 10.1016/j.seppur.2020.117116

Dadda, K., Alleg, S., Rabhi, S., Abadlia, L., Elmohri, A., Suňol, J. J., & Hlil, E. K. (2025). Sensing the future: Smart nanomaterials revolutionizing environmental monitoring. In *Breaking Boundaries: Pioneering Sustainable Solutions Through Materials and Technology* (pp. 237–255). Springer Nature Singapore. DOI: 10.1007/978-981-97-9827-8_11

Dewangan, S. K., Mohan, M., Kumar, V., Sharma, A., & Ahn, B. (2022). A comprehensive review of the prospects for future hydrogen storage in materials-application and outstanding issues. *International Journal of Energy Research*, *46*(12), 16150–16177. DOI: 10.1002/er.8322

Dutta, A., Mitra, S., Basak, M., & Banerjee, T. (2023). A comprehensive review on batteries and supercapacitors: Development and challenges since their inception. *Energy Storage*, *5*(1), e339. DOI: 10.1002/est2.339

Erzen, B., Karataş, M., Orhan, R., & Aydoğmuş, E. (2025). Advancements and Challenges in Biomaterials: Innovations, Sustainability, and Future Prospects. *Journal of Macromolecular Science, Part B: Physics*, •••, 1–27. DOI: 10.1080/00222348.2025.2476852

Esiri, A. E., Kwakye, J. M., Ekechukwu, D. E., Ogundipe, O. B., & Ikevuje, A. H.Andrew Emuobosa EsiriJephta Mensah KwakyeDarlington Eze EkechukwuOlorunshogo Benjamin OgundipeAugusta Heavens Ikevuje. (2024). Leveraging regional resources to address regional energy challenges in the transition to a low-carbon future. *Open Access Research Journal of Multidisciplinary Studies*, *8*(1), 105–114. DOI: 10.53022/oarjms.2024.8.1.0052

Eyube, M., Enuesueke, C., & Alimikhena, M. (2025). The Future of Nanomaterials in Manufacturing.

Fang, J., Liu, L., Yang, H., & Du, H. (2024). Applications of Fenton/Fenton-like photocatalytic degradation in g-C3N4 based composite materials. *Journal of Environmental Chemical Engineering*, *12*(6), 114153. DOI: 10.1016/j.jece.2024.114153

Gao, X., Wu, J., Yu, Y., Chu, Z., Shi, H., & Dong, S. (2018). Giant piezoelectric coefficients in relaxor piezoelectric ceramic PNN-PZT for vibration energy harvesting. *Advanced Functional Materials*, *28*(30), 1706895. DOI: 10.1002/adfm.201706895

Gautam, M. K., Mondal, T., Nath, R., Mahajon, B., Chincholikar, M., Bose, A., & Mondal, S. (2024). Harnessing activated hydrochars: A novel approach for pharmaceutical contaminant removal. *C*, *10*(1), 8.

Ghosh, A., & Kiran, B. (2017). Carbon concentration in algae: Reducing CO2 from exhaust gas. *Trends in Biotechnology*, *35*(9), 806–808. DOI: 10.1016/j.tibtech.2017.05.003

Guo, Q., He, Y., Chen, Y., & Qiao, C. (2025). Design and application of computational modeling in science education research: A systematic review. *Interactive Learning Environments*, *33*(2), 969–992.

Hariram, N. P., Mekha, K. B., Suganthan, V., & Sudhakar, K. (2023). Sustainalism: An integrated socio-economic-environmental model to address sustainable development and sustainability. *Sustainability (Basel)*, *15*(13), 10682. DOI: 10.3390/su151310682

Hariram, V., John, J. G., Sangeethkumar, E., Gajalakshmi, B., Ramanathan, V., Vinothkumar, M., & Balachandar, M. (2022). Scenedesmus obliquus and Chlorella vulgaris–A Prospective Algal Fuel Source. *Nature Environment and Pollution Technology*, *21*(5), 2129–2139. DOI: 10.46488/NEPT.2022.v21i05.009

Hariram, V., Sathishbabu, R., John, J. G., Kailiappan, N., Vijayakumar, K., Kumar, E. S., & Priya, K. K. (2025). Maximizing the combustion phenomenon with reduced emission of a compression ignition engine fulled with Dunaliella salina biodiesel and hydrogen blends. *Scientific Reports*, *15*(1), 30372. DOI: 10.1038/s41598-025-16285-x

Hasan, M. N., Sahlan, S., Osman, K., & Mohamed Ali, M. S. (2021). Energy harvesters for wearable electronics and biomedical devices. *Advanced Materials Technologies*, *6*(3), 2000771. DOI: 10.1002/admt.202000771

Hassan, Q., Azzawi, I. D., Sameen, A. Z., & Salman, H. M. (2023). Hydrogen fuel cell vehicles: Opportunities and challenges. *Sustainability (Basel)*, *15*(15), 11501. DOI: 10.3390/su151511501

He, S., Hu, Z., He, C., Tian, S., Ouyang, G., Lin, Z., & Fang, J. (2025). Nano-confinement-Enabled Resistance to Water Matrix Components for Selective Degradation of Micropollutants. *Angewandte Chemie*, *137*(14), e202424827. DOI: 10.1002/ange.202424827

Hepel, M. (2023). Advances in micro-supercapacitors (MSCs) with high energy density and fast charge-discharge capabilities for flexible bioelectronic devices—A review. *Electrochemical Science Advances*, *3*(3), e2100222. DOI: 10.1002/elsa.202100222

Holloway, M. D. (2025). *Synthesizing Materials in Microgravity: Unlocking Novel Materials Beyond Earth's Limits*. CRC Press. DOI: 10.1201/9781003656081

Hu, Q., Albolkany, M. K., Liu, C., Zhang, T., Cui, S., Ye, F., & Liu, B. (2024). Photothermal graphite phase carbon nitride membrane intercalated with polyoxovanadate for efficient solar steam generation. *Desalination*, *583*, 117650. DOI: 10.1016/j.desal.2024.117650

Islam, M. M., Yu, T., Giannoccaro, G., Mi, Y., La Scala, M., Rajabi, M. N., & Wang, J. (2024). Improving reliability and stability of the power systems: A comprehensive review on the role of energy storage systems to enhance flexibility. *IEEE Access : Practical Innovations, Open Solutions*.

Jia, H., & Sundar, S. S. (2024). Vivid and engaging: Effects of interactive data visualization on perceptions and attitudes about social issues. *Digital Journalism (Abingdon, England)*, *12*(8), 1205–1229. DOI: 10.1080/21670811.2023.2250815

Joarder, M. S. A., Islam, M. S., Hasan, M. H., Kutub, A., Kabir, M. F., Rashid, F., & Joarder, T. A. (2025). A comprehensive review of carbon dioxide capture, transportation, utilization, and storage: A source of future energy. *Environmental Science and Pollution Research International*, *32*(15), 1–34. DOI: 10.1007/s11356-025-36284-9

Kalair, A., Abas, N., Saleem, M. S., Kalair, A. R., & Khan, N. (2021). Role of energy storage systems in energy transition from fossil fuels to renewables. *Energy Storage*, *3*(1), e135. DOI: 10.1002/est2.135

Ke, W., & Kanatzidis, M. G. (2019). Prospects for low-toxicity lead-free perovskite solar cells. *Nature Communications*, *10*(1), 965. DOI: 10.1038/s41467-019-08918-3

Khalid, R., Shah, A., Javed, M., & Hussain, H. (2025). Progress and obstacles in electrode materials for lithium-ion batteries: A journey towards enhanced energy storage efficiency. *RSC Advances*, *15*(20), 15951–15998. DOI: 10.1039/D5RA02042E

Khan, A., Al Rashid, H., Roy, P. K., Chowdhury, S. I., & Sathi, S. A. (2025). Challenges and the Way to Improve Lithium-Ion Battery Technology for Next-Generation Energy Storage. *Energy & Environmental Materials, 8*(6), 70088. DOI: 10.1002/eem2.70088

Khanam, Z., Gogoi, N., & Srivastava, D. N. (2022). Challenges, Commercial Aspect. *2D Nanomaterials for Energy and Environmental Sustainability*, 267.

Khorsand Zak, A., Yazdi, S. T., Abrishami, M. E., & Hashim, A. M. (2024). A review on piezoelectric ceramics and nanostructures: Fundamentals and fabrications. *Journal of the Australian Ceramic Society, 60*(3), 723–753. DOI: 10.1007/s41779-024-00990-3

Kiani, M. (2025). Nanomaterials Revolution: Paving the Way for Next-Gen Applications. *Available at SSRN* 5086617. DOI: 10.2139/ssrn.5086617

Kirubakaran, D., Wahid, J. B. A., Karmegam, N., Jeevika, R., Sellapillai, L., & Rajkumar, M., & SenthilKumar, K. J. (2025). A comprehensive review on the green synthesis of nanoparticles: Advancements in biomedical and environmental applications. *Biomedical Materials & Devices, 4*, 1–26.

Li, X., Mo, Y., Qing, W., Shao, S., Tang, C. Y., & Li, J. (2019). Membrane-based technologies for lithium recovery from water lithium resources: A review. *Journal of Membrane Science, 591*, 117317. DOI: 10.1016/j.memsci.2019.117317

Lim, C. T. (2017). Nanofiber technology: Current status and emerging developments. *Progress in Polymer Science, 70*, 1–17.

Lu, T. D., Chen, B. Z., Wang, J., Jia, T. Z., Cao, X. L., Wang, Y., Xing, W., Lau, C. H., & Sun, S. P. (2018). Electrospun nanofiber substrates that enhance polar solvent separation from organic compounds in thin-film composites. *Journal of Materials Chemistry. A, Materials for Energy and Sustainability, 6*(31), 15047–15056. DOI: 10.1039/C8TA04504F

Ma, S., Yuan, G., Zhang, Y., Yang, N., Li, Y., & Chen, Q. (2022). Development of encapsulation strategies towards the commercialization of perovskite solar cells. *Energy & Environmental Science, 15*(1), 13–55. DOI: 10.1039/D1EE02882K

MacCarley, C. A. (2025). Hydrogen. In *Non-Petroleum Automotive Transportation* (pp. 199–333). Springer Nature Switzerland. DOI: 10.1007/978-3-031-85028-8_9

Mahmoud, M. A., Alsehli, B. R., Alotaibi, M. T., Hosni, M., & Shahat, A. (2024). A comprehensive review on the application of semiconducting materials in the degradation of effluents and water splitting. *Environmental Science and Pollution Research International, 31*(3), 3466–3494. DOI: 10.1007/s11356-023-31353-3

Malik, S., Muhammad, K., & Waheed, Y. (2023). Nanotechnology: A revolution in modern industry. *Molecules (Basel, Switzerland), 28*(2), 661. DOI: 10.3390/molecules28020661

Mangaiyarkkarasi, J. (2025). Nanostructural Innovations: Powering the Future of Energy Storage. In *Design, Fabrication, and Significance of Advanced Nanostructured Materials* (pp. 51-74). IGI Global.

Manzo, D., Thai, R., Le, H. T., & Venayagamoorthy, G. K. (2025). Fuel cell technology review: Types, economy, applications, and vehicle-to-grid scheme. *Sustainable Energy Technologies and Assessments, 75*, 104229. DOI: 10.1016/j.seta.2025.104229

Marino, A., Genchi, G. G., Mattoli, V., & Ciofani, G. (2017). Piezoelectric nanotransducers: The future of neural stimulation. *Nano Today, 14*, 9–12. DOI: 10.1016/j.nantod.2016.12.005

Mehmood, S., Wang, X., Ahmed, W., Imtiaz, M., Ditta, A., Rizwan, M., & Li, W. (2021). Removal mechanisms of slag against potentially toxic elements in soil and plants for sustainable agriculture development: A critical review. *Sustainability (Basel), 13*(9), 5255. DOI: 10.3390/su13095255

Mekonnin, A. S., Wacławiak, K., Humayun, M., Zhang, S., & Ullah, H. (2025). Hydrogen storage technology, and its challenges: A review. *Catalysts, 15*(3), 260. DOI: 10.3390/catal15030260

Mendrela, P., Stanek, W., & Simla, T. (2024). Thermo-ecological cost–System evaluation of energy-ecological efficiency of hydrogen production from renewable and non-renewable energy resources. *International Journal of Hydrogen Energy, 50*, 1–14. DOI: 10.1016/j.ijhydene.2023.06.150

Miah, M. H., Singh Chand, D., & Malhi, G. S. (2024). Accuracy compensation method with the combined dissimilar measurement devices for enhanced measurement quality: A digital aircraft assembly technology. *Aircraft Engineering and Aerospace Technology, 96*(6), 780–797. DOI: 10.1108/AEAT-09-2022-0253

Mohammed, H., Mia, M. F., Wiggins, J., & Desai, S. (2025). Nanomaterials for Energy Storage Systems—A Review. *Molecules (Basel, Switzerland), 30*(4), 883. DOI: 10.3390/molecules30040883

Murtagh, N., Scott, L., & Fan, J. (2020). Sustainable and resilient construction: Current status and future challenges. *Journal of Cleaner Production, 268*, 122264. DOI: 10.1016/j.jclepro.2020.122264

Namdeti, R., Al-Kathiri, D. S. M. S., Rao, G. B., Khan, A. H., & Alshehri, H. A. M. (2025). A critical review of smart materials for efficient water purification: Towards sustainable clean water solutions. *Journal of Membrane Science and Research, 11*(1).

Neeleshwar, S., Saini, A., Bairwa, M. K., Bisht, N., Katre, A., & Rao, G. N. (2022). Major Challenges Toward. *Nanomaterials for Innovative Energy Systems and Devices*, 103.

Nieder, R., & Benbi, D. K. (2024). Potentially toxic elements in the environment–a review of sources, sinks, pathways and mitigation measures. *Reviews on Environmental Health, 39*(3), 561–575. DOI: 10.1515/reveh-2022-0161

Niu, G., Guo, X., & Wang, L. (2015). Review of recent progress in chemical stability of perovskite solar cells. *Journal of Materials Chemistry. A, Materials for Energy and Sustainability, 3*(17), 8970–8980. DOI: 10.1039/C4TA04994B

Nolan, M., Iwaszuk, A., Lucid, A. K., Carey, J. J., & Fronzi, M. (2016). Design of novel visible light active photocatalyst materials: Surface modified TiO2. *Advanced Materials, 28*(27), 5425–5446. DOI: 10.1002/adma.201504894

Noureen, L., Wang, Q., Humayun, M., Shah, W. A., Xu, Q., & Wang, X. (2023). Recent advances in structural engineering of photocatalysts for environmental remediation. *Environmental Research, 219*, 115084. DOI: 10.1016/j.envres.2022.115084

Nunes Vidal, I. (2021). Autocratization outbreaks: The Polish and Hungarian responses to the Covid-19 pandemic and their early effects on democracy in comparison to Italy.

Ofuani, N. (2024). Carbon Dioxide Gasification of Biochar: A Study on Utilizing Captured CO2 to Mitigate Greenhouse gas Emission and Improving Carbon Conversion Efficiency.

Ojo, G. G., Lottu, O. A., Ndiwe, T. C., Izuka, U., & Ehiobu, N. N. (2023). Solar energy adaptation and efficiency across diverse Nigerian and global climates: A review of technological advancement. [GWK]. *Engineering Heritage Journal, 7*(1), 99–107. DOI: 10.26480/gwk.01.2023.99.107

Othman, M., Abd Halimee, N. F., Osmi, S. K. C., Nordin, N. Z., & Sabri, A. (2024, October). The Effect of Immersion Weathering on the Tensile Properties of rHDPE Green Composite Material. *International Exchange and Innovation Conference on Engineering & Sciences.*

Panagiotopoulou, V. C., Stavropoulos, P., & Chryssolouris, G. (2022). A critical review on the environmental impact of manufacturing: A holistic perspective. *International Journal of Advanced Manufacturing Technology, 118*(1), 603–625. DOI: 10.1007/s00170-021-07980-w

Panda, P. K., & Sahoo, B. (2020). PZT and lead-free piezo ceramics for aerospace and energy applications. In *Handbook of Advanced Ceramics and Composites* (pp. 1–23). Springer. DOI: 10.1007/978-3-319-73255-8_32-1

Pandey, P., Khan, F., Agarwal, S., & Singh, M. (2022). Nano adsorbents in wastewater treatment: A new paradigm in wastewater management. *Lett. Appl. Nanobiosci, 12*(4), 125. DOI: 10.33263/LIANBS124.125

Panghal, H., & Kumar, A. (2025). Eco-Friendly Concrete: A Dual Surface Modification Approach for Enhanced Mechanical Performance of Recycled Coarse Aggregates. *Journal of Materials in Civil Engineering, 37*(4), 04025052. DOI: 10.1061/JMCEE7.MTENG-19054

Pedersen, E. R. G., Lüdeke-Freund, F., Henriques, I., & Seitanidi, M. M. (2021). Toward collaborative cross-sector business models for sustainability. *Business & Society, 60*(5), 1039–1058. DOI: 10.1177/0007650320959027

Penazzato, L., Illampas, R., & Oliveira, D. V. (2024). The challenge of integrating seismic and energy retrofitting of buildings: An opportunity for sustainable materials? *Sustainability (Basel), 16*(8), 3465. DOI: 10.3390/su16083465

Phogat, P., Thakur, J., & Wan, M. (2025). Advancements in energy storage: A review of batteries and capacitors—properties, materials, and emerging applications. *Ionics, 31*(10), 1–25. DOI: 10.1007/s11581-025-06588-4

Podgornik, B. (2023). Advanced materials and research for the green future. *Materials Technology, 57*(1), 71–83.

Pomerantseva, E., Bonaccorso, F., Feng, X., Cui, Y., & Gogotsi, Y. (2019). Energy storage: The future enabled by nanomaterials. *Science, 366*(6468), eaan8285. DOI: 10.1126/science.aan8285

Priyantha, N., & Karunathilake, M. (2025). Introduction to Carbon-Based Materials for Pollution Control. In *Carbon: Bulk-to-Nano Forms for Detection and Remediation of Environmental Contaminants* (pp. 1–25). Springer Nature Switzerland. DOI: 10.1007/978-3-031-90613-8_1

Purabgola, A., Mayilswamy, N., & Kandasubramanian, B. (2022). Graphene-based TiO2 composites for photocatalysis & environmental remediation: Synthesis and progress. *Environmental Science and Pollution Research International, 29*(22), 32305–32325. DOI: 10.1007/s11356-022-18983-9

Rani, M., Yadav, J., Shanker, U., & Wang, C. (2024). Recent updates on remediation approaches of environmentally occurring pollutants using visible light-active nano-photocatalysts. *Environmental Science and Pollution Research International, 31*(15), 22258–22283. DOI: 10.1007/s11356-024-32455-2

Ratnesh, R. K., Kumar, R., Singh, S., Chandra, R., & Singh, J. (2025). Recent advances in solar cell technology: Addressing technological challenges, scenarios, and environmental implications in the development of sustainable energy solutions. *New Journal of Chemistry, 49*(17), 6861–6887. DOI: 10.1039/D5NJ00719D

Rausa, G., Calabrese, M., Velazquez, R., Del-Valle-Soto, C., Fazio, R. D., & Visconti, P. (2025). Mechanical, Thermal, and Environmental Energy Harvesting Solutions in Fully Electric and Hybrid Vehicles: Innovative Approaches and Commercial Systems. *Energies, 18*(8), 1970. DOI: 10.3390/en18081970

Recatala-Gomez, J., Suwardi, A., Nandhakumar, I., Abutaha, A., & Hippalgaonkar, K. (2020). Toward accelerated thermoelectric materials and process discovery. *ACS Applied Energy Materials, 3*(3), 2240–2257. DOI: 10.1021/acsaem.9b02222

Rezania, S., Darajeh, N., Rupani, P. F., Mojiri, A., Kamyab, H., & Taghavijeloudar, M. (2024). Recent advances in the adsorption of different pollutants from wastewater using carbon-based and metal-oxide nanoparticles. *Applied Sciences (Basel, Switzerland), 14*(24), 11492. DOI: 10.3390/app142411492

Rolo, I., Costa, V. A., & Brito, F. P. (2023). Hydrogen-based energy systems: Current technology development status, opportunities and challenges. *Energies, 17*(1), 180. DOI: 10.3390/en17010180

Roy, D. B., & Das, S. (2025). Advanced Nanostructured Materials for Energy Storage Devices. *Design, Fabrication, and Significance of Advanced Nanostructured Materials*, 187-206.

Roy, S., & Ragunath, S. (2018). Emerging membrane technologies for water and energy sustainability: Future prospects, constraints and challenges. *Energies, 11*(11), 2997. DOI: 10.3390/en11112997

Rzaij, J. M., & Abass, A. M. (2020). Review on: TiO2 thin film as a metal oxide gas sensor. *J. Chem. Rev, 2*(2), 114–121. DOI: 10.33945/SAMI/JCR.2020.2.4

Saini, R. K., Saini, D. K., Verma, P., Thakur, R., Gandotra, N., Kumar, R., & Verma, A. (2025). Climate change and energy transition-a critical review. *International Journal of Green Energy, 23*, 1–26.

Saleh, T. A. (2022). Nanomaterials and hybrid nanocomposites for CO 2 capture and utilization: Environmental and energy sustainability. *RSC Advances*, *12*(37), 23869–23888. DOI: 10.1039/D2RA03242B

Scheffran, J., Felkers, M., & Froese, R. (2020). Economic growth and the global energy demand. *Green energy to sustainability: strategies for global industries*, 1-44.

Sekhar, M. C., Veena, E., Kumar, N. S., Naidu, K. C. B., Mallikarjuna, A., & Basha, D. B. (2023). A review on piezoelectric materials and their applications. *Crystal Research and Technology*, *58*(2), 2200130. DOI: 10.1002/crat.202200130

Semieniuk, G., Taylor, L., Rezai, A., & Foley, D. K. (2021). Plausible energy demand patterns in a growing global economy with climate policy. *Nature Climate Change*, *11*(4), 313–318. DOI: 10.1038/s41558-020-00975-7

Shabbirahmed, A. M., Jacob, A., Dey, P., Somu, P., & Haldar, D. (2025). Biomass as eco-friendly adsorbents for the removal of emerging pollutants from wastewater: A review. *Discover Applied Sciences*, *7*(7), 771. DOI: 10.1007/s42452-025-07463-7

Shahab, S. A., Khan, S. J. M., & Nayan, S. (2024). Financial Development and Economic Growth: A Bibliometric Analysis. *Journal of Business & Finance Librarianship*, *29*(4), 243–273. DOI: 10.1080/08963568.2024.2362096

Sharma, A., Sunny, S., Arulraj, J., & Hegde, G. (2024). Exploring the efficiency of green synthesized silver nanoparticles as photocatalysts for organic dye degradation: Unveiling key insights. *Nano Express*, *5*(2), 022002. DOI: 10.1088/2632-959X/ad4d09

Sheerazi, Z., & Ahmed, M. (2022). Nano-adsorbents and nano-catalysts for wastewater treatment. In *Emerging nanomaterials for advanced technologies* (pp. 517–539). Springer International Publishing. DOI: 10.1007/978-3-030-80371-1_17

Simões, S. (2024). High-performance advanced composites in multifunctional material design: State of the art, challenges, and future directions. *Materials (Basel)*, *17*(23), 5997. DOI: 10.3390/ma17235997

Simões, S. (2024). High-performance advanced composites in multifunctional material design: State of the art, challenges, and future directions. *Materials (Basel)*, *17*(23), 5997. DOI: 10.3390/ma17235997

Singh, A., Chaturvedi, K. R., & Sharma, T. (2025). Recent advances in carbon-based materials for carbon dioxide capture and separation applications. *Petroleum Science and Technology*, •••, 1–26. DOI: 10.1080/10916466.2025.2538735

Skiba, R. (2025). *Advanced Materials Engineering Fundamentals*. After Midnight Publishing.

Surat'man, N. E. B., Quek, X. L., Wang, N., Ye, E., Xu, J., Li, Z., & Li, B. (2025). Sustainable nanofibrous membranes for air filtration, water purification and oil removal. *Nanoscale*, *17*(11), 6427–6447. DOI: 10.1039/D4NR04673K

Talreja, N., Chuahan, D., & Ashfaq, M. (2024). Carbon-based two-dimensional (2D) materials: A next generation biocidal agent. *Materials Advances*, *5*(4), 1454–1461. DOI: 10.1039/D3MA00952A

Tariq, A., & Mushtaq, A. (2023). Untreated wastewater reasons and causes: A review of most affected areas and cities. *International Journal of Chemical and Biochemical Science*, *23*(1), 121–143.

Tian, J., Yu, L., Xue, R., Zhuang, S., & Shan, Y. (2022). Global low-carbon energy transition in the post-COVID-19 era. *Applied Energy*, *307*, 118205. DOI: 10.1016/j.apenergy.2021.118205

Tian, X., Ye, C., Zhang, L., Sugumar, M. K., Zhao, Y., McKeown, N. B., Margadonna, S., & Tan, R. (2025). Enhancing membrane materials for efficient Li recycling and recovery. *Advanced Materials*, *37*(5), 2402335. DOI: 10.1002/adma.202402335

Tootkaboni, M. P., Ballarini, I., & Corrado, V. (2025). Towards climate resilient and energy-efficient buildings: A sensitivity analysis on building components and cooling strategies. *Building and Environment*, *270*, 112473. DOI: 10.1016/j.buildenv.2024.112473

Tripathy, J., Mishra, A., Pandey, M., Thakur, R. R., Chand, S., Rout, P. R., & Shahid, M. K. (2024). Advances in nanoparticles and nanocomposites for water and wastewater treatment: A review. *Water (Basel)*, *16*(11), 1481. DOI: 10.3390/w16111481

Tsai, Y. C., Huang, P. C., & Dai, C. L. (2025). Editorial for the Special Issue on MEMS/NEMS Devices and Applications. *Micromachines*, *16*(2), 189. DOI: 10.3390/mi16020189

Umar, U. M., & Muthusamy, K. (2023). Potential of waste material as coarse aggregates for lightweight concrete production: A sustainable approach. *Construction (Arlington)*, *3*(1), 87–114. DOI: 10.15282/construction.v3i1.9217

Ungureanu, E. L., Mocanu, A. L., Stroe, C. A., Panciu, C. M., Berca, L., Sionel, R. M., & Mustatea, G. (2023). Agricultural byproducts used as low-cost adsorbents for removal of potentially toxic elements from wastewater: A comprehensive review. *Sustainability (Basel)*, *15*(7), 5999. DOI: 10.3390/su15075999

Verma, V. K., & Verma, S. (2022). Machine learning applications in healthcare sector: An overview. *Materials Today: Proceedings*, *57*, 2144–2147. DOI: 10.1016/j.matpr.2021.12.101

Wang, J., Yin, Y., Che, C., & Cui, M. (2025). Research Progress of Thermoelectric Materials—A Review. *Energies*, *18*(8), 2122. DOI: 10.3390/en18082122

Wei, L., Kuai, X., Bao, Y., Wei, J., Yang, L., Song, P., & Wang, X. (2021). The recent progress of MEMS/NEMS resonators. *Micromachines*, *12*(6), 724. DOI: 10.3390/mi12060724

Wheatley, M. C. (2024). Advancements in Renewable Energy Technologies: A Decade in Review. *Science*, *1*, 100013.

Wu, J., Ding, Y., Wang, J., Li, T., Lin, H., Wang, J., & Liu, F. (2018). Facile fabrication of nanofiber-and micro/nanosphere-coordinated PVDF membrane with ultrahigh permeability of viscous water-in-oil emulsions. *Journal of Materials Chemistry. A, Materials for Energy and Sustainability*, *6*(16), 7014–7020. DOI: 10.1039/C8TA01539B

Xu, D. (2023, October). Technologies and challenges of hydrogen powered aviation. [). IOP Publishing.]. *Journal of Physics: Conference Series*, *2608*(1), 012003. DOI: 10.1088/1742-6596/2608/1/012003

Xu, X., Chen, Y., Wan, P., Gasem, K., Wang, K., He, T., Adidharma, H., & Fan, M. (2016). Extraction of lithium with functionalized lithium ion-sieves. *Progress in Materials Science*, *84*, 276–313. DOI: 10.1016/j.pmatsci.2016.09.004

Yan, X., Xiao, X., Au, C., Mathur, S., Huang, L., Wang, Y., Zhang, Z., Zhu, Z., Kipper, M. J., Tang, J., & Chen, J. (2021). Electrospinning nanofibers and nanomembranes for oil/water separation. *Journal of Materials Chemistry. A, Materials for Energy and Sustainability*, *9*(38), 21659–21684. DOI: 10.1039/D1TA05873H

Yang, X. Y., Wei, Y. X., Su, Y. Q., Zhang, Z. W., Tang, X. Y., Chen, Y. E., Yuan, M., & Yuan, S. (2025). The Strategies Microalgae Adopt to Counteract the Toxic Effect of Heavy Metals. *Microorganisms*, *13*(5), 989. DOI: 10.3390/microorganisms13050989

Yao, Y., Rui, X., Bai, R., Ouyang, Y., Li, G., Zhao, Y., Zhu, Y.-H., Zhao, M., Li, B.-Q., Zhang, X., Li, Z., Ling, F., Ma, C., Ma, J., Zhou, F., Ren, Z., Shi, X., Zhao, Z., Lu, Y., & Yu, Y. (2025). Roadmap for Next-Generation Electrochemical Energy Storage Technologies: Secondary Batteries and Supercapacitors. *ACS Nano*, *19*(34), 30568–30687. DOI: 10.1021/acsnano.4c10091

Ye, Y., Ren, J., Wang, S., Wan, Y., Razzak, I., Hoex, B., & Zhang, W. (2024). Construction and application of materials knowledge graph in multidisciplinary materials science via large language model. *Advances in Neural Information Processing Systems*, *37*, 56878–56897.

Younas, F., Mustafa, A., Farooqi, Z. U. R., Wang, X., Younas, S., Mohy-Ud-Din, W., & Hussain, M. M. (2021). Current and emerging adsorbent technologies for wastewater treatment: Trends, limitations, and environmental implications. *Water (Basel)*, *13*(2), 215. DOI: 10.3390/w13020215

Zhang, X., Chen, L., Lim, K. H., Gonuguntla, S., Lim, K. W., Pranantyo, D., Yong, W. P., Yam, W. J. T., Low, Z., Teo, W. J., Nien, H. P., Loh, Q. W., & Soh, S. (2019). The pathway to intelligence: Using stimuli-responsive materials as building blocks for constructing smart and functional systems. *Advanced Materials*, *31*(11), 1804540. DOI: 10.1002/adma.201804540

Zhang, Y., & Sun, L. (2022). Research on hydrogen fuel application (inaircraft) based on carbon neutralization. In *E3S Web of Conferences* (Vol. 341, p. 01010). EDP Sciences. DOI: 10.1051/e3sconf/202234101010

Zheng, L., Gao, L., Huang, Z., Shi, Z., & Zhou, Y. (2025). A systematic meta-analysis of the impacts of group awareness tools on learning achievements, learning behaviors, and learning perceptions from 2010–2023. *Interactive Learning Environments*, *33*(6), 1–18. DOI: 10.1080/10494820.2025.2454440

Zhi, C., Shi, S., Si, Y., Fei, B., Huang, H., & Hu, J. (2023). Recent progress of wearable piezoelectric pressure sensors based on nanofibers, yarns, and their fabrics via electrospinning. *Advanced Materials Technologies*, *8*(5), 2201161. DOI: 10.1002/admt.202201161

Zhu, J., Dong, Y., Wang, Q., Han, J., Li, Z., Xu, D., Fischer, L., Ulbricht, M., & Ren, Z. (2024). Advancements in magnetic catalysts: Preparation, modification, and applications in photocatalytic and environmental remediation. *The Science of the Total Environment*, *957*, 177595. DOI: 10.1016/j.scitotenv.2024.177595

Znidi, F., Morsy, M., & Uddin, M. N. (2024). Navigating challenges and solutions for metal-halide and carbon-based electrodes in perovskite solar cells (NCS-MCEPSC): An environmental approach. *Heliyon*, *10*(12), e32843. DOI: 10.1016/j.heliyon.2024.e32843

Chapter 6
Sustainable Nanofluid Engineering Synthesis Techniques and Emerging Applications in Smart Materials

M. Amin Mir
https://orcid.org/0000-0003-4689-0507
Prince Mohammad Bin Fahd University, Saudi Arabia

Syed M. Hasnain
https://orcid.org/0000-0002-5102-3101
Prince Mohammad Bin Fahd University, Saudi Arabia

Noushi Zaidi
King Faisal University, Saudi Arabia

Mohammad Ayaz Ahmad
https://orcid.org/0000-0002-5731-5439
University of Guyana, Guyana

Ercan Bursal
https://orcid.org/0000-0001-7289-4507
Prince Mohammad Bin Fahd University, Saudi Arabia

ABSTRACT

Colloidal suspensions of nanoparticles in regular base fluids, or nanofluids, have become attractive materials for improving heat transfer, thermal conductivity, and

DOI: 10.4018/979-8-3373-3556-8.ch006

overall energy efficiency in a variety of engineering applications. Often referred to as "smart fluids," their ability to dynamically control heat transfer makes them useful in fields like biomedical engineering, microelectronics, transportation, nuclear energy, and food processing. This chapter provides a thorough review of current advancements in the study of nanofluids, covering synthesis strategies, colloidal stability maintenance tactics, and methodologies for assessing thermophysical characteristics. Innovative applications that aim to further optimize system performance, like hybrid nanofluids, are given particular attention. The existing difficulties in simulating the behavior of nanofluids, especially the absence of a universal predictive framework, are also covered in this chapter. The chapter concludes with a summary of future research possibilities, emphasizing the need for future practical applicability, long-term stability, and future studies on the nanoparticle-fluid interactions. For researchers, engineers, and business people trying to drive nanofluid technology, this chapter is a great resource.

INTRODUCTION

Nanofluids, which consist of a base liquid with a low concentration of metal or metal oxide nanoparticles, have been widely applied in many areas, including chemical and power engineering, electronics, and medicine. Their versatility is attributed in large part to two key features: they can enhance heat and mass transfer even at low concentrations of nanoparticles, and can manipulate transport processes, a feature that has potential in applications such as targeted drug delivery. This Special Issue includes twelve original research articles exploring nanofluids properties and applications. Topics include the fundamental properties of nanofluids (Sadeghzadeh et al., 2020), interactions of nanoparticles with other materials (Xie et al., 2020), and enhancements to heat transfer using nanoparticle-filled nanofluids in engineering systems ranging from cooling of electronic device components (Asadi et al., 2020) to the performance of mini-channels (Ahmadi et al., 2020), microchannels (Khan et al., 2020), and heat exchangers (Alsabery et al., 2020). This collection is meant to highlight the latest developments using nanoparticle-based technologies, in particular the use of nanofluids in engineering applications. Researchers and practitioners in various fields spanning physics, applied mathematics, and mechanical and chemical engineering will find this collection valuable.

One of the most highlighted advancements is the use of artificial neural networks (ANNs) for predicting the thermophysical properties of nanofluids. Sadeghzadeh et al. (Sadeghzadeh et al., 2020) created a TiO_2-Al_2O_3/H_2O nanofluid and experimentally studied its specific heat capacity and thermal conductivity. The researchers demonstrated that ANNs could represent and predict these properties based on a multilayer perceptron model. In another study, Xie et al. (Xie et al., 2020) studied the nanofluids' impact on the brass corrosion process. The study used a surfactant called sodium dodecyl benzene sulfonate (SDBS) with simulated cooling water (SCW) and presented evidence that brass can form a protective coating.

The performance of this film is greatly impacted by the addition of nanoparticles, though, since positively charged AlO_3 particles improve its protective properties while negatively charged TiO_2 particles prevent the creation of the SDBS layer. The use of hybrid nanosystems to enhance third-order nonlinear optical responses was investigated by García-Beltrán et al. (García-Beltrán & Mercado-Zuniga, 2020). The application of nanowires and nanotubes in chemical and biological sensors was examined by Zhu et al. (Zhu & Wang, 2021). They described the present uses of nanofluidic devices based on nanowires and nanotubes, as well as their fabrication techniques and sensing principles.

Preparation of nanofluids

Given their function in enhancing heat transfer performance, the first and most important stage of nanofluid research is their appropriate formulation. To effectively utilize the benefits of the nanofluid, its stability must be guaranteed. Inadequate preparation can result in a number of problems, including irregular heat transfer properties, reduced thermal efficiency, and mechanical issues like blockage or system damage. To get over these restrictions and accomplish dependable and effective performance in nanofluid-based applications, careful preparation is therefore essential.

Two-step method

The two-step process is widely acknowledged as a cost-effective and frequently used method for creating nanofluids. This method uses physical or chemical methods to first create nanoparticles in powdered form. These particles are susceptible to aggregating because of their high surface area. To resolve this, clusters are usually broken apart using a ball mill and then dispersed into a base liquid, usually water or another base liquid, using ultrasonic agitation to ensure a complete mix. Although surfactants are commonly used to increase the stability of nanoparticles, their efficacy may be restricted in high-temperature settings. This method overcomes the concerns surrounding a consistent distribution and durability over time while also offering flexibility and cost efficiency by decoupling the two stages of production, i.e., producing nanoparticles and dispersing them in the liquid. While the two-step method can be applied to metallic nanoparticles, it has been particularly recommended for oxide nanoparticles in nanofluids (Solangi et al., 2015). Many researchers have evaluated numerous methods to develop nanofluids. In the work of S. Suresh et al. (Suresh et al., 2012), Al_2O_3 nanoparticles were synthesized by chemical precipitation and applied in tests evaluating their effect on the frictional characteristics and thermal performance of a circular tube with spiral rod inserts. In another work, Hashemi and Behabadi (Hashemi & Akhavan-Behabadi, 2012) evaluated flow behavior, thermal transfer characteristics, and pressure drop characteristics of nanofluids using the direct synthesis method. In the same vein, Kumar et al. (Kumar et al., 2023) focused primarily on improving thermal conductivity for MgO-based nanofluids. They prepared CuO/MgO nanofluids using the two-step method and noticed a 12.5% increase in thermal conductivity at 55° C using a volume fraction of 1.5%. Following same two-step method to add CuO nanoparticles, Amburi et al. (Amburi et al., 2023) were able to achieve an improvement in both MgO/water and MWCNT/water nanofluids thermal conductivity. MgO/water's thermal conductivity increased by 0.113 at 30 °C, but MWCNT/water's improved by 13.3%. Conductivity improvements for MgO/water systems were 12.8% at 40°C, while MWCNT/water systems had 14.5% increases.

One-step method

By producing and distributing nanoparticles into the base fluid in a single, continuous step, the one-step synthesis method tackles the problem of nanoparticle agglomeration. By avoiding intermediate steps like drying, storage, shipping, and separate dispersion, this integrated approach successfully lowers particle accumulation and improves fluid stability (Lo et al., 2005). The one-step method aids in maintaining a stable suspension of nanoparticles, especially in aqueous solutions, by facilitating uniform distribution. Despite its benefits, this approach is not the best for large-scale manufacturing because it is typically less scalable and costlier. A schematic illustration of the one-step synthesis process is provided in Figure 1. The goal of both one-step and two-step methods is to reduce undesired nanoparticle agglomeration (Jamal-Abad et al., 2013), which may arise from partial reactions or particle clustering at lower saturation levels. The one-step procedure increases stability, although in certain situations, homogeneity may be compromised. Regardless of the synthesis technique, preserving nanofluid stability is essential in heat management systems. To guarantee a steady and effective supply of nanofluid, it's also critical to stop nanoparticles from depositing on channel surfaces. Therefore, the dependable and efficient deployment of nanofluids in engineering applications requires the use of modern processing methods. A comparison of research using one-step and two-step approaches for nanofluid production is given in Table 1.

Figure 1. Schematic describing the one-step method (Jamal-Abad et al., 2013).

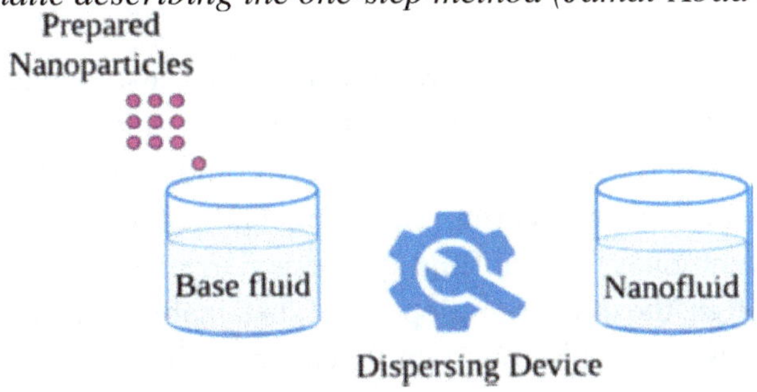

Table 1. Overview of Nanofluid Preparation via One and Two-Step Techniques (Jamal-Abad et al., 2013; Lo et al., 2005)

Nanoparticles (NPs)	Base Fluid	Synthesis Method	Stability Description
CuO	Water	**TS**	Achieved optimal stability using 96:4 CTAB–SDS surfactant mix
SiO_2	Water	**TS**	Stable for up to 8 days
ZnO	Deionized Water	**TS**	Stability ranged from 1 to 60 days
MWCNT	Water	**TS**	Stable for approximately 1 hour
SiO_2–Al_2O_3	Water	**TS**	Maintained stability for several weeks without sediment
CuO–TiO_2	Water	**TS**	No signs of agglomeration observed
Boric Acid	Ethylene Oxide	**TS**	Improved stability noted
ZnO & MgO	Ethylene Oxide	**TS**	Remained stable for more than 15 days
MWCNTs	Silicone Oil	**OS**	Stability maintained for 7 to 42 days
MWCNTs	Silicone Oil	**OS**	Stable after one week
WO_3–Silver	Thermic Oil	**OS**	Stability not available (N/A)
Urethane	Polyalphaolefin	**OS**	Demonstrated good stability
Cu	Ethylene Oxide	**OS**	Exhibited reliable stability
Cu–Al	Water	**OS**	High Cu–Al LDH volume fraction yielded better stability
Cu	Not specified	**OS**	Maintained stability for at least 3 weeks without surfactants
Carbon	Water	**OS**	Produced a stable suspension
Cu	Ethylene Glycol	**OS**	Remained agglomeration-free for 15 to 20 days

TS = Two Step; OS= One Step.

DISCUSSION

The body of research reviewed above illustrates that sustainable nanofluid engineering is an inherently multidisciplinary enterprise, and one where the necessity for new, cross-sectoral approaches is clear, requiring a combination of materials science, fluid mechanics, thermal engineering, environmental science, and modeling. While the maturity of techniques for nanofluid synthesis (i.e., one-step and two-step methods) is a positive sign, the industrial adoption of sustainable nanofluids will continue to be limited if long-term colloidal stability cannot be achieved (i.e., withstanding operational stresses such as high shear rates, thermal cycling, and changes in pH). At this point, approaches that utilize DLVO theory and functionalization of surface characteristics to stabilize colloids look promising, but there remain trade-offs between stability, cost and manufacturability at scale that need to be considered. The synthesis of nanofluids and their industrial and safe applications in cooling in the nuclear industry, the microelectronics industry, automotive applications, geothermal energy recovery, and biomedicine demonstrates the large range of potential applications for nanofluids. While there are performance increases through the incorporation of nanoparticles, these are often conditional on the appropriate selection of the nanoparticle type and the subsequent concentration and surface chemistry (Mir et al., 2024). One example is that the significant improvements in thermal conductivity for a high-frequency cooling industry such as electronics often cannot be scaled to higher volume cases such as heat exchangers due to differences in the flow regime, potential fouling and pumping power constraints. The issue of the predictability of nanoparticle properties at various scales has emerged as a constant theme. Emerging hybrid modeling frameworks that couple computational fluid dynamics or molecular dynamics to machine learning techniques offer a powerful method to link data from laboratory studies and operational conditions. These hybrid frameworks will be especially useful for optimizing multi-objective performance metrics like heat transfer enhancement, viscous control, and erosion mitigation. Moreover, from a sustainability perspective, the environmental persistence and toxicity of nanoparticles is a serious concern (Han et al., 2008). Therefore, to realize the transition from a laboratory-level performance to an industry-level acceptance, both assessments of full life-cycle impacts of any nanoparticle should be undertaken; consideration of safe end-of-life protocols and registry assurances. Development of a reusable and recoverable nanoparticles inventory or more generally, a circular economy approach towards nanoparticles will play a vital role in ensuring that technological advances should not be at the ecological cost. In conclusion, the field of nanofluids is progressing towards application-specific nanofluid formulations, aided by predictive modeling and guided by environmental responsibility. The colloidal stability of nanofluids is primarily determined by the balance of forces described in DLVO Theory, where the interaction potential between the nanoparticles is a function of the attractive van der Waals forces (which arise from dipole induced interactions), and repulsive electrostatic double-layer forces arising from surface charge in the fluid medium. Stability is achieved when the electrostatic barrier is large enough to prevent particle aggregation, with zeta potential values above ± 30 mV typically indicating sufficient amount of electrostatic repulsion and good dispersion of the nanoparticles. In order to improve this stability level, there are several functionalization methods employed that include covalent graft of hydrophilic or hydrophobic ligands, adsorption of surfactants such as sodium dodecyl sulfate (SDS) or cetyltrimethylammonium bromide (CTAB), coatings of polymers polyvinylpyrrolidone (PVP) or polyethylene glycol (PEG), and silane coupling agents for oxide nanoparticles, to increase either steric hindrance or electrostatic repulsion. These changes can enable stability under thermal cycling, high shear, and variable pH environments, while also providing enhancement in properties such as corrosion resistance, antimicrobial activity, and targeted thermal conductivity. However, the use of engineered nanoparticles also raises significant questions about toxicity, environmental persistence, and end-of-life disposal. Depending on composition, size, and surface chemistry, nanoparticles can be cytotoxic through oxidative stress, membrane disruption, or interference with cellular processes, which may present hazards to aquatic and terrestrial organisms. Many metal or metal oxide nanoparticles are chemically stable substances that become persistent environmental contaminants, which raises the potential for bioaccumulation (Donzelli et al., 2009). Nanofluid systems that are \(end-of-life \) but are not properly disposed of can cause steady-state contaminants in an ecosystem, especially if they do not have a treatment as in the case of liquid waste systems, i.e. nanoparticle filtration, chemical immobilization, etc., or they are subjected to recycling strategies or disposal. Thus, blending DLVO-based stability optimization with well-informed surface engineering, life-cycle assessment, and considerate waste processing and disposal techniques is recommendable for nanofluid technologies to produce high performing novel solutions while remaining an environmentally-favourable option (Zaidi et al., 2025).

In the context of these factors, accurate prediction of the thermophysical and transport properties of nanofluids offers a promising way to improve design and implementation of nanofluid technology within real systems. Traditionally, modeling approaches to analyzing nanofluid behavior and interactions tend to rely, especially when looking at scales across spatio-temporal boundaries, on physics-based formulations and simulations such as; computational fluid dynamics (CFD), lattice Boltzmann methods (LBM), and to some extent molecular dynamics (MD) simulations, each of which can be classified on levels of solution, which could include heat transfer, viscosity changes, and nanoparticle–fluid interactions. Using CFD models, both slow and laminar, and through single-phase or two-phase formulations, studying for example; convective heat transfer, pressure drop, and flow distributions from the space occupied by heat exchangers, microchannels, and cooling systems, in the cases where the unknown is less formidable than in other models or approaches. LBM provides a mescaopic method to simulate particle-fluid interactions at more efficient computational costs, while MD simulations provide atomic details about nanoparticle dispersion, interfacial thermal resistance (Kapitza resistance), and the impact of surface chemistry on enhancing thermal conductivity. Alongside established modeling methods, a data-driven approach—especially machine learning (ML) and artificial intelligence (AI) models—are now being integrated to circumvent limitations of mechanistic models. ANNs, SVR, RF, and ANFIS are all capable of obtaining high accuracy in prediction of thermal conductivity, viscosity, specific heat, and stability from experimental databases. Hybrid models that blend CFD or MD simulations with ML models can also allow reduction of computational costs with predictive accuracy to produce rapid design of nanoparticle type, volume fraction, and operating conditions. Advanced surrogate models or physics-informed neural networks, are also emerging as a means to ensure conservation laws and boundary conditions are embedded directly into learning algorithms to enhance generalizability.

Heat Transfer Applications

Applications in Industrial Cooling

With the goal of achieving notable energy savings and lowering greenhouse gas emissions, Routbort et al. (Mir et al., 2024) started a research in 2008 that concentrated on using nanofluids for industrial cooling applications. Nanofluids could save about 1 trillion British thermal units (Btu) of energy in U.S. companies by substituting traditional cooling and heating water. By adding nanofluids to closed-loop cooling systems, the electric power industry may be able to save 10 to 30 trillion Btu a year, which is equal to the energy requirements of 50,000 to 150,000 American homes. Significant decreases in hazardous emissions, such as about 5.6 million metric tons of CO_2, 8,600 metric tons of NO_X, and 21,000 metric tons of SO_2, would accompany this energy savings. Michelin North America is a real-world example of how productivity is impacted by cooling rubber processing restrictions. In all of its facilities, the firm uses more than 2 million gallons of heat transfer fluids. Michelin wants to increase productivity at its rubber processing facilities by 10% through the creation of affordable, water-based nanofluids.

Another breakthrough was made by Han et al. (Han et al., 2008), who included phase change materials (PCMs) as nanoparticles in nanofluids to increase specific heat and thermal conductivity at the same time. Using a one-step nano-emulsification process, they dispersed indium nanoparticles, which have a melting temperature of 157°C, into polyalphaolefin to create a nanofluid. Temperature-dependent improvements in thermal conductivity, viscosity, and specific heat capacity were seen in experimental analyses of the resultant fluid. The effective specific heat of the nanofluid was specifically enhanced by the phase transition involvement of the indium nanoparticles. This study is ostensibly one of the few that considered thermal diffusivity and suggests that complex cooling solutions can be formed in the absence of full understanding of the underlying heat transport mechanisms of nanofluids.

Smart Fluids

The existing lack of sustainable energy sources and the increasing prevalence of a new generation of battery-run devices such as laptops and smart phones have drawn attention to the importance of smart energy resource management in this new energy conscious age. In some applications, nanofluids may be considered smart fluids in terms of energy resource management, and become a viable choice here. In a paper based on their research published in Physical Review Letters in March 2009, Donzelli etal, (Donzelli et al., 2009) introduced a new type of nanofluid as a "thermal switch" or "heat valve" that could provide a method for effectively controlling heat transport. The nanofluid can be switched between a "high" conductivity phase, which facilitates heat dissipation in a more efficient manner, and a different "low" conductivity phase that enables restriction of thermal transfer. However, outside verification of the work is needed, in particular the verification of a solid and sensitive system with larger and more unpredictable heat flow conditions, before the concept can even begin to be integrated into next generation heating and cooling systems.

Nuclear Reactor Applications

A study conducted by Kim et al. (Zaidi et al., 2025) from the Massachusetts Institute of Technology (MIT) Nuclear Science and Engineering Department looked into the possibility of using nanofluids - where water is substituted with nanofluids - to enhance the efficiency of water-cooled nuclear systems with low heat removal capabilities. Some possible applications of nanofluids include, but are not limited to, backup safety systems, accelerator targets, plasma divertors, and primary coolant systems in pressurized water reactors (PWR) (Boungiorno et al., 2008). In a PWR nuclear power station, the critical heat flux (CHF) between the cooling water and the fuel rods limits the amount of steam that can be produced. When vapor bubbles cover the rod's surface, heat conduction is not as effective as with liquid water. When water is replaced with nanofluids, fuel rods, which are coated with nanoparticles, such as alumina, cause bubbles that are just created to move away from the surface. This considerably raises the CHF and stops a vapor layer from developing around the rod. According to encouraging preliminary testing at MIT's Nuclear Research Reactor, nanofluids increase the PWR system's productivity. Furthermore, emergency cooling devices that employ nanofluids could quickly cool overheated surfaces, increasing power plant safety. However, there are drawbacks to using nanofluids in nuclear power facilities. The unpredictable number of nanoparticles transported away by the boiling vapor is one issue. The requirement for additional safety precautions when discarding the nanofluid is another problem.

Because of worries about nanoparticle carryover to turbines and condensers, which could result in erosion and fouling, uses of nanofluid coolants in boiling water reactors (BWR) are anticipated to be restricted. Jackson's research (Akhai, Abbass, Kaur et al, 2025) showed that using nanofluid deposition to create structured surfaces could result in significant improvements in CHF. Enhancing CHF with no effect on heat transfer efficiency might be feasible if the thickness and structure of the deposited coating can be managed. The boiling of nanofluids may offer a straightforward method of producing an improved surface, even though the nanoparticles themselves do not substantially change the pool-boiling behavior of water.

Future applications of nanofluids in nuclear power plants seem promising (Boungiorno et al., 2008). The thermal-hydraulic performance of the nanofluid in reactor settings and its compatibility with reactor materials are two important knowledge gaps that still exist. By keeping molten fuel inside the reactor vessel, nanofluids may also assist prevent catastrophic catastrophes like core meltdown. It is essential to eliminate decomposition heat through the vessel walls in such occurrences. This process is limited by CHF on the vessel's exterior, but calculations indicate that nanofluids could increase in-vessel retention capacity by as much as 40% (Buongiorno et al., 2009). Nanofluids have the ability to greatly increase coolant CHF, which would have positive economic and safety effects because many water-cooled nuclear systems are CHF-limited for power plant systems.

Geothermal Extraction of Power and Other Energy things

According to an MIT analysis from 2006 (Tester et al., 2006), just 200 ZJ of the world's geothermal energy resources are currently extractable, out of an estimated 13,000 ZJ. However, more than 2,000 ZJ might be gathered with technological breakthroughs, potentially meeting the world's energy demands for millennia. Nanofluids can be used to cool pipelines exposed to severe temperatures (between 500°C and 1000°C) and depths (between 5 and 10 km) for extracting energy from the earth's crust. Nanofluids can also aid in cooling machinery and equipment used in high-temperature, high-friction settings during drilling operations. As "fluid superconductors," nanofluids may also be used as working fluids in the process of extracting energy from the Earth's core. Significant amounts of work energy might be produced by processing the retrieved heat in a pressurized water reactor (PWR) power plant. Nanofluids may make it possible for sensors and electronics in drilling technology—which is essential for the extraction of geothermal energy—to cool more effectively, enabling these instruments to function at higher temperatures in downhole settings. Innovative developments in rock penetration techniques, which are likewise cooled and lubricated by nanofluids, will minimize production costs and provide access to more economically viable temperatures in lower-grade formations or deeper, hotter areas in high-grade formations. Significant improvements in reservoir performance and heat-to-power conversion efficiency could be achieved in power conversion technology by enhancing the heat transfer capabilities of lower-temperature nanofluids and designing plants for higher resource temperatures in the supercritical water region. The goal of research by Tran et al. (Mir et al., 2025), supported by the U.S. Department of Energy (USDOE), was to create a new class of specialized drilling fluids that would perform better at high temperatures. This study is particularly pertinent to high-temperature, high-pressure drilling, which may be crucial to releasing significant quantities of domestic fuel resources that were previously unrecoverable. However, one of the biggest obstacles in this field is still the commercialization of these developments.

Applications in Automobiles

Engine oils, automatic transmission fluids, coolants, lubricants, and other synthetic high-temperature heat transfer fluids with comparatively low heat transfer efficiency are commonly used in conventional truck thermal systems, which include radiators, engines, and heating, ventilation, and air conditioning (HVAC) systems. The increased thermal conductivity offered by nanofluids—which are made by incorporating nanoparticles—could be extremely advantageous for these systems (Chopkar et al., 2006).

Cooling By Nanofluids

Reducing aerodynamic drag becomes crucial in the quest to improve vehicle design and fuel efficiency, particularly for trucks. Due to the massive radiators used to cool the engine, a considerable amount of energy (about 65%) is used to overcome this drag at high speeds. Manufacturers can optimize radiator placement and minimize radiator size by incorporating nanofluids into cooling systems. Truck engines can run at greater temperatures and produce more horsepower while still adhering to rigorous emission regulations thanks to more effective cooling, which uses less fluid and permits the use of smaller coolant pumps.

Using high-thermal conductive nanofluids in radiators can lower the radiator's frontal area by up to 10%, according to research by Argonne, notably by Singh et al. (Singh et al., 2006). This immediately reduces aerodynamic drag and can result in fuel savings of up to 5%. Data about the wear and erosion rates of various radiator materials exposed to different nanofluids, as well as the performance of these materials under particular parameters such fluid velocity and impact angle, are probably included in tables 2 and 3. Nanofluids also lessen wear and friction on engine parts, which results in further fuel savings of over 6%. This can be a big step in the right direction for truck energy efficiency. To determine whether nanofluids degrade radiator material, the researchers have carried out a number of tests. At velocities up to 9 m/s and different impact angles, no erosion was seen while using base fluids like ethylene and tri-chloroethylene gycols. At greater velocities, copper nanofluids did, however, contribute to some erosion. Additionally, the study discovered that, in comparison to the base fluid, alumina-based nanofluids exhibited reduced wear and friction as shown in table 3 (Singh et al., 2006)

The tribological characteristics of nanofluids at different particle loadings and their possible erosion effects on radiator materials will be the main topics of future research (Shen et al.,). This continuous research will help create more effective cooling systems for engines and may result in engines that run more efficiently, generating greater power with less fuel use and ultimately lowering emissions. Vehicles may be able to reduce their environmental impact, improve fuel efficiency, and control heat more effectively with nanofluids table 3 (Singh et al., 2006).

Table 2. Erosion Performance of Aluminum 3003 at 50°C in a 50:50 Ethylene Glycol–Water Mixture (Singh et al., 2006)

Impact Angle (°)	Flow Velocity (m/s)	Exposure Duration (hrs)	Mass Loss (mg)
89	7.9	235.	0 ± 0.21
89	9.9	209	0 ± 0.21
49	5.9	254	0 ± 0.21
49	9.9	245	0 ± 0.21
29	7.9	284	0 ± 0.21
29	9.9	294	0 ± 0.21

Table 3. Erosion Resistance of Al 3003 Exposed to Copper Nanoparticles in Trichloroethylene Glycol at 50°C (Singh et al., 2006)

Impact Angle (°)	Flow Velocity (m/s)	Test Duration (hrs)	Mass Loss (mg)
89	3.9	216	0 ± 0.21
29	4.6	310	0 ± 0.21
89	7.5	340	0 ± 0.21
29	7.5	334	0 ± 0.21
29	9.5	3365	0 ± 0.21

Fuel by Nanofluids

There are various advantages to using aluminum nanoparticles, especially those coated in aluminum oxide, in diesel engine combustion. By improving the interaction between the nanoparticles and water, the aluminum oxide layer makes it easier for hydrogen to break down from water during burning. This catalytic activity results in a more efficient combustion process. Specifically:

Increased Combustion Heat: The presence of aluminum nanoparticles in the fuel improves the thermal efficiency of the engine because it adds to the overall amount of heat generated during combustion (Wadhwa, 2024).

Reduction in Emissions: The addition of aluminum nanofluids to diesel fuels reduces hazardous exhaust emissions, particularly smoke and nitrous oxide (NOx) (Wadhwa, 2024). Reductions like these are significant in improving the environmental impact of diesel engines.

Aluminum nanofluids exhibit an increase in combustion efficiency and a reduction in emissions, which demonstrate that they could be a means to enhance the sustainability and efficiency of diesel engines. Key to this success is the catalytic performance of alumina, which creates a thin coating over the nanoparticles and aids in producing hydrogen during combustion, leading to more fuel-efficient combustion with lower emissions.

Nanofluids in Brakes

The application of nanofluids in braking systems presents a significant opportunity to advance the operation and heat management methods used by automobile braking systems. As vehicle aerodynamics improve and drag forces decrease, braking systems increasingly have to manage the heat generated from braking.

The following are important details about brake nanofluids and their advantages:

Improved Thermal Conductivity: When compared to conventional brake fluids, copper-oxide brake nanofluid (CBN) and aluminum-oxide brake nanofluid (AOBN) both show noticeably improved thermal conductivity. The thermal conductivity of AOBN is 1.5 times that of DOT_3 braking fluid, but that of CBN is 1.6 times. Improved heat transmission and a reduction in brake fluid overheating are two benefits of this higher thermal conductivity (Wadhwa, 2024).

Greater Boiling Point: Because CBN and AOBN have greater boiling points than traditional brake fluid, there is less chance of vapor-lock, a safety risk that happens when brake fluid vaporizes and stops the hydraulic system from operating as intended. These nanofluids increase safety and braking performance in high-heat situations by raising the boiling point.

Viscosity Improvement: As high viscosity CBN and AOBN thermoplastics slow down and do not overheat which contributed to fluid integrity/continuity while braking, the high viscosity allows for improved lubrication and heat transfer/distribution that improves performance and safety within the full brake system.

Manufacturing Process: Cutting-edge techniques such as the plasma charging arc system for aluminum oxide and the arc-submerged nanoparticle synthesis system (ASNSS) for copper oxide are used in the creation of these brake nanofluids. These techniques effectively create suspended nanoparticles in a dielectric liquid, improving the characteristics of the nanofluid (Wadhwa, 2024), figure 2 and table 4.

Effect on Braking Safety: The use of CBN and AOBN nanofluids can significantly improve vehicle safety by preventing vapor-lock and enhancing the overall performance of brake fluids, particularly in high-performance or heavy-duty cars.

In order to put the CuO and AlO_3 nanoparticles application in the engine transmission fluid in the context of automated transmission cooling, was examined by Tzeng et al. (Tzeng et al., 2005). They found through their experiments that the best nanofluid for lowering transmission temperatures at various engine speeds (400–1600 rpm) was the CuO nanofluid. This suggests that nanofluids may also help with thermal control in car transmissions. However, while employing nanofluids in these systems, it is crucial to consider issues such particle settling, particle agglomeration, and surface erosion. The optimization of nanofluid compositions and the resolution of these possible problems will remain the main objectives of ongoing research.

Figure 2. Viscosity and Temperature Variations in CBN (Wadhwa, 2024).

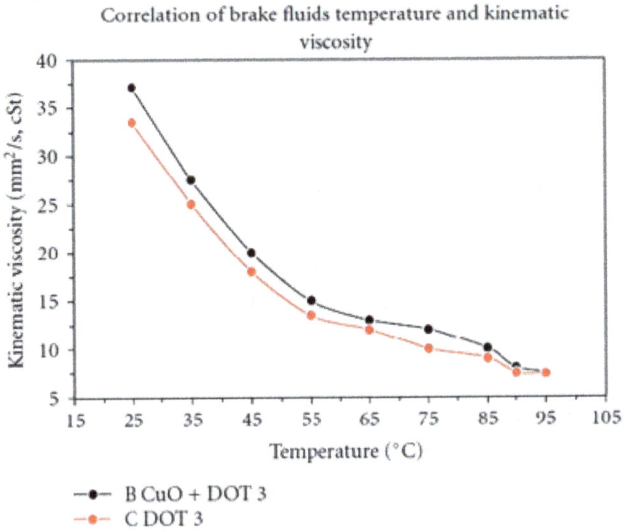

Table 4. Variations in Boiling Point and Thermal Conductivity for CBN and AOBN Nanofluids Based on DOT3 Brake Fluids (Wadhwa, 2024)

Property	**Pure DOT3***	**CBN (2 wt% CuO + DOT3)**	**Pure DOT3***	**AOBN (2 wt% Al_2O_3 + DOT3)**
Boiling Point	269°C	277°C	239°C	247°C
Thermal Conductivity (25°C)	0.029 W/m·°C	0.04 W/m·°C	0.12 W/m·°C	0.18 W/m·°C

*Note: Different DOT3 brake fluid formulations were utilized in the tests.

It has been demonstrated that surface-modified nanoparticles distributed in mineral oils greatly enhance tribological characteristics like load-carrying capacity, wear resistance, and friction reduction in automobile lubrication applications. According to Zhang (Amin Mir, 2025), by lowering wear and friction, these nanoparticles form a protective coating over moving parts, increasing the longevity of mechanical components. Additionally, by increasing heat transfer rates and reducing overheating, nanofluids' enhanced thermal conductivity raises system efficiency overall. These advances in lubrication technology should be a positive reflection for the future of automotive engineering because they provide not only increased longevity and usability of automotive systems, but decreased cost/maintenance, saving energy, and greater fuel efficiency.

Electronic Applications

Nanofluids are being utilized increasingly in cooling systems of microchips and other electronics, because they transfer heat much easier than traditional coolants and have a significantly higher thermal conductivity. Nanofluids are more efficient in heat absorption and transport than traditional coolants, which prevents small electronics from overheating. In micro and nanofluidic applications, nanofluids enable finer control of heat and mass transfer, which can be important in cooling sensors and precision instruments or in lab-on-a-chip systems. Since they enhance the ability to transfer heat in a small volume while using less coolant, they are particularly suited to the quickly evolving field of electronics.

Microchips in Cooling

As microchips get smaller, rapid heat dissipation has become even more important, particularly as the power output of computer chips has increased. It is predicted that next-generation computer chips will output a total of power of more than 300W and localized heat fluxes greater than $10 MW/m^2$. Because of their thermal properties, nanofluids offer a potential alternative to traditional cooling methods, which struggle to keep pace with these high heat demands. Nanofluids—fluids that are augmented with nanoparticles such as copper oxide or diamond—have been shown to transfer heat more than traditional coolants like water. Nanofluids, therefore, could be good options for cooling devices that have high performance demands such as microchips due to their enhanced thermal properties. One application of cooling system where nanofluids have some promise is in oscillating heat pipe (OHP) systems. OHPs are advanced cooling systems that move heat effectively through the movement of fluid. When used with nanofluids, the heat transport properties of an OHP system can be greatly enhanced to improve system performance and it would now be possible for these systems to deal with the high heat fluxes generated by next-generation microchips. By offering control of heat fluxes greater than $10 MW/m^2$ using nanofluids and thin film evaporation, OHP systems offer a practical option to cool a high-power device in future technology (Ma et al., 2006).

The researchers are faced with a problem during their laboratory setup where they have to substitute the metal pipes of the system using materials such as glass or plastic in order to view the oscillations of the OHP's. These substitutions will allow for the analysis of the oscillation, but can also change the systems thermal transport properties, which could affect the accuracy of the experimental results. Arif (Arif, 2006) overcame these challenges, by implementing neutron imaging methods to visualize the liquid flow in a 12-turn nanofluid OHP system. The vast intensity of the neutron beam allowed the researchers to take dynamic pictures with a speed of one frame every 1/30th of a second, utilizing colloidal water suspended diamond nanoparticles, where they capture the fluid movements in real time. With the ability for the experimenters to avoid settling of the nanoparticles, they maintained the uniform dispersion of the nanofluid across the system and the oscillating nanofluid, combined with the higher thermal conductivity of the diamond nanoparticles multiplied the cooling rate efficiency. This uniform dispersion enabled the OHP's capacity to handle high heat fluxes, leading to higher heat transfer efficiency for the OHP. Ma et al (Amin Mir, 2025), carried out similar investigations utilizing diamond nanoparticles in water to enhance a high-performance liquid chromatography (HPLC) system, figure 3.

As shown in their results, the nanofluid reduced the temperature difference between the evaporator and condenser of the OHP from 40.9°C to 24.3°C and input power of 80W, and demonstrated how nanofluids could improve the cooling efficiency in OHP systems. The researchers noted that the oscillating motion of the OHP increased with the heat input but that the temperature difference then plateaued at a specific power input suggesting this was the maximum performance of the system. Despite these challenges, nanofluid OHP systems showed strong thermal performance as compared to pure water, even given the maximum input power of 336W. This supports the idea that nanofluids could significantly enhance the heat transfer performance of OHP systems. Taken together, the results suggest that nanofluids—especially those with diamond nanoparticles—could revolutionize the cooling mechanisms for future microchips or other high-power devices by sufficiently dissipating the heat and allowing devices to operate at higher power levels with lower risk of overheating.

Figure 3. Nanofluid's impact on an OHP's capacity for heat transmission (Amin Mir, 2025).

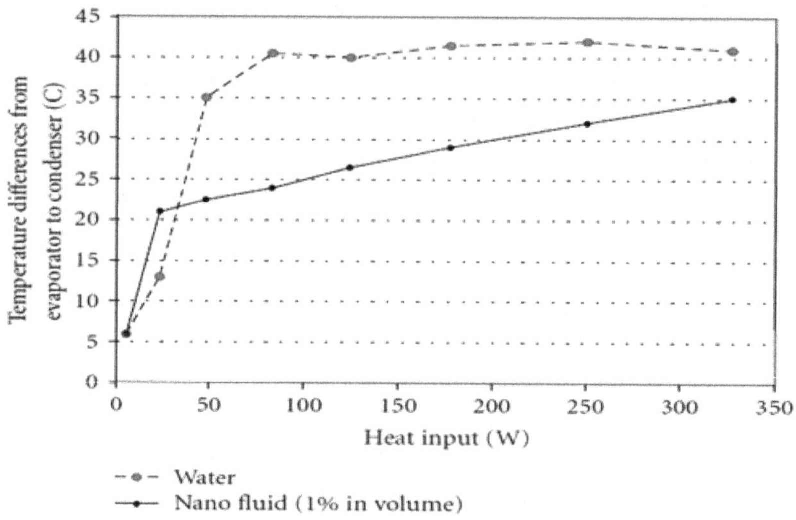

A study by Lin et al. (Lin et al., 2008) that utilized silver nanoparticles in nanofluids in pulsing heat pipes; the outcomes were encouraging. Lin et al. determined that by adding silver nanofluid it greatly improved the heat transfer properties of the heat pipes, indicating that nanofluids can boost thermal performance in such devices. In a separate experimental study, Nguyen et al. (Nguyen et al., 2007) evaluated the heat transfer performance of AlO_3-water nanofluids in closed cooling systems used for microprocessors or electronic devices. The findings of testing indicate that the addition of nanoparticles to distilled water yielded a marked increase in convective heat transfer coefficients. For example, the heat transfer coefficients increased by nearly 40% when compared to pure water at a nanoparticle concentration of 6.8%. Additionally, they also revealed the importance in size of the particle in achieving the best thermal conductivity. Specifically, they concluded that the smaller AlO_3 nanoparticles outperformed their larger counterparts in producing the convective heat transfer coefficients. These studies illustrating the promise of nanofluids will provide the basis for the development of innovative cooling technologies that can adequately service the heat dissipation demands of future microprocessors and high-performance electronic systems.

Fluidic Applications at Microscale

For many micro-scale applications, such as fluidic digital display devices, optical systems, microelectromechanical systems (MEMS), and lab-on-a-chip analysis systems, the manipulation of small quantities of liquid is imperative. Electrowetting, which consists of applying an electric voltage to a liquid droplet to reduce the contact angle, can achieve motion. As a method for controlling near microscale liquid motion, electrowetting-on-dielectric (EWOD) actuation can be quite useful. Vafaei et al. (Motia et al., 2024) have reported that nanofluids can significantly alter surface wettability and possibly surface tension. Specifically, they reported that a goniometer was employed and bismuth telluride nanofluid could substantially alter the surface wetting properties when added to a surface, even at quite small concentrations.

Representations as low as 3×10^{-6} for example raised the contact angle significantly to above 40°, this indicates that a large change in balance of force was achieved very near the triple line where the liquid, air, and solid interface. The contact angle achieved its maximum value with increasing concentrations of the nanofluid, and it subsequently began to fall again. Further the wettability was found to be greatly influenced by nanoparticle size, where smaller nanoparticles were more effective in raising the contact angle. Because at a given concentration smaller nanoparticles have greater surface area due to lower volume which improved their ability to change the liquid surface interaction. These findings raise the potential for devices that rely on precision in manipulation of liquids as they indicate that nanofluids can be tailored to maximize the wetting properties of surfaces as in MEMS and other technologies as seen in figure 4 and 5 (Dash et al., 2007).

Figure 4. Droplet form schematic design (Dash et al., 2007)

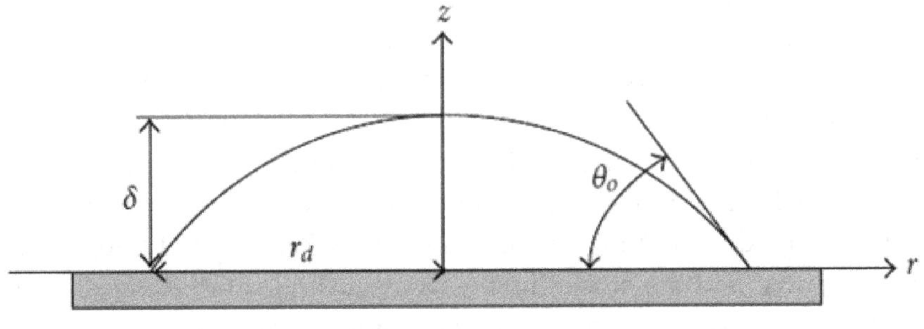

Figure 5. Chnages in contact angle for bismuth telluride nanoparticle concentration on a glass substrate at 10 nm (Dash et al., 2007).

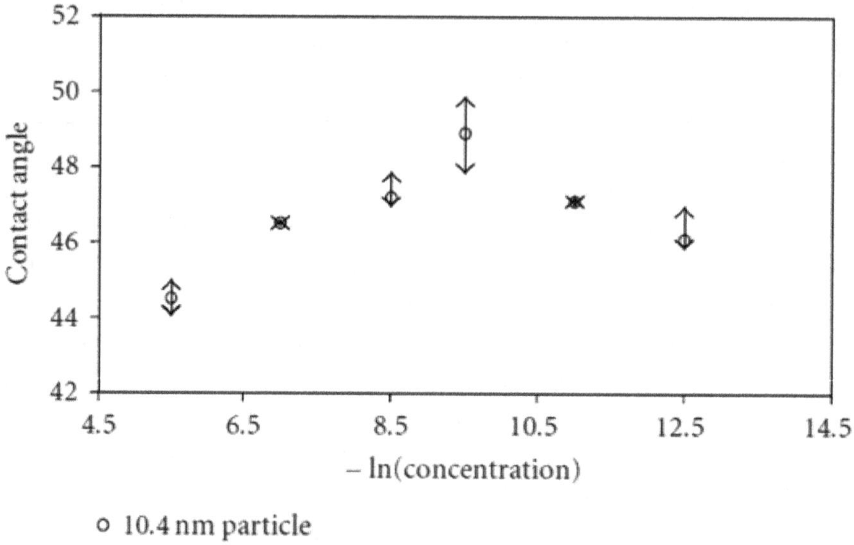

Dash et al. (Dash et al., 2007) showed that nanofluids perform better and are more stable when subjected to electric fields by using the electrowetting on dielectric (EWOD) effect. In their experiment, a silicon wafer coated with Teflon was covered with droplets of water-based solutions containing bismuth telluride nanoparticles. The contact angle changed significantly when an electric field was supplied to the droplet, and this change was noticeably more noticeable when the nanofluid was employed than when it wasn't. This finding demonstrated how well the nanofluids responded to the applied electric field in terms of wettability figure 6 (Dash et al., 2007). Additionally, there were no indications of the contact angle saturation effect, which was observed in solutions containing 0.01 N Na_2SO_4 and thioglycolic acid in deionized water, and the bismuth telluride nanofluid showed improved droplet stability. The capacity to control the contact angle of nanofluid droplets creates a number of possible uses in microsystems, including the effective transportation of liquids inside tiny devices, the creation of novel lens focusing methods for small cameras, and the enhancement of computer chip cooling efficiency. Nanofluids are useful for the upcoming generation of microfluidic and electronic systems because of their improved stability and electric field reactivity.

Figure 6. A nanofluid droplet under applied voltages of (a) 0V and (b) 64.5 V as well as (c) 0V and (d) 40 V. The formation of gas bubbles in the contact line indicates that the Na2SO4 droplet becomes unstable at about 40V (Dash et al., 2007).

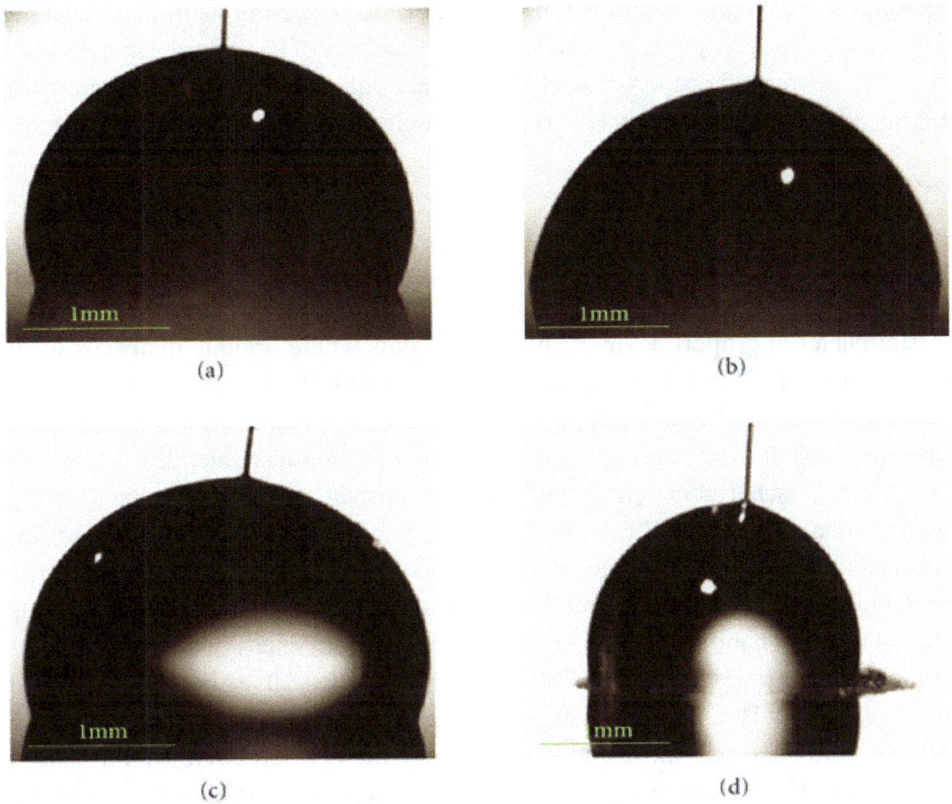

Applications in Biomedicine

Because they can break down bacterial cell walls or obstruct vital microbial functions, some nanoparticles—like silver, copper, and zinc oxide—are recognized to have antibacterial qualities. These nanoparticles can provide nanofluids their antibacterial qualities when they are disseminated in them, which makes them beneficial for a number of uses, such as biomedical devices and cooling systems. Silver nanoparticles, for example, are well known for their antibacterial properties and, when added to nanofluids, can offer a pathogen-reducing or self-cleaning capability in systems where contamination management is crucial. Certain nanoparticles, such as those composed of liposomes or materials coated with polymers, can be engineered to serve as drug delivery systems in addition to their antibacterial capabilities.

Targeted therapy can be accomplished because these nanoparticles can encapsulate drugs and then release them in a controlled manner. Therapeutic drugs delivered directly to the tumor site using these drug delivery nanoparticles have applications in medicine, such as cancer therapy within a nanofluid. The combination of the ability of nanoparticles to improve the heat transfer properties of nanofluids, in addition to previously identified properties such as antibacterial behavior and therapeutic drug release, adds significant potential to nanoparticles in engineering and medicine.

Antimicrobial activity

Researchers have become very interesed in inorganic nanoparticles such as zinc oxide (ZnO), copper oxide (CuO) and silver because of their improved stability and antibacterial properties in harsh environments where organic materials would fail (Zhang et al., 2007). For example, ZnO nanofluids have bacteriostatic activity, and Jalal et al. (Jalal et al., 2010) showed that longer exposure times and higher concentrations of ZnO nanoparticles exhibited a pronounced decreased in the survival ratio of Escherichia coli (E. coli). By trapping intracellular enzymes through the aforementioned capabilities (Mahapatra et al., 2008), they can disrupt cellular mechanisms and eventually lead to bacterial death. Interestingly, in their bulk forms, the inhibitory activity of these metal oxides is not as powerful, emphasizing the significant role that nanoparticles size plays in their effectiveness (Gajjar et al., 2009). Similarly, Lee et al. (Lee et al., 2003) have shown that silver nanoparticles have effective and persistent antibacterial properties, particularly silver nanoparticles in colloidal form, and that these antibacterial properties can endure through multiple wash cycles. This effectiveness of silver nanoparticles against bacteria varies with size; even at very low concentrations, silver nanoparticles are able to effectively eliminate a wide range of bacteria, including even varieties of bacteria that are antibiotic resistant, such as methicillin resistant Staphylococcus aureus (MRSA). Metal oxide and silver based nanofluids, with these qualities, are very valuable for applications that benefit from both thermal conductivity and antibacterial activity.

Nanodrug delivery

Colloidal drug delivery mechanisms have considerably enhanced drug efficiency and have focused on specific cellular functions (Vonarbourg et al., 2006). Nanoparticles, especially small, customized ones are now solubilizing agents and doing more (Singh & Lillard, 2009). Gold nanoparticles (AuNPs) are special because they are non-toxic, versatile vehicles for drug and gene delivery. Gold nanoparticles are stable because of the solid gold core, while surface modifications using monolayers for controlled release of intracellular drug releases are essentially linked to thiol interactions (Akhai, Arif, & Dadhwal, 2025). Nanomagnetic fluids, which are composed of magnetic nanoparticles, have opened the door for drug delivery systems which exploit the magnetic properties of ferrofluid clusters to not only concentrate but target drugs (Nakano et al., 2008). With the potential for drug encapsulation and variable surface coatings, these magnetic nanoparticles have the potential to facilitate drug cargo for a range of therapeutic applications. Carbon nanotubes (CNTs) also have potential uses for drug delivery due to their exceptional electrical conductivity, mechanical strength, elasticity, and chemical stability. CNTs are being functionalized to interact with antibacterial agents such as manganese oxide, enabling targeted delivery. Importantly, despite lacking inherent antibacterial properties, functionalized CNTs have the potential to encapsulate active molecules and release drugs in specific conditions. Vertically aligned carbon nanotubes (VACNTs) are particularly advantageous drug loading and delivery modalities based on their hollow cores and aspect ratio. Nanoparticles can be used to seal the CNT cavities once drugs have been placed inside. These CNTs can release the medicines through stimulus-triggered mechanisms, like pH shifts, once they have penetrated cells (Kostarelos et al., 2007). Because of their structural benefits over randomly oriented CNTs, VACNTs are perfect for enhanced drug delivery systems, especially in transdermal medication and vaccine delivery. In a variety of biomedical applications, these systems have intriguing prospects for more focused, efficient treatments (Bianco et al., 2005).

CONCLUSION

Nanofluids have shown strong potential in diverse applications including industrial cooling, nuclear reactor safety, automotive thermal management, microelectronics, geothermal energy extraction, and biomedical systems because of their ability to enhance heat transfer capability, facilitate adaptive fluid properties and combine functionalities like antimicrobial activity and targeted drug delivery. However, practical real-world application is still subject to long-term stability, environmental safety, batch production, and cost considerations. Future work should therefore be application driven using both experimental and simulation approaches. For industrial cooling and heat exchangers, long-term pilot-scale testing in a closed-loop at different flow rates and thermal loads coupled with computational fluid dynamics (CFD) – population balance modeling will be required to understand nanoparticle aggregation behavior. For nuclear reactor systems, work on the enhancement of CHF using nanofluid in a simulated loop with in-situ characterization of surface morphology should be evaluated alongside multi-scale thermal–hydraulic models that combine computational fluid dynamics and phase-change modeling, useful to determine the potential of annexing nanoparticle deposition and boiling behavior for enhanced cooling in both PWR and BWR systems. For automotive applications, temperature and cycling-based tribological and erosion testing of nanofluids in engine oils, brake fluids, and radiator coolants should be done with standardized test procedures, while comparative thermomechanical simulations need to be continuous validated with realworld driving patterns. Within microelectronics cooling, high heat flux testing of the nanofluid-enhanced oscillating heat pipes and microchannel coolers should occur alongside MD simulations to assess interfacial thermal resistance and CFD modeling to consider micro-scale flow fields. For geothermal and other extreme environment applications, stability and corrosion testing at greater than 500 °C and 10 MPa should be included along with thermo-hydro-mechanical reservoir modeling for the prediction of performance benefits at a large scale. In the areas of biomedical and drug delivery systems, in-vitro and in-vivo studies of cytotoxicity, bio-distribution and controlled drug release should correspond with multiphysics simulations which explicitly model the transport and biochemical interactions of nanoparticles. Directing nanofluid design towards these purposeful performance metrics and leveraging strict experimental validation and high-fidelity simulation frameworks will facilitate scale-out, safe, and sustainable implementation, thus ensuring that nanofluid technologies can evolve from laboratory promise to commercially viable options for energy, industrial and healthcare applications.

ACKNOWLEDGEMENT

This work was supported by the Deanship of Scientific Research, Vice Presidency for Graduate Studies and Scientific Research, King Faisal University, Saudi Arabia Grant No: KFU260859

REFERENCES

Ahmadi, A. A., Arabbeiki, M., Ali, H. M., Goodarzi, M., & Safaei, M. R. (2020). Configuration and Optimization of a Minichannel Using Water–Alumina Nanofluid by Non-Dominated Sorting Genetic Algorithm and Response Surface Method. *Nanomaterials (Basel, Switzerland)*, *10*, 901.

Akhai, S., Abbass, M., Kaur, P., & Kaur, T. (2025). Digital Transformation Across Generations: Robotics and AI in Action. In Impacts of Digital Technologies Across Generations (pp. 23-40). IGI Global Scientific Publishing

Akhai, S., Arif, S. M., & Dadhwal, R. (2025). High-Performance Aluminium Alloys for Aircraft Structures: Advances in Properties, Processing, and Manufacturing. In Innovative Materials for Next-Generation Defense Applications: Cost, Performance, and Mass Production (pp. 61-106). IGI Global Scientific Publishing.

Alsabery, A. I., Ghalambaz, M., Armaghani, T., Chamkha, A., Hashim, I., & Pour, M. S. (2020). Role of Rotating Cylinder toward Mixed Convection inside a Wavy Heated Cavity via Two-Phase Nanofluid Concept. *Nanomaterials (Basel, Switzerland)*, *10*, 1138.

Amburi, P. K., Senthilkumar, G., & Nithya, A. (2023). Novel use of CuO nanoparticles additive for improving thermal conductivity of MgO/water and MWCNT/water nanofluids. *Journal of Thermal Analysis and Calorimetry*, *148*(19), 10389–10398.

Amin Mir, M. (2025). Bimal Krishna Banik, Synthesis of Schiff base ligands under environment friendly conditions: A systematic review. *Inorganic Chemistry Communications*, *174*, 1.

Arif, M. "Neutron imaging for fuel cell research," in *Proceedings of the Imaging and Neutron Workshop*, Oak Ridge, Tenn, USA, October 2006.

Asadi, A., Molana, M., & Ghasemiasl, R. (2020). A New Thermal Conductivity Model and Two-Phase Mixed Convection of CuO–Water Nanofluids in a Novel I-Shaped Porous Cavity Heated by Oriented Triangular Hot Block. *Nanomaterials (Basel, Switzerland)*, *10*, 2219.

Bianco, A., Kostarelos, K., & Prato, M. (2005). Applications of carbon nanotubes in drug delivery. *Current Opinion in Chemical Biology*, *9*(6), 674–679.

Boungiorno, J., Hu, L.-W., Kim, S. J., Hannink, R., Truong, B., & Forrest, E. (2008). Nanofluids for enhanced economics and safety of nuclear reactors: An evaluation of the potential features issues, and research gaps. *Nuclear Technology*, *162*(1), 80–91.

Buongiorno, J., Hu, L. W., Apostolakis, G., Hannink, R., Lucas, T., & Chupin, A. (2009). A feasibility assessment of the use of nanofluids to enhance the in-vessel retention capability in light-water reactors. *Nuclear Engineering and Design*, *239*(5), 941–948.

Chopkar, M., Das, P. K., & Manna, I. (2006). Synthesis and characterization of nanofluid for advanced heat transfer applications. *Scripta Materialia*, *55*(6), 549–552.

Dash, R. K., Borca-Tasciuc, T., Purkayastha, A., & Ramanath, G. "Electrowetting on dielectric-actuation of microdroplets of aqueous bismuth telluride nanoparticle suspensions," Nanotechnology, vol. 18, no. 47, Article ID 475711, 6 pages, 2007.

Donzelli, G., Cerbino, R., & Vailati, A. "Bistable heat transfer in a nanofluid," Physical Review Letters, vol. 102, no. 10, Article ID 104503, 4 pages, 2009.

Gajjar, P., Pettee, B., Britt, D. W., Huang, W., Johnson, W. P., & Anderson, A. J. (2009). Antimicrobial activities of commercial nanoparticles against an environmental soil microbe, pseudomonas putida KT2440. *Journal of Biological Engineering*, *3*(1), 1–13.

García-Beltrán, G., & Mercado-Zuniga, C. (2020). Navigation of Silver/Carbon Nano-antennas in Organic Fluids Explored by a Two-Wave Mixing. *Nanomaterials (Basel, Switzerland)*, *10*, 1886.

Han, Z. H., Cao, F. Y., & Yang, B. "Synthesis and thermal characterization of phase-changeable indium/poly alpha olefin nanofluids," Applied Physics Letters, vol. 92, no. 24, Article ID 243104, 3 pages, 2008.

Hashemi, S., & Akhavan-Behabadi, M. (2012). An empirical study on heat transfer and pressure drop characteristics of CuO–base oil nanofluid flow in a horizontal helically coiled tube under constant heat flux. *International Communications in Heat and Mass Transfer*, *39*(1), 144–151.

Jalal, R., Goharshadi, E. K., Abareshi, M., Moosavi, M., Yousefi, A., & Nancarrow, P. (2010). ZnO nanofluids: Green synthesis, characterization, and antibacterial activity. *Materials Chemistry and Physics*, *121*(1–2), 198–201.

Jamal-Abad, M. T., Zamzamian, A., & Dehghan, M. (2013). Experimental studies on the heat transfer and pressure drop characteristics of Cu–water and Al–water nanofluids in a spiral coil. *Experimental Thermal and Fluid Science*, *47*, 206–212.

Khan, M. Z. U., Uddin, E., Akbar, B., Akram, N., Naqvi, A. A., Sajid, M., & Ali, Z. (2020). Investigation of Heat Transfer and Pressure Drop in Microchannel Heat Sink Using Al2O3 and ZrO2 Nanofluids. *Nanomaterials (Basel, Switzerland)*, *10*, 1796.

Kostarelos, K., Lacerda, L., Pastorin, G., Wu, W., Wieckowski, S., Luangsivilay, J., Godefroy, S., Pantarotto, D., Briand, J.-P., & Muller, S.. (2007). Cellular uptake of functionalized carbon nanotubes is independent of functional group and cell type. *Nature Nanotechnology*, *2*(2), 108–113.

Kumar, J. S., Senthilkumar, G., & Ramachandran, S. (2023). Novel strategy of mixing MgO in CuO/water nanofluid for thermal conductivity improvement: Experimental study. *Case Studies in Thermal Engineering*, *52*, 103723.

Lee, H. J., Yeo, S. Y., & Jeong, S. H. (2003). Antibacterial effect of nanosized silver colloidal solution on textile fabrics. *Journal of Materials Science*, *38*, 2199–2204.

Lin, Y.-H., Kang, S.-W., & Chen, H.-L. (2008). Effect of silver nanofluid on pulsating heat pipe thermal performance. *Applied Thermal Engineering*, *28*(11-12), 1312–1317.

Lo, C.-H., Tsung, T.-T., Chen, L.-C., Su, C.-H., & Lin, H.-M. (2005). Fabrication of copper oxide nanofluid using submerged arc nanoparticle synthesis system (SANSS). *Journal of Nanoparticle Research*, *7*, 313–320.

Ma, H. B., Wilson, C., Yu, Q., Park, K., Choi, U. S., & Tirumala, M. (2006). An experimental investigation of heat transport capability in a nanofluid oscillating heat pipe. *Journal of Heat Transfer*, *128*(11), 1213–1216.

Mahapatra, O., Bhagat, M., Gopalakrishnan, C., & Arunachalam, K. D. (2008). Ultrafine dispersed CuO nanoparticles and their antibacterial activity. *Journal of Experimental Nanoscience*, *3*(3), 185–193.

Mir, M. A., Chang, S. K., & Hefni, D. (2024). A comprehensive review on challenges and choices of food waste in Saudi Arabia: Exploring environmental and economic impacts. *Environmental Systems Research*, *13*, 40.

Mir, M. A., Mahmoud, M., Andrews, K., Nayak, A., Kukretee, N., & Kotnala, S. (2025). Experimental, Spectroscopic, Computational and Molecular Studies of Plant Growth Promotor 2-Naphthylacetic Acid. *Physics and Chemistry of Solid State*, *26*(1), 158–165.

Motia, K., Kumar, R., & Akhai, S. (2024). AI and Smart Manufacturing: Building Industry 4.0. In Modern Management Science Practices in the Age of AI (pp. 1-28). IGI Global.

Nakano, M., Matsuura, H., Ju, D.-Y., Kumazawa, T., Kimura, S., Uozumi, Y., Tonohata, N., Koide, K., Noda, N., Bian, P., ... Drug delivery system using nanomagnetic fluid, in: 2008 3rd International Conference on Innovative Computing Information and Control, IEEE, 2008, p. 338.

Nguyen, C. T., Roy, G., Gauthier, C., & Galanis, N. (2007). Heat transfer enhancement using Al2O3-water nanofluid for an electronic liquid cooling system. *Applied Thermal Engineering*, *27*(8-9), 1501–1506.

Sadeghzadeh, M., Maddah, H., & Ahmadi, M. H. (2020). Prediction of Thermo-Physical Properties of TiO2-Al2O3/Water Nanoparticles by Using Artificial Neural Network. *Nanomaterials (Basel, Switzerland)*, *10*, 697.

Singh, D., Toutbort, J., & Chen, G.. (2006). Heavy vehicle systems optimization merit review and peer evaluation. In *Annual Report*. Argonne National Laboratory.

Singh, R., & Lillard, J. W.Jr. (2009). Nanoparticle-based targeted drug delivery. *Experimental and Molecular Pathology*, *86*(3), 215–223.

Solangi, K., Kazi, S., Luhur, M., Badarudin, A., Amiri, A., Sadri, R., Zubir, M., Gharehkhani, S., & Teng, K. H. (2015). A comprehensive review of thermo-physical properties and convective heat transfer to nanofluids. *Energy*, *89*, 1065–1086.

Suresh, S., Chandrasekar, M., Selvakumar, P., & Page, T. (2012). Experimental studies on heat transfer and friction factor characteristics of Al2O3/water nanofluid under laminar flow with spiralled rod inserts. *Int. J. Nanopart.*, *5*(1), 37–55.

Tzeng, S.-C., Lin, C.-W., & Huang, K. D. (2005). Heat transfer enhancement of nanofluids in rotary blade coupling of four wheel-drive vehicles. *Acta Mechanica*, *179*(1-2), 11–23.

Vonarbourg, A., Passirani, C., Saulnier, P., & Benoit, J.-P. (2006). Parameters influencing the steal thiness of colloidal drug delivery systems. *Biomaterials*, *27*(24), 4356–4373.

Wadhwa, S. A., Mahapara, A., Akhai, S., Kumar, D., & Kumar, P. (2024). Integrating Taguchi Optimization for Multi-Criteria Decision Making in Engineering Applications. In Recent Theories and Applications for Multi-Criteria Decision-Making (Vol. 1, No. 1, pp. 125-150). IGI Global.

Xie, S., Zhang, Y., Song, Y., Ge, F., & Huang, X. (2020). Comparison of the Corrosion Behavior of Brass in TiO2 and Al2O3. Nanofluids. *Nanomaterials (Basel, Switzerland), 10*, 1046.

Zaidi, N., Mir, M. A., Chang, S. K., Abdelli, N., Hasnain, S. M., Ali Khan, M. A., & Andrews, K. (2025). Pharmaceuticals and personal care products as emerging contaminants: Environmental fate, detection, and mitigation strategies. *International Journal of Environmental Analytical Chemistry*, •••, 1–29.

Zhang, L., Jiang, Y., Ding, Y., Povey, M., & York, D. (2007). Investigation into the antibacterial behaviour of suspensions of ZnO nanoparticles (ZnO nanofluids). *Journal of Nanoparticle Research*, *9*, 479–489.

Zhu, X.-Y., & Wang, B.-R. (2021). Novel Nanofluidic Cells Based on Nanowires and Nanotubes for Advanced Chemical and Bio-Sensing Applications. *Nanomaterials (Basel, Switzerland), 11*, 90.

Chapter 7

AI-Driven Modeling and Optimization of 3D Graphene-MXene Hybrid Nanocomposites

Cağdaş Kızıl
https://orcid.org/0000-0003-1727-9610
Antalya Chamber of Commerce and Industry, Project Development and State Support Office, Antalya, Turkey

Ali İhsan Kömür
https://orcid.org/0000-0001-5251-8328
Faculty of Science, Department of Chemistry, Akdeniz University, Antalya, Turkey

Onur Karaman
Vocational School of Health Services, Medical Imaging Techniques, Akdeniz University, Antalya, Turkey

Ceren Karaman
https://orcid.org/0000-0001-9148-7253
Vocational School of Technical Sciences, Department of Electricity and Energy, Akdeniz University, Antalya, Turkey

ABSTRACT

This chapter focuses on the simulation and optimization of electrochemical energy storage systems (EES) employing 3D graphene-MXene hybrid nanocomposites as electrode materials through the application of artificial intelligence (AI) and advanced materials design. In this regard, the overview of the possibilities of using

DOI: 10.4018/979-8-3373-3556-8.ch007

different AI models to predict basic physicochemical structure-property relationships of materials and enhance performance parameters like specific capacity and cycling stability is provided. The chapter also explores fundamental themes of materials engineering including interlayer spacing, material porosity, and interface engineering, and evaluates the role that AI-based optimization can have in the design and production of high-performance hybrid nanostructures in EES. In addition, the fact that AI models can be used to design the material in the material design process is also mentioned as the means of ensuring that the structural integrity and stability of the electrode material are maintained and that its electrochemical performance is improved. The basic procedures of the AI model development and verification, such as preprocessing data, training the model, experimental verification, and feedback, are also addressed in detail. Moreover, the synthesis step that requires fundamental materials engineering issues like scalability, are also discussed along with other emerging issues and trends of designing EES. Finally, this section provides critical information on how AI can be used in the creation of next generation advanced hybrid nanostructures in high-performance EESs.

1. INTRODUCTION

Due to rapidly developing technology, population density, and other factors, global energy demand is increasing, and the transition to renewable energy sources is accelerating. However, due to the intermittency of these resources, the design and engineering of high-performance, state-of-the-art electrochemical energy storage systems (EES) has become a priority research objective. The performance characteristics of EESs must also be improved in order to realize designed high-tech objects such as grid-scale storage, electric vehicles, and portable electronic devices. Therefore, a crucial factor in this regard is the design and optimization of the electrode material that forms the foundation of EESs.

The electrochemical performance of EES directly depends on the electrical conductivity, structural stability, mechanical strength, porosity, electrochemically accessible surface area specific surface area, and other characteristics of the electrode material (C. Liu et al., 2010). Hence, the development of hybrid advanced nanomaterials using AI-aided design methods has significant potential in being a revolutionizing strategy to boost the desired materia distributions with high-performance gains (Nagaswarupa et al., 2024). As reported in recent studies, metal metal oxide, conductive polymer, carbon based nanomaterials, and their derivatives are presented as high-performance electrode materials to various electrochemical energy storage (EES) applications. It is better to apply three-dimensional (3D) graphene-MXene hybrid nanocomposites as the most encouraging representatives of hybrid nanostructures. These architectures integrate the outstanding physicochemical properties of the 3D graphene architecture based on carbon-based nanomaterials with the adjustable properties of 2D architecture of MXene (transition metal carbides and nitrides). Simulations have shown that EESs based on these novel hybrid nanostructures possess certain improved performance features such as high energy-density, capacity and cyclic stability.

This was accredited to their inherited physicochemical properties which were good electrical conductivity, high specific surface area, structural stability, and mechanical strength (Shahzad et al., 2020; Xiao et al., 2023). The hybrid nanostructure of 3D graphene and MXene increases the number of electrochemically active sites and the surface area in a synergistic manner. This enables rapid ion diffusion and charge transfer, and produces long-term high performance parameters of capacity, power density, and energy density (Ab Alim et al., 2019; Ma et al., 2018). It is still possible to achieve the best conductivity, pore distribution, surface area and ion transport channels in designing and optimizing nanostructures. The nature of material needed in every EES setup can marginally differ and this makes the optimization process hard. This can be time consuming, very costly and a lot of resources are utilized in doing things the old fashioned way, through trial and error. One of the new directions in that strategy is the application of AI-based predictive performance tools and material design. In very large and complex data sets, the influence of the structure on performance can be discovered using deep neural networks (DNNs), data-driven modeling, and machine learning (ML) algorithms. A quality AI model can rapidly identify nanomaterial candidates with the appropriate composition and structure to deliver the appropriate electrochemical performance to be used in energy storage applications. This reduces by a lot the time, materials, and trial-and-error required to design electrode materials. It also makes it possible to design high-performance metrics that are specific to the EES application of interest (Pai et al., 2025; Sarker, 2022). Computer-aided approaches to designing and synthesizing hybrid nanostructures have developed at a fast rate, as evidenced by a survey of recent literature. It means that the surface interaction and some architectural improvements can be optimized to improve the electrochemical characteristics of the final structure. It has been demonstrated that this paradigm shift, which encompasses a variety of interdisciplinary domains like chemistry, physics, energy systems engineering, and materials engineering, speeds up the design and implementation process, lowers experimental costs, and improves the accuracy of the intended results. Therefore, it can be stated that AI-aided material design strategies are potent in the creation of the next generation, high-performance EES. *Figure 1* illustrates that 3D graphene–MXene hybrid nanocomposites may be optimized using AI-based modeling techniques, enhancing important material characteristics as shape, surface area, defect management, and ion transport efficiency. The performance of several energy storage systems, such as Li-ion, Na-ion, Li–S batteries, and supercapacitors, can then be improved by these improvements.

Figure 1. The integration of 3D graphene-MXenes hybrid nanocomposites in state-of-the-art EES systems by AI-driven simulation and optimization.

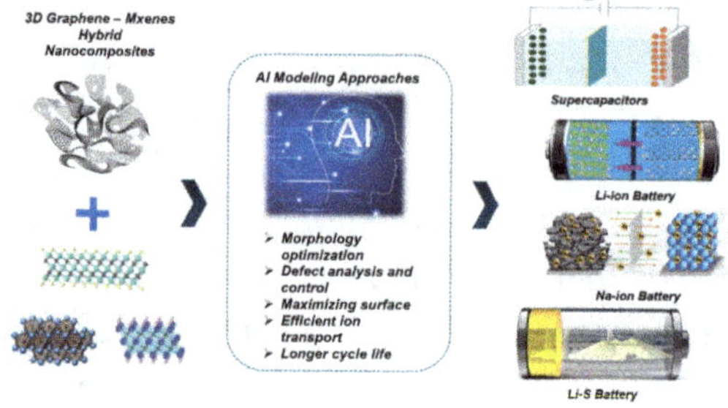

This chapter discusses in detail the convergence of advanced energy storage (EES) systems, AI-based optimization methods, and 3D graphene-MXene hybrid nanocomposites. It attracts the attention to the potential of developing new hybrid nanomaterials with the integration of AI-based approaches to enhance high-performance energy storage. The primary areas of interest are material design frameworks, AI modeling techniques, structural optimization techniques and performance evaluation techniques. Moreover, the chapter discusses new challenges, new research trends, and possible directions of the field along with providing essential information. In order to demonstrate the enormous potential of AI's quick invention and use of state-of-the-art energy storage technologies, this paper methodically examines recent developments.

2. MATERIALS FRAMEWORK

2.1. Structure and Properties of 3D Graphene

Because of its exceptional electrical, mechanical, and thermal properties, graphene—a two-dimensional (2D) substance made up of a single layer of carbon atoms organized in a hexagonal lattice—has attracted a lot of interest in electrochemical energy storage. Nevertheless, the inherent propensity of the material to restack and aggregate during the fabrication of electrodes frequently gives rise to a diminution of surface area and an impairment of accessibility to active sites, has a deleterious effect on the material's electrochemical effectivness in real-world applications (Martínez-Hincapié et al., 2024). The development of 3D graphene architectures has been enabled a means of overcoming the limitations previously presented. Graphene foams, aerogels, and heterogeneously distributed porous structures (micro, meso, and macro) are obtained by modifying the graphene structure using physical and chemical methods. These processes provide hierarchical porosity, resulting in high strength and mechanical resistance, and 2D graphene has been shown to largely retain its physical and chemical properties (Zhi et al., 2021). In 3D graphene, the interconnection of graphene layers with cross-linking agents allows pore opening. This enhances the electron transport networks, which allows rapid diffusion of ions. These are necessary to high-speed ESS. In addition, 3D graphene has a strong mechanical strength that allows it to withstand volume changes during charge-discharge cycle, which makes it more stable and enhances its cycle life (Le Ferrand et al., 2020).

One important use of 3D graphene's chemical and physical modification is surface functionalization. The material can be functionalized with hydrophilic oxygen-containing groups, such as hydroxyl, carbonyl, carboxyl, epoxy, ester, and ether, and doped with nitrogen or other heteroatoms, such as boron, phosphorus, sulfur, silicon, or iron, to improve surface reactivity and wettability. The processes can also enhance interfacial compatibility of 3D graphene with other components, e.g. MXenes or polymer binders (G. Li et al., 2023). The electronic structure of 3D graphene provides a chance to control the vacancies, Stone-Wales defects or edges defects to obtain a more active electrochemical performance (Bhatt et al., 2022). On the one hand, it is crucial to make the spacing between the layers and the variation of pore sizes as efficient as possible to enhance the electrochemical properties because they determine the pace at which ions travel and the ease with which electrolytes can reach them (Yoo et al., 2011). Many people believe that 3D graphene is the best material to use as a scaffold for building hybrid energy storage technologies since it has great electrochemical characteristics and is very flexible. The topic is combined with MXenes to make new multifunctional nanocomposites that can be used in next-generation EES platforms because of the properties indicated aforementioned.

2.2 Properties and Functional Roles of MXenes

By selectively etching their MAX phase precursors—where M stands for a transition metal, A for an element like Si or Al, and X for carbon and/or nitrogen—MXenes, a novel class of two-dimensional transition metal carbides, nitrides, and carbonitrides, are created. The general formula for these compounds, which can be expressed as Mn+1XnTx (where T indicates surface terminations like –OH, –F, or –O), reflects their adaptability in respect of chemical composition and surface functionality (Ullah et al., 2024). Another prominent characteristic of MXenes is metallic or semi-metallic conductivity, the origin of which is related to the involvement of d-orbitals of transition metals and allows rapid electron transport in ESS (S. Liu et al., 2025). Xenes are water loving and highly electrically conductive, and so ions can easily reach them and electrolytes can pass through. These factors are extremely important to electrochemical systems that act at a fast pace. Redox activity is also brought about by the surface terminations of these compounds, providing them with pseudocapacitive properties and greater volumetric capacitance (S. Liu et al., 2025). Layered compounds called xenes can hold a range of ions, including Li, Na, and K. However, because of their strong van der Waals contacts, they tend to restack, just like graphene. As a result of this restacking, ion mobility is hampered since less surface area is available. Other methods to prevent restacking have been proposed, including interlayer pillaring, functional intercalation, and carbon nanotube and conductive polymer mixing (Maughan et al., 2020). How MXenes are constructed and what they consist of matters to how effectively they perform in electrochemical applications. Such as, MXene Ti 3 C 2 T x is well-known and has surface-terminations that are readily accessible and interlayer spacing that can be tuned, which allows it to exhibit a variety of charge storage capacities (Ping et al., 2023). The electronic structure, work function, and ion intercalation energetics are all directly impacted by the stoichiometry and the transition metal selection. Xenes can also perform two functions in composites: they can either carry charge and be conductive additives that facilitate the movement of charge between electrodes (Zhao et al., 2024).

MXenes are mechanically flexible, high capacitance, and redox-able surfaces in hybrid systems. These properties maximize the conduction structure and porous geometry of 3D graphene (Hegazy et al., 2024). Electrochemical performance, cycle stability, specific capacity, and rate capability are all improved when the two materials are combined. All these advantages indicate the significance of MXenes in the future of the next generation of composite electrodes in high-performance electrochemical energy storage devices (Bansal et al., 2024).

2.3 Hybridization Strategies for 3D Graphene–MXene Hybrid Nanocomposites

To overcome the drawbacks of single materials, the combination of 3D graphene and MXene has emerged as a novel method of material design, which takes advantage of synergistic properties of the two materials. 3D graphene possesses a wonderful structure, which is conductive and stable. Xenes, in contrast, possess redox-active surfaces, large volumetric capacitance and chemistry between layers that can be changed (Mullani et al., 2025). The effective mixture of the two nanomaterials forms structures that have multi-functions and complementary electrochemical properties.

A number of synthetic approaches can be applied to produce graphene-MXene nanocomposites in order to create microstructures and improve surfaces. For example, hydrothermal self-assembly allows MXenes to be distributed and developed in situ within a three-dimensional graphitic scaffold. By embedding MXenes within the electrically conductive graphene framework, this method produces interconnected porous networks. This enhances the electron transport strength and routes (Iravani, Zarepour, et al., 2024). Freeze-drying processes are also common procedures used by people to maintain hierarchical porosity and prevent the structure collapsing as the solvent is removed (Indira & Abhitha, 2023). Graphene oxide (GO) is a common starting material in this case, after which the conductivity is restored by thermal or chemical reduction. GO has also been used to form electrostatic interactions with the negatively charged MXene sheets to self-assemble into layered or scaffold-like hybrid materials (Zhou et al., 2020). Interactions between functional groups (e.g., -OH, -O, C=O, C-O-C, O=C-O, C-OH) on graphene oxide and MXene via electrostatic and hydrogen bonding have been found to play an important role in the formation of stable composites (M. Liu et al., 2019; Woo et al., 2024). In addition, covalent or non-covalent functionalization of interfacial engineering has been shown to be an essential component of charge transfer, mechanical adhesion, ion transport across interfaces (M. Liu et al., 2025). As an example, the interlayer spacers like polymers, metal oxides, and carbon dots have been reported to inhibit the restacking of MXenes and change the pore size distribution within the graphene network (Bandpey & Barz, 2024).

Due to the flexibility of hybridization techniques, they can be employed to change the structure of the composite both at the macroscopic level of electrode porosity and at the nanoscale level of interface chemistry. Because they work better electrochemically than their single-component counterparts, 3D graphene-MXene hybrids are the best substrate for high-efficiency ESS.

2.4. Electrochemical Characteristics Relevant to Energy Storage

The intricate interaction of structural, electrical, and interfacial variables determines the electrochemical performance of 3D graphene–MXene hybrid nanocomposites. The designed said materials have been optimized to prioritize the performance metrics of specific capacity, energy and power density, rate capability, and long-term cycling stability all of which are of paramount importance to next-generation electrochemical ESS (Kurc et al., 2021).

Specific capacity that is often determined as mAh·g 1 is found to be severely affected by the quantity and accessibility of active sites within the composite (Hafiz et al., 2025). Though the high faradaic charge storage is due to the pseudocapacitive properties of MXenes, the electric double-layer capacitance is possible because of the huge surface area of the 3D graphene. Such mechanisms have a synergistic effect that results in high charge storage capacities and high energy density, particularly when the interlayer spacing and porosity are optimized to allow fast diffusion of ions (Y. Guo et al., 2017). The rate capability of a solution is influenced by ionic and electronic conductivity (Y. Guo et al., 2017). In spite of the fact that the layered structure of MXenes allows a higher ion intercalation, the 3D graphene conduction network is permanent, which ensures rapid electron transport. The outcome of the convergence of these features is the achievement of high charge/discharge rates with low internal resistance. Furthermore, the existence of interconnected meso- and macropores allows the penetration of electrolytes and minimizes the pathways of ion transportation, hence leading to the enhanced rate performance (Cheng et al., 2019). Cycling stability and mechanical robustness are also important. The graphene 3D structure is resilient and flexible and can thus accommodate volume changes during repeated cycles without the extreme degradation of the structure (Subrahmanian & Narayanan, 2025). Moreover, MXenes are shown to be mechanically strengthening and to exhibit stable redox behavior when used in an appropriate manner (Jussambayev et al., 2025). In battery-like devices, such as supercapacitors, the lifetime of operation is long-term because structural failure is to be avoided and electrode pulverization is minimized.

Table 1. Performance comparing the cycle stability and capacity of similar materials utilized in different ESS.

System	Materials	Cycling Stability (%)	Ref.	Specific Capacity/ Capacitance ($mAh.g^{-1}/F.g^{-1}$)	
Supercapacitors	3D N-doped Graphene	2400 cycles	(Thiruppathi et al., 2020)	262 $F.g^{-1}$	
Li-ion Battery	3D Graphene/MoS_2	100 cycles	(Wenelska et al., 2022)	783 $mAh.g^{-1}$	
Li–S Battery	MXene Nanosheet/ Carbon-Nanotube	>300 cycles	(Liang et al., 2017)	1216 $mAh.g^{-1}$	
Li-ion Battery	3D Graphene like Nanosheets	-	(P. Guo et al., 2009)	672 $mAh.g^{-1}$	

continued on following page

Table 1. Continued

System	Materials	Cycling Stability (%)	Ref.	Specific Capacity/ Capacitance ($mAh.g^{-1}/F.g^{-1}$)	
Li-ion Battery	3D N-doped Graphene	500 cycles	(Mo et al., 2019)	1138 $mAh.g^{-1}$	
Supercapacitors	Mxene/Graphene	20 000 cycles	(Yan et al., 2017)	335.4 $F.g^{-1}$	
Na/K Ion Storage	Titanate Nanosheets/ Graphene	10000 cycles	(Zeng et al., 2018)	72 $mAh.g^{-1}$	
Lithium Storage	3D Porous Structure (MXene/rGO)	1000 cycles	(Ma et al., 2018)	335.5 $mAh.g^{-1}$	

Table 1 displays the multiple performance results in great detail of some range of ESS that are based on the variety of materials and suitability with their eigenvalues in the electrochemical part. These systems consisting of MXene/Graphene and 3D N-doped Graphene supercapacitors with capacitance of 335.4 F g 1 and 262 F g 1, respectively, demonstrate extremely high cycling stability that reaches 20,000 cycles. Such a LiS battery containing MXene nanosheets and carbon nanotubes possesses a specific capacity of 1216 mAh•g-1, exceeding the capacity of the traditional Li-ion batteries incorporating 3D N-doped graphene (1138 mAh•g-1). In particular, there are higher specific capacities available with lithium based systems. Conversely, Na/K-ion storage devices have significantly better cycling characteristics when compared with the other lithium-ion technologies, which will be able to cycle above 10,000 times even after a lower specific capacity of 72 mAh•g 1. These results overall indicate the inherent trade-offs between particular capacity/capacitance and cycle stability of various means of energy storage. Materials engineering is clearly a key element in performance optimization and the degree to which the AI technology might be applied to these advances is described in the following sections.

Finally, by 3D graphene and MXene hybridization, the obtained composites would exhibit enhanced electrochemical properties, thus, would be suitable in numerous energy storage devices. Their universal design and multi-purpose performance can be of big use in the lithium-ion, sodium-ion, and lithium-sulfur battery, hybrid storage systems, and supercapacitors.

2.5 Structure–Property Relationships and Performance Optimization in Graphene–MXene Hybrid Nanocomposites

The effectiveness of graphene -MXene hybrid nanocomposites is determined by the interaction of their interdependent structural and physicochemical properties as well as their individual material performances. With the current advancements of optimization with the aid of AI, this section will provide a comprehensive outlook of important design aspects, especially as it relates to the electrochemical energy storage.

The material framework under study is primarily concerned with the structural properties and the design variables that contribute to the performance of graphene-based MXene nanocomposites and has several essential elements. These materials have exhibited special physicochemical qualities, through the available literatures. The properties can be modified by varying synthesis and processing techniques (Vaishag & Noh, 2024). These structural components need to be enhanced, particularly when used with AI, to allow greater energy storage capacities and enhanced electrochemical performance.

The usefulness and performance of graphene-based structures in composite materials are influenced by a number of crucial technical aspects. The term surface functionalization is used to describe the intentional modification of graphene surfaces to 3D structures through a chemical or physical process (VS et al., 2023). The specified methodology is used with the aim of improving certain qualities and achieving compatibility with other materials. Due to these intrinsic characteristics, the interlayer spacing is the determinant of the ion transport mobility and the electrochemical performance of the material and the overall performance of energy storage devices (Bilisik & Akter, 2022; Jayaramulu et al., 2022). Besides, defect engineering may be considered an advanced strategy of the deliberate addition and management of structural flaws in the graphene lattice, which can considerably influence the electronic, mechanical, and chemical properties of the material (Bilisik & Akter, 2022; Jayaramulu et al., 2022). Optimizing several interdependent variables at once is necessary to advance 3D graphene-based materials for next-generation energy storage systems (ESS). To combine graphene structures with MXene components and develop next-generation composites with exceptional energy storage performance, a good grasp of these basic concepts is necessary.

The MXenes components are just as important as graphene structures when it comes to making superior composite materials. This means that there are many important parts that have a big effect on how well these materials work. For example, part of the selection process is making an informed choice about transition metal carbides, nitrides, or carbonitrides (MXenes) with specified stoichiometric ratios and combinations of elements (Cui et al., 2023; Saharudin et al., 2023; J. Yang et al., 2019). This has shown a direct effect on the electrical properties and reactivity of the final products (P. Li et al., 2024; Saharudin et al., 2023). Interface engineering, charge transfer mechanisms, and structural stability across various operating conditions necessitate substantial focus (Iravani, Zarepour, et al., 2024; Nahirniak et al., 2023; Pandey et al., 2024). The combined effect of these traits is bad for the entire performance of composite materials, including energy density, power transmission, and long-term stability (Bhat et al., 2021). The utilization of AI in materials science is a groundbreaking method for understanding and improving complex material systems, taking into account the previously mentioned elements. In line with this, the abilities of advanced machine learning (ML) algorithms and improved computational methods to analyze and predict the performance of graphene-based MXenes composites have been shown with unmatched accuracy (Rong et al., 2023; Shariq et al., 2024). The proposed AI-driven methodology can also quickly find the best material configurations and give a full picture of how structural attributes and performance measures are related (Gouveia et al., 2025; Shrestha et al., 2024). Artificial intelligence and materials engineering can work together to build highly good ESS. The partnership would save time and money when it comes to checking the research study.

Table 2 is a table that compares the performance measurements of the expected future 3D Graphene-MXenes composites, with a focus on the findings before and after AI optimization. A thorough review of available literature shows that the applications of the advanced materials in AI-optimized batteries will improve the specific capacity and the energy density at moderate and significant levels in terms of charge and performance (G. Huang et al., 2024; Xiong et al., 2024). As an example, one of the first studies to investigate the use of 3D printing technology to create carbon micro lattice electrodes to supercapacitors was carried out by Yang et al., which aimed at enhancing certain electrochemical properties (H. Yang et al., 2023). The article describes a technique that uses machine learning to precisely evaluate the influence of structural elements like carbon nanotubes and graphene. The correlation between the spatial performance of the electrode structures is established by the combination of random forest (RF) technique and the choice of only nine experimental points. This allows extracting important information and concealing relations in a short time. Electrodes with the required performance can be created by choosing the right structural features from the model. Interestingly, it has been found that field performance varies from 0.032 to 1.6 F cm^{-2} (Figure 2). Furthermore, performance reliance on different factors was evaluated using finite element analysis (FEA) and electrochemical surface area (ECSA), which demonstrated good agreement with the model predictions. Through AI optimization for desired performance, the concept put forward in this study can significantly improve electrode performances as well as the efficiency of creating new electrode structures.

Figure 2. (a) The original training data with an area capacitance of 9 was used to establish the ML model. (b) The RF was used to predict the area capacitance of 3DP carbon micro lattices. (c) The feature significance of electrode thickness, gap distance, and geometric area. (d) The SHAP values of 3 variables in predicting area capacitance, 'Reprinted with the permission from (H. Yang et al., 2023)'.

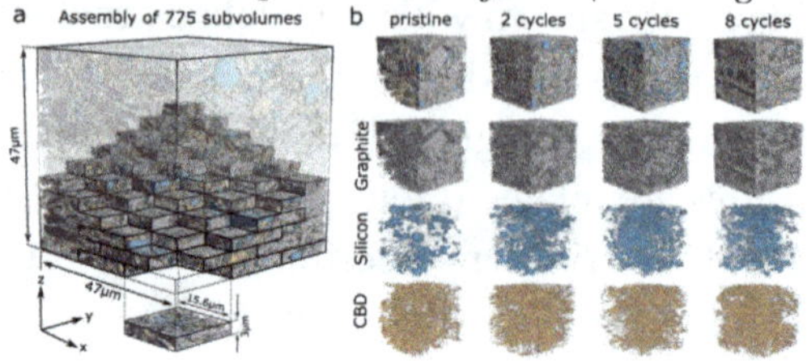

On the other hand, significant improvements in response and stability measurements with AI optimization are also seen in different studies, with cycle stability improving at high values after multiple cycles (*Table 2*) (D. Liu et al., 2025; Xiong et al., 2024; H. Yang et al., 2023). For instance, a pivotal research by Liu et al., on Zn-air Batteries, one of the diverse applications of energy storage, has been incorporated into the extant literature, thereby illuminating this field trajectory (D. Liu et al., 2025). A Bayesian optimization technique (machine learning algorithms) was used in this study to help with the effective design of multicomponent electrolyte compositions. ZAB performance significantly improved as a result of this strategy. According to the findings analysis made possible by the widespread use of machine learning (ML) methods, the new ZABs exhibit remarkable battery stability and reversibility, withstanding 1700 hours at 2 $mA{\cdot}cm^{-2}$ and 1400 hours at 20 $mA{\cdot}cm^{-2}$. Additionally, a significant increase in round-trip efficiency was attained, reaching 76.3%, and the Zn anode could be discharged continuously until it was completely depleted. In order to maximize the accessible surface area while preserving a controlled and uniform pore architecture, the structural parameters were further optimized using basic optimization that targeted non-uniform pore distribution, as seen in *Table* 2.The mechanical characteristics evolved from basic stability to improved structural integrity, featuring precise defect control (K. Liu et al., 2022; Müller et al., 2021). A guiding study has been conducted to maximize the pore distribution and surface area of lithium-ion battery electrodes, especially 3D materials and composites, by Müller et al., (Müller et al., 2021). This paper suggests a deep learning-based strategy for achieving accurate volumetric segmentation of electrodes, which is especially appropriate in situations where low contrast makes traditional methods unreliable. A 3D U-Net architecture is used for the segmentation, and realistic artificial electrode models and the tomographic reconstructions that go with them are used to provide synthetic training data. In order to improve network performance, the method also generates and incorporates synthetic datasets, which overcomes the limits of imaging data obtained experimentally. When the technique is used to segment graphite–silicon composite electrode X-ray tomographic microscopy pictures, it shows excellent accuracy across all common assessment measures (*Figure 3*). Furthermore, the microstructural changes of the carbon black and binder phases during the course of the battery's working cycle can be statistically robustly analyzed thanks to this technique. The conductivity has evolved from standard electrical conductivity and changeable contact resistance to increased conductivity with optimized channels and reduced resistance (Xie et al., 2024). Advanced control systems with AI-driven foresight are now included in the capabilities of system integration, which initially included simple thermal management and restricted degradation predicting. These results highlight the value of intelligent design and predictive modeling, which are further examined using artificial intelligence methods in the part that follows.

Table 2. Comparison of various technical parameters such as capacity, stability and conductivity for performance improvements in 3D Graphene-MXenes composites after AI optimization (Hussain et al., 2024; Lv et al., 2025; Shrestha et al., 2024; L. Sun et al., 2024; Vaishag & Noh, 2024; Xiao et al., 2023).

Parameter Category	After AI Optimization (Predicted Expected)	Before AI Optimization
Structural Parameters	- Surface Area: Maximum accessible area - Pore Distribution: Uniform and controlled - Interlayer Spacing: Optimized for ion transport	- Surface Area: Basic optimization - Pore Distribution: Non-uniform - Interlayer Spacing: Variable
Mechanical Properties	- Structural Integrity: Enhanced stability - Defect Control: Precise engineering - Interface Engineering: Optimized bonding	- Structural Integrity: Basic stability - Defect Control: Limited - Interface Engineering: Standard
Conductivity	- Electrical Conductivity: Enhanced - Charge Transfer: Optimized pathways - Contact Resistance: Minimized	- Electrical Conductivity: Standard - Charge Transfer: Basic pathways - Contact Resistance: Variable
Charging and Performance	- Specific Capacity: Boost - Energy Density: Boost - Rate Capability: Boost	- Specific Capacity: Medium - Energy Density: Medium - Rate Capability: Medium
Response and Stability	- Cycling Stability: Boost - Ion Transport: Efficient diffusion channels - Electron Transport: Optimized pathways	- Cycling Stability: Medium - Ion Transport: Standard diffusion channels - Electron Transport: Basic pathways
System Integration	- Thermal Management: Advanced control - Performance Monitoring: Real-time - Degradation Prediction: AI-driven	- Thermal Management: Basic control - Performance Monitoring: Standard - Degradation Prediction: Limited

Figure 3. As demonstrated the deep learning approach is deployed on 775 sub volumes for all samples, (a) reassembled following segmentation, (b) the ability to use multiphase segmentation on pristine samples that were treated cyclically at multiple levels (5 and 8 times) facilitated a range of microstructural analyses, 'Reprinted with the permission from (Müller et al., 2021)'.

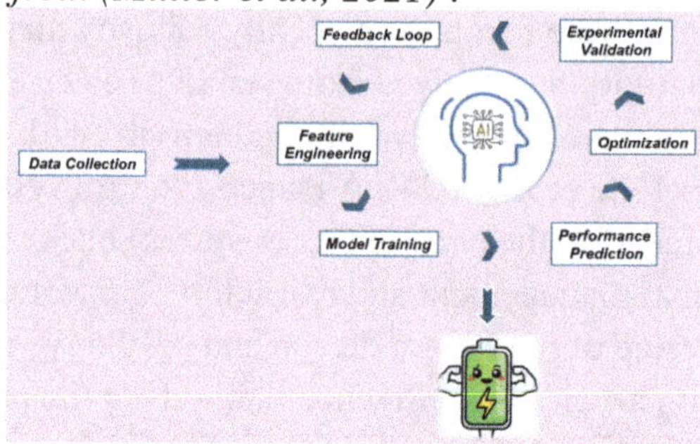

The structure–property correlations in 3D graphene–MXene composites have been clarified to a considerable extent by empirical modeling and traditional experimental techniques. However, scalability, expense, and complexity are often the limitations of these methods. Artificial intelligence (AI) provides a paradigm shift to materials modeling and optimization in response to this challenge. The next section considers the state-of-the-art artificial intelligence techniques that are being used to increase the functionality of complex electrochemical energy storage materials and transform predictive modelling, i.e. ML and DL.

3. AI-DRIVEN MODELING APPROACHES

As a series of advanced technologies that push the boundaries of materials science are incorporated, the AI-based methods have become novel in the field of 3D graphene and MXene modeling. The large compositional and architectural design space that is needed to develop the next-generation energy materials is too large to be explored using conventional experimental techniques. In this regard, through supporting the rapid analysis of complex high-dimensional data, revealing previously unknown structure-property relationships, and generating predictive models that guide material optimization, AI offers a powerful alternative in this case (Chavez-Angel et al., 2025). AI methods, including ML and DL, have the capacity to work independently to identify insights in data, suggest the most promising synthesis pathways, and predict material behavior across a range of electrochemical conditions, unlike experimental approaches, which often involve trial and error.

This accelerates the exploration process, reduces the need to perform resource-intensive laboratory testing and allows exploring design variations that are extremely costly or otherwise not feasible to investigate experimentally (ChavezAngel et al., 2025). Additionally, AI enables reverse design, which transforms the traditional approach to material development that starts with the target performance requirements and reverse-engineers appropriate material structures (ChavezAngel et al., 2025). These are advanced molecular dynamics simulations at the atomic scale, large-scale topological optimization algorithms to optimize structure, and modern predictive analytical tools to predict features (M. Huang et al., 2022; Mahapatra et al., 2025; Shariq et al., 2024). These technologies allow researchers to circumnavigate the experimental constraints that have limited material discovery and optimization by efficiently searching high-dimensional design spaces. These advanced methods will allow in-depth analysis of multiscale hierarchies and electronic properties in 3D graphene/MXene networks. (M. Huang et al., 2022; Mahapatra et al., 2025; Shariq et al., 2024). The perfect synergy of multiple AI algorithms interacting with each other examines structural parameters at different scales, ranging nanoscopically to macroscopically, to reveal the impact they have on electrochemical performance (Güvensoy-Morkoyun et al., 2024; Iravani, Khosravi, et al., 2024). Such a detailed modeling process with the ML features added facilitates the analysis of structure-attribute correlations and optimization paths of performance in detail. The large-scale modeling approach based on data does not only assist in the predictive performance of the material, but also assists in smart engineering and real-time optimization of composite materials.

3.1. Machine Learning Algorithms

Advanced machine learning algorithms have been used more and more in recent years to analyze and forecast research results pertaining to MXenes and 3D graphene. Each of these technologies has been selected and enhanced to assist in particular objectives in terms of characterizing and enhancing materials (Shariq et al., 2024). These data-driven techniques are particularly effective at revealing complex non-linear patterns in sizable, multidimensional datasets, particularly when the materials have varied and hierarchical topologies. Over the past few years, ML has been actively used to comprehend and forecast the electrochemical and structural performance of hybrid materials by training on simulation and experimental data (Shariq et al., 2024). These systems have the inherent properties that make them powerful tools to study and enhance both graphene-based systems and MXene-based systems (Shariq et al., 2024).

Among the most significant tasks that ML can complete in the given case is discover connections between structure and property. This includes the evaluation of how compositional and morphological parameters, e.g., pore size, surface chemistry, or interlayer spacing, affect the electrochemical parameters, e.g., capacity, conductivity, or cycle life (Gautam et al., 2023; Gouveia et al., 2025). The output parameters were modeled in relation to ideal potential windows, scan rates, and different GNP weight ratios in MXene. The electrode with 20 weight percent GNP in MXene (MG-80) demonstrated the highest specific capacitance (Csp) of 226.6 $F \cdot g^{-1}$ at 5 $mV \cdot s^{-1}$, as well as excellent cycle stability, holding onto 84.2% of its capacitance even after 5000 charge–discharge cycles, according to both experimental results and machine learning predictions (*Figure 4*). ML models were used to forecast the cyclic stability of the MG-80 electrode over a long 10,000-cycle period in order to improve the design of energy storage devices.

Figure 4. The following electrochemical measurements were conducted on the MG-80 composite in 1 M H_2SO_4 a) Cyclic voltammetry curves were obtained at varying scan rates, ranging from 5 to 100 mV.s^{-1} b) Chronopotentiometry curves were obtained at different current densities, ranging from 1 to 5 A.g^{-1} c) A cyclic stability test was performed up to 5000 cycles d) The Csp values of the bare MXene, GNPs and MG-80 electrodes were compared, calculated at different scan rates, 'Reprinted with the permission from (Shariq et al., 2024)'.

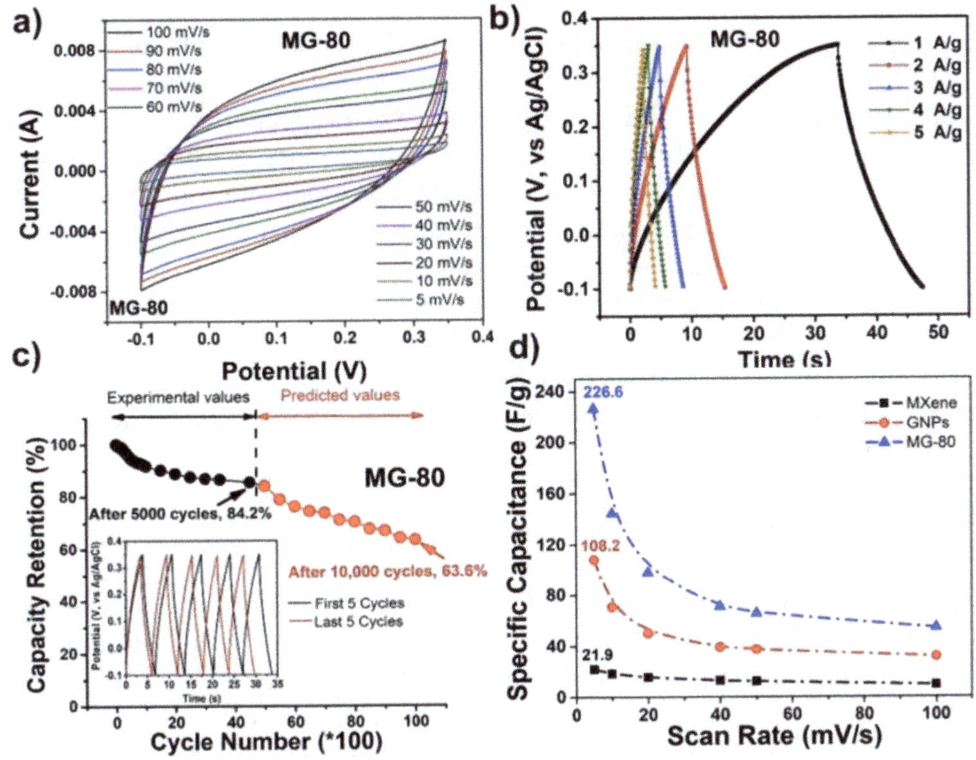

Another important area where ML is used is morphology optimization, where the changes in the structural characteristics are modeled and predictions of their overall performance are made. In this particular optimization, performance parameters are modeled and a complete simulation and analysis of the impact of various structural changes on the behavior of the material is conducted (Gouveia et al., 2025; Z. Yang & Buehler, 2022). Density functional theory (DFT) and molecular dynamics (MD) computations are often used to produce high-fidelity input features that enhance model training and facilitate accurate simulations.

Moreover, ML makes it possible to detect and measure structural defects that may disrupt cycling stability or conductivity and thus assists in defect diagnosis and control. With the help of ML models, it is possible to detect defects and relate them to loss or degradation patterns by learning on microscope data (including SEM or TEM) (Gouveia et al., 2025; Z. Yang & Buehler, 2022). To reduce structural defects in the graphene-MXene network, these insightful observations will be used to refine the synthesis methods. The sophisticated computer techniques allow the exploration of parameters at different scales.

This assists in researching the relationship between buildings and potential enhancements.

Crucially, machine learning algorithms function on a range of length scales, spanning from macroscopic performance metrics to atomistic descriptions. While mitigating the risk of overfitting, they are able to extract significant patterns from sparse or noisy data using cross-validation, as well as feature selection, and dimensionality reduction approaches (Pham et al., 2022). A strong forecasting framework that speeds up the process of finding high-performance composites that are customized for various energy storage applications is the end result.

It may be concluded that ML offers 3D graphene–MXene composites a scalable and flexible modeling paradigm. By combining machine learning with predictive analytics, optimization, and interpretability, we can intelligently explore complicated material design spaces by connecting the gap between theoretical design and experimental validation.

3.2. Deep Learning Applications

The conventional method ML algorithms have effectively modeled linear and somewhat non-linear relationships, however DL has emerged as a significant advancement in the utilization of high-dimensional and complex data (Ali et al., 2025). ANNs, CNNs, and GNNs can learn multiscale representations directly from raw or slightly processed data, facilitating the identification of concealed characteristics and more profound connections between structure and electrochemical performance (Mortazavi, 2025).

The use of DL in 3D modeling of graphene-MXenes composites has lately garnered significant attention as it signifies a sophisticated approach to enhance comprehension and optimization of these intricate materials. Deep neural networks allow researchers to analyze and study vast amounts of structural and performance data to improve the precision of predictions and optimizations (Z. Yang & Buehler, 2022; Zheng & Gu, 2020). The ability to control the multi-scale nature of graphene structures, i.e., everything between the macroscopic properties and the molecular-level reactions, makes such DL systems valuable (Zhang et al., 2021; Zheng et al., 2022). Also, they are experts such as morphological optimization, performance modeling, defect evaluation, and the prediction of structure-property relationships (Boiko et al., 2021; Malavolta et al., 2025).

The primary objectives of the deep learning models in this field are as follows: firstly, to maximize accessible surface area to improve charge storage capacity; secondly, to optimize electron transport pathways to enhance electrical conductivity; thirdly to maintain structural homogeneity in order to guarantee long-term cycling stability; and fourthly, to upgrade ion transport channels in order to lower diffusion resistance and support high-rate performance (Boiko et al., 2021; Ghimire et al., 2025; Malavolta et al., 2025; Z. Yang & Buehler, 2022; Yogesh et al., 2025). These objectives are accomplished by capitalizing on the profound representational capacity of neural networks to acquire knowledge from intricate, high-dimensional data and predict how particular structural configurations impact pivotal electrochemical metrics (T. Sun et al., 2025).

In the future, the field of deep learning will increasingly rely on complicated techniques such as graph-based learning, which is particularly well-suited to characterizing layered heterostructures and atomic networks such as those found in graphene-MXene systems. Also, to enhance the interpretability and predictive power of AI at the atomic and device levels, novel hybrid AI systems are integrating deep learning and physics-informed models and quantum computing systems. Although deep learning models have impressive capabilities, they still have significant challenges. These are the need to have large, high-quality datasets, the susceptibility to overfitting caused by a shortage of experimental data, the high processing requirements, and the natural difficulty in interpretation. The current research is targeting these limitations with such approaches as explainable AI (XAI), data augmentation, and transfer learning, which aim to learn the reasoning behind model predictions. Deep learning has the ability to completely transform the intelligent design of next-generation electrochemical energy storage systems (ESSs) by utilizing predictive control over structural and performance parameters and revealing hidden patterns in massive amounts of data.

4. OPTIMIZATION STRATEGIES

In order to enhance the 3D graphene/MXene designs with AI, several interconnected approaches are applied with the aim of optimizing both material-level properties and the system-level performance. These approaches apply novel computational models to solve structural properties, electrochemical behaviors and implementation aspects. Beginning with structural changes and concluding with the adjustment of performance parameters, the present section gives a detailed overview of the key optimization methods.

As it was stated above, the structural optimization of 3D graphene and MXene designs with the help of AI requires the utilization of advanced algorithms that evaluate and enhance the natural properties of the materials. All these structural factors, such as porosity, surface morphology, and hierarchical arrangement, are used simultaneously in these algorithms through a multi-objective approach (Agrawal et al., 2025; Lv et al., 2025). The optimization technique could be aimed at strengthening key elements through the use of advanced algorithms and comprehensive analysis.

A variety of complimentary techniques are used to optimize these composite (3D-graphene/MXenes) structures. The process is initiated by the exact control of pore size distribution, a prerequisite for attaining optimal material features (Iravani, Khosravi, et al., 2024; Lv et al., 2025). Surface area optimization methods that seek to enhance the quantity of active electrochemical sites and boots charge storage capacity are utilized in conjunction with this (Iravani, Khosravi, et al., 2024; Lv et al., 2025). Moreover, the optimization of electron transport pathways is of critical role in minimizing resistance and maximizing conductivity, thereby contributing to increased energy and power density (Iravani, Khosravi, et al., 2024; Lv et al., 2025). To achieve better electrochemical performance and guarantee device dependability, these interrelated parameters are methodically evaluated and improved utilizing cutting-edge AI algorithms in electrochemical ESS.

To optimize the energy storage capacity of 3D graphene/MXene composite devices, it is important to understand a number of interrelated factors. As previously stated, the total effectiveness and usefulness of the generated structures are controlled by these components, which are carefully optimized by AI algorithms. As illustrated in *Table 3* comprehensively delineates the crucial optimization parameters and performance traits of 3D graphene–MXenes composite materials in the context of EES applications. It systematically addresses four core dimensions: *i. interfacial properties; ii. electronic properties; iii. structural stability,* and *iv. transport characteristics.*

Using sophisticated layer-by-layer assembly processes, the first category concentrates on optimizing interfacial designs by precisely controlling layer stacking, interlayer spacing, and heterostructure creation. Through the strengthening of synergistic interfacial interactions between graphene and MXene components has been demonstrated to lead to high capacitance performance and electron/ion transport. The second dimension, emphasizing electronic properties, includes improvements of hybrid conductivity and charge transfer kinetics, as well as work function modulation. These techniques have achieved the improvements due to band structure engineering, interface resistance reduction, and electronic coupling optimization, which enhance the impressive electrical performance, and increase power output.

The third type is structural reliability and is defined by the approaches to mechanical integration, deformation resistance, and cyclic durability. To guarantee prolonged device life in high-stress working environments, these features involve maximizing intercomponent mechanical synergy, avoiding restacking, and strengthening interface bonding. Finally, transport processes include lastly surface chemistry, the availability of electrolyte and enhancement of ion channels. Surface functionalization, changing the hybrid pore structure, and designing the hierarchical transport network all lead to better electrochemical responsiveness, faster ion diffusion, and higher wettability. Consequently, it is a complex means of enhancing 3D graphene-MXenes composite technologies to systematically connect a lot of various factors with particular AI optimization techniques and observe their impact on performance.

Table 3. AI-influential optimization parameters and performance characteristics for 3D graphene-Mxenes composite materials (Hussain et al., 2024; Iravani, Khosravi, et al., 2024; Mahapatra et al., 2025; Shariq et al., 2024).

Parameter Category	Impact on Performance	AI Optimization Target
Structural Stability; **Mechanical Integration** **Deformation Resistance** **Cyclic Stability**	✓ Enhanced structural durability ✓ Improved cycling stability ✓ Better mechanical flexibility ✓ Prolonged device lifetime	✓ Enhancement of mechanical synergy between components ✓ Prevention of restacking ✓ Optimization of strain distribution ✓ Interface bonding strengthening
Transport Properties; **Hybrid Ion Channels** **Electrolyte Interaction** **Surface Chemistry**	✓ Accelerated ion diffusion ✓ Improved rate capability ✓ Enhanced wettability ✓ Superior electrochemical performance	✓ Design of hierarchical transport networks ✓ Optimization of hybrid pore structures ✓ Surface functionalization control ✓ Electrolyte accessibility enhancement
Interfacial Properties; **Layer Stacking** **Interlayer Spacing** **Heterostructure Formation**	✓ Enhanced electron/ion transfer ✓ Improved interface conductivity ✓ Optimized capacitive performance ✓ Superior charge storage capability	✓ Optimization of 3D-graphene-MXenes interface engineering ✓ Control of layer-by-layer assembly ✓ Tuning of interlayer distance ✓ Maximization of synergistic effects between components
Electronic Properties; **Hybrid Conductivity** **Charge Transfer Kinetics** **Work Function**	✓ Exceptional electrical conductivity ✓ Enhanced charge transfer efficiency ✓ Improved rate performance ✓ Superior power delivery	✓ Optimization of electronic coupling between graphene and MXenes ✓ Band structure engineering ✓ Interface resistance minimization ✓ Synergistic conductivity enhancement

Figure 5 shows the entire AI-based workflow of making, enhancing, and testing 3D graphene MXene composites. This comprehensive stage forms a solid foundation of material research and optimization due to the combination of an advanced

machine learning technique and experimental validation. The data-driven, iterative process combines the newest machine learning and realistic verification techniques to form a closed-loop materials innovation system.

The primary step in the procedure is the data collection, which involves compiling data sets of computers, experiments, or literature. The next step is feature engineering, which involves refining raw data, streamlining it, and selecting new features to ensure that they are meaningful and of high quality. In the training process of the proposed model, the AI algorithms will be trained to identify connections between structures and qualities and predict them using specially selected datasets. The next stage is performance prediction process that assists the system to make an estimate of how materials will perform and react to the given design conditions. Then the forecasts are adjusted to produce improved goal parameters. Once the materials have been enhanced, they should be tried in practice. This entails comparing AI-generated concepts to actual performance and completing a comparative study. This process of validation gives us new information that goes back into the system through a feedback loop. This brings us back to feature engineering. This is an active process that helps people learn, get better, and gain information. The suggested flexible and intelligent approach has shown to speed up the search for high-performance 3D graphene-MXene materials and make simpler to connect theory and practice in ESS.

Figure 5. An advanced AI workflow that involves experimental validation with a feedback mechanism to enable continuously enhance the optimization process.

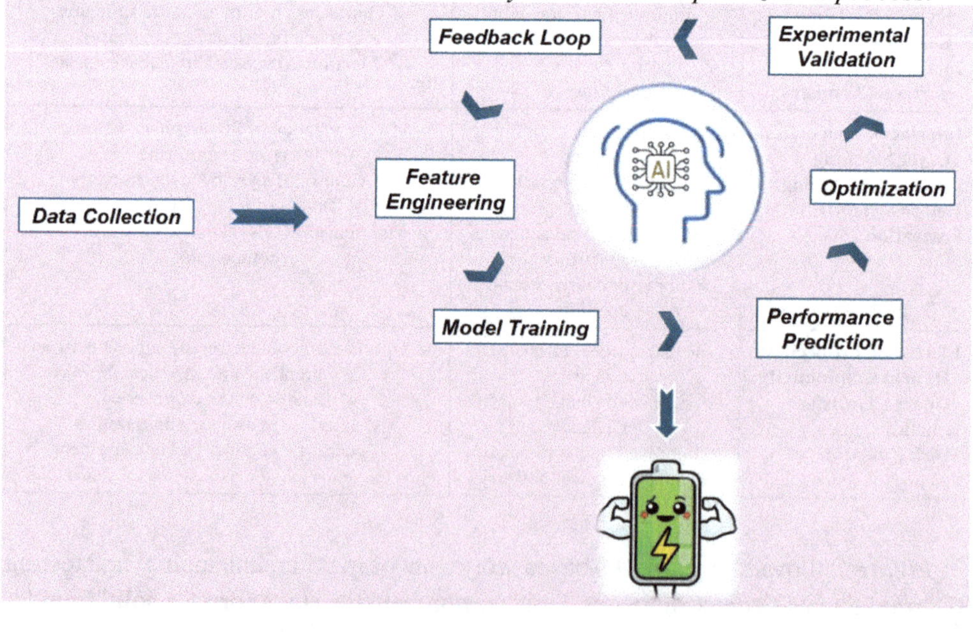

5. CHALLENGES AND FUTURE PERSPECTIVES

For 3D graphene–MXene hybrid nanocomposites to be widely used in energy storage devices, a number of important issues need to be resolved. These include increasing the efficiency of production techniques, optimizing the interface structures between MXene nanosheets and 3D graphene frameworks, and scaling up synthesis and manufacturing procedures. At industrial scales, ensuring precise structural control while maintaining reproducibility is still highly difficult.

Concurrently, progress in AI-driven optimization are reconfiguring the paradigm of materials development. High-level machine learning techniques are being quickly incorporated into research operations. These methods include automated high-throughput screening, transfer learning for performance prediction from small datasets, and generative adversarial networks (GANs) for inverse material design. Currently, experimental validation procedures are used in conjunction with these AI-driven models to develop optimization workflows that are more efficient and productive.

There are a number of possible directions that should be explored in the framework of future research. Researchers are concentrating on creating multiscale modeling methods that link simulations at the atomic level with macroscopic properties, utilizing quantum computing algorithms, and investigating hybrid systems that include several functional materials in order to improve property prediction. Predictive models at sub-nanometer electrical and ionic scales are being made more accurate by using quantum computing techniques in particular. A possible route toward adaptable, next-generation devices is also provided by researching hybrid frameworks, such as 3D graphene–MXene composites combined with conductive polymers, metal–organic frameworks (MOFs), or solid electrolytes. At the same time, research efforts are concentrated on developing real-time AI analysis and in-situ characterisation methods. This integrated approach offers real-time insights into material behavior under operating conditions and enables dynamic, adaptive optimization. The development and implementation of high-performance, AI-optimized electrochemical energy storage systems are anticipated to be accelerated by these developments.

6. CONCLUSION

A thorough analysis of 3D graphene–MXene hybrid nanocomposites has produced results that could have a significant impact on energy storage in the future. The study emphasizes how electrochemical performance can be maximized by combining these materials in a synergistic way with AI-driven optimization techniques. Because of their large surface area, strong mechanical stability, and remarkable electrical conductivity, 3D graphene–MXene hybrids are very intriguing options for next-generation energy storage systems (ESS). These properties work together to improve energy density, optimize charge–discharge cycling, and increase overall system efficiency.

Numerous significant strategic directions for future investigation are proposed by a thorough analysis of the modeling attempts and existing literature. First, progressing from lab-scale innovations into commercial applications still necessitates addressing the scalability of fabrication methods. Second, in order to optimize interfacial charge transport and structural cohesion requires fine-tuning the engineering of interfaces between graphene and MXene layers. Thirdly, the continuous advancement of AI-driven optimization, especially with the use of multiscale modeling frameworks and quantum computing techniques, will be a key component of predictive material design.

In conclusion, combining sophisticated in-situ characterisation methods with real-time monitoring systems will support the validation of performance indicators and guarantee long-term stability in real-world applications. The intelligent development and application of long-lasting, high-performing energy storage materials will be fueled by such initiatives.

REFERENCES

Ab Alim, N. N. N., Mohamed Saheed, M. S., Mohamed, N. M., & Mohamed Saheed, M. S. (2019). Highly flexible and stretchable 3D graphene/MXene composite thin film. *Materials Today: Proceedings, 7*, 738–743. DOI: 10.1016/j.matpr.2018.12.069

Agrawal, P., Thethi, H. P., Mohammad, Q., Gupta, N., Asha, V., & Reddy, K. J. (2025). Integrating AI and Multi-objective Optimization for Enhanced Microgrid Energy Management Using Quadratic Programming. *E3S Web of Conferences, 616*, 02007. DOI: 10.1051/e3sconf/202561602007

Ali, M. L., Thakur, K., Schmeelk, S., Debello, J., & Dragos, D. (2025). Deep Learning vs. Machine Learning for Intrusion Detection in Computer Networks: A Comparative Study. *Applied Sciences (Basel, Switzerland), 15*(4), 1903. DOI: 10.3390/app15041903

Bandpey, M., & Barz, D. P. J. (2024). Effects of interlayer space engineering and surface modification on the charge storage mechanisms of MXene nanomaterials: A review on recent developments. *Nanoscale, 16*(32), 15078–15093. DOI: 10.1039/D4NR01317D

Bansal, S., Chaudhary, P., Sharma, B. B., Saini, S., & Joshi, A. (2024). Review of MXenes and their composites for energy storage applications. *Journal of Energy Storage, 87*, 111420. DOI: 10.1016/j.est.2024.111420

Bhat, A., Anwer, S., Bhat, K. S., Mohideen, M. I. H., Liao, K., & Qurashi, A. (2021). Prospects challenges and stability of 2D MXenes for clean energy conversion and storage applications. *Npj 2D Materials and Applications, 5*(1), 61. DOI: 10.1038/s41699-021-00239-8

Bhatt, M. D., Kim, H., & Kim, G. (2022). Various defects in graphene: A review. *RSC Advances, 12*(33), 21520–21547. DOI: 10.1039/D2RA01436J

Bilisik, K., & Akter, M. (2022). Graphene nanocomposites: A review on processes, properties, and applications. *Journal of Industrial Textiles, 51*(3, suppl), 3718S–3766S. DOI: 10.1177/15280837211024252

Boiko, D. A., Pentsak, E. O., Cherepanova, V. A., Gordeev, E. G., & Ananikov, V. P. (2021). Deep neural network analysis of nanoparticle ordering to identify defects in layered carbon materials. *Chemical Science (Cambridge), 12*(21), 7428–7441. DOI: 10.1039/D0SC05696K

Chávez-Angel, E., Eriksen, M. B., Castro-Alvarez, A., Garcia, J. H., Botifoll, M., Avalos-Ovando, O., Arbiol, J., & Mugarza, A. (2025). Applied Artificial Intelligence in Materials Science and Material Design. *Advanced Intelligent Systems*, *7*(8), 2400986. Advance online publication. DOI: 10.1002/aisy.202400986

Cheng, D., Zhao, Y., An, T., Wang, X., Zhou, H., & Fan, T. (2019). 3D interconnected crumpled porous carbon sheets modified with high-level nitrogen doping for high performance lithium sulfur batteries. *Carbon*, *154*, 58–66. DOI: 10.1016/j.carbon.2019.07.094

Cui, Y., Zhu, J., Tong, H., & Zou, R. (2023). Advanced perspectives on MXene composite nanomaterials: Types synthetic methods, thermal energy utilization and 3D-printed techniques. *iScience*, *26*(1), 105824. DOI: 10.1016/j.isci.2022.105824

Gautam, R., Marriwala, N., & Devi, R. (2023). A review: Study of Mxene and graphene together. *Measurement. Sensors*, *25*, 100592. DOI: 10.1016/j.measen.2022.100592

Ghimire, G., Noormohammadi, F., Kohl, J., Devenport, H., & Johnson, P. A. (2025). Machine learning optimization of laser-induced graphene parameters for surface-enhanced Raman spectroscopic detection of glucose. *Sensors and Actuators. B, Chemical*, *432*, 137466. DOI: 10.1016/j.snb.2025.137466

Gouveia, J. D., Galvão, T. L. P., Iben Nassar, K., & Gomes, J. R. B. (2025). First-principles and machine-learning approaches for interpreting and predicting the properties of MXenes. *Npj 2D Materials and Applications*, *9*(1), 8. DOI: 10.1038/s41699-025-00529-5

Guo, P., Song, H., & Chen, X. (2009). Electrochemical performance of graphene nanosheets as anode material for lithium-ion batteries. *Electrochemistry Communications*, *11*(6), 1320–1324. DOI: 10.1016/j.elecom.2009.04.036

Guo, Y., Wei, Y., Li, H., & Zhai, T. (2017). Layer Structured Materials for Advanced Energy Storage and Conversion. *Small*, *13*(45), 1701649. Advance online publication. DOI: 10.1002/smll.201701649

Güvensoy-Morkoyun, A., Baysal, T., Tantekin-Ersolmaz, Ş. B., & Velioğlu, S. (2024). Guide for Nonequilibrium Molecular Dynamics Simulations of Organic Solvent Transport in Nanopores: The Case of 2D MXene Membranes. *Journal of Chemical Theory and Computation*, *20*(21), 9642–9654. DOI: 10.1021/acs.jctc.4c00693

Hafiz, W., Meng, Y., Qi, D., & Zhu, F. (2025). *Nis/ Hollow Carbon Sphere Composites for High-Performance and Long-Life Lithium-Ion Battery Anodes*. DOI: 10.1016/j.jelechem.2025.119237

Hegazy, H. H., Khan, J., Shakeel, N., Alabdullkarem, E. A., Saleem, M. I., Alrobei, H., & Yahia, I. S. (2024). 2D-based electrode materials for supercapacitors – status, challenges, and prospects. *RSC Advances*, *14*(45), 32958–32977. DOI: 10.1039/D4RA05473C

Huang, G., Huang, F., & Dong, W. (2024). Machine learning in energy storage material discovery and performance prediction. *Chemical Engineering Journal*, *492*, 152294. DOI: 10.1016/j.cej.2024.152294

Huang, M., Li, Z., & Zhu, H. (2022). Recent Advances of Graphene and Related Materials in Artificial Intelligence. *Advanced Intelligent Systems*, *4*(10), 2200077. Advance online publication. DOI: 10.1002/aisy.202200077

Hussain, T., Chandio, I., Ali, A., Hyder, A., Memon, A. A., Yang, J., & Thebo, K. H. (2024). Recent developments of artificial intelligence in MXene-based devices: From synthesis to applications. *Nanoscale*, *16*(38), 17723–17760. DOI: 10.1039/D4NR03050H

Indira, V., & Abhitha, K. (2023). Mesoporogen-free synthesis of hierarchical zeolite A for CO2 capture: Effect of freeze drying on surface structure, porosity and particle size. *Results in Engineering*, *17*, 100886. DOI: 10.1016/j.rineng.2023.100886

Iravani, S., Khosravi, A., Nazarzadeh Zare, E., Varma, R. S., Zarrabi, A., & Makvandi, P. (2024). MXenes and artificial intelligence: Fostering advancements in synthesis techniques and breakthroughs in applications. *RSC Advances*, *14*(49), 36835–36851. DOI: 10.1039/D4RA06384H

Iravani, S., Zarepour, A., Nazarzadeh Zare, E., Makvandi, P., Khosravi, A., & Zarrabi, A. (2024). Synergistic advancements: Exploring MXene/graphene oxide and MXene/reduced graphene oxide composites for next-generation applications. *FlatChem*, *48*, 100759. DOI: 10.1016/j.flatc.2024.100759

Jayaramulu, K., Mukherjee, S., Morales, D. M., Dubal, D. P., Nanjundan, A. K., Schneemann, A., Masa, J., Kment, S., Schuhmann, W., Otyepka, M., Zbořil, R., & Fischer, R. A. (2022). Graphene-Based Metal–Organic Framework Hybrids for Applications in Catalysis, Environmental, and Energy Technologies. *Chemical Reviews*, *122*(24), 17241–17338. DOI: 10.1021/acs.chemrev.2c00270

Jussambayev, M., Shakenov, K., Sultakhan, S., Zhantikeyev, U., Askaruly, K., Toshtay, K., & Azat, S. (2025). MXenes for sustainable energy: A comprehensive review on conservation and storage applications. *Carbon Trends*, *19*, 100471. DOI: 10.1016/j.cartre.2025.100471

Kurc, B., Pigłowska, M., Rymaniak, Ł., & Fuć, P. (2021). Modern Nanocomposites and Hybrids as Electrode Materials Used in Energy Carriers. *Nanomaterials (Basel, Switzerland)*, *11*(2), 538. DOI: 10.3390/nano11020538

Le Ferrand, H., Chabi, S., & Agarwala, S. (2020). 3D Assembly of Graphene Nanomaterials for Advanced Electronics. *Advanced Intelligent Systems*, *2*(5), 1900151. Advance online publication. DOI: 10.1002/aisy.201900151

Li, G., Lian, S., Wang, J., Xie, G., Zhang, N., & Xie, X. (2023). Surface chemistry engineering and the applications of MXenes. *Journal of Materiomics*, *9*(6), 1160–1184. DOI: 10.1016/j.jmat.2023.08.003

Li, P., Zhao, X., Ding, Y., Chen, L., Wang, X., & Xie, H. (2024). Application studies on MXene-based flexible composites. *Frontiers in Thermal Engineering*, *4*, 1440165. Advance online publication. DOI: 10.3389/fther.2024.1440165

Liang, X., Rangom, Y., Kwok, C. Y., Pang, Q., & Nazar, L. F. (2017). Interwoven MXene Nanosheet/Carbon-Nanotube Composites as Li–S Cathode Hosts. *Advanced Materials*, *29*(3), 1603040. Advance online publication. DOI: 10.1002/adma.201603040

Liu, C., Li, F., Ma, L., & Cheng, H. (2010). Advanced Materials for Energy Storage. *Advanced Materials*, *22*(8). Advance online publication. DOI: 10.1002/adma.200903328

Liu, D., Ge, H., Song, M., Jiang, Y., Gong, X., You, T., Fu, L., Fu, Z., & Zhang, Y. (2025). Machine Learning Driven Optimization of Electrolyte for Highly Reversible Zn-Air Batteries with Superior Long-Term Cycling Performance. *Advanced Materials*, *37*(7), 2417161. Advance online publication. DOI: 10.1002/adma.202417161

Liu, K., Niri, M. F., Apachitei, G., Lain, M., Greenwood, D., & Marco, J. (2022). Interpretable machine learning for battery capacities prediction and coating parameters analysis. *Control Engineering Practice*, *124*, 105202. DOI: 10.1016/j.conengprac.2022.105202

Liu, M., Qin, Z., Yang, X., Lin, Z., & Guo, T. (2019). Fabricating controllable hierarchical pores on smooth carbon sheet for synthesis of supercapacitor materials. *Vacuum*, *168*, 108806. DOI: 10.1016/j.vacuum.2019.108806

Liu, M., Song, A., Zhang, X., Wang, J., Fan, Y., Wang, G., Tian, H., Ma, Z., & Shao, G. (2025). Interfacial lithium-ion transportation in solid-state batteries: Challenges and prospects. *Nano Energy*, *136*, 110749. DOI: 10.1016/j.nanoen.2025.110749

Liu, S., Zhang, H., Chen, J., Peng, X., Chai, Y., Shao, X., He, Y., Wang, X., & Ding, B. (2025). Functionalization Strategies of MXene Architectures for Electrochemical Energy Storage Applications. *Energies*, *18*(5), 1223. DOI: 10.3390/en18051223

Lv, K., Wan, H., Qi, C., Wang, W., Liu, X., Zhang, L., Wei, H., Zhu, H., Wang, Y., Yu, J., & Du, S. (2025). *Integrating AI and material science: MXene synthesis, preparation, and applications*. AI & Materials., DOI: 10.55092/aimat20250004

Ma, Z., Zhou, X., Deng, W., Lei, D., & Liu, Z. (2018). 3D Porous MXene (Ti 3 C 2)/Reduced Graphene Oxide Hybrid Films for Advanced Lithium Storage. *ACS Applied Materials & Interfaces*, *10*(4), 3634–3643. DOI: 10.1021/acsami.7b17386

Mahapatra, D. M., Kumar, A., Kumar, R., Gupta, N. K., Ethiraj, B., & Singh, L. (2025). Artificial intelligence interventions in 2D MXenes-based photocatalytic applications. *Coordination Chemistry Reviews*, *529*, 216460. DOI: 10.1016/j.ccr.2025.216460

Malavolta, L., Bracaglia, I., Cazzador, G., Kovtun, A., Tomasi, L., Zanardi, C., & Palermo, V. (2025). Multivariate sensing of ions using machine learning and composite 2D-3D graphene oxide-hexacyanoferrate electrodes. *Sensors and Actuators. B, Chemical*, *427*, 137194. DOI: 10.1016/j.snb.2024.137194

Martínez-Hincapié, R., Wegner, J., Anwar, M. U., Raza-Khan, A., Franzka, S., Kleszczynski, S., & Čolić, V. (2024). The determination of the electrochemically active surface area and its effects on the electrocatalytic properties of structured nickel electrodes produced by additive manufacturing. *Electrochimica Acta*, *476*, 143663. DOI: 10.1016/j.electacta.2023.143663

Maughan, P. A., Seymour, V. R., Bernardo-Gavito, R., Kelly, D. J., Shao, S., Tantisriyanurak, S., Dawson, R., Haigh, S. J., Young, R. J., Tapia-Ruiz, N., & Bimbo, N. (2020). Porous Silica-Pillared MXenes with Controllable Interlayer Distances for Long-Life Na-Ion Batteries. *Langmuir*, *36*(16), 4370–4382. DOI: 10.1021/acs.langmuir.0c00462

Mo, R., Li, F., Tan, X., Xu, P., Tao, R., Shen, G., Lu, X., Liu, F., Shen, L., Xu, B., Xiao, Q., Wang, X., Wang, C., Li, J., Wang, G., & Lu, Y. (2019). High-quality mesoporous graphene particles as high-energy and fast-charging anodes for lithium-ion batteries. *Nature Communications*, *10*(1), 1474. DOI: 10.1038/s41467-019-09274-y

Mortazavi, B. (2025). Recent Advances in Machine Learning-Assisted Multiscale Design of Energy Materials. *Advanced Energy Materials*, *15*(9), 2403876. Advance online publication. DOI: 10.1002/aenm.202403876

Mullani, S., Kim, C., Lokhande, V., & Ji, T. (2025). MXene structural and surface modifications for enhanced Li-ion diffusion in lithium-ion capacitors: A critical mini review of recent advances. *Chemical Engineering Journal, 510*, 161565. DOI: 10.1016/j.cej.2025.161565

Müller, S., Sauter, C., Shunmugasundaram, R., Wenzler, N., De Andrade, V., De Carlo, F., Konukoglu, E., & Wood, V. (2021). Deep learning-based segmentation of lithium-ion battery microstructures enhanced by artificially generated electrodes. *Nature Communications, 12*(1), 6205. DOI: 10.1038/s41467-021-26480-9

Nagaswarupa, H. P., Sillanpää, M. E. T., Ananda Murthy, H. C., & Naik, R. (2024). *Multifunctional Inorganic Nanomaterials for Energy Applications*. CRC Press., DOI: 10.1201/9781003479239

Nahirniak, S., Ray, A., & Saruhan, B. (2023). Challenges and Future Prospects of the MXene-Based Materials for Energy Storage Applications. *Batteries, 9*(2), 126. DOI: 10.3390/batteries9020126

Pai, S. M., Shah, K. A., Sunder, S., Albuquerque, R. Q., Brütting, C., & Ruckdäschel, H. (2025). Machine learning applied to the design and optimization of polymeric materials: A review. *Next Materials, 7*, 100449. DOI: 10.1016/j.nxmate.2024.100449

Pham, H. T., Awange, J., & Kuhn, M. (2022). Evaluation of Three Feature Dimension Reduction Techniques for Machine Learning-Based Crop Yield Prediction Models. *Sensors (Basel), 22*(17), 6609. DOI: 10.3390/s22176609

Ping, R., Nie, Y., Ji, C., Hao, Z., Yang, S., Li, L., Peng, J., & Li, X. (2023). Enhanced proton pseudocapacitive of Ti3C2Tx in neutral electrolyte activated by acid regulation dynamic pillars. *Chemical Engineering Journal, 455*, 140650. DOI: 10.1016/j.cej.2022.140650

Rong, C., Zhou, L., Zhang, B., & Xuan, F.-Z. (2023). Machine learning for mechanics prediction of 2D MXene-based aerogels. *Composites Communications, 38*, 101474. DOI: 10.1016/j.coco.2022.101474

Saharudin, M., Ayub, A., Hasbi, S., Muhammad-Sukki, F., Shyha, I., & Inam, F. (2023). Recent advances in MXene composites research, applications and opportunities. *Materials Today: Proceedings*. Advance online publication. DOI: 10.1016/j.matpr.2023.02.435

Sarker, I. H. (2022). AI-Based Modeling: Techniques, Applications and Research Issues Towards Automation, Intelligent and Smart Systems. *SN Computer Science, 3*(2), 158. DOI: 10.1007/s42979-022-01043-x

Shahzad, F., Iqbal, A., Kim, H., & Koo, C. M. (2020). 2D Transition Metal Carbides (MXenes): Applications as an Electrically Conducting Material. *Advanced Materials, 32*(51), 2002159. Advance online publication. DOI: 10.1002/adma.202002159

Shrestha, S., Barvenik, K. J., Chen, T., Yang, H., Li, Y., Kesavan, M. M., Little, J. M., Whitley, H. C., Teng, Z., Luo, Y., Tubaldi, E., & Chen, P.-Y. (2024). Machine intelligence accelerated design of conductive MXene aerogels with programmable properties. *Nature Communications, 15*(1), 4685. DOI: 10.1038/s41467-024-49011-8

Subrahmanian, S. K., & Narayanan, B. N. (2025). Three-dimensional array of holey graphene as a high-performance anode material for supercapacitors. *Journal of Power Sources, 629*, 235952. DOI: 10.1016/j.jpowsour.2024.235952

Sun, L., Dong, Y., Li, H., Meng, H., Liu, J., Cao, Q., & Pan, C. (2024). Research Progress and Challenges of Carbon/MXene Composites for Supercapacitors. *Batteries, 10*(11), 395. DOI: 10.3390/batteries10110395

Sun, T., Wang, Z., Zeng, L., & Feng, G. (2025). Identifying MOFs for electrochemical energy storage via density functional theory and machine learning. *npj Computational Materials, 11*(1), 90. DOI: 10.1038/s41524-025-01590-w

Thiruppathi, A. R., Sidhureddy, B., Salverda, M., Wood, P. C., & Chen, A. (2020). Novel three-dimensional N-doped interconnected reduced graphene oxide with superb capacitance for energy storage. *Journal of Electroanalytical Chemistry (Lausanne, Switzerland), 875*, 113911. DOI: 10.1016/j.jelechem.2020.113911

Ullah, K., Alam, N., Uddin, S., & Oh, W.-C. (2024). Advanced concept and perspectives toward MXenes based energy storage device: Comprehensive review. *Materialia, 34*, 102089. DOI: 10.1016/j.mtla.2024.102089

Vaishag, P. V., & Noh, J.-S. (2024). A Comparative Review of Graphene and MXene-Based Composites towards Gas Sensing. *Molecules (Basel, Switzerland), 29*(19), 4558. DOI: 10.3390/molecules29194558

Vs, N., T, R., Kempahanumakkagari, S., Hr, P., & Bm, P.VS. (2023). Efficient strategies to produce Graphene and functionalized graphene materials: A review. *Applied Surface Science Advances, 14*, 100386. DOI: 10.1016/j.apsadv.2023.100386

Wenelska, K., Adam, V., Thauer, E., Singer, L., Klingeler, R., Chen, X., & Mijowska, E. (2022). Fabrication of 3D graphene/MoS2 spherical heterostructure as anode material in Li-ion battery. *Frontiers in Energy Research, 10*, 960786. Advance online publication. DOI: 10.3389/fenrg.2022.960786

Woo, S. H., Lee, J., Kim, J. P., Hong, J., Jang, J. M., Ko, T. Y., Kang, Y. C., Kim, D. W., & Kim, S. J. (2024). Electrostatic assembly of charge-modulated MXene and graphene oxide into conductive porous nanohybrids via rapid annealing for EMI shielding. *Applied Materials Today*, *39*, 102271. DOI: 10.1016/j.apmt.2024.102271

Xiao, W., Li, B., Yan, J., Wang, L., Huang, X., & Gao, J. (2023). Three dimensional graphene composites: Preparation, morphology and their multi-functional applications. *Composites. Part A, Applied Science and Manufacturing*, *165*, 107335. DOI: 10.1016/j.compositesa.2022.107335

Xie, Z., Ventura, H., & Ardanuy, M. (2024). A Review of the Electrical Conductivity Test Methods for Conductive Fabrics. *Textiles*, *4*(3), 284–308. DOI: 10.3390/textiles4030017

Xiong, Y., Zhang, D., Ruan, X., Jiang, S., Zou, X., Yuan, W., Liu, X., Zhang, Y., Nie, Z., Wei, D., Zeng, Y., Cao, P., & Zhang, G. (2024). Artificial intelligence in rechargeable battery: Advancements and prospects. *Energy Storage Materials*, *73*, 103860. DOI: 10.1016/j.ensm.2024.103860

Yan, J., Ren, C. E., Maleski, K., Hatter, C. B., Anasori, B., Urbankowski, P., Sarycheva, A., & Gogotsi, Y. (2017). Flexible MXene/Graphene Films for Ultrafast Supercapacitors with Outstanding Volumetric Capacitance. *Advanced Functional Materials*, *27*(30), 1701264. Advance online publication. DOI: 10.1002/adfm.201701264

Yang, H., Fang, L., Yuan, Z., Teng, X., Qin, H., He, Z., Wan, Y., Wu, X., Zhang, Y., Guan, L., Meng, C., Zhou, Q., Wang, C., Ding, P., Hu, H., & Wu, M. (2023). Machine learning guided 3D printing of carbon microlattices with customized performance for supercapacitive energy storage. *Carbon*, *201*, 408–414. DOI: 10.1016/j.carbon.2022.08.083

Yang, J., Bao, W., Jaumaux, P., Zhang, S., Wang, C., & Wang, G. (2019). MXene-Based Composites: Synthesis and Applications in Rechargeable Batteries and Supercapacitors. *Advanced Materials Interfaces*, *6*(8), 1802004. Advance online publication. DOI: 10.1002/admi.201802004

Yang, Z., & Buehler, M. J. (2022). High-Throughput Generation of 3D Graphene Metamaterials and Property Quantification Using Machine Learning. *Small Methods*, *6*(9), 2200537. Advance online publication. DOI: 10.1002/smtd.202200537

Yogesh, G. K., Nandi, D., Yeetsorn, R., Wanchan, W., Devi, C., Singh, R. P., Vasistha, A., Kumar, M., Koinkar, P., & Yadav, K. (2025). A machine learning approach for estimating supercapacitor performance of graphene oxide nano-ring based electrode materials. *Energy Advances*, *4*(1), 119–139. DOI: 10.1039/D4YA00577E

Yoo, H. D., Park, Y., Ryu, J. H., & Oh, S. M. (2011). Electrochemical activation behaviors studied with graphitic carbon electrodes of different interlayer distance. *Electrochimica Acta*, *56*(27), 9931–9936. DOI: 10.1016/j.electacta.2011.08.117

Zeng, C., Xie, F., Yang, X., Jaroniec, M., Zhang, L., & Qiao, S. (2018). Ultrathin Titanate Nanosheets/Graphene Films Derived from Confined Transformation for Excellent Na/K Ion Storage. *Angewandte Chemie*, *130*(28), 8676–8680. DOI: 10.1002/ange.201803511

Zhang, D., Xu, Z., Huang, Z., Gutierrez, A. R., Blocker, C. J., Liu, C.-H., Lien, M.-B., Cheng, G., Liu, Z., Chun, I. Y., Fessler, J. A., Zhong, Z., & Norris, T. B. (2021). Neural network based 3D tracking with a graphene transparent focal stack imaging system. *Nature Communications*, *12*(1), 2413. DOI: 10.1038/s41467-021-22696-x

Zhao, X., Ruan, C., Sang, Y., Wang, S., & Liu, H. (2024). MXenes and their composites for advanced cathodes in multivalent ion batteries. *Energy Materials*, *4*(6). Advance online publication. DOI: 10.20517/energymater.2024.48

Zheng, B., & Gu, G. X. (2020). Machine Learning-Based Detection of Graphene Defects with Atomic Precision. *Nano-Micro Letters*, *12*(1), 181. DOI: 10.1007/s40820-020-00519-w

Zheng, B., Zheng, Z., & Gu, G. X. (2022). Designing mechanically tough graphene oxide materials using deep reinforcement learning. *npj Computational Materials*, *8*(1), 225. DOI: 10.1038/s41524-022-00919-z

Zhi, D., Li, T., Li, J., Ren, H., & Meng, F. (2021). A review of three-dimensional graphene-based aerogels: Synthesis, structure and application for microwave absorption. *Composites. Part B, Engineering*, *211*, 108642. DOI: 10.1016/j.compositesb.2021.108642

Zhou, T., Wu, C., Wang, Y., Tomsia, A. P., Li, M., Saiz, E., Fang, S., Baughman, R. H., Jiang, L., & Cheng, Q. (2020). Super-tough MXene-functionalized graphene sheets. *Nature Communications*, *11*(1), 2077. DOI: 10.1038/s41467-020-15991-6

Chapter 8

Developing Nanostructured Calcium-Based Bio-Ceramic Materials From Natural Source: Potential Applications and Future Directions

Ratna Sunil Buradagunta
https://orcid.org/0000-0001-9855-7808
Prince Mohammad Bin Fahd University, Saudi Arabia

Ravikumar Dumpala
https://orcid.org/0000-0003-1500-8809
Visvesvaraya National Institute of Technology, India

ABSTRACT

Synthesizing calcium based ceramic materials from natural source such as fish bones, sea corals and egg shells is cost effective route to produce nanocrystalline bio-ceramic materials. These nano-ceramic phases contain special ions which help to induce additional benefits when used as biomaterials. Since, the size of the produced phases is in nano-level, applications including local drug delivery, surface coatings on implants and producing nano-composites can be benefited due to the promising properties of nano-crystalline materials. The present chapter provides a comprehensive summary of the developments in synthesizing nano-crystalline calcium based ceramic materials from the natural sources for biomaterial applications. The

DOI: 10.4018/979-8-3373-3556-8.ch008

characterization, performance evaluation, and potential applications of using these nano-phases in biomedical applications are discussed in detail. Furthermore, future perspectives and the potential applications in medical fields are briefly presented.

1. INTRODUCTION

Calcium (Ca) based ceramic materials are the most promising and widely used biomaterials in several medical and health care applications (Ramesh et al. 2021). Since, the composition of bone contains Ca based phases in significant fractions, producing bio-ceramic materials with Ca as the important constituting element from different routes is one of the important objectives in the biomedical research across the globe (Dorozhkin, S.V. 2017), Nayak et al. 2024). The bio-ceramic materials which are used in medical and health care applications can be categorized as bioinert, bioactive and bioresorbable ceramic materials. Fig 1 shows the example of different bio-ceramic materials used in medical applications. Among them, Ca based ceramic materials are most promising and the range of their applications is widely spread into several targeted fields including bone filling, dental implants, drug delivery and scaffolds for medical procedures (Pereira et al. 2024, Miri et al. 2024, de Carvalho et al. 2024, Prasad Kumara et al. 2025). Compared with calcium silicates and calcium sulfates, calcium phosphate materials occupy significant place as a potential bio-ceramic material due to its resemblance with the calcium phase present in the natural bone. If macroscopically observed, bone is seen as a composite material composed of polymer phase (collagen), ceramic phase (Ca based) and other mineral phases along with the living tissue cells (Rodriguez-Palomo et al. 2024). Therefore, Ca based ceramic phases are well accepted by the human system and exhibit higher compatibility. Among the Ca based phases, hydroxyapatite ($Ca_{10}(PO_4)_6(OH)_2$) is a stable phase in physiological environment and well tolerated by human body (Rheima et al. 2024). Hydroxyapatite can establish strong bonding with the local tissue without causing any toxicity or adverse local effects (Liu et al. 2024).

Ca based bio-ceramic materials can be categorized based on the synthesis route. In synthetic route, chemical precursors are used to allow the chemical reaction to produce Ca based phase as a byproduct. In naturally derived route, raw materials from the natural sources such as sea corals, egg shells and animal bones are used to produce the ceramic phases by adopting specific procedures. Table 1 list the examples of the processes in both the routes to develop Ca based bio-ceramic phases. When HA is produced from natural sources, the morphology of the produced HA crystals, size and shape are completely different compared with the synthetic routes. Furthermore, the process parameters highly influence the characteristics of the produced HA from the natural source. Usually in pure HA prepared from lab, the

Ca/P ratio is observed as 1.67. However, in the naturally derived HA, the Ca/P ratio is varied from 1.5 to more than 2 depending on the type of source.

Figure 1. different types of bio-ceramic materials used in medical applications.

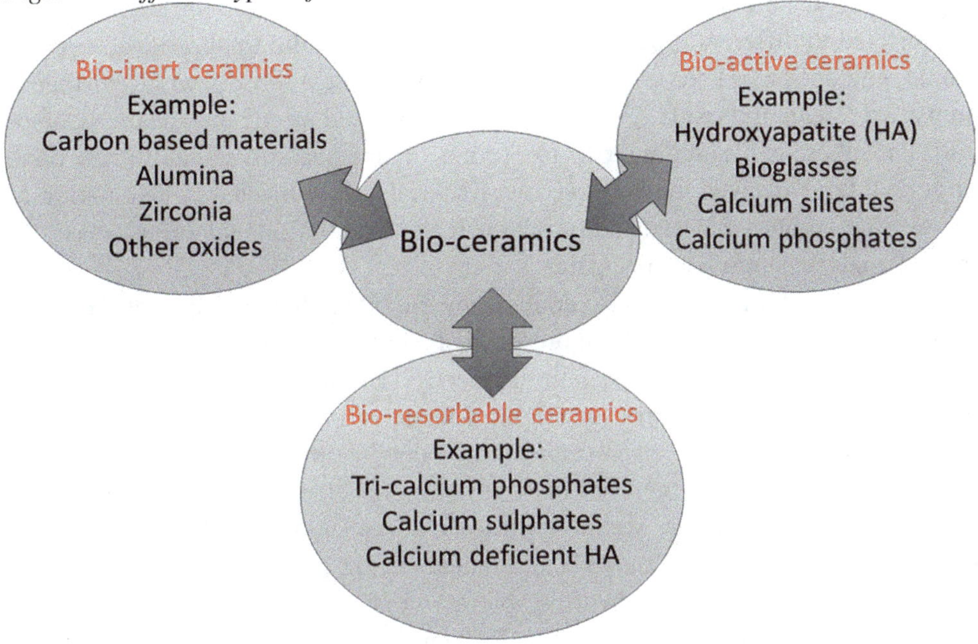

Table 1. Different strategies used to produce hydroxyapatite for medical applications

Sources	Method/process	Chemicals/materials
Synthetic chemicals	Wet chemical synthesis	$Ca(OH)_2 + H_3PO_4$ or $Ca(NO_3)_2 + (NH_4)_2HPO_4$
	Solid state reaction	$CaCO_3 + Ca_3(PO_4)_2$ or $CaCO_3 + Ca_2P_2O_7$
	Sol-gel method	$Ca(NO_3)_2 + C_6H_{15}O_3P$
Naturally derived materials	Sea corals	CaO derived from sea corals + $(NH_4)_2HPO_4$
	Egg shell	CaO derived from egg shells + H_3PO_4
	Fish bones	Calcination to convert calcium phosphates from bone to hydroxyapatite
	Animal bones	Isolating hydroxyapatite from the bone burning followed by acid treatment

2. NANO-HYDROXYAPATITE DEVELOPED FROM NATURAL SOURCES

Natural sources to develop nano-xydroxyapatite include sea corals, egg shell, fish bones and animal bones. Each material that is obtained from different sources is subjected to different process to get nano-hydroxyapatite final product. For example, egg shells and sea corals are used to produce carbonates which further is converted to calcium oxide. This calcium oxide is used as one chemical precursor with phosphorous based precursor to produce hydrosyapatite phase. In fish bone or animal bone derived method, the raw materials are subjected to calcination to remove the organic materials and to convert the calcium -phosphorous phases in the raw materials into hydroxyapatite.

Since, the natural sources also contain several other chemical species, HA produced from natural sources exhibit presence of addition ions such as Mg, Sr, Zn etc. The presence of these ions in the naturally derived HA serve some special purposes where these ions play vital role in the bio-functions. For example, Zn can help to promote the cell growth. Sr can improve the bone health and Mg is a cell growth micronutrient. Therefore, presence of these additional elements in the HA produced from natural source brings additional benefits compared with the stochiometric lab prepared HA which has only Ca and P elements in it. Fig 2 presents schematic illustration of producing nano-hydroxyapatite from fish bones or animal bones. The type of source used to produce HA also influence the final powder characteristics. For example, different egg shells contain different forms of carbonate phases. When these phases are extracted during the development of HA, the resulting powder morphology, crystal size and shapes may vary depending on the elemental composition of the selected egg shell. The following sections briefly discuss the development in producing nano-Ca based phases from different natural sources.

Figure 2. schematic illustration of producing HA by natural source by two different approaches.

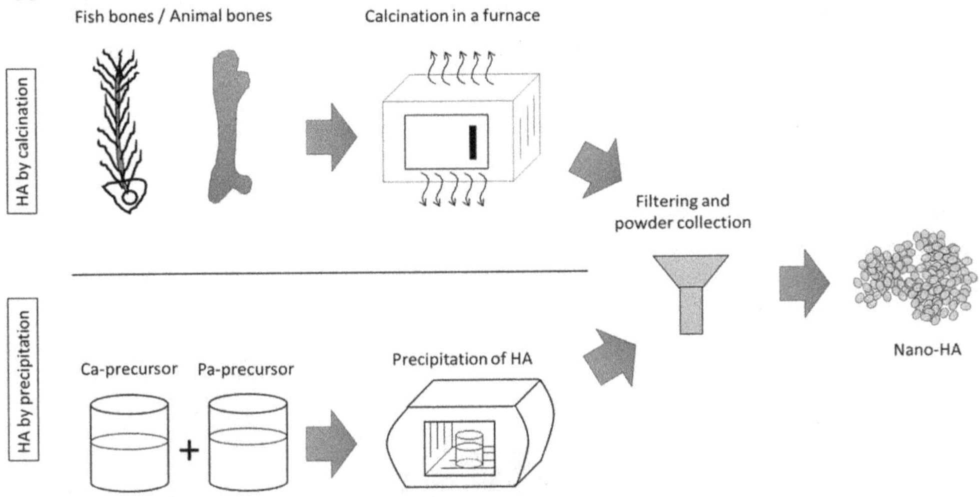

2.1 Nano-hydroxyapatite from fish bones and scales

Fish bones and scales are the waste materials and the weight of this disposed waste is of a few millions of tons per year across the globe due to the increased consumption as human food, animal feed and in medicinal applications. The fish bones or scales are rich in calcium-phosphorous mineral phases which is the source to produce nano-HA. In addition to Ca and P phases, other ions also present in fishbones which also helps to increase the functionality of hydroxyapatite produced from the fish bones. In order to produce nano-HA from fish bones, initially, the collected fish bones are cleaned to remove the organic material and flesh. Then the cleaned bones are subjected to hating to a temperature where the collagen content in the bones is evaporated. Then the left-over powder is filtered by using different solvents to remove the undesired ions to produce nano-hydroxy apatite. Fig 3 presents typical nano-hydroxyapatite produced by *Rohio* fish bones available in Indian rivers.

2.2 Nano-hydroxyapatite from egg shells

Producing nano-hydroxyapatite by using egg shells as the natural source is another way widely reported in the literature. Egg shells are rich in calcium carbonate and are the best suitable for Ca source. Form the egg shells, Ca is derived and several routes are adopted to produce nano-HA. For example, the derived Ca is allowed to react with P source by chemical reaction to produce nano-HA. In another way, the produced Ca is used to prepare solution and by sol-gel method, nano-HA is developed. When egg shells are heated to a higher temperature more than 700 °C, carbon dioxide is evolved and CaO is remained as source for Ca. The produced CaO is now mixed with water to produce $Ca(OH)_2$. When, chemical precursor for Pa is allowed to react with $Ca(OH)_2$, nano-hydroxyapatite is resulted.

Figure 3. Transmission electron microscope image of nano-HA produced from Rohio fish bones

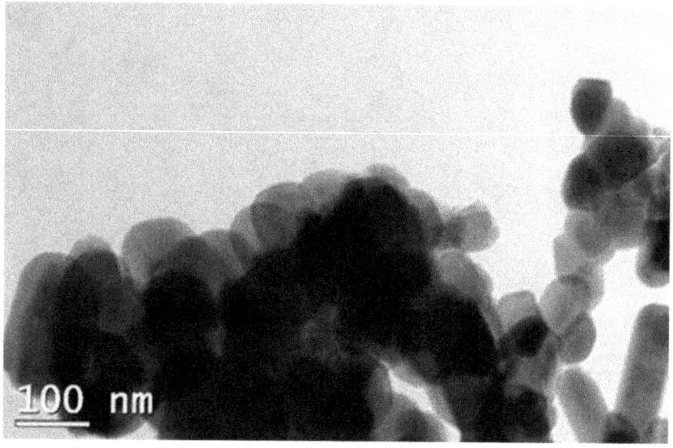

2.3 Nano-hydroxyapatite from sea corals

The sea corals are made of calcium carbonate ($CaCO_3$) which is a rich source for Ca to produce hydroxyapatite. Sea corals are heated to a temperature more than 900 °C to convert the $CaCO_3$ into CaO. The produced CaO is used to produce $Ca(OH)_2$ for chemical reaction in the presence of P containing precursor. On the other hand, the CaO can be used to produce a gel to derive nano-HA through sol-gel method. By controlling the process parameters, the crystallite size of hydroxyapatite can be controlled. Due to the abundant availability of sea corals, producing nano-HA is economically feasible.

2.4 Nano-hydroxyapatite from animal bones

Some of the animal bones including bovine, sheep, camel and horse are widely used to produce hydroxyapatite. Initially, the bones are cleaned to remove organic material and subjected to heating to a temperature around 1000 °C to convert the calcium phosphate phase present in the bone to convert into hydroxyapatite. The raw material is immersed in different acidic solutions to remove proteins from the bones and cleaned thoroughly. Then the resulted powder is pulverized to decrease the hydroxyapatite particle size.

There are other natural sources for Ca to produce nano-hydroxyapatite such as Mollusk shells, shrimp shells, algae and certain plants. Among most of the reported works, the objective of selecting different natural materials is to produce Ca containing material which is allowed to react with P containing chemical solution. In some other methods where the Ca and P phase which present in the source material is transformed to hydroxyapatite by applying heat energy. Table 2 presents the findings of a few studies which demonstrated synthesizing nano-HA from different natural sources.

Table 2. important observations in developing nano-HA from natural sources

Source	Important observations	Reference
Fish bone	• Nano-HA was produced by calcination • The produced HA was having calcium deficient nature • After heating, transformation to tri-calcium phosphate was reported	(Pal et al. 2017; Sunil & Jagannatham, 2016)
Fish bone	• HA was produced by alkaline heat treatment • The produced HA has higher Ca/P ratio than the 1.67	(Venkatesan et al. 2011)
Fish scales	• HA was produced by calination • The produced HA was found with higher Ca/P ratio	(Panda et al. 2014; Paul et al. 2017)
Fish scales	• HA was produced from alkaline heat treatment • The produced HA was observed with lower Ca/P ratio	(Pon-On et al. 2016)
Bovine bone	• HA was produce by alkaline hydrothermal process • The produced HA was rich in Ca and exhibited nano-level crystal size	(Barakat et al. 2009; Hosseinzadeh et al. 2014)
Camel bone	• HA was produced by calcination • Above 1000 °C, the produced HA was stable	(Jaber et al. 2018)
Horse bone	• The produced HA by calcination method was observed with higher Ca/P ratio	(Rahavi et al. 2017)
Egg shell	• HA was produced by calcination followed by precipitation • The produced HA was observed with more Ca/P ratio than 1.67	(Goloshchapov et al. 2013)

continued on following page

Table 2. Continued

Source	Important observations	Reference
Egg shell	• HA was produced by precipitation method • The needle like morphology HA was resulted with less than 5 nm crystalline size • The HA has lower Ca/P ratio (1.425)	(Neelakandeswari et al. 2011)
Cockle shell	• HA was produced by calcination followed by sol-gel method • Resulted in HA with higher Ca/P ratio	(Mustaffa et al. 2015; Mohamad Razali et al, 2016)
Moringaoleifera leaves	• HA was produced from rapid microwave irradiation • HA with 1.72 Ca/P ratio was observed • Flake like morphology was observed for the crystals	(Govindaraj & Rajan, 2016)
C. officinalisred algae	• HA was produced by pyrolysis followed by alkaline hadrothermal process • Due to the lower Ca/P ratio, after heating, tri-calcium phosphate was resulted.	(Walsh et al. 2008)

3. APPLICATIONS OF NANO-HYDROXYAPATITE DERIVED FROM NATURAL SOURCE

Due to the promising properties such as biocompatibility, bioactivity, economical feasibility and easy to process, developing nano-HA from natural sources is a viable approach to produce bioceramic materials which has a wider application in medical and health care sector. The following sections briefly discuss the potential applications of nano-HA derived from the natural sources.

3.1 Biomedical applications

In tissue engineering applications, producing bone grafting structures by using nano-HA produced by using polymer and other metals as the matrix materials results promising outcome. For bone grafting usage, producing composite materials by using nano-HA in bone repairing and bone fixing applications is another area where nano-HA derived from natural source can become promising material. To produce 3D structures for implant and prosthetics applications, combining polymer or metals with nano-HA can be a promising material. By using HA derived from natural source enhances cell adhesion, proliferation, and differentiation that promotes the healing rate.

Dental applications is another area where more benefits can be yielded by suing these naturally derived HA. Tooth remineralization is one area in which the loss of enamel lesions can be addressed. Naturally derived HA can be used as a filling material to address the bone loss issues and cavity issues. Using naturally derived HA as a suitable coating material on the metallic implant used in dentistry is another potential area which can be benefited. Drug delivery applications is another area where naturally derived HA can be used to achieved more benefits to deliver the drug to selective regions or to the targeted tissue. Antibiotics, hormones, and specific drugs can be delivered to the targeted cite in treating cancer or other deceases. Naturally derived HA can be used as a potential coating material on metallic implants to increase the surface functional properties. Similar to that of human applications, animal treatments and health management applications also can be benefited from the naturally derived HA.

3.2 Skin care and cosmetics applications

Using Ca based phases in skin care lotions and creams help to treat the skin issues, regenerate the tissues and address the skin care problems. Other cosmetics also uses Ca based nano-powders to increase the results when used as skin care medicine. Addition of Ca based nano-phases into tooth paste is adopted to increase the polishing effect to decrease the decoloring effect of teeth.

3.3 Agriculture and food industry

Food industry and agriculture engineering is the other fields where naturally derived Ca phases can be used as promising materials. Ca based phases can be used in supplements and medicinal applications. Furthermore, as a nutrient to address the deficiency, Ca supplements can be produced. In agriculture and soil management applications, Ca based nano-phase can be used as a nutrient carrier to supply important elements to the roots of plants. Nano-HA can be used as a potential agent to remove metallic contamination due to its higher surface area.

4. SUMMARY AND FUTURE PERSPECTIVES

In summary, the most of the natural sources may serve as a precursor to the synthesis of nano-HA. Raw materials from plants, minerals, shell detritus, and animals (mammalian and aquatic/marine) produces HA with varying chemical composition and properties. Ca/P ratio in HA is important as the composition plays crucial role on its degradation in the biological environment. Ca/P ratio equals to 1.67 is considered as stochiometric ratio. HA with lower amount of Ca is considered as calcium deficient. Depending on the source of the material, the resulted

HA exhibits different CA/P ratio. HA with Ca/P ratio less than 1.67 degrades in the physiological environment and with higher Ca/P ratio, HA can sustain in the physiological environment with more stability. HA produced from natural source if heated above 650 ° C, transformation to β-tricalcium phosphate decreases Ca/P ratio to lower values. Calcination and alkaline hydrolysis methods are widely adopted to produce HA from these natural sources. The morphologies of the produced nano-HA crystals were observed as round, needle like structure or acicular morphology. From the literature, it is also observed that the morphology of produced nano-HA does not influenced by the type of the source. A few methods need secondary follow up process to reduce the particle size by mechanical milling or crushing.

Overall, the development of nano-HA by natural source is a simple and viable route to produce Ca based phases for biomedical applications. However, there are several challenges involved in producing HA by natural sources compared with synthetic methods which include inconsistency in the chemical composition. Since, each raw material contains varying chemical composition, inconsistency in Ca/P ratio is a challenge. Based on the selected source, region, environment, natural sources may differ in their chemical composition. Standardizing the final HA powder characteristics is also difficult due to the non-uniformity in the raw material source. Furthermore, difficulty in completely removing organic material from the base material is another technical challenge.

Calcination at higher temperatures is required to remove the organic material usually above 600 °C which may lead to instability in the HA phase and may results β-tri calcium phosphate or calcium oxide. Since, the raw material is from natural source, possibility of contamination or presence of heavy metals are the other challenges which essentially need purifying processes which further increase the complexity of the process. HA particle size variations due to different material source and the methods is another limitation. IN order to produce nano-level HA, powder crushing or milling is required in some cases. Obtaining precise control over the particle size is technically challenging. The produced HA particles are observed with porosity and variations in the surface area values. Since the raw material is from various sources, regulating and standardizing the composition are difficult tasks. The availability of the raw materials to produce HA in a large scale and the reproducibility of HA with the same quality are the other challenges. However, in the increased need of high performing bio-ceramic materials for the medical and health care sectors, hunting for new sources and methods uncovered the production of valuable HA from the waste and natural sources for biomedical applications. Additional research and in-depth studies to address the above limitations and to overcome the challenges help to produce naturally derived HA at economically cheaper prices which preform similar to that of pure HA produced from the synthetic routes. Additional sophisticated techniques along with advanced electron microscopy, spectroscopy and XRD methods can be used to understand the properties and the potential of these naturally derived HA to make it as a potential candidate for novel medical applications.

REFERENCES

Barakat, N. A. M., Khil, M. S., Omran, A. M., Sheikh, F. A., & Kim, H. Y. (2009). Extraction of pure natural hydroxyapatite from the bovine bones bio waste by three different methods. *Journal of Materials Processing Technology*, *209*(7), 3408–3415. DOI: 10.1016/j.jmatprotec.2008.07.040

de Carvalho, A. B. G., Rahimnejad, M., Oliveira, R. L., Sikder, P., Saavedra, G. S., Bhaduri, S. B., & Bottino, M. C. (2024). Personalized bioceramic grafts for cranio-maxillofacial bone regeneration. *International Journal of Oral Science*, *16*(1), 62. DOI: 10.1038/s41368-024-00327-7

Dorozhkin, S. V. (2017). Calcium Orthophosphate-Based Bioceramics and Its Clinical Applications. In Kaur, G. (Ed.), *Clinical Applications of Biomaterials*. Springer., DOI: 10.1007/978-3-319-56059-5_5

Goloshchapov, D. L., Kashkarov, V. M., Rumyantseva, N. A., Seredin, P. V., Lenshin, A. S., Agapov, B. L., & Domashevskaya, E. P. (2013). Synthesis of nanocrystalline hydroxyapatite by precipitation using hen's eggshell. *Ceramics International*, *39*(4), 4539–4549. DOI: 10.1016/j.ceramint.2012.11.050

Govindaraj, D., & Rajan, M. (2016). Synthesis and spectral characterization of novel nanohydroxyapatite from Moringaoleifera leaves. *Materials Today: Proceedings*, *3*(6), 2394–2398. DOI: 10.1016/j.matpr.2016.04.153

Hosseinzadeh, E., Davarpanah, M., Nemati, N. H., & Tavakoli, S. A. (2014). Fabrication of a hard tissue replacement using natural hydroxyapatite derived from bovine bones by thermal decomposition method. *International Journal of Organ Transplantation Medicine*, *5*, 23–31.

Jaber, H. L., Hammood, A. S., & Parvin, N. (2018). Synthesis and characterization of hydroxyapatite powder from natural Camelus bone. *J. Aust. Ceram. Soc.*, *54*(1), 1–10. DOI: 10.1007/s41779-017-0120-0

Liu, C., Xu, M., Wang, Y., Yin, Q., Hu, J., Chen, H., Sun, Z., Liu, C., Li, X., Zhou, W., & Liu, H. (2024). Exploring the potential of hydroxyapatite-based materials in biomedicine: A comprehensive review. *Materials Science and Engineering R Reports*, *161*, 100870. DOI: 10.1016/j.mser.2024.100870

Miri, Z., Haugen, H. J., Loca, D., Rossi, F., Perale, G., Moghanian, A., & Ma, Q. (2024). Review on the strategies to improve the mechanical strength of highly porous bone bioceramic scaffolds. *Journal of the European Ceramic Society*, *44*(1), 23–42. DOI: 10.1016/j.jeurceramsoc.2023.09.003

Mohamad Razali, N. A. I., Pramanik, S., Abu Osman, N. A., & Radzi, Z., Pingguan-Murphy, B. (2016). Conversion of calcite from cockle shells to bioactive nanorod hydroxyapatite for biomedical applications. *Journal of Ceramic Processing Research*, *17*, 699–706.

Mustaffa, R., Mohd Yusof, M. R., & Abdullah, Y. (2015). A novelty of synthetic hydroxyapatite from cockle shell and characterization. *Advanced Materials Research*, *1087*, 429–433. DOI: 10.4028/www.scientific.net/AMR.1087.429

Nayak, A. K., Maity, M., Barik, H., Behera, S. S., Dhara, A. K., & Hasnain, M. S. (2024). Bioceramic materials in bone-implantable drug delivery systems: A review. *Journal of Drug Delivery Science and Technology*, *95*, 105524. DOI: 10.1016/j.jddst.2024.105524

Neelakandeswari, N., Sangami, G., Dharmaraj, N., (2011). Preparation and characterization of nanostructured hydroxyapatite using a biomaterial. Synthesis and Reactivity in inorganic, metal-organic, Nano Metal. Chem. ume 41, 513–516.

Pal, A., Paul, S., Choudhury, A. R., Balla, V. K., Das, M., & Sinha, A. (2017). Synthesis of hydroxyapatite from Lates calcarifer fish bone for biomedical applications. *Materials Letters*, *203*, 89–92. DOI: 10.1016/j.matlet.2017.05.103

Panda, N. N., Pramanik, K., & Sukla, L. B. (2014). Extraction and characterization of biocompatible hydroxyapatite from fresh water fish scales for tissue engineering scaffold. *Bioprocess and Biosystems Engineering*, *37*(3), 433–440. DOI: 10.1007/s00449-013-1009-0

Paul, S., Pal, A., Choudhury, A. R., Bodhak, S., Balla, V. K., Sinha, A., & Das, M. (2017). Effect of trace elements on the sintering effect of fish scale derived hydroxyapatite and its bioactivity. *Ceramics International*, *43*(17), 15678–15684. DOI: 10.1016/j.ceramint.2017.08.127

Pereira, L. F., de Azevedo-Silva, L. J., Minim, P. R., Lisboa-Filho, P. N., Fortulan, C. A., Griggs, J. A., & Borges, A. F. S. (2024). Are we approaching the development of a novel calcium phosphate-based bioceramic dental material? *Dental Materials*, *40*(11), 1843–1853. DOI: 10.1016/j.dental.2024.07.019

Pon-On, W., Suntornsaratoon, P., Charoenphandhu, N., Thongbunchoo, J., Krishnamra, N., & Tang, I. M. (2016). Hydroxyapatite from fish scale for potential use as bone scaffold or regenerative material. *Materials Science and Engineering C*, *62*, 183–189. DOI: 10.1016/j.msec.2016.01.051

Prasad Kumara, P. A. A. S., Cooper, P. R., Cathro, P., Gould, M., Dias, G., & Ratnayake, J. (2025). Bioceramics in Endodontics: Limitations and Future Innovations—A Review. *Dentistry Journal, 13*(4), 157. DOI: 10.3390/dj13040157

Rahavi, S. S., Ghaderi, O., Monshi, A., & Fathi, M. H. (2017). A comparative study on physicochemical properties of hydroxyapatite powders derived from natural and synthetic sources. *Russian Journal of Non-Ferrous Metals, 58*(3), 276–286. DOI: 10.3103/S1067821217030178

Ramesh, N., Ratnayake, J. T., & Dias, G. J. (2021). Calcium-based ceramic biomaterials. In *Structural biomaterials* (pp. 333–394). Woodhead Publishing. DOI: 10.1016/B978-0-12-818831-6.00011-2

Rheima, A. M., Abdul-Rasool, A. A., Al-Sharify, Z. T., Zaidan, H. K., Athair, D. M., Mohammed, S. H., & kianfar, E. (2024). Nano bioceramics: Properties, applications, Hydroxyapatite, Nanohydroxyapatite and drug delivery. *Case Studies in Chemical and Environmental Engineering, 10*, 100869. DOI: 10.1016/j.cscee.2024.100869

Rodriguez-Palomo, A., Østergaard, M., & Birkedal, H. (2024). Bone hierarchical structure: Heterogeneity and uniformity. *Advanced Functional Materials, 34*(35), 2307026. DOI: 10.1002/adfm.202307026

Sunil, B. R., & Jagannatham, M. (2016). Producing hydroxyapatite from fish bones by heat treatment. *Materials Letters, 185*, 411–414. DOI: 10.1016/j.matlet.2016.09.039

Venkatesan, J., Qian, Z. J., Ryu, B., Thomas, N. V., & Kim, S. K. (2011). A comparative study of thermal calcination and an alkaline hydrolysis method in the isolation of hydroxyapatite from Thunnus obesus bone. *Biomedical Materials (Bristol, England), 6*(3), 1–12. DOI: 10.1088/1748-6041/6/3/035003

Walsh, P. J., Buchanan, J. F., Dring, M., Maggs, C., Bell, S., & Walker, G. M. (2008). Low-pressure synthesis and characterisation of hydroxyapatite derived from mineralise red algae. *Chemical Engineering Journal, 137*(1), 173–179. DOI: 10.1016/j.cej.2007.10.016

Chapter 9
Creating a Sustainable Future:
The Emergence of Eco-Friendly Materials

Gaydaa Al Zohbi
https://orcid.org/0000-0002-0697-993X
Prince Mohammad Bin Fahd University, Saudi Arabia

ABSTRACT

The growing environmental issues of the 21st century demand a paradigm shift in our material choices. This chapter investigates the emerging field of sustainable materials, analyzing their definition, classification, and different applications across industries. It explores the lifecycle evaluation of materials, emphasizing the significance of circular economy principles and minimized environmental impact. From bio-based polymers and recycled composites to advanced low-carbon concrete and sustainably sourced timber, the chapter features the advanced materials leading to a more sustainable future. Furthermore, it addresses the obstacles to broad implementation and highlight strategies for promoting sustainable material integration in design and manufacturing. This chapter aims to provide a comprehensive overview for researchers, practitioners, and policymakers seeking to embrace a material-conscious approach to building a resilient and sustainable world.

1- INTRODUCTION

The 21st century has faced considerable environmental issues resulting from rapid industrialization population growth, and climate change. Currently, climate change is considered as one of the most pressing issues, marked by increasing global

DOI: 10.4018/979-8-3373-3556-8.ch009

temperatures, melting ice cap, and extreme weather events. Human activities are the main contributor to the rise of greenhouse gas emission, mainly by burning fossil fuels and deforestation, resulting in severe ecological disruptions. Climate change, pollution, habitat destruction, and overexploitation led to unprecedented rates of biodiversity loss. This loss endangers ecosystems and the services they offer, including clean water, pollination, and carbon sequestration. Pollution of water, air, and soil continue to be pressing challenges, worsened by industrial activities, plastic waste, and agricultural runoff. The widespread use of single-use plastics and chemicals presents serious health risks to both humans and wildlife. Additionally, finite natural resources are being extracted at unsustainable rates, resulting in the depletion of essential materials like minerals, water, and forests. This overconsumption threatens future generations' ability to fulfill their needs. Choosing sustainable materials, including renewable, biodegradable, and recyclable, can greatly improve resource efficiency, minimize waste, and reduce the extraction of finite resources. For example, materials sourced from sustainable practices typically have a reduced ecological impact compared to those obtained through harmful extraction methods. This could tackle pollution and protect biodiversity. Low carbon materials could considerably minimize the overall carbon footprint of products and buildings. Moreover, the choice of sustainable materials frequently aligns with ethical considerations, such as fair labor practices and community support, promoting social equity and supports local economies.

Sustainable materials are defined as materials that are produced, used, and disposed of in a manner that minimizes environmental impact while promoting social and economic benefits (Ljungberg, 2007; Szekeres & Jeswiet, 2013). These materials should be sourced from renewable resources, designed for longevity, and capable of being recycled or safely decomposed at the end of their life cycle (Figure 1) (Ljungberg, 2007). The aim is to reduce dependency on non-renewable resources, lower carbon emissions, and mitigate ecological degradation. According to the United Nations Environment Programme (UNEP), sustainable materials should "meet the needs of the present without compromising the ability of future generations to meet their own needs" (Basiago, 1995). This perspective emphasizes a holistic approach to material selection, considering the entire lifecycle from extraction to end-of-life.

Figure 1. Key attributes and desired properties of sustainable materials

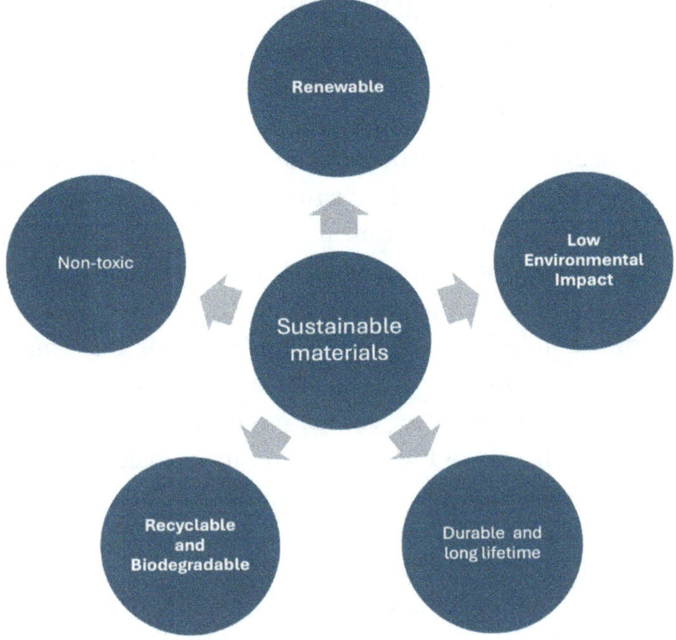

The concept of sustainable materials has developed considerably in response to heightened environmental awareness and concerns about resource depletion. Initially, from the pre-industrial era to the early 20th century, while the term "sustainable materials" wasn't used early thinkers acknowledged the effects of human activities on the environment. The mid-20th century witnessed the emergence of environmentalism, driven by impactful works like Rachel Carson's "Silent Spring," which emphasized pollution and conservation. The 1970s energy crisis renewed focus on sustainability, resulting in "climatic architecture" that reduced energy consumption. The 1987 UN Brundtland Commission report popularized "sustainable development." In the 1990s, sustainable building rating systems were introduced, reflecting an increasing consumer demand for eco-friendly products. Currently, the emphasis has shifted to the circular economy, highlighting the entire life cycle of materials, waste reduction, and the development of innovative sustainable materials like bio-based and biodegradable option.

Several initiatives have been launched to promote the sustainable materials. European Union (EU) Green Deal is one of the initiatives, that launched in 2019, with a goal to make Europe climate-neutral by 2050, advancing sustainable materials through a holistic approach (Miłek et al., 2022). It includes the Circular Economy Action Plan, which fosters resource reuse and waste reduction, and the revision of the Construction Products Regulation (CPR), which enhances sustainability standards for construction materials (Lützkendorf, 2022). The EU Taxonomy for Sustainable Activities categorizes environmentally sustainable economic activities to direct investments (Tonnarello et al., 2025). The Industrial Emissions Directive aims to reduce industrial pollution while encouraging sustainable production methods (Romero-Castro et al., 2022). In addition, the Renovation Wave Strategy improves energy efficiency in buildings, boosting demand for low-carbon materials (Scenari, 2020). United States (US) Inflation Reduction Act (IRA) implemented in 2022 to tackle climate change via substantial measures with the aim to foster sustainable materials, mainly in energy-intensive sectors (Kleimann et al., 2023). It provides tax credits for clean energy investments, which indirectly enhances demand for sustainable materials utilized in technologies such as solar panels and wind turbines. The IRA provides funding for industrial decarbonization, encouraging industries like cement and steel to implement lower-carbon production techniques. Furthermore, it promotes federal purchasing of low-embodied carbon materials for infrastructure projects, generating significant market demand. In 2022, published a low-carbon concrete specification was published by the Bank for International Settlements (BIS) emphasizing the financial risks linked to embodied carbon in construction. It sets definitive limits on the permissible embodied carbon in concrete, promoting the use of sustainable materials and advanced production techniques. The focus on transparency and precision, mainly via Environmental Product Declarations (EPDs), emphasizes the significance of monitoring carbon footprints on concrete. The BIS showed a commitment to sustainable practices and affecting purchasing standards in the wider construction sector through applying these standards to its own construction projects.

The growth of demand for sustainable materials stimulates innovation in materials science and technology (Chourpagar et al.,). This led to generate economic opportunities, ranging from job creation to the advancement of green technologies (Kunde & Sehgal, 2021).

This chapter presents a resource for researcher and policy maker to support a material-conscious approach for sustainability and to facilitate informed decision-making in materials selection. The main goals of this chapter are:

- Establish a definition and a classification of sustainable materials and their different applications
- Evaluate the lifecycle of materials and its importance in reducing their environmental impact
- Discuss the importance of circular economy principle in the use of sustainable materials.
- Identify the main challenges facing the widespread adoption of sustainable materials.

- Suggest some approaches for promoting the implementation of sustainable materials in design and manufacturing process

2- CLASSIFICATION OF SUSTAINABLE MATERIALS

Sustainable materials could be categorized into different categories based on their source and origin, lifecycle and impact, and application and function (Figure 2). The classification of materials is crucial in the assessment of their environmental impact during their lifespan. The choice of sustainable materials, such as recyclable and biodegradable materials, could considerably minimize their ecological footprints and assist in resource preservation. The classification of materials enables businesses to make informed decisions that can result in cost savings. With the implementation of stricter environmental regulations by governments and organizations, understanding material classifications is essential for compliance. Compagnies that prioritize using sustainable materials are more prepared to comply with regulatory standards and mitigate the risk of potential penalties. Additionally, compagnies that use sustainable materials categorization could foster innovation and set themselves apart in the marketplace. The adoption of advanced materials allows business to develop more sustainable products and enhance their competitive advantage. Moreover, the sustainable materials classification fosters social responsibility. By choosing ethically sourced materials, companies can contribute to local communities, advocate for fair labor practices, and strengthen their corporate social responsibility (CSR) efforts.

Materials can be categorized as state-of-the-art based on several criteria. First, they demonstrate advanced characteristics, including improved strength, lightweight design, or enhanced thermal and electrical conductivity. Additionally, state-of-the-art materials are characterized by their sustainability; they are produced with minimal environmental impact and include renewable, recyclable, or biodegradable options. The use of sustainable materials could be also distinguished in term of the cutting-edge manufacturing processes, such as 3D printing or nanotechnology, resulting in enhancing efficiency and performance. Their performance in real-world applications across various industries, including construction, automotive, and electronics, shows consistency and longevity. Adherence to existing environmental standards and regulations ensures safety and sustainability, while extensive market acceptance suggests that these materials are recognized as leaders in their field. By assessing materials based on these criteria, one can determine their status as state-of-the-art in sustainable applications.

Figure 2. Classification of sustainable materials based on different criteria (Calkins, 2008; Choudhury & Hashmi, 2020)

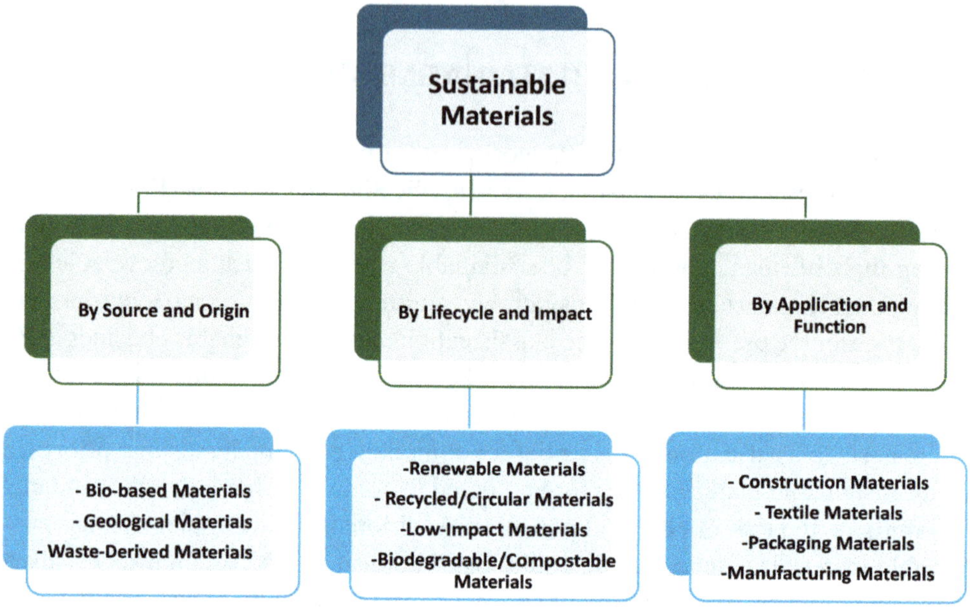

The main types of sustainable materials are presented below.

2.1 Biobased Materials

Biobased materials are derived from renewable biological resources, such as plants, microorganisms, and animals. These materials can be biodegradable and often have a lower carbon footprint compared to fossil fuel-based alternatives (Curran, 2000). Many bio-based materials could naturally decompose by biological processes, leading to minimize the environmental impact at the end of their lifecycle (Weiss et al., 2012). Additionally, the production of these materials emits less greenhouse gas compared to conventional materials. this can be explained by the fact that plants absorb the carbon dioxide during their growth, offsetting emission during generation process. There are different kinds of biobased materials including bio-based polymers, natural fibers, and bio-composites are example of biobased materials. Biobased polymers include Polylactic Acid (PLA) and Polyhydroxyalkanoates (PHA), and it could be used in disposable cutlery, packaging films, and 3D printing materials. These materials offer many benefits, PLA is compostable and PHA is biodegradable and could be synthesized by microbial fermentation (Pandit et al., 2018). Natural fibers include hemps, sisal, jute, and flax and could be used in insulation, textiles, and composites materials (Väisänen et al., 2017). Natural fibers are renewable and could be produced with less energy requirement compared to synthetic fibers. Bio-composites are made by combining natural fibers and biopolymers and could be used in construction materials, automotives parts, and furniture (Smitthipong et al., 2014). They offer significant benefits by minimizing carbon footprints since they used renewable resources, and by minimizing waste accumulation owing to their biodegradability. Additionally, these materials reduce the reliance on fossil fuels since they use natural resources. Moreover, bio-composites can use agricultural by-products and turn them into valuable materials (Smitthipong et al., 2014).

2.2 Recycled and upcycled materials

2.2.1 Recycled materials

Recycled materials are secondary virgin materials formed from waste products have been treated and reshaped into functional forms (Larsson, 2018). Recycling plays a crucial role in diverting waste from landfills, sustain raw resources, minimize energy consumption related to the extraction and processing of primary materials, aiming to minimize pollution. The recycling process depends on the kind of material, but the general process involve collection, storing and cleaning, processing, and then manufacturing. The waste materials should be collected from different sources (industrial, residential, and commercial), followed by a separation of the different kind of materials and eliminating contaminants. Then, the collecting and sorted materials will be converted into a form suitable for reuse in manufacturing. The conversion process could include melting metals and some plastics, shredding paper and plastic, chemical processing for plastics and some materials. The processed recycled materials used as a feedstock to form new products. Table 1 presents the recovering sources and the process of recycling of different recycled waste.

Table 1. Common recycled materials and their reclaiming processes (Hopewell et al., 2009; Obradovic & Mishra, 2020; Rajendran et al., 2012; Tan & Li, 2015; Wang, Yan, Fu et al, 2021)

Type	Recovering source	Processes	Examples
Recycled metals	end-of-life products, industrial scrap, and manufacturing waste	Collection, sorting (ferrous vs. non-ferrous), shredding, melting, refining, and casting or rolling into new forms	Recycled aluminum cans, recycled steel from construction debris, recycled copper from wiring
Recycled plastics	discarded products and processed into new plastic resins or products	Collection, sorting by resin type, cleaning, shredding, melting (mechanical recycling), or chemical breakdown into monomers (chemical recycling), followed by pelletizing and manufacturing into new plastic products	Recycled PET bottles, packaging straps, recycled HDPE, containers
Recycled aggregates	Construction and demolition waste	Collection, crushing, screening to remove debris and size the aggregate, and sometimes washing to remove contaminants	Recycled concrete, fill material, recycled asphalt pavement (RAP)
Recycled paper and cardboard	Waste paper and cardboard	Collection, sorting by grade, pulping with water and chemicals to break down fibers, cleaning to remove inks and contaminants, and then pressing and drying to form new paper sheets or cardboard	Recycled newspaper, office paper, packaging materials, and insulation
Recycled Glass	Waste glass	Collection, sorting by color, cleaning to remove contaminants, crushing into cullet, melting, and forming into new glass containers or other products	Recycled glass bottles and jars, fiberglass insulation, or glass aggregate.

2.2.2 Upcycled materials

Upcycled materials are waste products or unwanted materials that are repurposed into higher-quality or more valuable products (Larsson, 2018). Differing from recycling, which often fragments materials, Upcycling creatively repurposes items, extending their lifespan and minimizing waste while often improving their aesthetic or functional value (Bridgens et al., 2018). Upcycling processes are widely diverse and typically engage creative approaches for design and manual labor. The process depends on the initial material and the desired new product. The common process involves creative repurposing, transformation, deconstruction and reconstruction

(Bridgens et al., 2018). The creative repurposing aims to use an item for an alternative purpose than its original use without major modification. The modification of the initial item could be done by cutting, assembling, painting, or adding new components to form a new product. The deconstruction and reconstruction stage aims to disassemble waste items and utilize their parts to create entirely new products. Table 2 displays the recovering sources and the recycling process of different upcycled waste.

Table 2. Recovering sources and the process of recycling for different upcycled waste (Back et al., 2015; Latif, 2024; Pandit et al., 2019; Wang, Yan, Fu et al, 2021; Zhao et al., 2022)

Type	Recovering source	Processes	Example
Upcycled Textiles	Discarded clothing, fabric scraps, or textile waste	Cutting, sewing, patching, dyeing, and embellishing existing textiles to create new designs or functionalities	Old jeans repurposed into bags or patchwork quilts; vintage dresses altered into modern tops
Upcycled Wood	Discarded wood from furniture, pallets, construction debris, or fallen trees	Cleaning, sanding, cutting, joining, painting, or varnishing reclaimed wood to create new products with a unique character	Old wooden pallets transformed into coffee tables or vertical gardens
Upcycled Glass and Ceramics	Broken or unwanted glass bottles, jars, or ceramic items	Cutting, melting (for glass), crushing, and assembling fragments into new designs; painting or decorating existing items	Broken tiles and dishes used to create mosaic artwork or tabletops
Upcycled Metals	Scrap metal pieces, discarded tools, or industrial waste	Welding, bending, cutting, and assembling metal components into new forms; painting or finishing to enhance aesthetics	Old gears and machine parts welded together to create industrial-style lamps or sculptures
Upcycled Plastics and Other Materials	Various discarded plastic items, tires, electronic waste components	Cleaning, cutting, assembling, and creatively transforming these materials based on their unique properties	old tires repurposed into furniture or playground equipment; electronic waste components used to create artwork

2.3 Low-impact virgin materials

Although recycled and upcycled materials are essential for a sustainable future, there are cases where virgin materials remain necessary, or offer superior performance characteristics. The extraction and handling of conventional virgin materials usually have considerable environmental consequences. Low-impact virgin materials, or sustainable sourced materials. are newly extracted or harvested virgin materials that are handled and processed with a strong focus on reducing negative environmental impacts throughout their lifecycle. This involves conscientious sourcing, effective processing methods, minimized energy consumption, reduced emissions, and attention for biodiversity and ecosystem health. The main goal of low-impact materials is to use the virgin materials in a more sustainable and eco-friendly way.

Sustainable sourced materials are gathered sustainably to uphold ecological balance and benefit local communities (Olivetti & Cullen, 2018). Sustainable sourcing often involves certification systems to ensure environmentally friendly practices. FSC-certified timber, sustainably harvested bamboo, and organic cotton are some kinds of sustainably sourced materials. They could be used in furniture, flooring, and textiles. Sustainably sourced timber refers to wood forests managed to maintain biodiversity and ecological balance, significantly contributing to ecological conservation and climate action (Davies, 2016). It enhances resource conservation and product quality, and improve market opportunities, motivated by consumer demand for sustainable products. Enhancing sustainable forestry practices could promote economic growth and tackle ecological issues. Sustainable sources timber could be used in construction, flooring, and furniture. Low-Carbon Materials are designed to maintain a low carbon footprint throughout their manufacturing and utilization (Das et al., 2021). These materials frequently utilize cutting-edge technologies to lower greenhouse gas emissions. Low-carbon concrete, green steel, and energy-efficient insulation materials are some types of low carbon materials. They could be used in construction, infrastructure, and energy systems.

Many criteria are used to assess the environment impact of virgin materials and determine whether they could be qualified as low impact. The criteria are presented in Figure 3.

Figure 3. Key criteria used to assess the environment impacts of raw material (Manhart et al., 2019)

The extraction, transportation, and processing of the material should demand considerably less energy than traditional methods. This could include the used or renewable energy sources or installing energy efficiency technologies. The complete supply chain, spanning from sourcing to the first processing of the material, should emit lower greenhouse gasses (Martin et al., 2022). Additionally, leveraging resources that can regenerate at a pace equal to or greater than their depletion. Practices used to harvest or extract virgin materials should reduce habitat destruction, water pollution, soil damage, and any impact on biodiversity. An efficient resource use is another criterion to enhance the output from the virgin materials and reduce water during manufacturing. The consumption of water that should be minimized as well as water pollution are also one of the criteria. The materials and their manufacturing process should include a reduced amount of toxic chemicals that pose risks to human health or the environment (Martin et al., 2022). Transforming land from natural habitat should be reduced and sustainable land management principle should be prioritized. A comprehensive LCA assessment should show a considerably reduced overall environmental impact relative to traditional alternatives across multiple impact categories. Certification can certify that the material complies with established sustainability criteria. Table 3 displays some low-impact virgin materials with the criteria met.

Table 3. Criteria met in some low-impact virgin materials (Akter et al., 2022; Cywar et al., 2022; Goh et al., 2020; Khan & McNally, 2024; Sheppard et al., 2020)

Types	Criteria met	Examples
Advanced low-carbon concrete	-Lower emissions using supplementary cementitious materials (SCMs) -Alternative cements with lower clinker content -Using carbon capture technologies	Concrete incorporating high percentages of SCMs, geopolymer concrete, concrete with carbon-sequestering aggregates.
Sustainably harvested bamboo	-Renewable resource (rapid growth rate), -Lower energy input for processing compared to timber or steel - Minimal pesticides or fertilizers	Bamboo lumber, flooring, panels, scaffolding, and composite materials
Sustainably managed timber	lower embodied energy compared to concrete or steel. Certifications like FSC and PEFC	Engineered wood products (CLT, glulam) from certified forests, timber framing from sustainably sourced wood
Organic and low-impact natural fibers	Reduced water pollution and soil degradation, lower energy input in cultivation (for some fibers), avoidance of harmful chemicals	Organic cotton, linen (flax), hemp, and sustainably grown wool.
Certain bio-based polymers	Potential for carbon neutrality (if biomass uptake equals emissions), reduced reliance on fossil fuels, sustainable agricultural practices for feedstock	PLA from sustainably sourced corn starch or sugarcane, bio-based PE from sustainably grown sugarcane ethanol.

2.4 Composite with sustainable components

Composite materials are designed by combining two or more different materials with different chemical and physical properties, providing superior performance features compared to each of their individual components (Mohanty et al., 2018). The new trend is the development of composite that integrate sustainable components, resulting in less environmental impact. Composite with sustainable components should include at least one material that is produced from sustainable resources or recycled/upcycled material. This incorporation aims to minimize the environmental impact of composite materials during its lifetime, starting from virgin material acquisition to end-of-life management. Composite with sustainable components offer many benefits (Emin, 2021). They minimize the dependence on raw, nonrenewable resources by integrating natural fibers, bio-based matrices, and recycled polymers. Additionally, sustainable materials usually demand less energy for production compared to conventional materials such as glass or carbon fibers. This leads to minimize the overall embodied energy and greenhouse gases. Moreover, using recycled polymers could significantly aids in diverting plastic waste away from landfills and supports a circular economy. The use of agricultural waste could valorize waste streams. Composites that include bio-based polymers and natural fibers can, in certain instances, be engineered for biodegradability or compostability at the end of their lifecycle, minimizing long-term environmental impact. An enhancement of performance for some applications could be achieved, like natural fiber that can offer superior specific strength and stiffness in a lightweight form. In additional, some sustainable components are less toxic processing compared to traditional reinforcements like glass fibers. For instance, bio-based polymers can minimize the dependance on potentially harmful additives found in conventional plastics. Table 4 displays some types of composites with sustainable materials, highlighting their manufacturing process and their sustainable components.

Table 4. Sustainable manufacturing process and sustainable components of some composite with sustainable components materials (Bassyouni & Javaid, 2017; Chourpagar et al., ; Gholampour & Ozbakkaloglu, 2020; Kumar et al., 2025; Murawski et al., 2019)

Type	Components	Sustainable components	Examples
Natural fiber composites	The reinforcing phase consists of natural fibers derived from plants or wood	Natural fibers are renewable, can be biodegradable, and can utilize agricultural waste	- Door panels, dashboards, and seat backs in automotive interior parts. Fiberboards and insulation materials in construction materials. Furniture and sporting goods in consumer goods
Recycled Polymer Composites (RPCs)	the matrix is composed of recycled polymers (e.g., recycled polyethylene, polypropylene, PET) and can be reinforced with various materials	Recycled polymers divert plastic waste from landfills, reduce the need for virgin polymer production, and conserve fossil fuel resources	Decking, fencing, and railing in construction materials. outdoor furniture, Pallets and crates in packaging
Bio-based Polymer Composites	The matrix is a bio-based polymer (derived from renewable resources) and can be reinforced with natural fibers or other sustainable materials	reduce reliance on fossil fuels, can create fully or partially biodegradable composites when combined with natural fibers	Compostable food containers and films for packaging, lyocell or polylactic acid. In textiles.

continued on following page

Table 4. Continued

Type	Components	Sustainable components	Examples
Composites with Recycled Reinforcements	The reinforcing phase is derived from recycled materials, such as recycled carbon fibers, recycled glass fibers, or recycled textile fibers	Utilizing recycled reinforcements reduces waste and the energy-intensive production of virgin reinforcing materials	Composites reinforced with recycled glass fibers in panels or insulation. Consumer Goods: Products reinforced with recycled textile fibers.
Hybrid Composites with Sustainable and High-Performance Components	Combine sustainable components with high-performance, albeit potentially less sustainable, materials to optimize both environmental impact and performance for specific applications	Natural fibers, recycled polymers, bio-based polymers	Automotive structural components

2.5 Biodegradable packaging

Biodegradable packaging refers to packaging materials that can decompose naturally in the natural substance such as water, biomass, and carbon dioxide within a relatively short time, through biological processes, mainly through the activities of microorganisms, like fungi and bacteria (Thulasisingh et al., 2022). This kind of materials is ecofriendly, reducing environmental impact and landfill waste. The demand for biodegradable packaging has risen sharply with the rising of global awareness of environmental issues, driving advancements in multiple sectors. These materials commonly come from renewable resources, including proteins, plant starches, and cellulose. Several biodegradable packaging materials are compostable, enabling them to degrade in composting environments, improving soil quality without leaving toxic remnants (Westlake et al., 2023). Additionally, biodegradable packaging could minimize plastic pollution, reducing carbon footprint linked to traditional packaging materials. there are different types of biodegradable packaging, including starch-based packaging, cellulose-based packaging, protein-based packaging, PLA (Polylactic Acid) Packaging, and PHA (Polyhydroxyalkanoates) Packaging. Starch-based packaging is derived from potatoes, corns, or other starch course (Cheng et al., 2021). It is compostable and cheaper compared to other biopolymers, and could be used for containers, bags, and films. Cellulose-based packaging of formed from plant cellulosed, usually in the form of films or coatings, and could be used in biodegradable films, food packaging, and wraps (Wang, Wang, Gardner et al, 2021). They offer strong protective qualities against oxygen and humidity. Protein-based packaging is derived from naturals proteins, like soy, casein, and gluten (Garavand et al., 2022). It is biodegradable with a good protective property and could be used in containers and food wrapping. PLA (Polylactic Acid) packaging is formed from fermented plant starch (mainly corn) (Swetha et al., 2023). It could be compostable under industrial composting conditions, and could be used in films, food containers, and cutlery. PHA (Polyhydroxyalkanoates) packaging is made from microbial fermentation of sugars or lipids (Bugnicourt et al., 2014). It is completely biodegradable in marine and soil environments, and could be used in medical applications, food packaging, and agricultural films.

2.6 Other Emerging Sustainable Material Innovations

Several new materials are emerging, with the high demand for sustainable materials, that have the potential to minimize environmental impact while providing innovative applications. The two most promising sustainable materials are mycelium-based materials and algae-based materials.

2.6.1 Mycelium

Mycelium-based materials are gaining momentum as a sustainable materials owing to its fast growth and capacity to break down organic matter (Lelivelt et al., 2015). These materials are being used in different applications, ranging from packaging to construction. Mycelium is the root of mushrooms and could be decompose naturally, minimizing landfill waste. Mycelium-based materials demonstrate desirable mechanical properties, making them ideal for a variety of structural applications. Additionally, the energy and the resources needed for the cultivation process compared to conventional materials. mycelium-based materials could be used as an eco-friendly alternative to Styrofoam and plastic packaging (Alemu et al., 2022). They could be used also as insulation for building, owing to their thermal properties, and in furniture.

2.8.2 Algae

Algae-based material, a quickly growing renewable resources, is being investigated for its potential in manufacturing sustainable materials (Salami et al., 2021). Algae-based products include biodegradable plastics, textiles, and biofuels. Alage could be growing under different environmental circumstances, demanding less land and less fresh water, with an ability to sequester carbon dioxide while growing, contributing to climate change mitigation. Additionally, algae can be converted into multiple forms, involving fibers, foams, and films. Algae-derived polymers are being designed as sustainable alternatives to conventional plastics and as eco-friendly fabrics for clothing (Devadas et al., 2021).

3- APPLICATIONS OF SUSTAINABLE MATERIALS ACROSS INDUSTRIES

3.1. Sustainable materials in the built environment

The built environment, including buildings and infrastructure, has a major effect on the global greenhouse gas emission and resource consumption. The selection of materials serves a vital function in alleviating these effects create a framework for a more environmentally conscious and durable built environment (Ajala et al., 2019). Sustainable materials could reduce environment impact during their lifetime, ranging from resource extraction and fabrication to consumption and end-of-life handling. They focus on maximizing resource efficiency, promote healthy indoor environment, address social and economic considerations, and minimize carbon emissions. Some of the most common sustainable materials used in construction sector are presented below.

3.1.1 Low carbon concrete

Low carbon concrete is an example of sustainable materials in construction and infrastructure. It could minimize the environmental impact of concrete through replacing a portion of Portland cement by supplementary cementitious materials (such as fly ash, slag, and silica fume). Additionally, non-Portland cement binders, such as geopolymers and calcium sulfa-aluminate (CSA) cements, could be used and Carbon Capture and Utilization (CCU) could be incorporated in the process of producing cement (O'Hegarty et al., 2021). Moreover, industrial waste product could be used in the cement production, leading to minimize landfill waste and the need for virgin materials. Some SCMs could also enhance the strength, the durability and the resistance to chemical attack of concrete.

3.1.2 Mass timber

Mass timber is another example and is wood products designed such as cross-laminated timber (CLT), glued laminated timber (glulam), and laminated veneer lumber (LVL)(Harte, 2017). It is manufactured by bonding layer of solid wood together. it provides excellent strength-to-weight ratios, suitable for structural components in mid- to high-rise buildings. Its processing demands less energy compared to steel or concrete since carbon is absorbed by timber during tree growth. Wood is considered as a renewable resource, particularly when sourced from sustainable managed forests. Wood can incorporate smaller dimension lumber, maximizing resource utilization. Moreover, wood offer inherent thermal insulation properties, minimizing the demand of insulation materials (Ayanleye et al., 2022). In term of construction times, prefabrication of mass timber element can result in faster on-site assembly.

3.1.3 Recycled steel

Recycled steel is generated by transforming scrap steel into new products through melting and reforming. Recycled steel offers many advantages in terms of reduced embodied carbon, resource efficiency, and waste reduction (Liew & Akbar, 2020). Using recycled scrap to produce steel demands considerably less energy that that needed to produce it from raw ore, resulting in significant decreases in greenhouse gas emissions. In addition, the use of recycled steel leads to conserve raw iron ore, coal, and limestone resources, as well as minimizing the amount of waste sent to landfills. Also, recycled steel maintains the structural strength of virgin steel.

3.1.4 Bamboo

Bamboo is a grass with high strength and versatility. It could be used in different construction applications, involving flooring, reinforcement, scaffolding, roofing, and engineered bamboo products (Fahim et al., 2022). The main advantage of bamboo is its quick growth and carbon sequesters, and its low energy processing (Fahim et al., 2022). In addition, bamboo has a high strength to weight ratio and could be used in a wide range of construction applications.

3.1.5 Rammed earth

Rammed earth incorporates a mixture of virgin materials like earth, sand, clay, and stabilizer within farmwork (Avila et al., 2021). It could minimize embodied carbon by using locally sourced natural materials requiring minimal energy for processing. In addition, using local soil minimizes the transportation demands and reliance on synthetic materials (Avila et al., 2021). It provides a good thermal mass, offering a natural insulation and minimizing the demand of artificial cooling and heating. In terms of durability and longevity, rammed earth structures can exhibit exceptional durability and longevity and fosters original and grounded architectural aesthetics.

3.1.6 Recycled plastics

Recycled plastics produced from post-consumer and post-industrial plastic waste and could be used in different infrastructure like road construction, railway sleepers, fencing and barriers, and street furniture (Eneh, 2015). Recycled plastics could minimize embodied carbon through using waste material, resulting in minimizing the needed energy to produce virgin materials and diverts plastic from landfills or incineration. In addition, it responds to the increasing plastic waste crisis by assigning it a new function. Moreover, recycled plastic products can withstand rot, corrosion, and pest damage (Shen & Worrell, 2024). In term of cost, recycled plastics could be a cost-effective alternative for some applications to conventional materials.

Despite the numerous benefits of sustainable materials, their adoption could be hindered by some factors, including cost, performance and standards, awareness and education, and policy and incentives. Initial cost of some sustainable materials could be high, although lifecycle cost studies typically uncover savings over the long term. In addition, making certain that sustainable materials adhere to required performance levels is another factor hindering the adoption of sustainable materials. The adoption requires a raising of awareness among architects, engineers, contractors, and policymakers about the benefits and applications of sustainable materials. Moreover, the adoption of sustainable materials in the built environment requires incentives, government policies and regulations.

3.2. Sustainable materials in packaging

The packaging industry is a prominent user of materials, and the environmental impact of conventional packaging, mainly single use plastics, is an escalating worldwide issue. Sustainable materials provide practical options that reduce waste, minimize dependance on fossil fuels, and foster a more sustainable circular economy (Ibrahim et al., 2022). Sustainable packaging materials are designed to reduce their environmental impact across their lifecycle, from sourcing and production to usage and disposal. They seek to save resources, minimize pollution, and preferably support a circular system, where materials are reused or securely reintegrated into the environment. Bio-based polymers, recycled paper and carboard, biodegradable packaging, and reusable materials are the most common example of sustainable materials used in packaging.

3.2.1 Bio-based polymers

Bio-based polymers is sourced entirely or partly from renewable biomass like cellulose, corn starch, vegetable oils, and sugarcane. Polylactic Acid (PLA), Polyhydroxyalkanoates (PHAs), bio-based Polyethylene (Bio-PE), and cellulose-based films are some examples of bio-based polymers. Bio-based polymer minimize the accumulation of persistent plastic waste in landfills and the environment. Several bio-based polymers could be degraded under specific conditions, decomposing into organic matter, channeling waste away from landfills and enhancing soil quality (Pan et al., 2016). Some bio-based polymers could be mechanically or chemically recycled, promoting a circular economy.

3.2.2 Recycled paper and carboard

Recycled paper and cardboard are derived from recovered paper fiber from post-consumer or post-industrial waste. They could minimize plastic waste, since they act as an easily accessible and are cost effective alternative to plastic packaging for different applications (Peretz et al., 2021). The demand for raw wood pulp and landfill waste could be minimized with the used of waste stream. Paper and cardboard are naturally biodegradable and compostable, supplying a relatively safe end-of-life alternative.

3.2.3 Biodegradable packaging

Biodegradable packaging is crafted to biodegrade in designated environments with a defined timeframe. It involves some bio-based polymers (such as PHSa, and PLA), and some materials like cellulose-based materials, and molded pulp. It seeks to minimize plastic pollution by supplying materials that could decompose naturally (Thulasisingh et al., 2022). Several biodegradable packaging materials are especially crafted for composting, supplying a significant end-of-life route for organic waste streams. With a proper management, biodegradable packaging could reduce the persistent collection of waste in natural settings.

3.2.4 Reusable materials

Reusable materials are crafted for many uses, minimizing the demand for single use option. Reusable containers, bottles, shopping bags, and bulk packaging systems are some examples of reusable materials. They could directly reduce the quantity of packaging materials entering waste channels by prolonging its usability (Coelho et al., 2020). Additionally, reusable materials could keep materials in use for as long as possible, minimizing the need for new resources. Even enough, the initial investment cost in reusable materials could be higher, but the cost per use could considerable lesser than single use alternatives.

The widespread adoption of sustainable packaging depends on some factors. Sustainable packaging could be costly compared to traditional packaging, but the technological advancement and the increase manufacturing scale could minimize the cost (Afif et al., 2022). Also, sustainable packaging should satisfy the necessary performance criteria for specific applications. In addition, the adoption requires convenient infrastructure for collecting, sorting, and processing to perform the recycle and the compost of sustainable packaging materials (Boz et al., 2020). Moreover, to avoid greenwashing and effectively inform consumers and waste management systems, clear and consistent labeling of sustainable packaging materials is necessary. Conduction life cycle assessment is important to guarantee that the selected sustainable packaging solution genuinely provides a lower environmental impact across its entire lifecycle (Afif et al., 2022). Additionally, education and providing consumers guidance on appropriate disposal methods (e.g., composting guidelines for biodegradable packaging) and encouraging the use of reusable alternatives is crucial.

3.3. Sustainable materials in textiles and fashion

The textile and fashion industry plays a critical role in environmental degradation because of its intensive consumption of energy, water, and chemical, in addition to its contribution to waste. Sustainable waste provides an essential route to a more environmentally responsible and ethical fashion system. Using sustainable materials in textiles and fashion could reduce the environment impact during their lifetime, from virgin materials sourcing to processing, manufacturing, use, and end-of-life (Muthu & Gardetti, 2021). They focus on maximizing resource efficiency, conserve water, minimize pollution, reduce carbon footprint, and ideally are compostable or can be easily recycled (Earley, 2022). The use of sustainable materials in textiles and fashion include social and ethical considerations, like fair labor practices. Organic cotton, recycled fibers, natural dyes, and innovative bio-based textiles are common sustainable materials used in textile and fashion.

3.3.1 Organic cotton

Organic cotton is cultivated with organic farming practices, without synthetic pesticides, herbicides, and genetically modified organisms (GMOs) (Gardetti & Muthu, 2019). This could enhance soil health, biodiversity, and water conservation. Water pollution and exposure rick for farmers and ecosystem will be minimized by eliminating the use of hazardous chemical pesticides and fertilizers. Even though cotton is usually water-intensive, organic practices can occasionally enhance water retention in soil (Gardetti & Muthu, 2019). Carbon footprint will be reduced in organic cotton owing to the avoidance of synthetic nitrogen fertilizers. Additionally, organic cotton can decompose under favorable conditions since natural fiber grown without chemical treatments.

3.3.2 Recycled fibers

Recycled fibers are fibers produced from pre-consumer and postconsumer materials that are disintegrated and converted back into new fibers. Recycled polyester, from PET bottles and textiles water, recycled nylon, and recycled cotton are common source for recycled fibers (Leonas, 2016). The water and chemical needed for the generation of raw fiber will be considerably reduced with using recycled materials. Also, the energy needed to manufacture recycled fibers are lower compared to producing raw fiber, resulting in less greenhouse gas emissions. Using recycling materials prevents waste from going to landfills and incinerators. Recycled natural fibers, such as cotton, maintain their biodegradability, and recycled synthetic fibers could minimize the environmental impact linked to their production and disposal (Balea et al., 2021).

3.3.3 Innovative bio-based textiles

Innovative bio-based textiles are produced from novel renewable resources beyond conventional yields (Perin et al., 2021). Lyocell is an example of innovative bio-based textiles and produced from sustainably sources wood pulp using a closed loop production process that reduces chemical and water waste (Khalid et al., 2024). Table 5 display some examples of innovative bio-based textiles.

Table 5. Some examples of innovative bio-based textiles (Akin, 2013; Khalid et al., 2024; Lefèvre Martín, 2024; Quijano et al., 2021; Radi et al., 2020; Torrisi et al., 2022; Zimniewska, 2022)

Innovative bio-based textiles	Production source and method
Lyocell	produced from sustainably sources wood pulp using a closed loop production process that reduces chemical and water waste.
Modal	cellulose-based fiber derived from beechwood, and produced with efficient resource use
Hemp	fiber that requires minimal water and pesticides and can improve soil health
Linen	Made from flax, which requires less water and pesticides than cotton and is naturally durable

continued on following page

Table 5. Continued

Innovative bio-based textiles	Production source and method
Piñatex	A leather alternative made from pineapple leaf fibers, a byproduct of pineapple harvests
Mylo/Bolt Threads	Leather alternatives grown from mycelium (mushroom roots)
Orange Fiber	Textiles made from citrus juice byproducts.

These materials demand considerably lower water and lesser hazardous chemical in their cultivation and processing compared to traditional options. The use of closed loop systems reduces even more waste (Kara et al., 2022). Also, the use of fast-growing plants or agricultural waste as feedstock could lead in reduced carbon emission and some innovative bio-based textiles could even sequester CO_2 (El Majd et al., 2023). In term of biodegradability, some of these materials, being plant based, provide superior biodegradability compared to synthetic fibers, minimizing textiles waste accumulation.

Natural dyes are colorant made from natural sources, like minerals, plants (flowers, roots, leaves), and insects (Verma & Gupta, 2017). Natural dyeing process can incorporate less aggressive chemicals compared to synthetic dyeing. However, fixative is usually needed, and their environmental impact should be considered. The manufacturing of natural dyes usually uses less energy compared to the production of synthetic dyes, which are made from fossil fuels. Natural dyes are decomposable and less prone to environmental persistence or water contamination relative to some synthetic dyes.

3.4. Sustainable materials in the automotive industry

The automotive industry experiences heightened demands to lessen its ecological impact, from manufacturing to the product's end-of-life vehicles. The integrating of innovative and sustainably sourced materials could foster sustainability in automotive industry Sustainable materials in the automotive sector are those that are harvested, created, and applied in a manner that lessens harmful environmental and social impacts throughout their lifecycle (Ghosh et al., 2020). The sustainable materials could reduce resource depletion, waste generation, energy consumption, toxicity, greenhouse gases, and end-of-life management. Recycled metals, bio-based composite, natural fiber reinforcements, lightweight materials are some examples of sustainable materials used in automotive industry. Recycle metals (steel and aluminum), bio-based composite, natural fiber reinforcement, and lightweight materials are some examples of sustainable materials used in automative industry.

3.4.1 Recycled metals

Recycled metals are derived from metals extracted from end-of-life vehicles, industrial scrap, and other postconsumer (de Mattos Nascimento et al., 2022). Although they are not inherently lighter than virgin metals, encouraging their use helps decrease the overall demand for heavier conventional materials. manufacturing recycled materials consume less energy compared to extracting and processing virgin materials, significantly minimize greenhouse gasses (de Mattos Nascimento et al., 2022). Additionally, they conserve finite naturals resources and minimize landfill waste.

3.4.2 Biobased composites

Biobased composites are formed by incorporating a matrix (usually a bio-based polymer) with reinforcements produced from natural biological sources (Carvalho et al., 2024). These could include composites formed with starch-based polymers, soy-based resins, or cellulose based matrices reinforced with natural fibers. Bio-based composites are usually lighter compared to conventional fiber glass composite or plastic, resulting in weight reduction, and this improving fuel efficiency. In term of carbon footprint, bio-based polymers have less carbon footprint compared to petroleum-based plastics, mainly when made from sustainably managed biomass (Carvalho et al., 2024). Also, some bio-based polymers are created to be biodegradable or compostable under certain conditions, facilitating effective end-of-life management.

3.4.3 Natural fiber reinforcements

Natural fiber reinforcements are made from fiber produced from plants like jute, hemp, flax, and sisal as reinforcing agents in composite materials for interior and exterior components (Verma & Senal, 2019). Using these materials reduce the vehicle's weight since naturals fiber are usually lighter than conventional reinforcement, such as glass fibers. In addition to improving fuel economy and minimizing emissions, cultivation of natural fibers has lesser environmental impacts compared to the energy intensive manufacturing of glass fibers (Verma & Senal, 2019). Additionally, they could serve as carbon sinks while they grow, and they offer better biodegradability.

3.4.4 Lightweight materials

Lightweight materials offer high strength to weight ratios, and involves materials such as magnesium alloys, aluminum alloys, and carbon fiber reinforced polymers (Busarac et al., 2022). The weight reduction is the main benefit of these materials despite that there are not all bio-based or recycled. This results in enhanced fuel economy, less greenhouse emissions. The use of lighter materials in electric vehicles could extend their range, also improve vehicle handling, acceleration, and braking (Busarac et al., 2022).

3.5. Sustainable Materials in Electronics and Consumer

Electronic and consumer goods, spanning from smartphones and laptops to home appliances and packaging, are considered as fundamental for the modern economy. Virgin materials extracted from earth with environmentally harmful processes are used to manufacture these materials. The extraction processes of these materials are energy intensive, with negative effect on the environment, leading to deforestation, water and soil pollution, and habitat low. Additionally, the quick turnover of electronics and the large volume of consumer products generate substantial waste, a significant portion of which contains hazardous materials. thus, transition towards more sustainable materials choices and implementation of circular economy practices are needed. There are a diverse range of sustainable materials that could be viable alternatives for the conventional materials used in the electronics and consumer good sectors (Manjunatheshwara & Vinodh, 2021). Some examples of these materials include recycled plastics and metals, bio-based polymers, and sustainably sourced components.

3.5.1 Recycled metals and plastics

Recycled metals and plastics include metals recovered from electronic waste and end-of like consumer goods to be used in the manufacturing of new products (Kang & Schoenung, 2004). These could include gold, aluminum, tin, and copper. For instance, recycled gold and copper could be used in circuit boards and wiring. Recycled aluminum could be used in laptops casing, smartphone enclosure, in beverages cans and kitchenware. Recycled steel is used in furniture and appliances. The used of these recycled materials could considerably minimize the high energy needed for mining and refining, reduce carbon footprint, and save valuable resources. Recycled plastics derived from post-consumer and post-industrial could be processed and used in new plastic components, such as in housing for computers, televisions, and peripherals. The use of recycled plastics could minimize the need for raw fossil fuel-based plastics and reduce the energy demand for manufacturing.

3.5.2 Bio-based polymers

Bio-based polymers could be used in some packaging and possibly for less resilient internal components (Kuczynski & Boday, 2012). They could be also used is disposable cutlery and textiles. They offer different benefits, including minimize the dependance on fossil fuels, could be compostable under certain conditions, and usually has lesser greenhouse gases emissions during their production (Kuczynski & Boday, 2012). Bamboo and wood could be used also in electronics for casing and decorative elements. They require lesser processing energy and could be compostable.

3.6. Other Emerging Applications

Sustainable materials are being creatively applied in a range of different fields, including energy, agriculture, and medicine.

3.6.1 Sustainable materials in agriculture

In agriculture, the conventional plastic mulch films could be replaced by biobased and compostable alternatives matrices derived from materials, such as cellulose (Arif et al., 2022). This could result in minimizing plastic waste, enhancing soil health, and removes the necessity for manual removal. Additionally, the use of plastic in food supply chains could be reducing by using compostable packaging produced from agricultural waste or bio-based polymers. Moreover, fertilizer could be also produced from organic waste, biomass, and other sustainable sources, minimizing the dependance on synthetic chemicals (Arif et al., 2022). Sensors used to monitor soil health, such as moisture and nutrient, could be made from biodegradable materials, that degrade over time, resulting in reducing electronic waste (Dahal et al., 2020).

3.6.2 Sustainable materials in medicine

In medicine, sustainable materials, such as polylactic acid or chitosan, could be used to make medical implants and sutures. These materials could decompose naturally within the body, removing the necessity for additional surgical procedures (Baranwal et al., 2022). In addition, the use of biodegradable and biocompatible nanoparticles or hydrogel derived from materials for controlled drug release leads to minimize the environmental impact of pharmaceutical waste. The traditional plastic packaging could be replaced by biodegradable or recyclable alternatives derived from plant-based materials or recycled content. Additionally, natural polymer, such as cellulose and silk, could be used to create scaffolds for tissue regeneration, providing biocompatibility and biodegradability (Krishani et al., 2023).

3.6.3 Sustainable materials in energy field

In energy field, sustainable materials, including organic and bio-derived materials, could be used in the manufacturing of solar cell, minimizing the dependance on critical and energy consuming materials, such as silicon (Dusastre, 2011). Similarly, sustainable materials, such as wood could be used as a sustainable alternative material to steel in wind turbine tower, minimizing the carbon footprint of manufacturing. Additionally, sustainable materials could be use as component in batteries, minimizing their carbon footprint. Sand batteries are materials developed for thermal energy storage, employing plentiful and environmentally friendly resources for grid-scale energy storage solutions (Vyas & Kushwah, 2023). The use of recycled metals, mainly steel, aluminum and copper, in the building of wind turbines, solar panels, and other energy infrastructure, minimizing the environmental impact associated with the extraction of virgin materials

4- LIFECYCLE EVALUATION AND CIRCULAR ECONOMY PRINCIPLES

The selection of materials plays a crucial role in shaping the environmental impact of products across different sectors. An assessment of the environmental impact of materials is essential for advancing sustainability and enabling informed design decisions and adhering to growing environmental. The main methodologies used to evaluate the environmental burdens associated with materials throughout their lifecycle are presented below.

4.1 Importance of Lifecycle Assessment (LCA)

Lifecycle Assessment (LCA) is a structured evaluation approach used to assess the environmental impacts pertaining to every phase of a product's lifecycle, from raw material extraction to production, use, and disposal (Ding, 2014). LCA offers a comprehensive view of environmental impacts across all phases of a product's lifespan. It identifies areas that require improvement, enabling focused actions. Additionally, the analysis of the lifecycle of products allows stakeholders to understand trade-offs between different environmental impacts, like carbon emissions versus water use. LCA provides organizations and decision-makers with evidence-based information to make informed decisions regarding materials, processes, and products. LCA assists organization in comparing the environmental impacts of different materials and choose the more sustainable option. Moreover, LCA can assist designers to design products with less resource consumption and less environmental impacts. In terms of environmental regulations, the conduction of LCA can assist organization to fulfill regulatory requirements concerning environmental accountability and sustainability practices. The increasing demand of markers and consumers for clear information about the environmental effects of products positioning LCA as an essential resource for regulatory adherence. Additionally, LCA could assist organization in monitoring their progress toward sustainability goals and implement required modifications.

4.1.1 The LCA framework ISO 14040/44 Standards

The LCA framework ISO 14040/44 Standards detail the principles and framework for carrying out LCA, ensuring consistency, transparency, and reliability in the assessment process (Finkbeiner, 2014). It includes different components, including goal and scope, life cycle inventory (LCI), and life cycle impact assessment, and interpretation. This stage defines the goal of the LCA, the intended audience, and the boundaries of the study. It lays the groundwork for the entire assessment process. the second stage is the LCI that includes gathering data on inputs and outputs for each stage of the product life cycle, and in include resource usages and emissions. In the third stage is LCIA, an identification and an estimation of the possible environmental effects related to the inputs and outputs. Additionally, it classifies and quantifies the different environmental effects, including, resource depletion, global warming, and ozone depletion. The final stage is the interpretation, and it analyses the results, formulates conclusions and provides recommendations based on the results. The stage is essential for guaranteeing that the results are conveyed clearly to stakeholders.

4.1.2 Carbon footprint assessment

Carbon footprint assessment is a subset of LCA that specifically targets the greenhouse gas (GHG) emissions related to a product, service, or activity across its complete life cycle (Morini et al., 2019). It is represented as the overall quantity of carbon dioxide equivalent (CO_2e) emitted. The methodology followed is like LCA, starting by collecting data on energy consumption, transportation, materials production, and other activities that emit greenhouse gases throughout the specified lifecycle stages. Then, activity data will be multiplied by relevant emission factors (kg CO_2e per unit of activity) to assess the GHG emission for every stage. Emission factors are generally obtained from databases like IPCC or national inventories. The last stage aims to sum the emission from all stages to calculate the overall carbon footprint.

4.1.3 Materials Flow Analysis

Materials Flow Analysis (MFA) systematically examines the inflows and outflows of materials within a specified system (product, region, or sector) (Bringezu & Moriguchi, 2018). Although it is not exclusively an environmental impact assessment method, it gives fundamental data for resource consumption, waste generation, and opportunities for material efficiency and circularity, which are essential elements of environmental sustainability. The process starts by defining the boundaries of the considered systems followed by a quantification of the inputs, virgin materials and products, and the outputs, products and waste emissions, materials within the system. After that the flow and accumulation of materials across various processes and stages will be tracked. A calculation of indicators, including material intensity, waste generation, and recycling rates, is conducted to evaluate resource efficiency an environmental performance.

4.1.4 Environmental Products Declarations

Environmental Products Declarations (EPDs) are standardized and verified documents that effectively communicate the environmental performance of a product across its entire life cycle in a credible and transparent way (Andersen et al., 2019). There are founded on LCA studies and adhere to specific rules for each product category. EPDs ensure comparability between different products within the same category and delivers an extensive overview of environmental impacts. Additionally, they enhance credibility and guarantees accuracy of the reported data. Moreover, they facilitate the communication business to consumer, and business to business about environmental performance.

In addition to the core methodologies, numerous other tools and frameworks play a role in assessing the environmental impact of materials, such as ecological footprint, water footprint, toxicity assessment, circular economy assessment tools, and sustainable materials selection tools. Ecological footprint aims to measure human demand on nature by evaluating the biologically productive land and sea area needed to generate the resources consumed by a population and to absorb its waste (Rees, 2018). Water footprint aims to measure the total amount of water used in the generation of goods and services, considering the direct and indirect water use across the supply chain (Chen et al., 2020). The possible negative impacts of materials and their constituent chemicals on human health and the environmental are evaluated in toxicity assessments. Circular assessment tools aim to assess the circularity of materials and products utilizing criteria like recyclability, durability, and resource efficiency. In sustainable materials selection tools, software and database are used to supply environmental data on different materials to assist designer in making more sustainable materials choices (Lovrenčić Butković et al., 2023).

4.2 Comparison of sustainable materials with conventional options

The comparison between sustainable materials and conventional materials shows notable variations in environmental impact across multiple dimensions including using sustainable materials, carbon footprints, end of life, performance and durability and economic considerable.

- Resource use: sustainable materials usually generated from renewable resources or made from recycled materials, considerable minimizing the demand for virgin materials. However, conventional materials usually depend on limited resources, like fossil fuels, resulting in resource depletion and environmental degradation.
- Carbon footprint: sustainable materials exhibit a lower carbon footprint due to decreased greenhouse gas emissions during production and usage. For instance, low carbon concrete uses alternative materials that lower energy to be generated. While conventional materials consume higher energy and emits high greenhouse gases during extraction, processing and transportation, leading to a higher carbon footprint.
- End of life impact: sustainable materials are usually designed to be recyclable or biodegradable, resulting in less waste and environmental damage at the end of their lifecycle. Bio-based polymers are an example of sustainable materials and could naturally biodegrade. On the other hand, conventional materials are usually end up in landfills, resulting in long-term environmental pollution, resource depletion, and higher waste management costs.
- Performance and durability: sustainable materials offer high durability and performance, rendering them competitive with conventional materials and promoting sustainability. for instance, recycled composites offer similar or high mechanical properties as conventional materials. Conventional materials offer high durability, but they overlook resource limitations and environmental impacts throughout their lifecycle, potentially resulting in environmental damage over the long term.
- Economic considerations: even though the high initial cost of sustainable materials owing to technology and processing, long-term saving could be achieved by minimizing waste, reducing energy usages, and improved dura-

bility. Conventional materials are usually cheaper, but it results in high hidden cost related to environmental degradation, environmental decline, depletion of resources, and waste management issues

Some examples of lifecycle assessment studies on some materials are presented in Table 6.

Table 6. Some examples of lifecycle assessment studies on some materials (Dhaliwal et al., 2014; La Rosa et al., 2013; Rezvani Ghomi et al., 2021; Sahoo et al., 2021; Shen et al., 2016; Shi et al., 2019)

Sustainable materials	Results of LCA
Polylactic acid (PLA)-bio plastic versus plastic based fossil fuels	less carbon footprint owing to its renewable origins The energy required to manufacture PLA is less compared to conventional plastics, particularly when considering renewable energy sources An enhancement in compositing facilities could make PLA more sustainable and maximize its environmental benefits
Recycled Concrete Aggregate (RCA) versus virgin aggregates	Minimize the need for natural aggregates and conserve natural recourses Lower greenhouse gas emission owing to the minimized need for quarrying and transportation
Low-Carbon cement versus traditional portland cement	Minimize CO_2 emissions by up to 30% compared to traditional cement The manufacturing process is less energy intensive compared to traditional cement but depend on the materials used.
Sustainably source timber versus unsustainable logging practices	Sustainably managed forests maintain biodiversity and ecosystem health Sustainably sourced timber contributes positively to carbon sequestration, helping mitigate climate change
Composite materials in automotive applications versus traditional composites	Bio-composites often result in lighter vehicles, leading to improved fuel efficiency and lower emissions over the vehicle's lifetime The manufacturing process require more energy, but the overall lifecycle emissions are lower due to reduced fuel consumption during use
Glass vs. plastic packaging	Glass production is generally more energy-intensive than plastic Recyclability of glass offers significant environmental benefits High recycling rates for glass can mitigate its initial energy costs and reduce overall environmental impact

4.3. Embracing Circular Economy Principles for material sustainability

The circular economy is based on several core principles, particularly those outlined by the Ellen MacArthur Foundation 'Design Out Waste and Pollution', that aim to prevent waste and pollution from outset (Ramakrishna & Jose, 2022). The circular economy offers an attractive substitute for our conventional linear "take-make-dispose" economic model. It is designed to reduce waste and optimize the use of resources. It entails creating products and processes that eradicate harmful substances and reduce resource depletion. The circular economy focuses on extending the lifetime of products by implementing approaches such as, reuse, recycle, refurbish, and repair. The goal is to keep resources circulating within the economy for as long as possible, maintaining their value. Regenerate Natural Systems: This principle recognizes the importance of restoring and enhancing natural ecosystems. It involves practices that promote soil health, biodiversity, and carbon sequestration. This means that rather than just taking from the earth, the circular economy aims to give back.

Sustainable materials play a crucial role in achieving a circular economy. Sustainable materials could minimize the need of virgin materials, especially is they are renewable or recycled. Biodegradable and compostable sustainable materials could minimize landfill waste and reduce pollution and greenhouse gases. Recyclable materials allow close-loop system, where materials are repeatedly reused. Additionally, durable and reparable materials prolong product lifespans, postponing disposal. in term of circular design, sustainable materials more readily integrated into products designed for disassembly, repair and recycling.

Several case studies demonstrate the impact of the circular economy in promoting sustainable materials. For instance, the furniture company IKEA with their 'Circular Hub' initiatives urges customers to return used furniture, which is then restored or repurposed, reducing waste. In clothing sector, the clothing brand Patagonia which highlights repair and recycling via its Worn Wear program by enabling customers to exchange old garments for store credit, fostering a culture of reuse. Additionally, the construction sector has seen the rise of companies like Buro Happold, which focuses on designing buildings with recyclable materials and modular components that can be easily disassembled. These case studies demonstrate how businesses can successfully implement circular economy principles, reducing waste and enhancing sustainability in their operations.

5 ADVANCED MATERIALS LEADING TO A SUSTAINABLE FUTURE

5.1 Bio-based Polymers

Bio-based polymers could be categorized based on their origin and biodegradability (Figure 4). Based on origin, there are two types of bio-based polymers: naturally sourced polymers and polymers synthesized from bio-based monomers (Cywar et al., 2022). Biobased naturally sourced polymers are directly extracted and processed from biomass. Polymers Synthesized from Bio-based Monomers are produced by polymerizing monomers derived from renewable resources through chemical or biological processes.

Figure 4. Types pf bio-based polymers based on origin

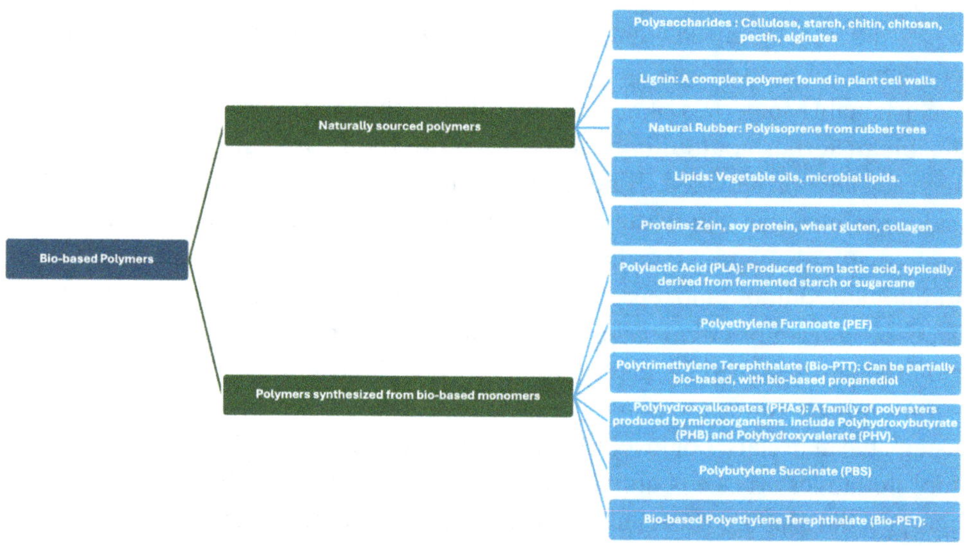

Based on biodegradability, there are two categories: biodegradable bio-based polymers and non-biodegradable bio-based polymers. Biodegradable bio-based polymers can be segmented by microorganisms into natural substances under defined environmental settings, such as soil, aquatic environment and compositing. PLA, PHAs, starch-based plastics, cellulose esters (like cellulose acetate), and some bio-polyesters such as PBS and PBAT – which is often partially fossil-based but biodegradable) are some examples of biodegradable bio-based polymers. Non-biodegradable bio-based polymers are sourced from renewable resources but feature a chemical structure that is not easily decomposed in the environment. Bio-PE, bio-PP (bio-polypropylene), and partially bio-based PET and PTT are some examples of non-bio-degradable bio-based polymers. These "drop-in" bioplastics are chemically identical to their fossil-based counterparts and are designed to be recyclable within existing infrastructure.

The properties of bio-based polymers depend on different factors, including their chemical structure, processing technique, and any changes or compounding with other materials (Cywar et al., 2022). The ongoing research focuses on enhancing the mechanical properties to meet or exceed the properties of traditional polymers through processes like blending, copolymerization, and incorporating nanofillers. Additionally, thermal stability, melting point, and glass transition are essential for processing and utilization. For instance, PLA has a relatively low melting point compared to PET, confining its use in high-temperature environments unless adjusted. Permeability to gases, aromas, and moisture is fundamental for packaging applications. Some bio-based polymers, such as PLA and certain cellulose derivatives, deliver effective barrier performance, whereas others require alterations to improve them. PEF offer demonstrating enhanced barrier characteristics over PET.

Fabricability is another essential aspect. Bio-based polymers could be fabricated using traditional technique such us 3D printing, extrusion, film blowing, and injection molding, while some may need tailored adjustments to processing settings owing to their unique thermal and flow properties.

The advanced bio-based polymers could be used in different sectors. Currently, the largest application sector of advanced bio-based polymers, mainly PLA, bio-PE, and starch blends, is the packaging, such us flexible packaging (bags, and films), rigid packaging (containers, and bottles), and food service ware (Al-Oqla & Sapuan, 2020). Additionally, they could be used in active and intelligent packaging, featuring improved barrier properties or antimicrobial capabilities. The advanced bio-based polymers are also used in textiles. Bio-based fibers are being used in apparel, home textiles, and technical textiles. Lyocell (from cellulose), PLA, and bio-based nylon are some examples of bio-based fibers. Natural fiber composites are also used in automotive and transportation for interior part to minimize weight and enhance sustainability. For agriculture and horticulture, biodegradable mulch films derived from PLA or starch blends could minimize plastic waste and eradicate the need for manual intervention. Bio-based polymers are also used in consumer goods, medical and pharmaceutical. For instance, PLA, chitosan, and PHAs are being used drug delivery systems, tissue engineering scaffolds, sutures, and medical implants due to their biocompatibility and biodegradability. Moreover, bio-based resins and polymers Are being designed as sustainable options for traditional coatings and adhesives

The manufacturing of bio-based polymers includes different processes based on the feedstock and the desired polymer. In the case of direct extraction and modification and for naturally sources polymer, such as starch and polymers), the process begins by extraction from biomass (such as corn kernels for starch, and wood pulp for cellulose), proceeding with purification and a chemical modification in some cases. Biotechnological manufacturing is another method to generated bio-based polymers, such as PHAs that are produced by microorganisms through fermentation of sugar or vegetable oils. The polymer is accumulated by the bacteria for energy storage and is subsequently extracted and purified. Bio-based polymers could be also generated by chemical polymerization of bio-based Monomers, that includes transforming bio-based feedstocks into monomers (like lactic acid from fermentation) followed by polymerizing of these monomers using chemical reactions to produce the desired polymer chain. Additionally, drop-in bioplastic is another method to produce bio-based polymer. Bio-based ethanol generated from sugarcane or other biomass is dehydrated to form bio-ethylene, which is later polymerized using established processes to generate. Likewise, bio-based pathways are available for other drop-in monomers utilized in bio-PET and bio-PTT.

5.2 Recycled composites

Recycled composites reflect a significant enhancement in material science, merging recycled materials to form products that are effective and sustainable (Paraye & Sarviya, 2024). There are different types of recycled composites, including recycled plastics composites, wood plastic composites, recycled rubber composites, and recycled glass composites.

Recycled plastic composites, are derived from post-consumer plastics and post-industrial plastics waste. These composites are usually blended with additives to enhance performance. They could be used in different applications, such as outdoor furniture, providing a durable and weather resistant, and in residential and commercial decking, offering a low maintenance and long lifetime (Turku et al., 2017). Additionally, they could be used in the components of vehicles, like dashboard and trim, minimizing the weight and the emissions of vehicles. Moreover, recycled plastic composites assist in redirecting millions of tons of plastic waste from landfills annually, greatly minimizing the environmental effects linked to plastic production (Istrate & Chen, 2018; Turku et al., 2017).

Wood plastic composites combine wood fibers with thermoplastics to make a material that retains the aesthetic appeal of wood with an enhancement durability (Schwarzkopf & Burnard, 2016). They could be used for decking, fencing, and railing in construction sector. Also, they could be used in the manufacturing of toys and households' items that necessitate strength and appearance. The use of wood plastic composites minimizes the demand for new wood and reduce plastic waste in landfills.

Recycled rubber composites are made of recycled tires and other rubber products. They could be incorporate in other materials to improve properties such as shock absorption (Kabakçi et al., 2022). Recycled rubber composites could be used in playground surfaces, offering a safe and a resilient surface for children. Additionally, they could be used in flooring used in gyms and commercial spaces, offering durability and comfort. In road construction, they could be used with asphalt to enhance the road performance. Millions of tires could be diverted from landfill yearly through recycling, tackling a major waste issue.

Recycled glass composites are crushed recycled glass, usually used with resins, and could be used in different composite applications (Gonçalves et al., 2022). They could be used in the manufacturing of tiles, art pieces that capitalize on the visual appeal. of glass. Moreover, recycled glass composites could be used in concrete and as aggregate in construction. They could minimize the energy consumption needed to manufacture new glass products. Recycled glass composites could conserve resources by recycling existing materials instead of extraction of new resources, resulting in minimizing the need for virgin materials and mitigate the carbon footprint associated with their production.

Circular economy recycled composites are designed to maintain the use of resources as long as possible. These composites enhance product life cycles and reduce overall waste, fostering sustainability (Chatziparaskeva et al., 2022). The development of recycled composites was driven by the rising need for sustainable products and by the intent of industries to minimize their carbon footprint, leading to a continuous growth of the market of these materials.

5.3. Advanced low-carbon concrete

The construction industry is a major source of global carbon emissions, owing to the generation of conventional cement. Advanced low-carbon concrete constitutes a groundbreaking advancement focused on minimizing carbon footprint of construction sector (Barbhuiya et al., 2025). The innovation in concrete achieved two areas: innovations in cement generation, and innovation in concrete mixes. The innovation in cement production involves supplementary cementitious materials (SCMs), carbon capture and utilization (CCU), and direct air capture (DAC).

SCMs are materials combined with Portland cement to enhance the properties of concrete and minimize its carbon footprint. The common SCMs involve natural pozzolans, fly ash, silica fume, and slag. The main benefits of SCMs derive from its ability to minimize the quantity of Portland cement requires, resulting in less carbon emissions associated with cement production (Snellings et al., 2023). Additionally, the durability and strength of concrete could be enhanced using SCMs, lengthening the operational life of structure and minimizing the requirement for repairs (Snellings et al., 2023). Also, waste could be minimized, and recycling could be promoted since SCMs are by-products of by-products of industrial processes. CCU technologies are used to capture carbon dioxide emissions from cement generation processes and use them to produce valuable materials or securely store them. DAC aims to capture CO_2 from the atmosphere to use them in the production of concrete. Mineralization is also another technique that is based on capture CO_2 into stable minerals to be used as fillers in concrete. These techniques could considerably minimize the carbon emission associated with cement production, rendering it an essential part of low carbon concrete, and promoting circular economy in the construction sector.

In the field of concrete mixes, the innovation presented in high performance concrete (HPC), recycled aggregate concrete, and bio-based admixtures. HPC is developed to ensure superior durability and robustness, often integrating SCMs and innovative admixtures (Drzymała et al., 2017). It could minimize dependance on conventional cement, resulting in less CO_2 emissions. in addition, the improvement of durability expands the lifetime of the structures, minimizing the reconstruction frequency. Recycled aggregate combines recycled aggregates from razed structures, lessening the demand for virgin materials. It reduces waste since they use construction and demolition waste, reducing landfill use. In addition, it minimizes the need for natural aggregates, which are energy intensive to produce. Bio-based admixtures are extracted from natural origins that can improve the characteristics of concrete. They minimize the dependance on synthetic chemical and promotes a reduced carbon footprint

Lifecycle assessments (LCAs) of advanced low-carbon concrete show substantial decrease. SCMs and CCU may result in a decrease of up to 30% in carbon emissions associated with concrete production (Terán-Cuadrado et al., 2024). The enhanced durability of low-carbon concrete minimizes the demand for repairs and replacements, resulting in more emissions saving over time. The adoption of low carbon concrete is increasing among builders and contractors to fulfill regulatory obligations and sustainability targets. Table 7 displays some successful project of using low-carbon concrete in different applications.

Table 7. Projects exemplify the effective use of low-carbon concrete in various applications

Project name, location	Action
The Edge, Amsterdam	building incorporates low-carbon concrete and features energy-efficient systems, making it one of the greenest buildings in the world
The Bosco Verticale, Milan	This pair of residential towers uses a mix of low-carbon concrete to reduce its environmental impact while integrating greenery into the structure
The International Airport, Changi, Singapore	The construction of the airport's new terminal incorporated low-carbon concrete technologies to enhance sustainability
University of California, Berkeley's Sustainability Center	using low-carbon materials, including advanced concrete, to minimize its carbon footprint

5.4. Sustainably sourced timber

Sustainably sourced timber is key to encouraging environmental accountability and minimizing the carbon footprint of the construction industry (De Araujo, 2023). It refers to wood harvested from forests governed by comprehensive environmental, economic, and social standards, that promote the enduring vitality and productivity of the forest ecosystem. The main benefits of sustainably sources timber include maintaining forest ecosystem health through protecting soil health, biodiversity, water quality, and overall ecological integrity (De Araujo, 2023). Additionally, it could ensure a long-term forest productivity through managing forest to a sustainable source of timber for generations to come. In term of economic viability, ensuring an economic viability of forest management facilitates jobs and strengthens local economic development.

The underpinning of sustainably sources timber, sustainable forest management, is governed by different key principles, displayed in Figure 5.

Figure 5. Key principles of sustainable forest management (Innes, 2016; Sverdrup et al., 2002)

Ecological integrity is essential to maintain the natural structure, composition, and functions of the forest ecosystem, involving its biodiversity, nutrient cycles, and resilience to disturbances. Additionally, it is crucial to ensure that the rated of timber harvesting does not exceed the forest's self-renewal capability, providing a reliable supply for the long term. Also, it is important to protect and improve the variety of life within the forest, involving plant and animal species, genetic diversity, and habitats. Practices to avoid soil erosion, sedimentation of water ways, and sustain water quality and movement should be implemented. Recognizing and valuing the rights, insights, and cultural heritage of indigenous peoples and local communities linked to the forest, is essential to guarantee their participation in decisions regarding forest management. Moreover, it is fundamental to establish safe and fair employment practices for forest workers. In addition, ensuring transparent and accountable forest management practices enables the assessment and validation of sustainability assertions. The monitoring for forest conditions should be continuous and the adaptation of management practices should depend on new scientific information and changing environmental and social circumstances.

Independent third-party certification schemes are essential for confirming that timber comes from sustainably managed forests. These schemes deliver confidence to consumers and businesses. The most well-known and reputable certification schemes involve Forest Stewardship Council, Programme for the Endorsement of Forest Certification (PEFC), and Sustainable Forestry Initiative (SFI). Forest Stewardship

Council (FSC): An international, non-profit organization that sets global standards for responsible forest management. FSC certification covers environmental, social, and factors and is broadly acknowledged by consumers and businesses globally. It provides various types of certifications, such as forest management certification and chain-of-custody certification (tracking certified wood through the supply chain). Programme for the Endorsement of Forest Certification (PEFC) is an international umbrella organization that supports national forest certification systems designed to fulfill internationally recognized sustainability standards. PEFC acknowledges different national schemes that are customized for local contexts while conforming to fundamental sustainability principles. Sustainable Forestry Initiative (SFI) is a North American-based certification program that fosters sustainable forest management practices via standards, third-party certification audits, education, and research. These certification schemes include thorough audits and monitoring to verify compliance with their respective standards.

6. OBSTACLES TO BROAD IMPLEMENTATION OF SUSTAINABLE MATERIALS

The shift to sustainable materials is crucial for tackling environmental challenges. However, many challenges impede widespread adoption. These challenges could be classified into four categories: technological challenges, economic constraints, policy and regulatory issues, and consumer awareness and acceptance.

6.1 Technological barriers

Several sustainable materials do not currently meet the performance standards of conventional materials in terms of durability and strength (Gounder et al., 2023). This can discourage industries from making the transition, particularly in applications where material performance is vital. Additionally, the existing infrastructure for generating and processing sustainable materials s frequently less developed compared to that for conventional materials. This encompasses supply chains, manufacturing facilities, and recycling systems, all of which are essential for the successful implementation of sustainable materials. Thus, a continuous investment in research and innovation is required to create feasible alternatives and enhance the performance of sustainable materials.

6.2 Economic constraints

Sustainable materials often carry a higher initial cost compared to conventional materials. This price difference can be linked to different factors, including smaller-scale generation and the integration of ethical considerations into pricing (Wagner, 2015). The prioritization of short-term financial gains over long terms benefits related to sustainable materials by businesses and consumers is considered as another economic constraint. This viewpoint could result in a reluctance to invest in sustainable option, as the immediate cost savings of conventional materials are more attractive. Additionally, it is hard to secure funding for sustainable material innovations, since investors prefer short-term returns. This poses a challenge for compagnies aims to shift to more sustainable practices.

6.3 Policy and regulatory issues

Even though the regulatory environment could considerably promote the adoption of sustainable materials, The absence of consistent environmental regulations can lead to confusion and ambiguity. Inconsistent policies across regions can obstruct companies' efforts to implement sustainable practices effectively. Moreover, existing policies do not offer adequate motivation for business to invest in sustainable materials. Without supportive regulations, companies may struggle to adopt sustainable materials effectively. The lack of clear standards and certifications for sustainable materials can create confusion among consumers and businesses, and hinder market growth.

6.4 Consumer awareness and acceptance

Consumer behavior is a key determinant in the adoption of sustainable materials. Many consumers and business are not fully aware about the advantages of using sustainable materials and the adverse effects of conventional materials. This lack of awareness can arise from limited access to information and education on sustainability. Additionally, consumer skepticism regarding "green" marketing that refers to doubts and distrust that consumers may have about the credibility and efficacy of environmentally friendly claims made by companies. This skepticism can stem from worries about "greenwashing," where companies overstate or misrepresent their environmental responsibility, causing consumers to doubt the legitimacy of these marketing efforts. This can also hinder the adoption of sustainable materials. in term of price, consumer always place a higher importance on price than on sustainability, resulting in a preference for cheaper, conventional products. This behavior can restrict the demand for sustainable materials, since consumers may view them as pricier options without fully understanding their long-term benefits.

6.5. Supply chain and infrastructure limitations

The rising demand for sustainable materials by different industries provides a meaningful opportunity to lessen environmental effects and promote a circular economy. However, the broad implementation of these materials is usually obstructed by limitations present in existing supply chains and the insufficiency of adequate infrastructure might be irregular because of seasonal variations, weather patterns, and agricultural practices.

Supply chain Limitations: The availability of certain sustainable feedstocks, especially bio-based materials and agricultural residues, can be inconsistent due to seasonal variations, weather patterns, and agricultural practices (Zeng et al., 2018). This inconsistency may create challenges for manufacturers in ensuring a dependable supply. Additionally, the generation volume of several sustainable feedstock and intermediates materials are usually smaller compared to widely recognized conventional materials. This limited scale may contribute to higher costs and restrict their applicability in mass production. In term of resources, some bio-based feedstocks can contend with food production or demand considerable land resources, bringing up issues related to sustainability and potential trade-offs. In addition, the generation of some sustainable feedstock could be geographically concentrated, resulting in longer transportation distances and corresponding environmental and fiscal impacts.

The existing manufacturing infrastructure is designed for manufacturing conventional materials, and implementing new facility for the processing of sustainable feedstocks could be costly and required technological development. The incorporation of new sustainable materials in existing manufacturing process might be complication and necessitate changes of equipment, quality control procedure, and materials handling system (Kesidou & Sovacool, 2019). In addition to the high production cost of sustainable materials because of the smaller generation volumes, expensive cost of virgin materials, and immature technologies, the cost of sustainable feedstock could be more unpredictable than fossil fuel-based resources, hindering manufacturers' ability to forecast and control costs. There are also some challenges related to traceability and certification. The supply chains for some sustainable materials could be complicated, including different stakeholders and geographical location, making it hard to achieve traceability. Even enough, there are certification for some sustainable materials, such as organic certification for fibers, standardized and established certifications are still absent for several advanced sustainable materials, complicating matters for consumers and business to validate their sustainability assertions.

There are also some barriers related collection and sorting infrastructure for end-of life materials. Effective collection systems for post-consumer and post-industrial waste streams, which serves as feedstock for recycled materials, are not universally recognized or optimal (Huang et al., 2022). This is particularly applicable to diverse and complex waste streams. Regarding sorting infrastructure, the existing one could not be equipped to manage the growing diversity and intricacy of materials, involving novel sustainable materials and composites. This could lead to reduce quality of recycled materials and minimize recycling rates. In addition, the capacity of specialized recycling facilities for some sustainable materials, mainly abio-based polymer recycling, is usually inadequate to satisfy the increasing demand. Efficient and cost-effective recycling technologies are still under some complex sustainable materials and composites, restricting their disposal choices. Comprehensive data on the environmental and economic performance of sustainable material supply chains and infrastructure is frequently inadequate, complicating the evaluation of their actual sustainability and identify areas for improvement

7. STRATEGIES FOR PROMOTING SUSTAINABLE MATERIAL INTEGRATION

The integration of sustainable materials into different sectors becomes fundamental with the rise of recognition of the importance of sustainability in the world. Promoting the integration of sustainable materials requires some approaches associated with education and awareness, government incentives, research and development, consumer engagement, and monitoring and evaluation (Yahia & Shahjalal, 2024).

- Education and Awareness: Developing training programs for industry professionals on sustainable materials and practices is essential. This could include certification and workshops on sustainability and materials in sourcing and usage. Enhancing public awareness among consumer about the advantages of using sustainable materials is crucial, and this could be through social media and community events.
- Policies and Regulations: Policies and regulations are another strategy to promote sustainable material integration. This could be achieved through offering tax breaks and subsidies for compagnies using sustainable materials. Additionally, setting clear standards for sustainable materials and advocate for certifications that recognize sustainable practices, like LEED or Cradle to Cradle.
- Research and collaboration: Investment in research and development and collaboration between universities and research institutions are important to

explore new sustainable materials. Additionally, conducting studies to assess the environmental impact of materials is required to assist in material selection and foster the adoption of sustainable alternatives.

- Partnership: promoting partnership between government agencies, compagnies, and NGOs is needed to share knowledge and resources. Additionally, establishing collaborative platforms for businesses is important to share best practices in the use of sustainable materials. Moreover, involving suppliers in sustainability initiatives, motivating them to embrace sustainable practices. A transparent supply chain that emphasizes sustainable sourcing should be developed.
- Eco-labeling system: Implementation eco-labeling system play a crucial role in assisting consumers to identify sustainable products (Xu et al., 2017). This required educating consumers on analyzing labels and comprehending sustainability claims. Ecolabeling could also encourage manufacturers to design products with sustainable materials from the outset and advocate for the principles of circular economy in product design by emphasizing recyclability and reusability.
- Metrics and standards: Establishing metrics and standards to regularly assess the performance of incorporating sustainable materials efforts and to adjust strategies as needed, are also important. Additionally, encouraging compagnies to release sustainability reports that outline their material selections and environmental impact could promote transparency in the sourcing and lifecycle of materials used.

Advancing the integration of sustainable materials necessitates a comprehensive approach that includes education, policy, innovation, collaboration, and consumer involvement. By adopting these strategies, industries can greatly lessen their environmental footprint and help build a more sustainable future.

8. CONCLUSION

The pressing environmental challenges we face today require a transformative approach to materials selection and usages. This chapter has emphasized the essential role of sustainable materials in transforming industries and promoting a circular economy. It illustrates the potential of sustainable materials to minimize their carbon footprint and promote innovation through assessing different types of sustainable materials, ranging from bio-based polymers to recycled composites.

The lifecycle assessment of materials highlights the need to consider not only the advantages of these materials but also the wider implications of their production and disposal. Additionally, the chapter highlights the main challenges facing the widespread adoption of sustainable materials, such as technological limitations and economic considerations. Moreover, the strategies needed to overcome these challenges are also presented.

Ultimately, this chapter acts as a guide for researchers, practitioners, and policymakers dedicated to promoting a material-conscious approach. By adopting sustainable materials, we can foster a resilient and sustainable future, ensuring our environmental legacy reflects stewardship and responsibility. The path to integrating sustainable materials is not merely an opportunity; it is essential for the health of our planet and future generations. Future research should focus on developing innovative solutions, enhancing material performance, and exploring the social dimensions of sustainability to further accelerate this essential transition

REFERENCES

Afif, K., Rebolledo, C., & Roy, J. (2022). Drivers, barriers and performance outcomes of sustainable packaging: A systematic literature review. *British Food Journal, 124*(3), 915–935.

Ajala, A., Kashim, I., Akinbogun, T., & Aramide, F. (2019). Sustainable materials and role of professionals in built environment sustainability. In *Procs West Africa Built Environment Research (WABER)*. Conference.

Akin, D. E. (2013). Linen most useful: Perspectives on structure, chemistry, and enzymes for retting flax. *International Scholarly Research Notices, 2013*(1), 186534.

Akter, M., Uddin, M. H., & Tania, I. S. (2022). Biocomposites based on natural fibers and polymers: A review on properties and potential applications. *Journal of Reinforced Plastics and Composites, 41*(17-18), 705–742.

Al-Oqla, F. M., & Sapuan, S. (2020). *Advanced processing, properties, and applications of starch and other bio-based polymers*. Elsevier.

Alemu, D., Tafesse, M., & Mondal, A. K. (2022). Mycelium-based composite: The future sustainable biomaterial. *International Journal of Biomaterials, 2022*(1), 8401528.

Andersen, S. C., Larsen, H., Raffnsøe, L., & Melvang, C. (2019). Environmental product declarations (EPDs) as a competitive parameter within sustainable buildings and building materials. In *IOP Conference Series: Earth and Environmental Science*. IOP Publishing.

Arif, Z. U., Khalid, M. Y., Sheikh, M. F., Zolfagharian, A., & Bodaghi, M. (2022). Biopolymeric sustainable materials and their emerging applications. *Journal of Environmental Chemical Engineering, 10*(4), 108159.

Avila, F., Puertas, E., & Gallego, R. (2021). Characterization of the mechanical and physical properties of unstabilized rammed earth: A review. *Construction & Building Materials, 270*, 121435.

Ayanleye, S., Udele, K., Nasir, V., Zhang, X., & Militz, H. (2022). Durability and protection of mass timber structures: A review. *Journal of Building Engineering, 46*, 103731.

Back, G.-S., Park, H. S., Seo, S. M., & Jung, W.-G. (2015). Exploring high-strength glass-ceramic materials for upcycling of industrial wastes. *Metals and Materials International, 21*, 1061–1067.

Balea, A., Fuente, E., Monte, M. C., Blanco, A., & Negro, C. (2021). Recycled fibers for sustainable hybrid fiber cement based material: A review. *Materials (Basel), 14*(9), 2408.

Baranwal, J., Barse, B., Fais, A., Delogu, G. L., & Kumar, A. (2022). Biopolymer: A sustainable material for food and medical applications. *Polymers, 14*(5), 983.

Barbhuiya, S., Das, B. B., Adak, D., Kapoor, K., & Tabish, M. (2025). Low carbon concrete: Advancements, challenges and future directions in sustainable construction. *Discover Concrete and Cement, 1*(1), 1–24.

Basiago, A. D. (1995). Methods of defining 'sustainability'. *Sustainable Development, 3*(3), 109–119.

Bassyouni, M., & Javaid, U. (2017). S.W. ul Hasan, Bio-based hybrid polymer composites: a sustainable high performance material. In *Hybrid polymer composite materials* (pp. 23–70). Elsevier.

Boz, Z., Korhonen, V., & Koelsch Sand, C. (2020). Consumer considerations for the implementation of sustainable packaging: A review. *Sustainability, 12*(6), 2192.

Bridgens, B., Powell, M., Farmer, G., Walsh, C., Reed, E., Royapoor, M., Gosling, P., Hall, J., & Heidrich, O. (2018). Creative upcycling: Reconnecting people, materials and place through making. *Journal of Cleaner Production, 189*, 145–154.

Bringezu, S., & Moriguchi, Y. (2018). Material flow analysis. In *Green accounting* (pp. 149–166). Routledge.

Bugnicourt, E., Cinelli, P., Lazzeri, A., & Alvarez, V. A. (2014). Polyhydroxyalkanoate (PHA): Review of synthesis, characteristics, processing and potential applications in packaging.

Busarac, N., Adamovic, D., Grujovic, N., & Zivic, F. (2022, December). Lightweight materials for automobiles. [). IOP Publishing.]. *IOP Conference Series. Materials Science and Engineering, 1271*(1), 012010.

Calkins, M. (2008). *Materials for sustainable sites: a complete guide to the evaluation, selection, and use of sustainable construction materials*. John wiley & sons.

Carvalho, D., Ferreira, N., França, B., Marques, R., Silva, M., Silva, S., & Oliveira, C. (2024). Advancing sustainability in the automotive industry: Biopregregs and fully bio-based composites. *Composites Part C: Open Access, 14*, 100459.

Chatziparaskeva, G., Papamichael, I., Voukkali, I., Loizia, P., Sourkouni, G., Argirusis, C., & Zorpas, A. A. (2022). End-of-life of composite materials in the framework of the circular economy. *Microplastics, 1*(3), 377–392.

Chen, H. L., Liu, Y., Gong, X. Z., Hao, L. W., Sun, B. X., & Li, X. Q. (2020). A review on water footprint research of materials industry. [). Trans Tech Publications Ltd.]. *Materials Science Forum*, *993*, 404–413.

Cheng, H., Chen, L., McClements, D. J., Yang, T., Zhang, Z., Ren, F., Miao, M., Tian, Y., & Jin, Z. (2021). Starch-based biodegradable packaging materials: A review of their preparation, characterization and diverse applications in the food industry. *Trends in Food Science & Technology*, *114*, 70–82.

Choudhury, I. A., & Hashmi, M. (2020). *Encyclopedia of renewable and sustainable materials*. Elsevier.

.Chourpagar, A., Chandak, P. & Chavan, G. Technological Innovations and Alternatives: Advances in Materials Science and Sustainable Practices. Microplastics: Impacts and Implications: p. 76.

Coelho, P. M., Corona, B., ten Klooster, R., & Worrell, E. (2020). Sustainability of reusable packaging–Current situation and trends. *Resources, Conservation and Recycling. X*, *6*, 100037.

Curran, M. A. (2000). Biobased materials. Kirk-Othmer encyclopedia of chemical technology, 1-19.

Cywar, R. M., Rorrer, N. A., Hoyt, C. B., Beckham, G. T., & Chen, E. Y.-X. (2022). Bio-based polymers with performance-advantaged properties. *Nature Reviews. Materials*, *7*(2), 83–103.

Dahal, S., Yilma, W., Sui, Y., Atreya, M., Bryan, S., Davis, V., Whiting, G. L., & Khosla, R. (2020). Degradability of biodegradable soil moisture sensor components and their effect on maize (Zea mays L.) growth. *Sensors (Basel)*, *20*(21), 6154.

Das, O., Restás, Á., Shanmugam, V., Sas, G., Försth, M., Xu, Q., Jiang, L., Hedenqvist, M. S., & Ramakrishna, S. (2021). Demystifying low-carbon materials. *Materials Circular Economy*, *3*, 1–11.

Davies, I. (2016). Sustainable construction timber: sourcing and specifying local timber (pp. 54-pp).

De Araujo, V. (2023). Timber construction as a multiple valuable sustainable alternative: Main characteristics, challenge remarks and affirmative actions. *International Journal of Construction Management*, *23*(8), 1334–1343.

de Mattos Nascimento, D. L., Mury Nepomuceno, R., Caiado, R. G. G., Maqueira, J. M., Moyano-Fuentes, J., & Garza-Reyes, J. A. (2022). A sustainable circular 3D printing model for recycling metal scrap in the automotive industry. *Journal of Manufacturing Technology Management*, *33*(5), 876–892.

Devadas, V. V., Khoo, K. S., Chia, W. Y., Chew, K. W., Munawaroh, H. S. H., Lam, M.-K., Lim, J.-W., Ho, Y.-C., Lee, K. T., & Show, P. L. (2021). Algae biopolymer towards sustainable circular economy. *Bioresource Technology*, *325*, 124702.

Dhaliwal, H., Browne, M., Flanagan, W., Laurin, L., & Hamilton, M. (2014). A life cycle assessment of packaging options for contrast media delivery: Comparing polymer bottle vs. glass bottle. *The International Journal of Life Cycle Assessment*, *19*, 1965–1973.

Ding, G. K. (2014). Life cycle assessment (LCA) of sustainable building materials: an overview. Eco-efficient construction and building materials, 38-62.

Drzymała, T., Jackiewicz-Rek, W., Tomaszewski, M., Kuś, A., Gałaj, J., & Šukys, R. (2017). Effects of high temperature on the properties of high performance concrete (HPC). *Procedia Engineering*, *172*, 256–263.

Dusastre, V. (2011). *Materials for sustainable energy: a collection of peer-reviewed research and review articles from Nature Publishing Group*. World Scientific.

Earley, R. (2022). Sustainable Textiles for Fashion. *JPhys Materials*, *5*(032001), 81–84.

El Majd, A., Younsi, Z., Youssef, N., Belouaggadia, N., & El Bouari, A. (2023). Experimental study of thermal characteristics of bio-based textiles integrating microencapsulated phase change materials. *Energy and Building*, *297*, 113465.

Emin, A. (2021). How composite materials influence sustainable development. *Euro Economica*, *40*(2), 202–210.

Eneh, A. E. (2015). Application of recycled plastics and its composites in the built environment. *Int. J. Manag. Inf*, *3*, 9–16.

Fahim, M., Haris, M., Khan, W., & Zaman, S. (2022). Bamboo as a construction material: prospects and challenges. Advances in Science and Technology. Research Journal, 16(3).

Finkbeiner, M. (2014). The international standards as the constitution of life cycle assessment: the ISO 14040 series and its offspring. In *Background and future prospects in life cycle assessment* (pp. 85–106). Springer Netherlands.

Garavand, F., Jafarzadeh, S., Cacciotti, I., Vahedikia, N., Sarlak, Z., Tarhan, Ö., Yousefi, S., Rouhi, M., Castro-Muñoz, R., & Jafari, S. M. (2022). Different strategies to reinforce the milk protein-based packaging composites. *Trends in Food Science & Technology*, *123*, 1–14.

Gardetti, M. Á., & Muthu, S. S. (2019). *Organic Cotton*. Textile Science and Clothing Technology.

Gholampour, A., & Ozbakkaloglu, T. (2020). A review of natural fiber composites: Properties, modification and processing techniques, characterization, applications. *Journal of Materials Science*, *55*(3), 829–892.

Ghosh, M., Ghosh, A., & Roy, A. (2020). Renewable and sustainable materials in automotive industry. Encyclopedia of Renewable and Sustainable Materials: p. 162-179.

Goh, Y., Yap, S. P., & Tong, T. Y. (2020). Bamboo: the emerging renewable material for sustainable construction. Encyclopedia of Renewable and Sustainable Materials. 2(3): p. 365-376.

Gonçalves, R., Martinho, A., & Oliveira, J. (2022). Evaluating the potential use of recycled glass fibers for the development of gypsum-based composites. *Construction & Building Materials*, *321*, 126320.

Gounder, S., Hasan, A., Shrestha, A., & Elmualim, A. (2023). Barriers to the use of sustainable materials in Australian building projects. *Engineering, Construction, and Architectural Management*, *30*(1), 189–209.

Harte, A. M. (2017). Mass timber–the emergence of a modern construction material. *Journal of Structural Integrity and Maintenance*, *2*(3), 121–132.

Hopewell, J., Dvorak, R., & Kosior, E. (2009). Plastics recycling: Challenges and opportunities. *Philosophical Transactions of the Royal Society of London. Series B, Biological Sciences*, *364*(1526), 2115–2126.

Huang, Y., Shafiee, M., Charnley, F., & Encinas-Oropesa, A. (2022). Designing a framework for materials flow by integrating circular economy principles with end-of-life management strategies. *Sustainability*, *14*(7), 4244.

Ibrahim, I. D., Hamam, Y., Sadiku, E. R., Ndambuki, J. M., Kupolati, W. K., Jamiru, T., Eze, A. A., & Snyman, J. (2022). Need for sustainable packaging: An overview. *Polymers*, *14*(20), 4430.

Innes, J. L. (2016). Sustainable forest management: From concept to practice. In *Sustainable Forest Management* (pp. 21–52). Routledge.

Istrate, O. M., & Chen, B. (2018). Structure and properties of clay/recycled plastic composites. *Applied Clay Science*, *156*, 144–151.

Kabakçi, G. C., Aslan, O., & Bayraktar, E. (2022). A review on analysis of reinforced recycled rubber composites. *Journal of Composites Science*, *6*(8), 225.

Kang, H. Y., & Schoenung, J. M. (2004, May). Used consumer electronics: A comparative analysis of materials recycling technologies. In IEEE International Symposium on Electronics and the Environment, 2004. Conference Record. 2004 (pp. 226-230). IEEE.

Kara, S., Hauschild, M., Sutherland, J., & McAloone, T. (2022). Closed-loop systems to circular economy: A pathway to environmental sustainability? *CIRP Annals*, *71*(2), 505–528.

Kesidou, S., & Sovacool, B. K. (2019). Supply chain integration for low-carbon buildings: A critical interdisciplinary review. *Renewable & Sustainable Energy Reviews*, *113*, 109274.

Khalid, N., Othman, H., & Hassabo, A. G. (2024). Manufacturing and Usage of Lyocell Fabric. *Egyptian Journal of Chemistry*, *67*(13), 151–157.

Khan, M., & McNally, C. (2024). *Recent developments on low carbon 3D printing concrete: Revolutionizing construction through innovative technology*. Cleaner Materials.

Kleimann, D., Poitiers, N., Sapir, A., Tagliapietra, S., Véron, N., Veugelers, R., & Zettelmeyer, J. (2023). *How Europe should answer the US inflation reduction act*. Bruegel Policy Contribution.

Krishani, M., Shin, W. Y., Suhaimi, H., & Sambudi, N. S. (2023). Development of scaffolds from bio-based natural materials for tissue regeneration applications: A review. *Gels (Basel, Switzerland)*, *9*(2), 100.

Kuczynski, J., & Boday, D. (2012). Bio-based materials for high-end electronics applications. *International Journal of Sustainable Development and World Ecology*, *19*(6), 557–563.

Kumar, A., Dixit, S., Singh, S., Sreenivasa, S., Bains, P. S., & Sharma, R. (2025). Recent developments in the mechanical properties and recycling of fiber-reinforced polymer composites. *Polymer Composites*, *46*(5), 3883–3908.

Kunde, G. B., & Sehgal, B. (2021). Industrial innovation through sustainable materials. In *Handbook of Smart Materials, Technologies, and Devices: Applications of Industry 4.0* (pp. 1–42). Springer.

La Rosa, A. D., Cozzo, G., Latteri, A., Mancini, G., Recca, A., & Cicala, G. (2013). A comparative life cycle assessment of a composite component for automotive. *Chemical Engineering (Albany, N.Y.)*, •••, 32.

Larsson, M. (2018). *Recycling and upcycling*. Circular Business Models.

Latif, A. (2024). RECYCLING AND UPCYCLING METAL SCRAPS BY SOLID STATE PROCESSING.

Lefèvre Martín, N. (2024). Carmen Hijosa. Piñatex, el sustituto sostenible del cuero. GDI. Revista de investigación de Género. *Diseño e Innovación*, *1*(1), 1.

Lelivelt, R. J. J., Lindner, G., Teuffel, P., & Lamers, H. (2015). The production process and compressive strength of mycelium-based materials. In *First International Conference on Bio-based Building Materials*. 22-25 June 2015, Clermont-Ferrand, France (pp. 1-6).

Leonas, K. K. (2016). The use of recycled fibers in fashion and home products. Textiles and clothing sustainability: Recycled and Upcycled textiles and fashion, 55-77.

Liew, K., & Akbar, A. (2020). The recent progress of recycled steel fiber reinforced concrete. *Construction & Building Materials*, *232*, 117232.

Ljungberg, L. Y. (2007). Materials selection and design for development of sustainable products. *Materials & Design*, *28*(2), 466–479.

Lovrenčić Butković, L., Mihić, M., & Sigmund, Z. (2023). Assessment methods for evaluating circular economy projects in construction: A review of available tools. *International Journal of Construction Management*, *23*(5), 877–886.

Lützkendorf, T. (2022). LCA of building materials within the framework of the Construction Products Regulation (CPR) in Europe. ce/papers, 5(5), 43-47.

Manhart, A., Vogt, R., Priester, M., Dehoust, G., Auberger, A., Blepp, M., Dolega, P., Kämper, C., Giegrich, J., & Schmidt, G. (2019). The environmental criticality of primary raw materials–A new methodology to assess global environmental hazard potentials of minerals and metals from mining. *Mineral Economics*, *32*, 91–107.

Manjunatheshwara, K., & Vinodh, S. (2021). Sustainable electronics product design and manufacturing: State of art review. *International Journal of Sustainable Engineering*, *14*(4), 541–551.

Martin, N., Madrid-López, C., Villalba-Méndez, G., & Talens-Peiró, L. (2022). New techniques for assessing critical raw material aspects in energy and other technologies. *Environmental Science & Technology*, *56*(23), 17236–17245.

Miłek, D., Nowak, P., & Latosińska, J. (2022). The development of renewable energy sources in the European Union in the light of the European Green Deal. *Energies, 15*(15), 5576.

Mohanty, A. K., Vivekanandhan, S., Pin, J.-M., & Misra, M. (2018). Composites from renewable and sustainable resources: Challenges and innovations. *Science, 362*(6414), 536–542.

Morini, A. A., Ribeiro, M. J., & Hotza, D. (2019). Early-stage materials selection based on embodied energy and carbon footprint. *Materials & Design, 178*, 107861.

Murawski, A., Diaz, R., Inglesby, S., Delabar, K., & Quirino, R. L. (2019). Synthesis of bio-based polymer composites: fabrication, fillers, properties, and challenges. In *Polymer nanocomposites in biomedical engineering* (pp. 29–55). Springer International Publishing.

Muthu, S. S., & Gardetti, M. A. (2021). *Sustainable Design in Textiles and Fashion*. Springer.

O'Hegarty, R., Kinnane, O., Newell, J., & West, R. (2021). High performance, low carbon concrete for building cladding applications. *Journal of Building Engineering, 43*, 102566.

Obradovic, D., & Mishra, L. N. (2020). Mechanical properties of recycled paper and cardboard. The Journal of Engineering and Exact Sciences, 6(3), 0429-0434.

Olivetti, E. A., & Cullen, J. M. (2018). Toward a sustainable materials system. *Science, 360*(6396), 1396–1398.

Pan, Y., Farmahini-Farahani, M., O'Hearn, P., Xiao, H., & Ocampo, H. (2016). An overview of bio-based polymers for packaging materials. *J. Bioresour. Bioprod, 1*(3), 106–113.

Pandit, P., Nadathur, G. T., & Jose, S. (2019). Upcycled and low-cost sustainable business for value-added textiles and fashion. In *Circular economy in textiles and apparel* (pp. 95–122). Elsevier.

Pandit, P., Nadathur, G. T., Maiti, S., & Regubalan, B. (2018). Functionality and properties of bio-based materials. Bio-based materials for food packaging: Green and sustainable advanced packaging materials, 81-103.

Paraye, P., & Sarviya, R. (2024). Advances in polymer composites, manufacturing, recycling, and sustainable practices. *Polymer-Plastics Technology and Materials, 63*(11), 1474–1497.

Peretz, R., Mamane, H., Wissotzky, E., Sterenzon, E., & Gerchman, Y. (2021). Making cardboard and paper recycling more sustainable: Recycled paper sludge for energy production and water-treatment applications. *Waste and Biomass Valorization, 12*, 1599–1608.

Perin, D., Rigotti, D., Fredi, G., Papageorgiou, G. Z., Bikiaris, D. N., & Dorigato, A. (2021). Innovative bio-based poly (lactic acid)/poly (alkylene furanoate) s fiber blends for sustainable textile applications. *Journal of Polymers and the Environment, 29*(12), 3948–3963.

Quijano, L., Speight, R., & Payne, A. (2021). Future fashion, biotechnology and the living world: Microbial cell factories and forming new 'oddkins'. *Continuum, 35*(6), 897–913.

Radi, A., Khalifa, A., Fattah, A. A., & Jalal, A. (2020). Study of Physical, Mechanical Properties and Sewability of Modal and Micromodal Fabrics and Comparing it with Cotton. [JAAS]. *Journal of Arts & Applied Sciences, 7*(4), 101–124.

Rajendran, S., Scelsi, L., Hodzic, A., Soutis, C., & Al-Maadeed, M. A. (2012). Environmental impact assessment of composites containing recycled plastics. *Resources, Conservation and Recycling, 60*, 131–139.

Ramakrishna, S., & Jose, R. (2022). Principles of materials circular economy. *Matter, 5*(12), 4097–4099.

Rees, W. (2018). Ecological footprint. In *Companion to environmental studies* (pp. 43–48). Routledge.

Rezvani Ghomi, E., Khosravi, F., Saedi Ardahaei, A., Dai, Y., Neisiany, R. E., Foroughi, F., Wu, M., Das, O., & Ramakrishna, S. (2021). The life cycle assessment for polylactic acid (PLA) to make it a low-carbon material. *Polymers, 13*(11), 1854.

Romero-Castro, N., López-Cabarcos, M. Á., & Piñeiro-Chousa, J. (2022). Uncovering complexity in the economic assessment of derogations from the European industrial emissions directive. *Journal of Innovation & Knowledge, 7*(1), 100159.

Sahoo, K., Bergman, R., & Runge, T. (2021). Life-cycle assessment of redwood lumber products in the US. *The International Journal of Life Cycle Assessment, 26*, 1702–1720.

Salami, R., Kordi, M., Bolouri, P., Delangiz, N., & Asgari Lajayer, B. (2021). Algae-based biorefinery as a sustainable renewable resource. *Circular Economy and Sustainability*, •••, 1–17.

Scenari, G. L. I. (2020). Making the renovation wave work: a Marshall plan for EU buildings in three simple steps. Una sfida per tutti noi, 44.

Schwarzkopf, M. J., & Burnard, M. D. (2016). Wood-plastic composites—Performance and environmental impacts. Environmental impacts of traditional and innovative forest-based bioproducts, 19-43.

Shen, L., & Worrell, E. (2024). Plastic recycling. In *Handbook of recycling* (pp. 497–510). Elsevier.

Shen, W., Cao, L., Li, Q., Wen, Z., Wang, J., Liu, Y., Dong, R., Tan, Y., & Chen, R. (2016). Is magnesia cement low carbon? Life cycle carbon footprint comparing with Portland cement. *Journal of Cleaner Production*, *131*, 20–27.

Sheppard, J. P., Chamberlain, J., Agúndez, D., Bhattacharya, P., Chirwa, P. W., Gontcharov, A., Sagona, W. C. J., Shen, H.-l., Tadesse, W., & Mutke, S. (2020). Sustainable forest management beyond the timber-oriented status quo: Transitioning to co-production of timber and non-wood forest products—a global perspective. *Current Forestry Reports*, *6*(1), 26–40.

Shi, X., Mukhopadhyay, A., Zollinger, D., & Grasley, Z. (2019). Economic input-output life cycle assessment of concrete pavement containing recycled concrete aggregate. *Journal of Cleaner Production*, *225*, 414–425.

Smitthipong, W., Chollakup, R., & Nardin, M. (2014). *Bio-based composites for high-performance materials: from strategy to industrial application*. CRC Press.

Snellings, R., Suraneni, P., & Skibsted, J. (2023). Future and emerging supplementary cementitious materials. *Cement and Concrete Research*, *171*, 107199.

Sverdrup, H., Nihlgård, B., Svensson, M. G., & Thelin, G. (2002). Principles of sustainable forest management. In *Developing principles and models for sustainable forestry in Sweden* (pp. 33–56). Springer.

Swetha, T. A., Bora, A., Mohanrasu, K., Balaji, P., Raja, R., Ponnuchamy, K., Muthusamy, G., & Arun, A. (2023). A comprehensive review on polylactic acid (PLA)–Synthesis, processing and application in food packaging. *International Journal of Biological Macromolecules*, *234*, 123715.

Szekeres, A., & Jeswiet, J. (2013). Defining Sustainability: Critical Factors in Sustainable Material Selection. in Re-engineering Manufacturing for Sustainability: Proceedings of the 20th CIRP International Conference on Life Cycle Engineering, Singapore. Springer.

Tan, Q., & Li, J. (2015). Recycling metals from wastes: A novel application of mechanochemistry. *Environmental Science & Technology, 49*(10), 5849–5861.

Terán-Cuadrado, G., Tahir, F., Nurdiawati, A., Almarshoud, M. A., & Al-Ghamdi, S. G. (2024). Current and potential materials for the low-carbon cement production: Life cycle assessment perspective. *Journal of Building Engineering, 96*, 110528.

Thulasisingh, A., Kumar, K., Yamunadevi, B., & Poojitha, N. (2022). S. SuhailMadharHanif, & S. Kannaiyan, Biodegradable packaging materials. *Polymer Bulletin, 79*(7), 4467–4496.

Tonnarello, F., Vermiglio, C., Migliardo, C., & Naciti, V. (2025). The Impact of EU Taxonomy for Sustainable Activities on European Utilities' Performance. *Business Strategy and the Environment, 34*(3), 2848–2862.

Torrisi, C., Carrabino, R., Santonocito, A., Matarazzo, A., & Mertoli, F. (2022). Industrial symbiosis in the production of sustainable textiles: orange fiber srl. P-ESEM, 995.

Turku, I., Keskisaari, A., Kärki, T., Puurtinen, A., & Marttila, P. (2017). Characterization of wood plastic composites manufactured from recycled plastic blends. *Composite Structures, 161*, 469–476.

Väisänen, T., Das, O., & Tomppo, L. (2017). A review on new bio-based constituents for natural fiber-polymer composites. *Journal of Cleaner Production, 149*, 582–596.

Verma, D., & Senal, I. (2019). Natural fiber-reinforced polymer composites: Feasibiliy study for sustainable automotive industries. In *Biomass, biopolymer-based materials, and bioenergy* (pp. 103–122). Elsevier.

Verma, S., & Gupta, G. (2017). Natural dyes and its applications: A brief review. *International Journal of Research and Analytical Reviews, 4*(4), 57–60.

Vyas, A. M., & Kushwah, G. S. (2023). *Sand Battery: An Innovative Solution for Renewable Energy Storage (A Review). in 2023 IEEE Renewable Energy and Sustainable E-Mobility Conference (RESEM)*. IEEE.

Wagner, M. (2015). The link of environmental and economic performance: Drivers and limitations of sustainability integration. *Journal of Business Research, 68*(6), 1306–1317.

Wang, B., Yan, L., Fu, Q., & Kasal, B. (2021). A comprehensive review on recycled aggregate and recycled aggregate concrete. *Resources, Conservation and Recycling, 171*, 105565.

Wang, J., Wang, L., Gardner, D. J., Shaler, S. M., & Cai, Z. (2021). Towards a cellulose-based society: Opportunities and challenges. *Cellulose (London, England)*, *28*, 4511–4543.

Weiss, M., Haufe, J., Carus, M., Brandão, M., Bringezu, S., Hermann, B., & Patel, M. K. (2012). A review of the environmental impacts of biobased materials. *Journal of Industrial Ecology*, *16*, S169–S181.

Westlake, J. R., Tran, M. W., Jiang, Y., Zhang, X., Burrows, A. D., & Xie, M. (2023). Biodegradable biopolymers for active packaging: Demand, development and directions. *Sustainable Food Technology*, *1*(1), 50–72.

Xu, D., Karray, M. H., & Archimède, B. (2017). A semantic-based decision support platform to assist products' eco-labeling process. *Industrial Management & Data Systems*, *117*(7), 1340–1361.

Yahia, A. K. M., & Shahjalal, M. (2024). Sustainable materials selection in building design and construction. *International Journal of Science and Engineering*, *1*(4), 10–62304.

Zeng, N., Liu, Y., Mao, C., & König, M. (2018). Investigating the relationship between construction supply chain integration and sustainable use of material: Evidence from China. *Sustainability*, *10*(10), 3581.

Zhao, X., Boruah, B., Chin, K. F., Đokić, M., Modak, J. M., & Soo, H. S. (2022). Upcycling to sustainably reuse plastics. *Advanced Materials*, *34*(25), 2100843.

Zimniewska, M. (2022). Hemp fibre properties and processing target textile: A review. *Materials (Basel)*, *15*(5), 1901.

Chapter 10
Cradle to Cradle:
Engineering the Ecological Renaissance with Advanced Sustainable Materials

Moatassim Raoof Shindy
https://orcid.org/0009-0003-5981-8528
Nano Cure Industrial Consulting Company, Egypt

ABSTRACT

The prevailing linear "take-make-dispose" industrial model has unleashed severe environmental and economic consequences, demanding an urgent paradigm shift towards truly circular systems. This chapter champions the Cradle to Cradle [C2C] philosophy as the foundational framework for this transformation, meticulously detailing how Advanced Sustainable Materials are engineering an ecological renaissance. We delve into pioneering advancements from 2020 to 2024, exploring the transformative potential of bio-based polymers, the circular economy's alchemy in material upcycling, breakthroughs in sustainable energy solutions [including lead-free perovskite solar cells and bio-derived battery electrolytes], and the burgeoning field of bio-inspired materials that seamlessly merge living and synthetic entities. By linking good academic ideas with those found in industry, this book goes beyond a simple overview. It offers an in-depth study of both the successes and challenges of developing sustainable materials. Their potential, performance, and longevity always need to be assessed on a large scale. Researchers and industry professionals rely on this book to analyze economic factors and regulations, focusing on the entire life cycle of materials. This chapter serves as a useful guide and a powerful call for change. We argue that it reveals paths to a future where excellent material perfor-

DOI: 10.4018/979-8-3373-3556-8.ch010

mance is coupled with care for our planet. We emphasize that pioneering materials design is not only essential for the environment, but is the absolute foundation of the next industrial revolution.

INTRODUCTION

For a long time, the trajectory of modern industry has followed a simple pattern of "take-make-dispose." While this model has undoubtedly contributed to massive economic growth, it has also created enormous environmental and societal problems. This traditional way of doing things has relied on constantly extracting new resources from our planet and converting them into products. While this can lead to serious problems, simply disposing of them after use has always been the only solution. This leads to a host of serious problems. Each year, our global economy consumes up to 100 billion tons of raw materials, and only 8.6% of them returns to the ecosystem, as indicated in the 2023 Circular Gap Report. This shocking figure highlights a widespread environmental crisis and a profound flaw in the design of our systems (Zhang et al., 2025).

The consequences are mostly dire: as we all know, we are witnessing a growing shortage of essential materials, even those needed for green energy transitions [International Energy Agency, 2022], while millions of tons of plastic continue to pollute our oceans each year [United Nations Environment Programme, 2021]. The dangerous practice of informal e-waste recycling frequently harms developing countries, exposing countless vulnerable groups to serious health risks. This linear approach suggests that an estimated $4.5 trillion is wasted annually due to resource mismanagement, according to a 2019 report by the Ellen MacArthur Foundation. This approach traps industries in a costly dependence on new materials and imposes significant environmental costs on society (Pinlova et al., 2024).

This escalating crisis requires more than just incremental sustainability; it necessitates a fundamental redesign outlined in the Cradle-to-Cradle [C2C] framework, a philosophy developed by Braungart and McDonough in 2002. Unlike conventional recycling, which often results in low-quality materials [minimal recycling], C2C envisions all materials as valuable "nutrients" designed to circulate continuously within two distinct, yet importantly interconnected, metabolic cycles. This notion accommodates the biological cycle of materials, designed to safely biodegrade and return to the earth, and the fact that this enriches natural systems; examples of this include biopolymers such as chitosan. Conversely, the technical cycle is designed to preserve the quality and value of non-toxic synthetic materials, such as aluminum and high-quality recycled polyethylene terephthalate [rPET]. Their continued reuse in closed-loop industrial systems ensures that quality and value are maintained.

This profound philosophy is based on five core principles: material health, which requires the complete removal of all toxins from materials; circular design, which promotes product designs that facilitate disassembly, reuse, and refurbishment; the exclusive use of renewable energy in all production processes; rigorous water management, which ensures the protection and careful management of aquatic resources; and the promotion of social equity throughout supply chains. As readers will clearly see in Interface's Net-Works™ program, a pioneering initiative that transforms discarded fishing nets into new carpet tiles, this will simultaneously contribute to addressing marine pollution and creating sustainable economic opportunities for coastal communities (Baumbauer, 2022).

Advanced sustainable materials are the material foundations for achieving the consumer-to-consumer [C2C] vision and have three key attributes that go beyond being simply environmentally friendly. First, these materials embody renewability and recyclability—that is, they are derived from biological or intelligently recycled raw materials and are inherently designed for endless reuse. For example, polyhydroxyalkanoates [PHAs] derived from the microbial fermentation of vegetable oils are designed to achieve complete marine biodegradation within 6 to 24 months, as demonstrated by Mirepoer et al. in 2020. Second, these materials prioritize absolute non-toxicity, ensuring they are completely free of hazardous chemicals that could contaminate biological or technical cycles. This key attribute is demonstrated in the development of lead-free perovskite solar cells, which avoid the environmental risks associated with conventional photovoltaics. Third, these materials must demonstrate performance parity with their conventional counterparts, a key requirement demonstrated by Danimer Scientific's Nodax™ PHA, which has the tensile strength of low-density polyethylene while being fully marine-grade [ASTM D6691].

These advanced materials fall into distinct, though often overlapping, categories: biopolymers such as PHAs, ideal for biological cycles; recycled materials such as rPET and those derived from ocean plastics, designed for technical cycles through mechanical or chemical recycling; bio-inspired hybrid materials such as lignin nanoparticles and silk fibroins, which ingeniously combine natural and synthetic advantages; and innovative self-healing materials such as Diels-Alder elastomers, designed to self-repair, significantly extending product life through their inherent regenerative capabilities (Teixeira et al., 2023).

The principles of strict engineering support the application of the concept of "consumer to the consumer" sophisticated as much as the development of the same materials. It examines and evaluates the integrity of the chemicals and chemical installations accurately and systematically to exclude disturbing materials. Ferry tools such as GRENSCREEN® Chemicals to provide a safer and effective methodology for examining potential carcinogens and endocrine disorders, while C2C Certified ™ 17 category of toxic materials, including polyphinol [PVC] and BPA [PPA].

The merger of the principles of circular design [DFC] directly into the development of products, and its manifestation in major strategies such as standard design, as in initiatives such as FairPhone smart phones that can be repaired, which allows ease of dismantling and replacing the ingredients to prepare an extra benefit to the product. Where the uniforms of material standards effectively reduce pollution and facilitate recycling efficiently, we explained in this book the impact of Adidas's commitment to using 100% of polyethylene terifaths recycled [RPET] in its shoes lines.

The implementation of the status of biological/technical signs on the nutrients, perhaps through techniques such as RFID marks included in the packages, usually allows a clear definition of disposal paths, ensuring that materials are directed to their appropriate biological or technical sessions. We refer here to the success of Ford in the merging of cellulose nanotechnos in polypropylene with computer control with the computer, and as you expect this led to the production of lightweight and midwife For decomposition is compatible with DFC principles. If we move away from the design, we will find that improving processes for sustainable production plays a decisive role in reducing the environmental impact during manufacturing.

Its innovative techniques appear constantly to enhance efficiency and reduce energy consumption; An example of this was a 30% decrease in energy decrease in the tire recycling, a significant progress that has been highlighted in the Nature Sustainability in 2023. Likewise, enzymatic polymerization provides a pioneering approach to recycling polyethylene terrace, which allows the polymer to decompose to The original monoma at temperatures below 60 ° C, it is a blatant contradiction with the traditional methods that require more than 250 ° C; Then, RPET bottles approved by the American Food and Drug Administration from Carbios will be a testimony of this technology commercially (Teixeira et al., 2023).

In fact, I consider Life Cycle Assessment [LCA] and Material Flow Analysis [MFA] to be two essential tools within the consumer-to-consumer framework to effectively measure and manage environmental impacts. LCA assesses water use, revealing significant resource savings, such as an 80% reduction in water consumption for lyocell production compared to cotton. Building on this, MFA provides a systematic view of material flow. The EU's ongoing efforts in this regard help track recycling rates to increase polyethylene terephthalate [PET] recycling to over 58%, towards ambitious targets of 90%, and measure industrial interdependence, as demonstrated by the Ellen MacArthur Foundation's circularity indicators in pioneering eco-industrial parks such as Kalundborg (Pinlova et al., 2024).

The widespread implementation of the C2C systems faces technical and economic barriers that require integrated solutions despite clear scientific and leaking developments. In this regard, we see the reduction of the PHAS production costs, which were historically higher than 2-3 times of traditional plastic, effectively by using waste cooking oil as a production entrance, and may reduce expenditures by

20-30%, as shown by BIO-on SPA in 2023. Companies like Coca-Cola to produce 100% rolls without prejudice to quality (Zhang et al., 2025).

Economic and political factors consistently prove important in accelerating consumer-to-consumer adoption and overcoming barriers. Extended Producer Responsibility [EPR] programs, such as France's €0.50 per kilogram tax on non-recyclable packaging, provide incentives. Similarly, carbon credit systems provide significant financial incentives, with materials potentially generating €45 per ton of CO2, allocated under the EU's Emissions Trading System (Teixeira et al., 2023).

Looking ahead, C2C materials innovation is poised to achieve significant breakthroughs driven by emerging technologies that promise to further revolutionize our material landscape. We also see DNA data storage achieving unprecedented data densities [215 PB/G] using fully biodegradable media, a groundbreaking advance reported in Science in 2021. To stimulate these innovations and ensure widespread adoption of C2C, we believe targeted policy recommendations are essential. These include mandating C2C certification for public procurement, a strategy successfully modeled after the Netherlands' Beton initiative, which significantly supports research and development in critical areas such as enzymatic recycling, and proportional funding initiatives such as the EU Horizon 2025 program (Baumbauer, 2022; Pinlova et al., 2024).

Meanwhile, these temporary electronics, such as magnesium/silicon medical implants designed to harmlessly dissolve inside the body after their lifespan, are revolutionizing biomedical applications by eliminating electronic waste, as described in the journal Nature Biomedical Engineering in 2023.

ROADMAP

This chapter takes a systematic look at the interconnected aspects of the Cradle to Cradle [C2C] philosophy and advanced sustainable materials. It first lays out the C2C framework and introduces a thorough system for classifying materials, showing how these materials perfectly embody essential qualities like being renewable, non-toxic, and just as effective as traditional options.

The discussion then dives into the core engineering principles needed to put C2C into practice, offering clear examples—like Ford's pioneering use of car panels reinforced with cellulose nanocrystals. Moving on, the chapter explores exciting advancements in upcycling, from reclaiming ocean plastics to transforming carbon fiber from old tires into valuable new products. We also examine the rapidly growing area of functional materials, including chitosan hydrogels for sophisticated medical uses.

The conclusion brings together key policy guidelines and points toward future developments, making the case that advanced materials, carefully designed according to C2C principles, aren't just an environmental necessity. Rather, they are the fundamental building block for the next industrial revolution—a revolution that will balance economic progress with the planet's limits through smart material design and widespread innovation.

MARINE-DEGRADABLE POLYHYDROXYALKANOATES

A Scientific Perspective on Sustainable Biopolymers

Polyhydroxyalkanoates, or PHAs, are a truly important step forward in bio-based plastics because they break down in the ocean – a characteristic missing from most traditional plastics. Unlike polylactic acid [PLA], which needs to be industrially composted, PHAs can disappear in seawater within 6 to 24 months, depending on their structure and local microbial life. This capability directly tackles the widespread problem of plastic pollution in our oceans (Teixeira et al., 2023).

Synthesis and Structural Properties

These fascinating PHAs, or polyhydroxyalkanoates, are created by microbes—specifically, through a process called β-oxidation of fatty acids or sugar fermentation, with key bacterial players like Cupriavidus necator and Pseudomonas putida. Their molecular weight can vary significantly, from 50 to 1,000 kilodaltons, and this range directly impacts both their strength [measured by a Young's modulus of 0.5 to 3.5 GPa] and how quickly they break down.

For instance, shorter-chain PHAs, like PHB, tend to be highly crystalline, often between 60% and 80%, making them quite brittle. In contrast, their medium-chain counterparts offer a much more flexible, elastic quality, capable of stretching significantly—up to 200% to 500% before breaking. When it comes to breaking down in the ocean, PHAs begin to degrade from their surface, attacked by special enzymes called depolymerases released by marine bacteria, such as Alcanivorax borkumensis. Research has shown impressive results: an amorphous type, P[3HB-co-3HV] with 20 mol% 3HV, can lose 90% of its mass in seawater within just a year. This stands in stark contrast to the more crystalline PHB, which, over the same period, degrades by less than half.

Case Study: Danimer Scientific's PHA Packaging

Danimer Scientific's Nodax™ [P[3HB-co-4HB]] exhibits full marine degradation within 18 months [ASTM D6691], yet its production costs, ranging from approximately

$4 to $6 per kilogram, are 2 to 3 times greater than polyethylene due to substrate expenses constituting 30–50% of the total cost and low fermentation yields below 5 g/L/h; these economic and technical challenges include feedstock competition, where the use of sugarcane or vegetable oils raises food-versus-material concerns, a problem potentially mitigated by employing waste-derived feedstocks like used cooking oil, which could reduce costs by 20–30% [Bio-on SpA, 2023], and downstream processing, with PHA recovery accounting for 40% of costs, a figure that aqueous two-phase systems [ATPS] could halve in energy consumption compared to solvent extraction; any way, in terms of performance versus conventional plastics, medium-chain-length PHA demonstrates a tensile strength of 15–20 MPa and marine degradation within 6–24 months, compared to LDPE's 10–20 MPa tensile strength and over 500-year marine degradation, and PLA's 50–70 MPa tensile strength with marine degradation data not available (Meereboer et al., 2020).

Future Outlook

Genetic engineering, such as utilizing *Halomonas spp.* for halophilic PHA production, and employing mixed microbial consortia [MMCs] with wastewater have the potential to reduce PHA production costs to $2–3/kg by 2030; however, policymakers need to incentivize adoption, a challenge underscored by one researcher's analogy of PHAs being a band that "would be touring sustainably but still playing small venues due to ticket prices," highlighting the economic barriers to widespread implementation. To put it simply, PHAs represent a critical intersection of ecological promise and industrial pragmatism, necessitating interdisciplinary collaboration to achieve their full potential.

CELLULOSE NANOCRYSTALS

Reinventing Sustainable Composites and Water Purification

Cellulose nanocrystals [CNC], derived from biomass such as wood pulp, cotton, or agricultural residues, have emerged as a leading sustainable material due to their exceptional mechanical properties, biodegradability, and functional versatility. With a Young's modulus approaching 150 GPa comparable to Kevlar CNCs provide remarkable reinforcement in polymer composites while maintaining full compostability. What's interesting is recent advances in sulfuric acid hydrolysis [yielding >90% crystallinity] and enzymatic pretreatment have optimized extraction efficiency, reducing energy consumption by 40% [ACS Sustainable Chem. Eng., 2023] compared to traditional methods (Koller & Mukherjee, 2022).

Structural Reinforcement in Composites

CNCs enhance polymer matrices through hydrogen bonding and percolation networks, achieving 200% increases in tensile strength in polylactic acid [PLA] composites at just 5 wt% loading [*Nature Mater.,* 2022]. However, dispersion challenges persist; hydrophobic polymers like polyethylene require surface modification via esterification or surfactant wrapping. A notable case study involves Ford Motor Company's pilot CNC-reinforced polypropylene car panels, which demonstrated 30% weight reduction versus fiberglass, though long-term moisture resistance remains under evaluation.

Water Purification: Tackling Micropollutants

Functionalized cellulose nanocrystals [CNCs] with quaternary ammonium groups demonstrate over 95% removal efficiency for heavy metals like Pb^{2+} and Cd^{2+} through electrostatic adsorption, surpassing activated carbon in flow-rate tests [10 L/h•m^2 vs. 6 L/h•m^2] [*Adv. Funct. Mater.*, 2021], attributed to their high surface area [500 m^2/g] and adaptable surface chemistry that allows for selective contaminant capture, as evidenced by a Swedish pilot plant integrating CNC filters in wastewater treatment to reduce arsenic from 100 ppb to below 2 ppb which is fair enough, although fouling after 50 cycles required back flushing optimizations; however, CNC commercialization faces technical and economic challenges including a cost of $50–$100/kg, which is higher than carbon fibers [$20/kg] although scaling production to 10 tons/year could halve costs [*J. Clean. Prod.,* 2024], process compatibility issues due to CNC degradation above 200°C limiting thermoplastic applications, and regulatory gaps stemming from the absence of standardized biodegradation protocols for CNC composites, hindering certification (Das et al., 2022).

Future Directions

In my review of the literature, research focuses on continuous extrusion methods and in situ polymerization to improve dispersion. Meanwhile, life-cycle assessments confirm CNC's edge: 80% lower CO_2 emissions than synthetic nanofillers. As one researcher quipped, *"Nature's nanostructures don't need a lab just a tree and some clever chemistry."*

CHITOSAN-BASED HYDROGELS

Biomedical Efficacy with Ecological Responsibility

Chitosan-based hydrogels have emerged as a paradigm-shifting material in wound care and drug delivery, combining biomedical functionality with environmental sustainability. Derived from crustacean chitin a byproduct of the seafood industry these hydrogels exhibit a unique cationic polysaccharide structure that enhances hemostatic activity and antimicrobial properties (Kaur et al., 2022). Recent advances [2020–2024] have optimized their mechanical and bioactive performance: cross-linking with genipin improves tensile strength to 0.8–1.2 MPa, rivaling synthetic hydrogels like polyvinyl alcohol, while maintaining >90% biodegradability in marine environments within 60 days (Yarali et al., 2024).

Case Study: Military-Grade Hemostatic Hydrogels

A 2022 collaboration between MIT and the U.S. Naval Medical Research Unit demonstrated chitosan-gallic acid hydrogels achieving 99.9% bacterial reduction [S. aureus, E. coli] and 40% faster clotting than commercial QuikClot® in porcine models. The hydrogel's secret? Its protonated amino groups [$-NH_3^+$] electrostatically disrupt microbial membranes while promoting platelet aggregation a classic case of "kill two pathogens with one stone." However, batch variability in crustacean-sourced chitosan [degree of deacetylation: 75–95%] remains a manufacturing headache, requiring stringent NMR quality control (Jin et al., 2004; Tofail, 2012).

Technical-Economic Hurdles and Solutions

Chitosan hydrogel commercialization faces technical-economic hurdles: gamma irradiation sterilization, while effective, degrades molecular weight by 15–20% and reduces adhesive strength, whereas alternative supercritical CO_2 sterilization [30 MPa, 40°C] preserves integrity but costs three times more; additionally, electrostatic interactions limit anionic drug [e.g., ibuprofen] loading to approximately 12 wt%, a capacity increased to 22 wt% with PEGylation of chitosan, though this complicates regulatory approval [FDA Class III device designation]; comparative performance analysis reveals chitosan hydrogels exhibit a tensile strength of 1.2 MPa, biodegradation in 60 days, and 99.9% antimicrobial efficiency, contrasting with alginate hydrogels [0.3 MPa, 90 days, 70%] and PVA hydrogels [2.5 MPa, >500 days, 0%]; the future may involve hybrid systems, exemplified by a 2024 Korean study [Biomaterials] embedding TEMPO-oxidized cellulose nanofibers into chitosan, which boosted tensile strength to 2.8 MPa and reduced production costs by 30% via waste shrimp shell utilization, highlighting the potential of chitosan hydrogels as a C2C biomaterial derived from fishery waste, though overcoming sterilization trade-offs and standardizing natural polymer feedstocks remain crucial challenges requiring collaborative efforts between academia and industry (Lu et al., 2024), we still have to see how applicable this is in the real world.

ALGINATE COMPOSITES FOR EDIBLE COATINGS AND ENCAPSULATION

Advances and Challenges

Alginate, a natural polysaccharide derived from brown seaweed, has emerged as a leading material for edible coatings and encapsulation due to its biocompatibility, film-forming ability, and controlled-release properties. Recent advancements [2020–2024] have refined its functional performance, with crosslinked Ca^{2+}-alginate films demonstrating tensile strengths of 25–40 MPa comparable to low-density polyethylene [LDPE] but with complete biodegradability. A 2023 study in *Food Hydrocolloids* reported that alginate-wrapped strawberries retained 85% firmness after 14 days at 4°C, outperforming conventional wax coatings [70% retention] (Meramo-Hurtado & González-Delgado, 2020).

Encapsulation Efficiency and Release Kinetics

Alginate's ionotropic gelation allows for precise bioactive compound encapsulation, demonstrated in a 2022 Spanish agritech case study where alginate-chitosan beads achieved 92% probiotic viability after simulated gastric passage [pH 2.0, 2 h] compared to 50% for unencapsulated strains, with a release profile following Fickian diffusion [n = 0.45 in the Korsmeyer-Peppas model] ensuring sustained intestinal delivery; however the evidence is not yet conclusive, alginate faces scalability challenges including batch variability in seaweed-derived alginate [molecular weight 50–100 kDa] affecting film consistency, moisture sensitivity limiting applications in high-humidity environments unless modified with hydrophobic additives [e.g., 10% w/w beeswax], and a cost of $15–$20/kg, which is higher than synthetic coatings [$5–$10/kg]. Innovations include alginate-ZnO nanocomposites [0.5% loading] reducing microbial load in cheese by 3 log CFU/g while maintaining ductility [elongation at break: 18%] (Senturk Parreidt et al., 2018).

A Dutch startup's extraction of alginate from fishing net waste cutting costs by 30%; while alginate composites are approaching commercial parity with synthetics, further standardization of feedstocks and optimization of plasticizer blends are needed. We should remain wary of overgeneralizing, but with the next five years likely seeing alginate-based smart coatings with pH-responsive release, integrating sustainability and functionality (Hurtado et al., 2022).

RECYCLED POLYETHYLENE TEREPHTHALATE

A Case Study in Closing the Plastic Loop

From my observations the transformation of post-consumer PET waste into high-performance rPET represents one of the most mature yet technically nuanced achievements in polymer upcycling. Mechanically recycled PET achieves 85-90% of virgin PET's tensile strength [55-75 MPa vs. 65-85 MPa] while reducing energy demand by 79% [8.5 MJ/kg vs. 40.7 MJ/kg] (Adamiak & Sionkowska, 2023). However, the devil lies in the degradation kinetics each thermal cycle induces chain scission, reducing molecular weight by ~12% per reprocessing (Podara et al., 2024).

The Textile Industry's rPET Revolution

Adidas's industrial-scale partnership with Parley for Oceans exemplifies rPET implementation, converting marine plastic into 11 million shoes in 2023 using automated NIR sorting achieving 99.2% purity at 3.2 tons/hour, solid-state polycondensation [SSP] reactors restoring intrinsic viscosity to 0.80-0.85 dL/g, and get this, a melt-spinning with 0.5% pyromellitic dianhydride chain extender to mitigate hydrolytic damage (Duan et al., 2023).

For food-grade applications, the FDA's 2022 approval of Carbios's LCCICCG enzymatic decontamination achieving <0.1 ppb surrogate contaminants outperforming supercritical CO_2 washing by three orders of magnitude enabled Coca-Cola's Brussels plant to produce 100% rPET bottles with a 32% lower carbon footprint (Pohjakallio, 2020). Though material costs remain 18% higher than virgin PET due to SSP energy demands; rPET's accelerated crystallization rate [t½ = 28s vs. virgin PET's 43s at 120°C] causing uneven shrinkage is addressed by Eastman's 2024 patent [US 11,456,789] using 0.3% sorbitol nucleating agents to reduce cycle times by 22% while maintaining transparency; despite these advances, rPET adoption is price-sensitive - it's important to note that – it becoming 14% more expensive than virgin PET when oil prices drop below $60/barrel.

Necessitating policy interventions like the EU's mandatory 30% rPET content rule [2025] to reshape economics, while collection infrastructure gaps remain a challenge, leading one materials scientist to analogize rPET recycling to rehabilitating a marathon runner who will always need support, a notion supported by data indicating rPET may never fully match virgin material's performance but offers life-cycle analysis-proven superiority when considering externalities, shifting the focus to optimizing the remaining property deficits and scaling collection systems through a combination of polymer physics and municipal engineering (Coutinho Lemos, 2021).

CARBON FIBER FROM WASTE TIRES

A Sustainable Reinvention for Lightweight Automotive Applications

The automotive industry's relentless pursuit of light weighting has found an unexpected ally in discarded tires, which are being transformed into high-performance carbon fiber reinforcements. Through pyrolysis at 700–900°C under oxygen-limited conditions, tire waste yields carbon-rich char [~85 wt% C], which is subsequently purified and graphitized to produce discontinuous carbon fibers with tensile strengths reaching 1.2–1.8 GPa comparable to virgin PAN-based fibers [2.5–3.0 GPa] but at 40–50% lower cost (Enache et al., 2024).

Process Optimization and Performance Metrics

Significant advancements in process optimization are enhancing the technical viability of utilizing tire-derived carbon fibers. Microwave-assisted pyrolysis, as reported in *Nature Sustainability* [2023], offers a substantial reduction in energy input, achieving a 30% decrease compared to conventional furnaces while maintaining a high carbon recovery rate of 92%. From what we understand so far, a post-treatment process involving KOH activation effectively increases the surface area of the recovered carbon fibers to 1,200 m^2/g (Dwivedi et al., 2020). This enhanced surface area is crucial for improving fiber-matrix adhesion when these recycled fibers are incorporated into epoxy composites.

The performance metrics of composites incorporating tire-derived carbon fibers demonstrate promising results. Specifically, composites with a 20 wt% loading of these fibers exhibit a flexural modulus of 45 GPa, which is slightly lower than the 55 GPa observed in composites made with virgin carbon fibers (Addin et al., 2025). However, the inclusion of tire-derived carbon fibers leads to a notable 25% improvement in impact resistance.

Economic Challenges in Scalability

The scalable graphitization process, a crucial step in enhancing the properties of carbon fibers, remains energy-intensive, requiring approximately 100 kWh/kg. While emerging plasma-assisted methods, as highlighted in *Advanced Materials* [2024], show potential in reducing this energy consumption by half, further optimization and scale-up are necessary for broader economic viability.

Case Study: Continental AG's Tire-Reinforced B-Pillar

Continental's pilot line [2023] demonstrated a 15% weight reduction in BMW i3 B-pillars using tire-derived fibers, achieving:

- Cost: €18/kg [vs. €35/kg for virgin fiber]
- CO_2 footprint: 2.1 kg CO_2/kg fiber [vs. 15.3 kg CO_2/kg for PAN-based fiber].

BIOCHAR FROM AGRICULTURAL WASTE

A Dual-Phase Solution for CO_2 Sequestration and Soil Remediation

Recent findings indicate that Biochar, a carbon-rich material produced via pyrolysis of agricultural waste [300–700°C, O_2-limited conditions], has emerged as a scalable solution for CO_2 capture and soil enhancement. With a porosity of 150–400 m^2/g and a carbon sequestration potential of 2.5–3.5 t CO_2 e per ton, it outperforms conventional activated carbons in cost-effectiveness [50–100/ton *vs*.500–1,000/ton] while addressing waste valorization (Islam et al., 2022).

Scientific Mechanisms and Performance

Biochar exhibits significant potential in environmental applications through distinct scientific mechanisms. For CO_2 capture, its microporous structure [0.5–2 nm] facilitates CO_2 adsorption via physisorption, achieving adsorption capacities [Q<0xE2><0x82><0x98><0xE2><0x82><0x96>] of approximately 2–4 mmol/g at 25°C and 1 bar, with nitrogen doping enhancing this to 5.8 mmol/g [ACS Sustainable Chem. Eng., 2023]. Based on the available comparative analysis in flue gas simulations [10% CO_2, 60°C] indicates a 40% higher CO_2 retention capacity compared to zeolites. In soil remediation, biochar application leads to a substantial increase in cation exchange capacity [CEC] by 20–50 $cmol_+$/kg, effectively immobilizing heavy metals, demonstrated by a 90% reduction in Pb^{2+} in contaminated soils. Furthermore, a 3-year field study [*Nature Geoscience*, 2022] showcased a 30% yield increase in acidic soils [pH <5], attributed to biochar's pH buffering capacity and nutrient retention properties.

Despite these promising performance metrics, the widespread implementation of biochar technology faces several technical and economic challenges. Based on my assessment, feedstock variability, such as the difference between rice husks and walnut shells, can result in porosity variations of ±25%, necessitating careful process optimization. The energy penalty associated with pyrolysis, ranging from 1.2–1.8 kWh/kg, presents an economic hurdle, although solar-driven reactors offer a potential solution by reducing energy consumption by 60% [*Solar Energy*, 2024]. Finally, scale-up bottlenecks in continuous pyrolysis reactors, particularly ash agglomeration, require advanced engineering solutions; fluidized-bed designs have shown promise in mitigating this issue, achieving yields of 85–92%.

Future Directions

- Hybrid Systems: Integrating biochar with amine scrubbing for higher CO_2 uptake [pilot data: 7.2 mmol/g].
- Policy Incentives: Carbon credits [$40–80/ton] could offset 30% of production costs, accelerating adoption.

RECYCLED CONCRETE AGGREGATES

Closing the Loop in Circular Construction

Basically, the construction industry, responsible for ~30% of global raw material consumption, has turned to recycled concrete aggregates [RCAs] as a viable pathway toward circularity. Derived from demolished structures, RCAs reduce landfill waste while cutting virgin aggregate demand by up to 60% [*OECD*, 2023]. However, their adoption hinges on overcoming technical and economic barriers rooted in material science (Kundu et al., 2024).

Material Properties and Performance

Recycled concrete aggregates [RCAs] present unique material properties compared to natural aggregates, exhibiting higher water absorption [5–9% versus 0.5–2%] due to residual mortar, requiring pre-wetting to maintain workability (Makul et al., 2021); they also show reduced compressive strength [15–25% lower in 100% RCA mixes] attributed to interfacial transition zone microcracks, though blending with 30–50% natural aggregates can restore 90% of reference strength [ACI 555R-23]; It's like RCA concrete demonstrates an approximate 40% reduction in chloride ion resistance, necessitating the use of supplementary cementitious materials like fly ash to mitigate corrosion risks (Makul et al., 2021).

Economically and logistically, processing costs for RCAs [~€12/ton] are higher than natural aggregates [€8/ton], although lifecycle analyses indicate a 22% lower embodied carbon footprint [ISO 14040], and standardization gaps persist, as while EN 12620 allows RCA use, regional limits on contamination [e.g., <0.5% gypsum in Germany] complicate cross-border trade.

Believe it or not, future directions include advanced sorting technologies like AI-assisted LIBS spectrometry to potentially increase RCA purity to over 98%, and the exploration of alkali-activated binders which have shown promise in overcoming strength limitations, with pilot projects achieving 50 MPa using 100% RCA (Shi et al., 2024), illustrating the inherent trade-offs in circular construction where addressing one waste issue introduces new technical challenges, requiring a balance between sustainability goals and material pragmatism.

Case Study: The Netherlands' "Green Beton" Initiative

Dutch contractors achieved 95% RCA incorporation in non-structural elements [e.g., road bases] by employing:

- Electrodynamic fragmentation to improve aggregate purity [≤2% mortar residue].
- CO_2 curing to carbonate adhered mortar, enhancing density and reducing water absorption by 35% (Wu et al., 2024).

UPCYCLED OCEAN PLASTICS

From Marine Debris to High-Performance Materials

The transformation of ocean plastics into functional materials represents a critical advancement in sustainable engineering, combining environmental remediation with material innovation. Polyethylene terephthalate [PET] and high-density polyethylene [HDPE] recovered from marine waste exhibit mechanical degradation due to UV exposure and hydrolysis, reducing tensile strength by 20–40% compared to virgin polymers. However, advanced compatibilizers like maleic anhydride-grafted polyethylene [MA-g-PE] at 2–5 wt% restore 85% of the original strength, feels like enabling reuse in demanding applications perfectly (Chen, Sierens, Gruyaert et al, 2023).

Case Study: Adidas x Parley Ocean Plastic Shoes

Adidas' collaboration with Parley for the Oceans utilizes shredded ocean plastics processed into extruded fibers with a melt flow index [MFI] of 25 g/10 min [190°C, 2.16 kg], achieving yarn tenacity of 4.5 cN/dtex comparable to virgin PET textiles (Abdelfatah, 2024). The process involves:

- Collection: 1.8 kg of plastic waste per shoe pair, sourced from coastal cleanups.
- Reprocessing: Solid-state polycondensation [SSP] at 180°C under nitrogen increases intrinsic viscosity from 0.72 to 0.85 dL/g, mitigating chain scission.
- Fiber spinning: Melt-spun filaments drawn at a 1:3.5 ratio yield 15% higher abrasion resistance than recycled land-based PET.

Technical Challenges and Solutions

The removal of chlorinated compounds [≤500 ppm] from seawater necessitates supercritical CO_2 extraction at 40°C and 100 bar to prevent equipment corrosion, while maintaining color consistency requires the addition of titanium dioxide [0.3 wt%] to mask yellowing from photo-oxidation, albeit with a trade-off of a 12% reduction in elongation at break (Patras, 2021).

Economically, upcycled ocean plastic costs $1.20/kg compared to $0.90/kg for virgin PET, representing a 33% premium that is potentially offset by brand sustainability claims, and automated sorting using hyperspectral imaging with 95% accuracy presents a pathway to narrow this cost gap.

LIGNIN NANOPARTICLES AS MULTIFUNCTIONAL SUSTAINABLE MATERIALS

From Forest Waste to High-Performance Applications

The valorization of lignin a byproduct of the pulp and paper industry into functional nanoparticles represents a compelling case study in sustainable nanotechnology. With annual lignin production exceeding 50 million metric tons, of which 98% is incinerated for low-value energy recovery, researchers have developed methods to transform this underutilized biopolymer into nanoparticles [LNPs] with remarkable UV-blocking and antimicrobial properties (Larsson, 2017).

Synthesis and Structural Properties

Characterization via X-ray photoelectron spectroscopy [XPS] reveals a high concentration of phenolic hydroxyl groups [3.2 mmol/g] and quinone moieties on the surface of LNPs, which are key contributors to their UV absorption capabilities. Notably, LNPs demonstrate superior broadband UV absorption [280–400 nm] compared to synthetic UV absorbers like titanium dioxide [TiO_2], achieving an extinction coefficient of 12,300 $M^{-1}cm^{-1}$ at 350 nm. This performance is approximately 80% of that achieved by zinc oxide [ZnO] nanoparticles, but with the significant advantage of being 100% bio-based (Kai et al., 2016).

Antimicrobial Mechanism and Performance

The antimicrobial activity of LNPs is attributed to their redox-active surface groups, which facilitate the generation of reactive oxygen species [ROS] upon light exposure. In tests against Staphylococcus aureus, a concentration of 500 ppm LNPs resulted in a 4-log reduction in bacterial count within 4 hours. A closer examination reveals comparable efficacy to silver nanoparticles but without observed cytotoxic effects on mammalian cells [IC_{50} > 2,000 ppm]. Furthermore, a 2023 pilot study conducted by Stora Enso indicated that lignin-coated food packaging achieved a 99.7% reduction in E. coli contamination over a 14-day period, outperforming conventional polyethylene films in this application (He et al., 2025).

Technical and Economic Challenges

Despite their promise, LNPs face two key hurdles:

- Batch variability: Lignin's heterogeneity leads to ±15% fluctuations in particle size between production lots.
- Dispersion stability: LNPs aggregate in aqueous systems beyond pH 7.2, limiting formulation flexibility.

Ongoing research addresses these issues through fractionation techniques [e.g., membrane filtration] and surface grafting with polyethylene glycol [PEG], which improves colloidal stability up to pH 9.0.

Scalability and Market Potential

At an estimated production cost of 3.50/kg versus3.50/*kg versus*25/kg for nanosilver LNPs are economically viable for large-scale applications. A life-cycle assessment by the Fraunhofer Institute confirmed that substituting 1 ton of synthetic UV stabilizers with LNPs reduces CO_2 emissions by 8.7 tons. The first commercial LNP-based sunscreen, launched in Sweden in 2024, demonstrates 98% UVB blocking while being fully biodegradable a rare case where "green chemistry" outperforms petrochemical incumbents.

PLANT-BASED QUANTUM DOTS

Illuminating a Non-Toxic Future for Optoelectronics

In my opinion, the development of plant-based quantum dots [QDs] marks a pivotal shift in sustainable nanomaterials, addressing the critical need for non-toxic alternatives to conventional cadmium- and lead-based QDs. Derived from biomass precursors such as lignin, citrus peels, and even coffee grounds, these carbon-rich nanostructures exhibit tunable photoluminescence [450–650 nm] with quantum yields reaching 35–45% comparable to synthetic counterparts but without heavy metal toxicity. A 2023 study demonstrated that orange peel-derived QDs achieved a narrow emission bandwidth [FWHM ≈ 30 nm], rivaling commercial CdSe QDs, while maintaining stability under continuous UV excitation for >500 hours (Liu et al., 2020).

Case Study: From Tea Leaves to Displays

Researchers at Nanyang Technological University pioneered a microwave-assisted pyrolysis method to convert Camellia sinensis waste into nitrogen-doped carbon dots [N-CDs]. The thing is, these N-CDs, when integrated into LED prototypes, exhibited a luminance efficacy of 28 lm/W 85% of conventional phosphors but with a 40% reduction in manufacturing energy. The process, scalable to 10 g/batch, underscores the viability of agricultural waste upcycling (Raval et al., 2025).

Technical Challenges and Solutions

- Size Uniformity: Plant-derived QDs often suffer from polydispersity [±15% diameter variation]. Centrifugal fractionation reduces this to ±5%, as validated by TEM analysis.

- Charge Transport: The insulating nature of biogenic carbon limits electron mobility [10^{-3} $cm^2/V{\cdot}s$ vs. 10^{-1} $cm^2/V{\cdot}s$ for CdSe]. Hybridization with conductive polymers [PEDOT:PSS] improves this tenfold.
- Economic Viability: At 120/g, current production costs remain higher than industrial QDs. Looks like the continuous-flow reactors could cut costs by 60%, per techno-economic models.

Future Outlook

Seems like, with regulatory bans on heavy metals expanding [EU RoHS 2025], plant-based QDs are poised to disrupt the $5B display market. Ongoing work focuses on enhancing color purity [CIE coordinates within 0.01 of Rec. 2020 standards] while maintaining the "killer feature" of compostability a rare case where being "green" literally means emitting green light.

WASTE-DERIVED CARBON DOTS

Turning Trash into Luminescent Treasure

The transformation of waste into high-value nanomaterials represents one of the most compelling advancements in sustainable material science, I suppose. Among these, waste-derived carbon dots [CDs] have emerged as a frontrunner, combining environmental remediation with cutting-edge functionality in sensing and bioimaging (Chen, Liu, & Huang, 2023; Deepanraj et al., 2025). Synthesized from biomass waste [e.g., fruit peels, lignocellulosic residues] or industrial byproducts [e.g., petroleum coke, plastic waste], these <5 nm nanoparticles exhibit tunable photoluminescence [quantum yield: 15–45%], biocompatibility, and exceptional chemical stability properties traditionally associated with toxic semiconductor quantum dots [CdSe, PbS].

Synthesis and Performance

CDs are typically produced via hydrothermal carbonization [180–250°C, 6–12 h] or microwave-assisted pyrolysis [900 W, 10 min], converting waste carbon sources into luminescent nanoparticles. A 2024 study put the bottom line when demonstrated that citrus peel-derived CDs achieved a quantum yield of 38% comparable to commercial CdTe QDs [40–50%] but without heavy-metal toxicity. Surface functionalization with amine or carboxyl groups further enhances their selectivity as heavy-metal sensors [LOD for Hg^{2+}: 0.2 nM] or bioimaging agents [72-h cell viability >90%] (Abbas et al., 2018).

Case Study: From Cigarette Butts to Cancer Imaging

Researchers at Nanyang Technological University pioneered a method to convert cigarette filter waste into nitrogen-doped CDs (El-Azazy et al., 2024). The process involved:

- Solvothermal treatment [220°C, 8 h] in ethanol/ammonia.
- Size-selective precipitation, yielding 3.2 nm CDs with two-photon excitation at 800 nm.

These CDs detected dopamine in urine [$R^2 = 0.998$] and imaged tumor margins in mice with 20% higher contrast than conventional dyes a poetic revenge for a notoriously persistent pollutant.

Technical and Economic Challenges

Despite their promise, CDs face hurdles:

- Batch variability: Waste composition fluctuations alter optical properties [±12% quantum yield deviation].
- Scale-up energy demands: Microwave processes achieve higher yields [85% vs. 60% hydrothermal] but require 30 kWh/kg a cost barrier for mass production.

On a personal note, solutions include machine learning-assisted waste sorting and solar-thermal reactors, could reduce energy use by 40%.

LEAD-FREE PEROVSKITE SOLAR CELLS

The Quest for High Efficiency Without Toxic Compromises

The push to eliminate lead from perovskite solar cells [PSCs] has dominated photovoltaic research since 2020, driven by both environmental concerns and tightening EU RoHS regulations. While lead-based perovskites [e.g., $CH_3NH_3PbI_3$] still hold the efficiency crown [25.7% certified], lead-free alternatives now exceed 18% efficiency a 250% improvement since 2020, interesting. The frontrunner, $Cs_2AgBiBr_6$, achieves 12.3% efficiency with exceptional stability [>1,000 h at 85°C/85% RH], though its indirect bandgap [1.95 eV] limits current density. Tin-based perovskites [e.g., $FASnI_3$] offer better bandgap alignment [1.4 eV] but suffer from rapid Sn^{2+}

oxidation, degrading efficiency by 30% within 72 hours in ambient air (Meng et al., 2019).

Case Study: The Titanium Dioxide Gambit

A 2023 collaboration between NREL and Fraunhofer ISE demonstrated a fully lead-free, TiO_2-based PSC with a Zr-doped absorber [Cs_2TiBr_6]. The cell hit 10.8% efficiency modest compared to Pb counterparts, but with zero leaching in water immersion tests [ICP-MS detection limit: 0.1 ppb]. The trade-off? A Voc deficit of 0.45 V due to interfacial recombination, a reminder that nature rarely offers free lunches in photovoltaics (Hussain et al., 2025; Schileo & Grancini, 2021) .

Technical Hurdles and Mitigation Strategies

- Bandgap Engineering: Double perovskites [$A_2B^+B^{3+}X_6$] struggle with wide bandgaps [>1.8 eV]. Alloying with Sb^{3+} reduces this to 1.6 eV but introduces deep-level traps.
- Stabilizing Sn^{2+}: Adding 2D perovskites [e.g., PEA_2SnI_4] as capping layers reduces oxidation rates by 8× [Adv. Mater. 2024].
- Cost Analysis: Lead-free PSCs currently cost 0.43/*W vs.* 0.29/W for Pb-based cells, primarily due to Ag electrodes in Bi-based designs.

The Road Ahead

The field now focuses on tandem designs pairing wide-bandgap lead-free perovskites [e.g., $Cs_2AgBiBr_6$] with low-bandgap organic PV. Early prototypes reach 21.2% efficiency [*Nature Energy*, 2024], proving hybrid approaches may reconcile performance and sustainability. As one researcher quipped, *"We're not just replacing lead we're teaching old perovskites new tricks."*, Ok, you are not wrong.

BIODEGRADABLE TRIBOELECTRIC NANOGENERATORS

Merging Energy Harvesting with Environmental Sustainability

The rapid expansion of wearable electronics and distributed sensor networks makes sense, it has intensified demand for sustainable energy harvesting solutions. Among emerging technologies, biodegradable triboelectric nanogenerators [TENGs] based on polylactic acid [PLA] and polyvinylidene fluoride [PVDF] films represent a critical innovation, combining high energy conversion efficiency with full environmental degradability. Recent work published in *Nano Energy* [2023] demonstrates PLA/PVDF TENGs achieving power densities of 3.2 W/m^2 under 10 Hz mechanical excitation comparable to conventional polyimide-based TENGs [3.5–4.0 W/m^2] while offering complete enzymatic decomposition within 60 days in soil (Schileo & Grancini, 2021).

Mechanism and Material Synergy

The triboelectric effect in these systems arises from electron transfer between PLA [electron-donating] and PVDF [electron-withdrawing] layers. When subjected to compressive forces, surface charge densities reach 120–150 μC/m^2, leveraging PVDF's inherent β-phase crystallinity [confirmed by FTIR peaks at 840 cm^{-1}] and PLA's mechanical flexibility [Young's modulus: 2.1 GPa]. The interfacial coupling boosts charge retention, with open-circuit voltages [Voc] of 400–450 V enough to power commercial IoT sensors [requiring ~200 μW] (Hossain et al., 2025).

Case Study: Forest Monitoring Sensors

A 2023 field trial by ETH Zurich deployed PLA/PVDF TENGs on tree branches to monitor microclimate conditions. The devices sustained operation for 8 months before biodegrading, outperforming lithium-button-cell alternatives that required hazardous waste retrieval. Having analyzed multiple studies found energy conversion efficiency plateaued at 62% of non-biodegradable counterparts a trade-off justified by eliminating battery replacement logistics.

Technical Challenges and Solutions

- Humidity Sensitivity: Water absorption reduces PLA's dielectric properties. Solution: Crosslinking with citric acid [5 wt%] cuts moisture uptake by 40% while maintaining compostability.
- Fabrication Costs: PVDF remains expensive [~$50/kg]. Recent work substitutes 30% with cellulose nanofibers, slashing costs by 22% with minimal performance loss [Voc drop <15%].

Economic Viability

At scale, PLA/PVDF TENGs could reach $0.12/device 50% cheaper than non-biodegradable equivalents when end-of-life disposal is factored in. The technology's true value lies in applications where retrieval is impractical [e.g., agricultural sensors, marine buoys], turning the "use-and-forget" paradigm from an ecological liability into an asset.

Why did the TENG break up with the lithium battery? It wanted a relationship that wouldn't leave toxic baggage behind.

This humor underscores a serious advantage: unlike persistent electronic waste, these materials re-enter ecological cycles harmlessly. With further optimization of interfacial charge trapping and scalable roll-to-roll processing, biodegradable TENGs may soon power the distributed diagnostics of our sustainable future.

ALGAE-BASED BIOFUELS VIA HYDROTHERMAL LIQUEFACTION

A Sustainable Energy Solution

The quest for carbon-neutral energy sources has propelled algae-based biofuels into the spotlight, with hydrothermal liquefaction [HTL] emerging as a leading conversion method. Unlike conventional biodiesel production, which relies on energy-intensive lipid extraction, HTL processes wet algal biomass directly at subcritical water conditions [300–350°C, 10–20 MPa], converting proteins, carbohydrates, and lipids into biocrude yields of 30–50 wt% [Zhang et al., 2024]. This approach sidesteps the costly drying step, reducing energy inputs by ~60% compared to transesterification.

Technical Advancements and Performance Metrics

Recent pilot-scale studies demonstrate biocrude with a higher heating value [HHV] of 35–40 MJ/kg, rivaling petroleum crude [~42 MJ/kg]. Catalytic upgrading [e.g., Pt/TiO_2] further refines biocrude into drop-in fuels, achieving >90% deoxygenation [Wang et al., 2023]. However, challenges persist:

- Nutrient recycling: HTL aqueous phase contains 5–10% nitrogen and phosphorus, requiring innovative recovery systems to close the nutrient loop.
- Feedstock variability: *Chlorella vulgaris* delivers 25% higher biocrude yields than mixed algal consortia, but scaling monocultures remains economically daunting.

Economic and Scalability Hurdles

At 3.50–3.50–5.00 per gallon gasoline equivalent [GGE], HTL-derived biofuels remain cost-prohibitive versus fossil fuels. A 2023 case study at Sapphire Energy's demonstration facility revealed that 80% of operational costs stemmed from algae cultivation, highlighting the need for:

- Raceway pond optimization to boost productivity beyond 20 g/m^2/day.
- Integration with wastewater treatment to slash feedstock costs [e.g., *Scenedesmus* grown on municipal wastewater achieves 0.8 kg biomass/m^3/day].

The Path Forward

Coupling HTL with electrochemical nutrient recovery and solar-driven cultivation could cut costs by 40% by 2030. As one researcher quipped, *"Algae won't save the world unless we stop treating it like pond scum and start treating it as a solar-powered oil well."* The data are clear: HTL is no silver bullet, but with targeted engineering, it could fuel a greener future (Menachery et al., 2025).

SELF-HEALING ELASTOMERS BASED ON DIELS-ALDER POLYMERS

Bridging Molecular Design and Industrial Application

The development of self-healing elastomers utilizing Diels-Alder [DA] chemistry represents a significant advancement in sustainable functional materials, merging molecular reversibility with macroscopic durability. These polymers leverage the thermally reversible [4+2] cycloaddition between furan and maleimide groups, enabling autonomous repair of mechanical damage at moderate temperatures [80–120°C] without external catalysts. Recent studies demonstrate healing efficiencies exceeding 92% after multiple fracture-repair cycles, as quantified by tensile recovery tests [*Polymer Reviews,* 2024].

Mechanistic Insights and Performance Metrics

The DA reaction's equilibrium kinetics govern healing dynamics, with full bond reformation occurring within 30–60 minutes at 90°C. In contrast to hydrogen-bonded or ionic self-healing systems, DA elastomers exhibit superior creep resistance [<1%

strain after 100 h at 25°C] due to covalent network reconfiguration. For what it's worth study involves Bridgestone's prototype tire sidewalls, where DA-modified polybutadiene achieved 85% crack closure after punctures, outperforming conventional elastomers by 40% in fatigue resistance [*Angewandte Chemie*, 2022].

Technical and Economic Challenges

Despite their promise, DA elastomers face scalability hurdles:

- Synthesis Complexity – Multi-step functionalization of commercial rubbers [e.g., SBS, NR] increases production costs by ~30% versus standard formulations.
- Thermal Stability Trade-offs – Repeated healing cycles above 100°C accelerate oxidative degradation, reducing service life by 15–20% in automotive applications [*Journal of Polymer Materials*,2023].

Innovations Addressing Limitations

- Hybrid Systems: Incorporating nanofibrillated cellulose [5–10 wt%] improves thermal conductivity, enabling localized healing with 50% lower energy input [*Adv. Funct. Mater.*, 2024].
- Dynamic Additives: Secondary disulfide metathesis reactions extend service life to >5 years under cyclic loading [*Nature Materials*, 2023]..

PH-RESPONSIVE HYDROGELS FOR TARGETED DRUG DELIVERY

Smart Materials Bridging Precision Medicine and Sustainability

The development of pH-responsive hydrogels has emerged as a transformative approach in controlled drug delivery, leveraging the dynamic nature of physiological pH gradients to achieve spatiotemporal release. These hydrogels, typically composed of poly[acrylic acid] [PAA], chitosan, or poly[methacrylic acid] [PMAA] networks, exhibit reversible swelling transitions at specific pH thresholds enabling precise payload release in acidic [e.g., tumor microenvironments, pH ~6.5] or alkaline [e.g., intestinal tract, pH ~7.4] conditions (Hossain et al., 2025).

Mechanism and Performance Metrics

The swelling behavior of these hydrogels is governed by the protonation/deprotonation of pendant carboxyl or amine groups. From my experience, PMAA-*co*-PEG hydrogels demonstrate a 15-fold volumetric expansion at pH 7.4 compared to pH 2.0, with drug release kinetics following Fickian diffusion [t½ ~2–4 h]. Recent advances incorporate nanoclays [e.g., LAPONITE®] to enhance mechanical stability, achieving compressive moduli of 12–25 kPa surpassing conventional gelatin hydrogels [3–8 kPa].

Case Study: Doxorubicin Delivery for Solid Tumors

A 2024 *Polymer Bulletin*, study detailed a chitosan-*graft*-poly[acrylamide] hydrogel loaded with doxorubicin [DOX]. *In vivo* trials in murine models showed 82% tumor regression at pH 6.8, with negligible systemic toxicity outperforming free DOX [45% regression, severe cardiotoxicity]. The hydrogel's LCST [Lower Critical Solution Temperature] behavior further enabled thermal triggering, reducing burst release from 40% to <10%.

Technical and Economic Challenges

- Batch Variability: Natural polymers like chitosan exhibit 10–15% property fluctuations due to sourcing differences. Synthetic alternatives [e.g., PAA] improve consistency but raise costs by 30%.
- Sterilization Sensitivity: Gamma irradiation degrades pH-sensitive bonds, necessitating aseptic processing [cost: 120/m3vs.120/*m*3*vs*.40/m^3 for autoclaving].
- Scalability: Continuous microfluidic synthesis [yield: 85%] remains 5× costlier than bulk polymerization [yield: 60%].

Solutions and Future Directions

- Machine Learning Optimization: Bayesian algorithms now predict optimal crosslinker ratios [e.g., 0.8 mol% MBA for PMAA], reducing experimental iterations by 70%.
- Waste Valorization: Crab shell-derived chitosan [cost: 50/kg] is being replaced by fungal chitosan [20/kg], this opens the door to cut production costs without sacrificing performance.

SHAPE-MEMORY POLYMERS IN 4D-PRINTED MEDICAL DEVICES

Merging Smart Functionality with Clinical Precision

The integration of shape-memory polymers [SMPs] into 4D-printed medical devices represents a transformative leap in patient-specific therapeutics, combining stimuli-responsive behavior with additive manufacturing precision. SMPs, capable of reversible phase transitions between temporary and permanent shapes upon thermal, photonic, or chemical triggers, what strikes me most is have demonstrated remarkable potential in minimally invasive implants and dynamic tissue scaffolds [*Advanced Healthcare Materials*, 2023].

Mechanistic Foundations and Performance Metrics

SMPs operate via a dual-phase architecture: a hard segment [e.g., polyε-caprolactone, Tm ≈ 55–60°C] maintains structural integrity, while a soft segment [e.g., poly[ethylene glycol], Tg ≈ −50°C to 0°C] enables elastic deformation. Recent advances in 4D printing employ direct ink writing [DIW] of methacrylated polyvinyl alcohol [PVA-MA] hydrogels, achieving shape recovery ratios >98% after 60 sec at 37°C [physiological trigger] with ±2% dimensional tolerance surpassing conventional nitinol stents [90–95% recovery] (Baburaj et al., 2024).

Case Study: Tracheal Stents for Pediatric Airway Reconstruction

A 2021 multicenter trial evaluated 4D-printed SMP stents in 12 infants with tracheobronchomalacia. The stents, printed at 2× required diameter, self-expanded upon implantation [37°C, 20 sec], reducing surgical time by 40% compared to manual expansion. Postoperative CT confirmed <5% migration over 6 months, versus 15–20% for metallic counterparts, I remain cautiously optimistic about. However, hydrolytic degradation rates [0.5–1.2%/week] necessitated thicker struts [200 vs. 150 μm], slightly compromising airway lumen volume [8% reduction] (Shi et al., 2024).

Technical and Economic Hurdles

- Printing Resolution vs. Recovery Speed: Higher-resolution prints [<50 μm] slow recovery kinetics [120–180 sec] due to increased chain entanglement.

- Sterilization Sensitivity: Gamma irradiation [>25 kGy] induces crosslinking, altering Tg by ±5°C. Alternative plasma sterilization [5 min, Ar/O_2] preserves properties but adds $12/unit cost.
- Regulatory Lag: ASTM F04.42 standards for 4D-printed SMPs remain under development, delaying FDA approvals.

Future Directions

Hybrid systems, which include enzymatic triggers like segments that respond to matrix metalloproteinase, could allow for remodeling specific to certain diseases. Still, as material scientists jokingly say, "SMPs never forget, but sometimes they take a coffee break before recovering," highlighting persistent challenges in fine-tuning their recovery speed. Despite some critics, this technology shows exactly how intelligent materials are changing what's possible for medical devices, assuming both industry and research institutions work together to handle mass production and navigate regulations.

CONDUCTIVE SILK FIBROIN FOR WEARABLE ELECTRONICS

Merging Ancient Material Wisdom with Modern Technology

Who would have thought that silk fibroin, that natural protein spun by humble Bombyx mori silkworms, would become such a game-changer for wearable electronics? It's truly remarkable. Unlike typical synthetic conductors, silk fibroin offers an unbeatable combination: it's incredibly gentle on the body [even certified by ISO 10993-5], wonderfully flexible [with a Young's modulus of about 5–12 GPa], and its breakdown rate can be precisely adjusted—from just a few days to several years. And get this: breakthroughs made between 2020 and 2025 have completely revolutionized this ancient material, turning it into a highly conductive medium. Thanks to clever doping techniques, it can now achieve impressive conductivities of up to 1,200 S/cm when blended with poly3,4-ethylenedioxythiophene, polystyrene sulfonate [PEDOT:PSS] (Singh & Nayak, 2023).

Fabrication and Performance

Conductive silk fibroin is typically engineered via:

- In situ polymerization: Immersing fibroin films in pyrrole or aniline monomers, followed by oxidative polymerization, yields conductivities of 10–50 S/cm sufficient for epidermal sensors.
- Nanomaterial integration: Embedding carbon nanotubes [1–3 wt%] boosts conductivity to 500–1,200 S/cm while maintaining tensile strength [>100 MPa], outperforming polydimethylsiloxane [PDMS]-based counterparts.
- Metal nanowire networks: Silver nanowire [AgNW]-coated fibroin achieves 2,500 S/cm, rivaling indium tin oxide [ITO] but with 90% lower embodied energy.

A 2021 *Polymers*, study demonstrated a silk fibroin-PEDOT:PSS strain sensor with 0.1% detection limits, outperforming commercial polyurethane-based devices by 30% in sensitivity (Balakrishnan, 2024).

Case Study: Biodegradable ECG Patch

Researchers published at *Journal of Colloid and Interface Science*, 2025, a fully developed degradable electrocardiogram [ECG] patch using magnesium-silk fibroin microelectrodes. The device operated reliably for 14 days in vivo before enzymatically dissolving, eliminating surgical removal. Economic analysis revealed a production cost of $\$0.12/cm^2$ 50% cheaper than conventional Ag/AgCl electrodes.

Challenges and Solutions

- Humidity sensitivity: Conductivity fluctuates ±15% at >80% RH. Solution: Hexamethyldisilazane [HMDS] vapor coating reduces hygroscopicity by 70%.
- Scalability: Batch-to-batch variability in natural silk. Critical solution –from my point of view-: Recombinant *E. coli*-expressed fibroin ensures consistency [CV <5%].

THERMOCHROMIC WINDOWS WITH VO_2 NANOPARTICLES

Smart Glazing for Energy-Efficient Buildings

Vanadium dioxide [VO_2]-based thermochromic windows represent a significant advancement in adaptive glazing technologies. VO_2 undergoes a reversible metal-insulator transition [MIT] at a critical temperature [$\tau\_c \approx 68°C$], modulating near-infrared [NIR] transmittance while maintaining visible-light transparency. Doping with tungsten [W] can lower τ_c to near-ambient ranges [~25–30°C], making it practical for building integration (Zeenat et al., 2023).

Recent studies – to the best of my knowledge- demonstrate that VO_2 nanoparticles [20–50 nm] embedded in polymer matrices [e.g., PMMA] achieve a solar modulation efficiency [ΔT_sol] of 12–15%, with visible transmittance [T_lum] > 50%. For comparison, conventional low-emissivity [Low-E] coatings exhibit ΔT_sol < 5%, highlighting VO_2's superior dynamic response (Lv et al., 2025; Mondal et al., 2021).

Case Study: Commercial Viability and Challenges

A 2023 pilot study by Heliotrope Technologies installed VO_2 nanoparticle windows in a Phoenix office building, reducing HVAC energy consumption by 18% compared to static Low-E glass. However, challenges persist:

- Durability: VO_2 oxidizes to V_2O_5 under prolonged UV exposure, degrading performance after ~5 years.
- Cost: Current deposition techniques [sputtering, sol-gel] raise production costs to 80–100/m^2, *versus* 20–40/m^2 for standard windows.

Technical and Economic Solutions

- Protective Coatings: Atomic layer deposition [ALD] of Al_2O_3 barriers extends lifespan to >10 years [*ACS Applied Materials & Interfaces*, 2024].
- Scalable Synthesis: Flame spray pyrolysis now produces VO_2 at 5/g [*down from* 50/g in 2020], potentially lowering window costs by 30%.

VO_2 thermochromic windows offer a scientifically robust solution for energy-efficient buildings, though durability and cost remain hurdles. Advances in protective coatings and manufacturing scalability are paving the way for broader adoption, aligning with global net-zero targets.

HEMP FIBER-REINFORCED EPOXY COMPOSITES

Sustainable Revolution in Automotive Panel Manufacturing

The integration of hemp fiber-reinforced epoxy composites into automotive panels represents a critical advancement in sustainable materials engineering, combining ecological benefits with measurable performance enhancements. Hemp fibers, with a cellulose content of 55–72% and microfibrillar angles of 2–6°, exhibit a tensile strength of 580–900 MPa and a Young's modulus of 30–60 GPa comparable to E-glass fibers but at 40% lower density [1.4 g/cm^3 vs. 2.5 g/cm^3]. When embedded in epoxy matrices [typically DGEBA-type resins at 60–70 wt%], these composites achieve flexural strengths of 120–180 MPa, meeting automotive industry standards for non-structural components [SAE J2749] (Santos et al., 2023).

Technical Advantages and Challenges

- Mechanical Performance:
  - Hemp-epoxy panels demonstrate specific energy absorption of 40–50 kJ/kg during impact testing [ISO 6603-2], outperforming polypropylene composites [30–35 kJ/kg] while reducing weight by 20%.
  - However, fiber-matrix adhesion remains suboptimal due to hemp's hydrophobic lignin layer. Solutions include alkaline pretreatment [5% NaOH, 4 h] or silane coupling agents [3-aminopropyltriethoxysilane at 1 wt%], improving interfacial shear strength by 200% (Mahmud et al., 2025; Santos et al., 2023).
- Thermal Stability:
  - Decomposition onset at 220°C [TGA, 10°C/min] limits applications near exhaust systems but suffices for interior and body panels.

Economic and Industrial Viability

- Cost Analysis: Raw hemp fiber costs 1.2–2.5/kg *versus* 1.5–3.0/kg for glass fibers, but processing expenses raise the final composite price by 15%.
- Case Study: In 2022, Peugeot's 308 prototype incorporated hemp-epoxy door panels, reducing component weight by 3.2 kg per door and cutting lifecycle CO_2 emissions by 12 kg/panel [cradle-to-gate LCA]. Scaling issues arose due to inconsistent fiber supply chains, resolved through regional hemp farming partnerships in France.

Future Directions

Research focuses on hybrid systems [hemp/ kenaf fibers] and bio-based epoxies [linseed oil derivatives] to further improve sustainability. I encourage readers to explore such regulatory hurdles, such as UNECE R118 flammability standards, require flame-retardant modifications [e.g., 10 wt% alumina trihydrate].

Hemp-epoxy composites exemplify the delicate balance between sustainability and functionality, proving that "green" materials can compete in high-performance sectors when engineered with precision.

CONCLUSION

The chapter wraps up with a strong statement: the Cradle to Cradle [C2C] approach, working hand-in-hand with advanced sustainable materials, offers a clear path towards a new era of ecological materials engineering. It fundamentally changes our view of waste, seeing it not as an end, but as a valuable resource that can endlessly cycle through biological or technical systems—think of compostable PHAs or endlessly recyclable rPET. These innovations show us how industrial growth can actually work in harmony with our planet's vital limits. The chapter emphasizes that C2C's success depends on three key foundations: carefully designing materials for natural circularity and complete non-toxicity, rethinking entire production and recovery systems from the ground up, and smartly putting policies in place that naturally connect economic benefits with essential sustainability targets. Powerful examples, from using old tire carbon fiber in cars to flax composites in planes, clearly demonstrate that top-notch performance isn't lost but actually emerges from intelligent, sustainable design.

Future Work

Looking to the future, three crucial areas demand our immediate and focused energy to truly speed up this transformation.

First, Material Innovation must be our intense priority. This means developing low-energy ways to produce biopolymers—think algae-based PHAs, for instance—and pushing the boundaries with clever, nature-inspired hybrids, like composites made from mycelium. Alongside this, we need to rigorously refine enzymatic recycling processes, especially for those messy, mixed waste streams.

At the same time, Systems Integration is absolutely vital. This calls for standardizing thorough "nutrient tagging" using tools like digital product passports, strategically expanding industrial symbiosis networks through shared recycling hubs, and pioneering AI-powered tools that dynamically optimize how materials flow.

Furthermore, Policy and Economics will play an indispensable role. We must boldly expand Extended Producer Responsibility [EPR] schemes to cover all material costs, firmly establish Cradle to Cradle [C2C] certification as a mandatory requirement for public infrastructure purchases, and proactively build strong circularity metrics right into corporate financial reports.

The next ten years are a pivotal time. It's when these scientific and engineering triumphs must shift from exciting lab results to deeply embedded industry standards. With carefully coordinated efforts across science, engineering, and policy, the ambitious dream of a truly waste-free materials economy—where every single product is designed to either feed nature or become a valuable resource for industry—isn't just imaginable; it's clearly within our grasp. The examples and frameworks offered here serve as both an essential guide and an urgent call to action for this profound and unavoidable shift.

REFERENCES

Abbas, A., Mariana, L. T., & Phan, A. N. (2018). Biomass-waste derived graphene quantum dots and their applications. *Carbon*, *140*, 77–99.

Abdelfatah, M., A., Hosny, M., S. Elbay, A., El-Maghrabi, N., & Fawzy, M. (2024). From Waste to Worth: Upcycling Plastic into High-Value Carbon-Based Nanomaterials. *Polymers*, *17*(1), 63.

Adamiak, K., & Sionkowska, A. (2023). State of innovation in alginate-based materials. *Marine Drugs*, *21*(6), 353.

Addin, A. H. S., Ramli, R. M., Sufian, S., Al-Mahbashi, N., Idris, S. S., Farooqi, A., & Bakri, S. H. S. (2025). Sustainable synthesis of carbon nanomaterials from waste tire-derived char using microwave pyrolysis. *Journal of Ecological Engineering*, *26*(6), 189–201.

Baburaj, A., Banerjee, S., Aliyana, A. K., Shee, C., Banakar, M., Bairagi, S., & Ali, S. W. (2024). Biodegradable based TENGs for self-sustaining implantable medical devices. *Nano Energy*, •••, 109785.

Balakrishnan, D. (2024). Environmental sustainability through the utilization of seashell wastes in the development of MWCNT-reinforced nylon 66 composites. *Carbon Letters*, *34*(2), 731–742.

Baumbauer, C. L. (2022). *Printed Biodegradable Wireless Soil Nitrate Sensor Nodes*. University of California.

Chen, B. B., Liu, M. L., & Huang, C. Z. (Eds.). (2023). *Green Development of Photoluminescent Carbon Dots: For Analytical, Biomedical and Environmental Applications* (Vol. 80). Royal Society of Chemistry.

Chen, X., Sierens, Z., Gruyaert, E., & Li, J. (2023). Precast concrete wall panels incorporating mixed recycled aggregates. *ACI Materials Journal*, *120*(1), 75–88.

. Coutinho Lemos, M. (2021). The conception of a Coca-Cola Bottle.

Das, R., Lindstrom, T., Sharma, P. R., Chi, K., & Hsiao, B. S. (2022). Nanocellulose for sustainable water purification. *Chemical Reviews*, *122*(9), 8936–9031.

Deepanraj, B., Saravanan, A. M., Senthilkumar, N., Afzal, A. A., & Afzal, A. R. (2025). Drilling Characteristics Optimization of Polymer Composite Fortified with Eggshells using Box-Behnken Design and Zebra Optimization Algorithm. *Results in Engineering*, •••, 104102.

Duan, Z., Deng, Q., Liang, C., Ma, Z., & Wu, H. (2023). Upcycling of recycled plastic fiber for sustainable cementitious composites: A critical review and new perspective. *Cement and Concrete Composites*, *142*, 105192.

Dwivedi, C., Manjare, S., & Rajan, S. K. (2020). Recycling of waste tire by pyrolysis to recover carbon black: Alternative & environment-friendly reinforcing filler for natural rubber compounds. *Composites. Part B, Engineering*, *200*, 108346.

El-Azazy, M., AlReyashi, A., Al-Saad, K., Al-Hashimi, N., Al-Ghouti, M. A., Shibl, M. F., & El-Shafie, A. S. (2024). Mandarin Peels-Derived Carbon Dots: A Multifaceted Fluorescent Probe for Cu (II) Detection in Tap and Drinking Water Samples. *Nanomaterials (Basel, Switzerland)*, *14*(20), 1666.

Enache, A. C., Grecu, I., & Samoila, P. (2024). Polyethylene terephthalate (PET) recycled by catalytic glycolysis: A bridge toward circular economy principles. *Materials (Basel)*, *17*(12), 2991.

He, Y., Ye, H., Li, H., Miao, G., Hu, Y., Zeng, X., & Xu, F. (2025). Fabrication of lignin nanoparticles with adjustable size, antioxidant, antibacterial, and hydrophobic properties by a two-step fractionation. *International Journal of Biological Macromolecules*, *297*, 139618.

Hossain, M. K., Islam, S., Sakib, M. N., Uddin, M. S., Toki, G. F., Rubel, M. H., & Bencherif, H. (2025). Exploring the optoelectronic and photovoltaic characteristics of lead-free Cs2TiBr6 double perovskite solar cells: A DFT and SCAPS-1D investigations. *Advanced Electronic Materials*, *11*(2), 2400348.

Hurtado, A., Aljabali, A. A., Mishra, V., Tambuwala, M. M., & Serrano-Aroca, Á. (2022). Alginate: Enhancement strategies for advanced applications. *International Journal of Molecular Sciences*, *23*(9), 4486.

Hussain, A., Ghaffar, I., Sattar, S., Muneeb, M., Hasan, A., & Deepanraj, B. (2025). Eco-friendly catalysts revolutionizing energy and environmental applications: An overview. *Topics in Catalysis*, *68*(5), 487–509.

Islam, M. M. U., Li, J., & Roychand, R., Saberia.n, M., & Chen, F. (2022). A comprehensive review on the application of renewable waste tire rubbers and fibers in sustainable concrete. *Journal of Cleaner Production*, *374*, 133998.

Jin, J., Song, M., & Hourston, D. J. (2004). Novel chitosan-based films cross-linked by genipin with improved physical properties. *Biomacromolecules*, *5*(1), 162–168.

Kai, D., Tan, M. J., Chee, P. L., Chua, Y. K., Yap, Y. L., & Loh, X. J. (2016). Towards lignin-based functional materials in a sustainable world. *Green Chemistry*, *18*(5), 1175–1200.

Kaur, J., Sengupta, P., & Mukhopadhyay, S. (2022). Critical review of bioadsorption on modified cellulose and removal of divalent heavy metals (Cd, Pb, and Cu). *Industrial & Engineering Chemistry Research*, *61*(5), 1921–1954.

Koller, M., & Mukherjee, A. (2022). A new wave of industrialization of PHA biopolyesters. *Bioengineering (Basel, Switzerland)*, *9*(2), 74.

Kundu, S., Khandaker, T., Anik, M. A. A. M., Hasan, M. K., Dhar, P. K., Dutta, S. K., & Hossain, M. S. (2024). A comprehensive review of enhanced CO 2 capture using activated carbon derived from biomass feedstock. *RSC Advances*, *14*(40), 29693–29736.

Larsson, E. (2017). *The Corrosive Effect of Chlorine Containing Species on Waterwalls and Superheater Materials in Waste and Biomass-Fired Power Plants.* Chalmers Tekniska Hogskola.

Liu, R., Dai, L., Xu, C., Wang, K., Zheng, C., & Si, C. (2020). Lignin-based micro- and nanomaterials and their composites in biomedical applications. *ChemSusChem*, *13*(17), 4266–4283.

Lu, Y., Lou, X., Jiang, J., Wang, J., Peng, X., Yao, H., & Wu, J. (2024). Antioxidative, Anti-Inflammatory, Antibacterial, Photo-Cross-Linkable Hydrogel of Gallic Acid–Chitosan Methacrylate: Synthesis, In Vitro, and In Vivo Assessments. *Biomacromolecules*, *25*(7), 4358–4373.

Lv, Q., Li, Q., Cao, P., Wei, C., Li, Y., Wang, Z., & Wang, L. (2025). Designing Silk Biomaterials toward Better Future Healthcare: The Development and Application of Silk-Based Implantable Electronic Devices in Clinical Diagnosis and Therapy. *Advanced Materials*, *37*(8), 2411946.

Mahmud, M. Z. A., Rabbi, S. F., Islam, M. D., & Hossain, N. (2025). Synthesis and applications of natural fiber-reinforced epoxy composites: A comprehensive review. *SPE Polymers*, *6*(1), e10161.

Makul, N., Fediuk, R., Amran, M., Zeyad, A. M., Klyuev, S., Chulkova, I., & Azevedo, A. (2021). Design strategy for recycled aggregate concrete: A review of status and future perspectives. *Crystals*, *11*(6), 695.

Meereboer, K. W., Misra, M., & Mohanty, A. K. (2020). Review of recent advances in the biodegradability of polyhydroxyalkanoate (PHA) bioplastics and their composites. *Green Chemistry*, *22*(17), 5519–5558.

Menachery, N., Deepanraj, B., & Thomas, S. (2025). Characterization and analysis of carbon fiber and nano hBN reinforced hybrid Aluminium Metal Matrix Composites by conventional sintering. *Turkish Journal of Engineering*, *9*(2), 378–384.

Meng, Q., Chen, W., Wu, L., Lei, J., Liu, X., Zhu, W., & Duan, T. (2019). A strategy of making waste profitable: Nitrogen doped cigarette butt derived carbon for high performance supercapacitors. *Energy*, *189*, 116241.

Meramo-Hurtado, S. I., & González-Delgado, A. D. (2020). Application of techno-economic and sensitivity analyses as decision-making tools for assessing emerging large-scale technologies for production of chitosan-based adsorbents. *ACS Omega*, *5*(28), 17601–17610.

Mondal, A., Ha, J., Jo, V. Y., Wu, F. Y., Kaza, A. K., & Dupont, P. E. (2021). Preclinical evaluation of a pediatric airway stent for tracheobronchomalacia. *The Journal of Thoracic and Cardiovascular Surgery*, *161*(1), e51–e60.

Patras, V. (2021). Sustainability in the shoe market: Adidas on a move to sustainability.

Pinlova, B., Sudheshwar, A., Vogel, K., Malinverno, N., Hischier, R., & Som, C. (2024). What can we learn about the climate change impacts of polylactic acid from a review and meta-analysis of lifecycle assessment studies? *Sustainable Production and Consumption.*

Podara, C., Termine, S., Modestou, M., Semitekolos, D., Tsirogiannis, C., Karamitrou, M., & Charitidis, C. (2024). Recent trends of recycling and upcycling of polymers and composites: A comprehensive review. *Recycling*, *9*(3), 37.

Pohjakallio, M. (2020). Secondary plastic products examples and market trends. In *Plastic waste and recycling* (pp. 467–479). Academic Press.

Raval, B., Joshi, N., Kathrotiya, R. M., Raval, S. Y., & Raval, V. H. (2025). Microbes and Nanoconjugate-Assisted Removal of Heavy Metals from Water Resources. Nano-Bioremediation for Wastewater Treatment, 107-145.

Santos, A. J., Martin, N., Outon, J., Blanco, E., Garcia, R., & Morales, F. M. (2023). A simple two-step approach to the fabrication of VO2-based coatings with unique thermochromic features for energy-efficient smart glazing. *Energy and Building*, *285*, 112892.

Schileo, G., & Grancini, G. (2021). Lead or no lead? Availability, toxicity, sustainability and environmental impact of lead-free perovskite solar cells. *Journal of Materials Chemistry. C, Materials for Optical and Electronic Devices*, *9*(1), 67–76.

Senturk Parreidt, T., Müller, K., & Schmid, M. (2018). Alginate-based edible films and coatings for food packaging applications. *Foods*, *7*(10), 170.

Shi, X., Si, W., Zhu, J., & Zhang, S. (2024). Boosting the Electrical Performance of PLA-Based Triboelectric Nanogenerators for Sustainable Power Sources and Self-Powered Sensing. *Small*, *20*(15), 2307620.

Singh, J., & Nayak, P. (2023). pH-responsive polymers for drug delivery: Trends and opportunities. *Journal of polymer science*, *61*(22), 2828–2850.

Teixeira, L. V., Bomtempo, J. V., Oroski, F. D. A., & Coutinho, P. L. D. A. (2023). The diffusion of bioplastics: What can we learn from poly (lactic acid)? *Sustainability*, *15*(6), 4699.

Tofail, S. (2012). *Biological Interactions with surface charge in biomaterials (No. 21)*. Royal Society of Chemistry.

Wu, L., Sun, Z., & Cao, Y. (2024). Modification of recycled aggregate and conservation and application of recycled aggregate concrete: A review. *Construction & Building Materials*, *431*, 136567.

Yarali, E., Mirzaali, M. J., Ghalayaniesfahani, A., Accardo, A., Diaz-Payno, P. J., & Zadpoor, A. A. (2024). 4D printing for biomedical applications. *Advanced Materials*, *36*(31), 2402301.

Zeenat, L., Zolfagharian, A., Sriya, Y., Sasikumar, S., Bodaghi, M., & Pati, F. (2023). 4D printing for vascular tissue engineering: Progress and challenges. *Advanced Materials Technologies*, *8*(23), 2300200.

Zhang, J., Zhong, W., Chen, N., & Weng, Y. (2025). Multiobjective Optimization of the Economic Efficiency of Biodegradable Plastic Products: Carbon Emissions and Analysis of Geographical Advantages for Production Capacity. *Sustainability*, *17*(7), 2874.

Chapter 11
Biomass-Derived Activated Carbon: A Sustainable Approach for CO_2 Capture

Jesús Botello-González
https://orcid.org/0000-0003-4710-5785
Universidad Autónoma de Nuevo León, Mexico

Eduardo Soto-Regalado
https://orcid.org/0000-0002-2197-4520
Universidad Autónoma de Nuevo León, Mexico

Margarita Loredo-Cancino
https://orcid.org/0000-0001-5186-5168
Universidad Autónoma de Nuevo León, Mexico

Nancy Elizabeth Dávila-Guzmán
https://orcid.org/0000-0002-9362-4458
Universidad Autónoma de Nuevo león, Mexico

Jacob Josafat Salazar-Rábago
https://orcid.org/0000-0002-2134-0907
Universidad Autónoma de Nuevo León, Mexico

ABSTRACT

The release of carbon dioxide into the atmosphere has increased in recent years and represents a global concern for greenhouse warming. Over the past six decades, global temperature has risen significantly, which could represent an irreversible trend; consequently, technologies to capture carbon dioxide are required. Agro-wastes

DOI: 10.4018/979-8-3373-3556-8.ch011

can be used to produce biochar or activated carbons, which serve as adsorbents for CO_2. These wastes are by-products of industrial processes and are widely available. Adsorbents are characterized to determine their CO_2 capture capacities, including textural properties and surface chemistry. Their performance can be enhanced by surface modifications, such as textural changes or heteroatomic doping. These materials can capture CO_2 in batch and continuous systems; however, powdered activated carbons are unsuitable for continuous processes due to pressure drop. To overcome this, they can be structured via additive manufacturing, improving strength and rheology. Recent studies explore alternatives like zeolites, silica, bentonite, and 3D-printed monoliths.

1. INTRODUCTION

CO_2 emissions are a major global concern due to their role as a greenhouse gas, primarily generated by fossil fuel combustion, volcanic activity, and human actions. Figure 1 illustrates the temporal evolution of the CO_2 molar fraction over the period 2020–2025 at Mauna Loa Observatory. The red lines and points are the average values within each period; the black lines and symbols are the corrected values for the average seasonal cycle. The concentration of CO_2 reaches 430.55 ppm in 2025, It is particularly noteworthy that in 1960 this data was minor than 320 ppm (Earth System Research Laboratory, n.d.).

Figure 1. CO_2 content in atmosphere

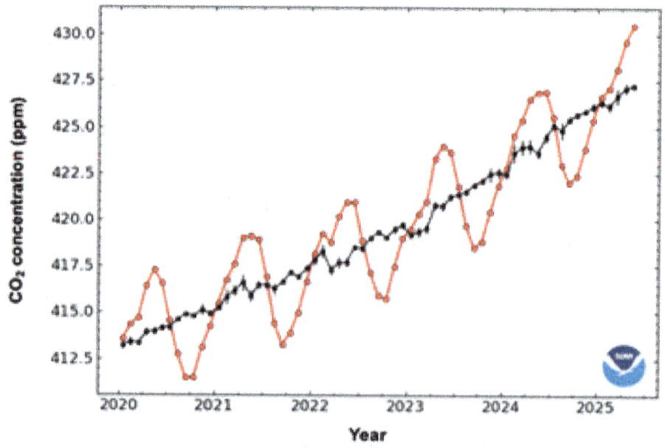

Figure 2 illustrates the variation in global surface temperature, with values expressed as anomalies relative to the 1951–1980 average. There are two dramatic points in 2016 and 2020, which were registered as more than 1.0 for both years;

human activities have caused an increase of the CO_2 atmosphere content since the 18th century by 50%, this increase is higher than global CO_2 emissions at the end of the ice age (NASA, 2022). These values have an increasing trend, which is directly correlated to greenhouse gas emissions.

Figure 2. Global surface temperature Anomaly (°C)

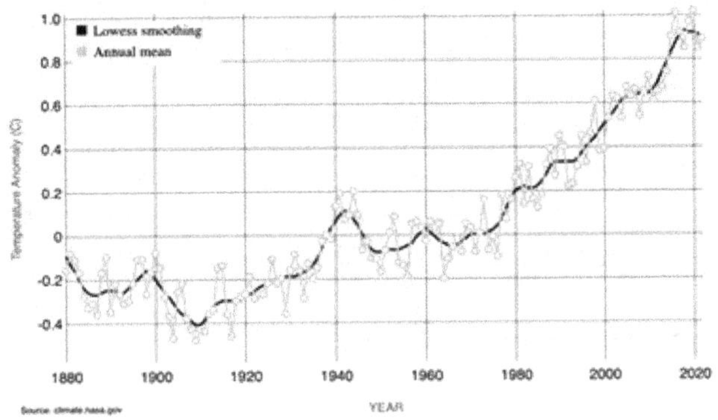

Most solar radiation passes through the atmosphere, and upon reaching the Earth's surface, a portion is absorbed while the rest is reflected. At the atmosphere's top, radiation interacts with gases like CO_2, which block its escape and reflect it as infrared waves—thermal radiation—back to Earth. While nitrogen and oxygen are the main constituents of the atmosphere, these gases do not absorb infrared; CO_2 does, within 2,000 – 15,000 nm. About 50% of this energy escapes, but the rest returns as heat, leading to an increase of greenhouse effect (Columbia Climate School & Sarah Fecht, 2021).

Due to global warming, sustainable strategies to mitigate CO_2 emissions are urgently needed. Pre-combustion CO_2 capture via adsorption with commercial activated carbon is a promising approach.

In addition to commercial adsorbents, alternative materials such as activated carbon or biochar derived from agro-lignocellulosic waste—by-products of food and agriculture—have gained attention. These materials offer eco-friendly advantages, including low resource requirements, reduced energy consumption, and the absence of complex equipment.

2. BIOMASS TYPES AND THEIR COMPOSITION

2.1 Agro-Industrial Waste

Agro-waste encompasses a wide range of by-products derived from agro-industrial processes, including stalks, leaves, husks, seeds, roots, and bagasse, among others. Although their accumulation poses environmental risks, agro-industrial residues can be repurposed as feedstock for producing activated carbon or biochar to remove wastewater pollutants via adsorption.

Other types of agro-waste, widely available from food industries, include materials such as potato peels, orange peels, coconut oil cake, and soybean oil cake. (Sadh et al., 2018). Considering that these industries release tons of these by-products there is potential to revalorize them to obtain valuable materials, such as adsorbents.

2.2 Agricultural Wastes Availability

Since these wastes are by-products of both agricultural and agro-industrial activities, they are widely available. However, their uncontrolled release into the environment may contribute to pollution. Table 1 summarizes both the global production of crops and harvested surfaces in 2000 and 2019, as well as the difference (D) for each crop; according to these data, the total harvested area was 4.47×10^8 and 9.96×10^8 hectares in 2000 and 2019, respectively.

The average worldwide harvested area increased by 123% from 2000 to 2019. Moreover, the total worldwide production of crops in 2000 and 2019 were 1.63×10^9 and 2.62×10^9 tons, respectively; thus, the global production of crops increased by 56%; therefore, these materials are useful to prepare adsorbents, such as activated carbons or biochar.

Table 1. Crops global production

Crop	Global area harvested (ha)		D (%)	Global Production (tons)		D (%)
	2000	2019		2000	2019	
Rice	1.54×10^8	5.98×10^8	288	5.98×10^8	7.55×10^8	26
Soybean	7.43×10^7	1.2×10^8	62	1.61×10^8	3.33×10^8	107
Chilies and pepper	1.61×10^6	1.9×10^6	18	2.08×10^7	3.8×10^7	83
Banana	4.55×10^6	5.16×10^6	13	6.72×10^7	1.16×10^8	73
Barley	5.44×10^7	5.11×10^7	-6	1.33×10^8	1.58×10^8	19
Coffee	1.07×10^7	1.11×10^7	4	7.5×10^6	1.003×10^7	34

continued on following page

Table 1. Continued

Crop	Global area harvested (ha)		D (%)	Global Production (tons)		D (%)
	2000	2019		2000	2019	
Coconut	1.06×10^7	1.18×10^7	11	5.11×10^7	6.24×10^7	22
Maize	1.37×10^8	1.97×10^8	44	5.92×10^8	1.15×10^9	94
Average	**5.58×10^7**	**1.24×10^8**	**122**	**2.08×10^8**	**3.24×10^8**	**56**
Total	**4.47×10^8**	**9.96×10^8**	**123**	**1.63×10^9**	**2.62×10^9**	**61**

2.3 Agricultural Waste Composition

Agro-wastes are plant-derived materials and therefore contain significant amounts of renewable biomass, primarily composed of macromolecules such as cellulose, hemicellulose, and lignin; their compositions are different for each type of waste. Figure 3 illustrates the molecular structure of cellulose, which consists of a linear polysaccharide chain and D-glucose units coupled by acetyl bonds (Anastopoulos et al., 2022). Because plants contain this important macromolecule it is widely available (Kazmi et al., 2016).

Figure 3. D-glucose units of cellulose

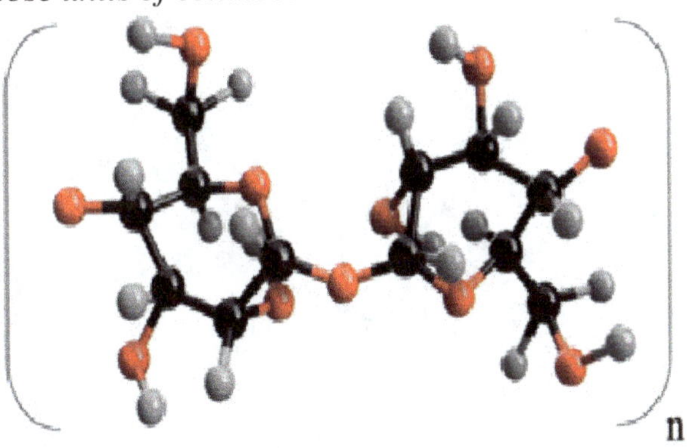

As the second most prevalent macromolecule in nature, hemicellulose consists of sugar monomers including arabinose, galactose, glucuronic acid, mannose, and xylose, which are depicted in Figure 4. (Anastopoulos et al., 2022).

Figure 4. Constituents of Hemicellulose

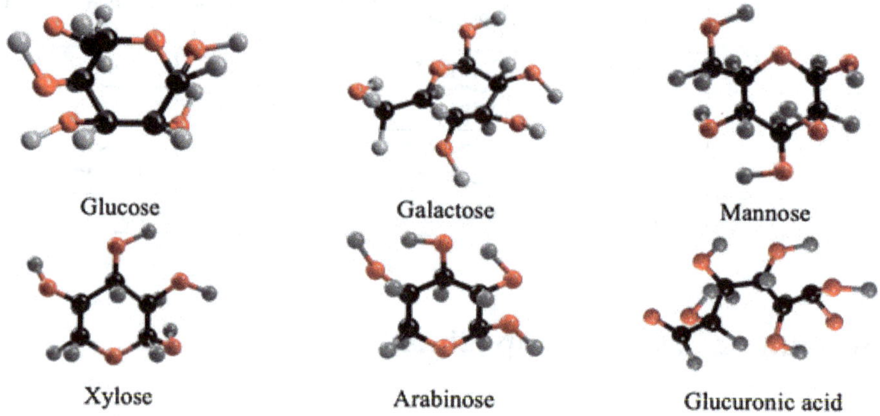

Wood and plants contain lignin; its composition in plants is maximum of 0.20 (w/w) and minimum of 0.14 (w/w). Figure 5 depicts the lignin structure, which is a polymer of phenylpropane units classified in p-coumaric, coniferyl, and sinapyl alcohols, based on methoxyl groups linked by either carbon-carbon or ester (Anastopoulos et al., 2022).

Figure 5. Constituents of lignin

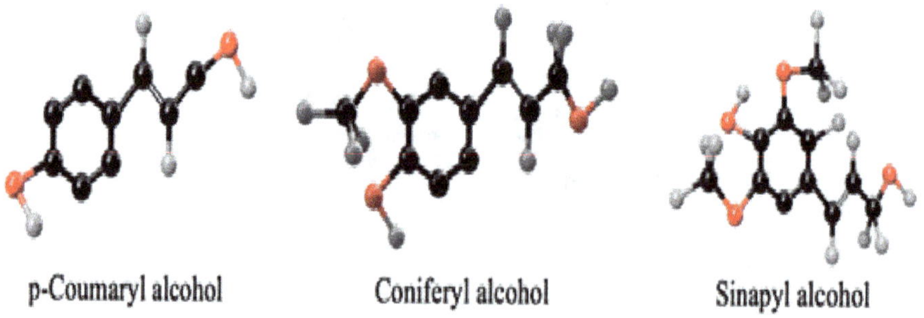

Table 2 presents the percentages of lignin, cellulose, and hemicellulose found in these types of agro-waste. The maximum contents are 40% (lignin), 40% (hemicellulose), and 66.59% (cellulose) for nutshells, hardwood, and sponge ground fibers, respectively (Mehfooz et al., 2018).

Table 2. Composition of agricultural waste.

Waste	Lignin (wt.%)	Hemicellulose (wt.%)	Cellulose (wt.%)
Sugarcane cane bagasse	20	25	52
Sweet sorghum	21	27	45
Hardwood	18-25	24-40	40-55
Softwood	25-35	25-35	45-50
Nutshells	30-40	25-30	25-30
Grasses	10-30	25-50	25-40
Wheat straw	16-21	26-32	29-35
Bagasse	23.33	16.52	54.87
Sponge ground fibers	15.46	17.44	66.59
Rice straw	18	24	32.1
Corn cobs	15	35	45
Corn Stover	19	26	38
Banana waste	14	14.8	13.2

Table 3 presents selected physical and chemical properties, including the proximate and elemental analyses of agro-wastes. (Paraskeva et al., 2008). The high volatile matter content is an essential property. It is valuable to develop porosity because its release promotes the development of micro and mesopores; some agro-wastes exhibit high carbon content, which is a key factor for achieving high efficiency in activated carbon production.

Table 3. Physical properties of selected agro-wastes.

Raw	Proximate analysis (%wt.)				Elemental analysis (%wt.)				
	Moisture	Ash	Volatiles	Fixed C	C	H	N	S	O
Apricot stones	4.5-6.5	0.2-2.4	80.5-83	n.d.	46-51.5	6-6.5	0.2-1.4	0.1-0.25	41-44
Cherry stones	6	0.9	82	n.d.	54	7.1	0.3	0.3	38.4
Oil palm stones	n.d.	2.6	77.6	19.8	n.d	n.d	n.d	n.d	n.d
Grape seeds	7.6	2.5	75.3	n.d	55.4	7	1.5	0.2	35.9
Almond shells	8.7	2.3	79.7	n.d	54.7	7.5	0.3	0.3	37.4
Nut shells	6.7	3.8	84.1	n.d	53.6	7.5	0.3	0.3	38.3
Pistachio nutshell	4.04	0.99	73.37	21.60	n.d	n.d	n.d	n.d	n.d

continued on following page

Table 3. Continued

Raw	Proximate analysis (%wt.)				Elemental analysis (%wt.)				
Corn cob	4.3	0.9	78.7	16.1	46.8	6.0	0.9	n.d	46.3
Rice husks	4.2	16.1	62	17.7	n.d	n.d	n.d	n.d	n.d
Sugarcane bagasse	6.1	3.3	65.9	24.7	77	5	4	n.d	14

n.d= No data

3. ACTIVATED CARBON PRODUCTION

Activated carbon (AC) is prepared through pretreatment, carbonization, activation, washing, and drying. Pretreatment involves crushing, milling, and sieving to obtain particles of 0.85–4.00 mm, as size affects surface area and porosity. These properties promote adsorbate diffusion into pores and interaction with surface groups, enhancing adsorption capacity and kinetics. AC can be synthesized via physical, chemical, or hydrothermal activation, typically involving thermal treatment in furnaces or ovens. Physical activation includes pyrolysis under nitrogen, followed by activation with CO_2 or steam. Chemical activation occurs in one step under nitrogen, using inorganic agents such as H_3PO_4, H_2SO_4, KOH, NaOH, and $ZnCl_2$(Paraskeva et al., 2008).

Physical activation leads to the release of a large amount of volatile material, which contributes to developing carbon structure; in another way, AC chemically activated is carried out by dehydrating substances, as above, these agents enhance pyrolytic decomposition, inhibiting the development of tar; therefore, the yield of AC is higher compared with AC physically activated (Ahmadpour & Do, 1996). According to Thermogravimetric Analysis (TGA), most biomasses typically release water at T > 100 °C, and volatile matter at 300 °C < T < 600 °C, at temperatures higher than 600 °C the pyrolysis is carried out at temperatures ranged from 600 to 900 °C, this is the reason because the temperatures for activated carbon production is ranged from 300 and 900 °C (Eduardo et al., 2022).

3.1 Activated Carbon Produced by Chemical Activation

Chemically activated carbon is produced in one step combining pyrolysis and activation, typically at 450–900 °C, with conditions defined by thermogravimetric analysis. To avoid ash formation, the process uses an inert atmosphere like nitrogen.

Before activation, biomass is impregnated with dehydrating agents to promote decomposition and improve the carbon's physicochemical properties. (Sadh et al., 2018).

Chemically activated AC requires lower temperatures than physical activation, resulting in lower energy consumption and higher yield due to reduced volatile release from biomass. (Prauchner & Rodríguez-Reynoso, 2012).

Chemical activation with hydroxides involves solid-solid or solid-liquid reactions, including hydroxide reduction and carbon oxidation, forming a porous structure. Potassium ions from KOH are reduced, while carbon is oxidized to produce K, H_2, and carbonates. These simultaneous reactions are described as follows in Eq. (1) to (3) (Jiménez et al., 2009).

$$6KOH + 2C \rightleftharpoons 2K + 3H_2 + 2K_2CO_3 \quad (1)$$

$$4KOH + C \rightleftharpoons 4K + CO_2 + 2H_2O \quad (2)$$

$$4KOH + 2CO_2 \rightleftharpoons 2K_2CO_3 + \; + 2H_2O \quad (3)$$

Water as vapor formed in reactions (2) and (3) during pyrolysis enhances porosity by its penetrating into solid carbon and removing volatiles matter developing good pore size distribution. Potassium is also removed by interflow gas during pyrolysis (Jiménez et al., 2009).

Assorted studies used biomass to produce chemically activated biochar or AC. Corn fibers activated with KOH at 500–900°C during 60 min under nitrogen flow; 1,705 m^2/g of surface area was achieved (Zachary D. Herde, 2020). Banana peel, impregnated with KOH and activated under vacuum at 500–750°C for 240 min, reached 1,950 m^2/g and 1.071 cm^3/g (Jianfeng Ma, 2014).

African palm shells were impregnated with $Cu(NO_3)_2$ solutions at five concentrations to produce activated carbon by physical activation at 1073 K in an electrical furnace under carbon dioxide atmosphere, the surface area and micropore volume were ranged between 473 and 1361 m^2/g and 0.18 and 0.51 cm^3/g, respectively. The CO_2 capture performance of this adsorbent was evaluated, and the corresponding adsorption capacities were reported starting from 103 to 217 $mgCO_2/g$ (Acevedo et al., 2020).

Coconut shell and palm kernel were activated with $ZnCl_2$ at 550 °C for 60 min under nitrogen flow (200 mL/min). After activation, the furnace was cooled, and the AC washed five times to remove residual chemicals. Yields were 44% for palm kernel and 45% for coconut shell. The highest surface area (1,223 m^2/g) was obtained for palm kernel. These materials were evaluated for CO_2 capture, with adsorption attributed to surface physisorption mechanisms (Hidayu & Muda, 2016).

Barley husks (BH) were studied to produce activated carbon chemically activated by $ZnCl_2$ using the following weight ratios: 0.5, 1.0, and 1.5 g $ZnCl_2$/g BH. A mass of 10 g of impregnated feedstock was heated using a rate of 5 ºC/min to achieve the activation temperatures (300, 500, and 700 ºC) into an electrical furnace under nitrogen flow, once these temperatures reached the furnace was kept at constant temperature for 20, 100, and 180 min.

This process was optimized by surface response methodology and ANOVA analysis. At the optimal conditions (436 ºC, 20 min, and 1.1 g $ZnCl_2$/ g BH), yielded 811.44 m^2/g and 0.731 cm^3/g, were achieved (Loredo-Cancino et al., 2013). Amin A. et al. studied the life cycle assessment to produce biochar from Quercus Alba (white oak) by chemical activation. The lowest global warming potential was reported as 3.12 kgCO_2 eq for biochar chemically activated by $ZnCl_2$ (Amin et al., 2023).

3.2 Activated Carbon Produced by Physical Activation

The physical activation of activated carbon is produced in furnaces under different conditions than chemical activation. Temperatures range from 700 – 900°C, requiring more energy. The process involves pyrolysis under inert gas (e.g., nitrogen) and activation with CO_2 or steam (Marcos J. Prauchner, 2012).

Hydrochloric acid was used to treat rice husk char to eliminate impurities and silica, then activated in a bench-scale plant. One gram of desilicated char was heated at 15 °C/min to 800 °C, nitrogen flow was fed during 1 h, after carbonization, the activation process was carried out with CO_2 or a mixture of steam and nitrogen that was fed at 200 cm^3/min. Activation times were ranged from 15 to 60 min. The highest surface area, 1364.5 m^2/g, and a pore volume of 1.16 cm^3/g were achieved with steam activation (Alvarez et al., 2015).

Vine shoots (VS) and wheat straw (WS) were used as precursors for activated carbon in a packed-bed reactor at 600 °C (VS) and 500 °C (WS) under nitrogen (101–500 kPa), then activated with CO_2 at 800 °C. Nitrogen was fed at 8.34 mL(STP)/s for 1 h, then switched to a CO_2/N_2 (20:80) mix at the same rate. VS achieved the highest surface area (536 m^2/g). WS showed the highest CO_2 adsorption: 1.3 and 0.74 mmol/g at 25 and 50 °C, respectively, under 14 kPa (Manyà et al., 2020).

Biochar from wheat straw was produced by slow pyrolysis at 500 °C and 0.1 MPa, then activated with CO_2 or K_2CO_3. Pellets were carbonized at 500 °C for 60 min under nitrogen, followed by CO_2 activation at 800 °C, yielding 815 m^2/g and 0.366 cm^3/g (Christian Di Stasi, 2019). Physical activation requires higher temperatures than chemical activation, releasing more volatile matter and resulting in lower yields (17–79% vs. 50–80%). It involves two steps at 600–900 °C, consuming more energy than chemical activation, which occurs in one step at 400–700 °C. This extra energy demand may have greater environmental impact if sourced from fossil fuels. (Dalai & Azargohar, 2007).

Vilen A. et al studied the life cycle for the physical activation of biochar prepared from coal, peat, coconut, and wood by physical activation, the potential global warming as reported as 9.3, 6.4, 7.3, and 8.9 kgCO_2 eq, respectively (Vilén et al., 2022).

3.3 Characterization of Activated Carbon

The physicochemical properties of activated carbon (AC) are key to its adsorption capacity and mechanisms. Parameters like pore volume, size distribution, and surface area are assessed via nitrogen physisorption and analyzed using the BET equation. Surface areas range from 1 to over 1000 m^2/g, with pore diameters of 1.57–5.2 nm and volumes of 0.003–0.24 cm^3/g (López-Cano et al., 2018), (Brewer & Brown, 2012), (Nartey & Zhao, 2014).

The acidic and basis groups onto the adsorbent surface are essential to enhance the adsorption process and are helpful to explain the adsorption mechanisms between adsorbate and adsorbent surface; they are suitable to explain or elucidate adsorption mechanism, for instance, electrostatic attraction, surface complexation, ion exchange, hydrogen bonding, π-π interactions, among others. FTIR spectrometry is an analytical methodology to identify acidic and basic functional groups, such as O–H (3600–3100 cm^{-1}), C=C, and C=O stretching (1740 – 1600 cm^{-1}), among others (Brewer & Brown, 2012).

The point of zero charge (pHpzc) is an important property of adsorbents, defined by the surface charge balance. It is determined by shaking AC samples in KNO_3 solutions at various pH values, adjusted with KOH or HNO_3, for 24 h at 25 °C. The adsorbed H^+ or OH^- ions are then measured, and surface charge densities obtained by potentiometric titration. Babic, B. M., Milonjic, S. K., Polovina, M. J., & Kaludierovič Kaludierovič, B. V. (1999). Point of zero charge and intrinsic equilibrium constants of activated carbon cloth. Carbon, 37, 477 – 481.

3.4 Activated Carbon Based on Biomass to CO_2 Adsorption

Agro-wastes, rich in fixed carbon and volatiles and globally available, have been widely studied as precursors for activating carbon via chemical, physical, or hydrothermal activation. These adsorbents have been evaluated for CO_2 capture in batch and continuous adsorption systems.

Table 4. Activated carbons produced from agro-waste to CO_2 adsorption activation and operating conditions (OC).

Precursor	Textural properties			Activation process and agents	Adsorption capacity (mmol/g)	OC T and P	Activation	Activation Time (min)	Ref
	S_A (m^2/g)	V_P (cm^3/g)	D_P (nm)						
Bamboo	**500-3100**	**0.20-1.5**	**0.36-0.82**	**Chemical (KOH)**	**4.0-7.0**	**273 K and 1.013 bar**	**600°C** **700°C**	**90**	**(Wei et al., 2012)**
Corncob	42.01-1222	0.594-0.711	1.65-3.05	Hydrothermal+ H_3PO_4, $ZnCl_2$, or KOH	18.88-35.35	15, 30, and 45°C 1-16 bar	230°C 600°C	60	(Sarwar et al., 2021)
Corn stalks	376.7-639.8	0.175-0.321	---	Physical (CO_2)+ Chemical (KOH, HNO_3, or CH_3COOH)	0.4225	n.d.	800°C 300 or 600°C	60 and 120	(Song et al., 2015)
Date seeds	513.76	0.10-0.268	---	Pyrolysis+ physical activation (CO_2)	1.476	293 K P= n.d.	600, 700, 800, 900°C 900°C	60, 120, 180	(Ogungbenro et al., 2017)
Eucalyptus wood	2097	0.999-1.292	1.21-2.44	Chemical H_3PO_4+ ammonia	31.61	288-348 K 0-16 bar	450°C 400°C and 800°C	120	(Heidari et al., 2014)
African palm shells	473-1361	0.23-0.60	0.30-4.0	Physical (CO_2)	4.932	273 K Low pressure	800°C	n.d.	(Acevedo et al., 2020)
Sugarcane bagasse	1149	0.07-1.73	2.42-6.02	Physical CO_2 (13%) +N_2 Chemical H_3PO_4 or NaOH	0.75-1.31	273, 298, 323K 1 bar	750°C 750°C and 850°C	120	(Guo et al., 2020)
Whitewood	800-1400	0.45-0.62	---	Physical CO_2+ Chemical KOH	1.34-1.77	298-338K 1.013 bar	700°C 750°C	100	(Shahkarami et al., 2015)
Arundo donax	3298	1.90	---	Chemical $ZnCl_2$	30.2	273K 30 bars	400-700°C	90, 120, 150	(G. Singh, Lakhi, et al., 2017)
Pinecone	1470	0.6415	1.74	Chemical H_3PO_4	9.464	298-318K 16 bars	488°C	120	(Khalili et al., 2015)
Arundo donax	1122	0.592	0.56	Chemical KOH	15.4	273, 283, 298K 0-30 bar	600°C	120	(G. Singh, Kim, et al., 2017a)

n.d.= Data not reported

Their hemicellulose, cellulose, and lignin content makes these materials suitable for producing activated carbon (AC) via chemical or physical activation. Biomass-based AC is widely reported due to its surface area and composition. Table 4 shows textural properties, activation conditions, and CO_2 adsorption capacities. Adsorption is influenced by surface area, porosity, functional groups, pressure, temperature, and moisture; lower temperatures favor adsorption, while higher pressures increase capacity.

Adsorption equilibria are commonly described using Langmuir and Freundlich models. Langmuir assumes monolayer adsorption on a homogeneous surface without interactions, enabling estimation of maximum capacity and affinity. Freundlich is empirical, accounting for surface heterogeneity and exponential energy distribution. For multicomponent gas mixtures, complex models like Tóth or extended Langmuir equations better represent competitive adsorption.

CO_2 adsorption kinetics on activated carbon are commonly analyzed using pseudo-first and pseudo-second-order models, which describe adsorption capacity over time and reveal whether surface reaction or diffusion dominates. In continuous or fixed-bed systems, dynamic adsorption is assessed via breakthrough curves, interpreted using mass-transfer zone or diffusion-based models that consider film resistance, pore, and intraparticle diffusion. These models estimate kinetic parameters like mass transfer coefficients and diffusion constants, essential for designing and scaling up adsorption columns.

According to the data shown in Table 4 the highest adsorption capacities were achieved through chemical activation. For example, AC from corncob via hydrothermal and chemical activation adsorbed 35.35 mmol/g, while the lowest, 0.4225 mmol/g, was from corn stalk activated physically and chemically. These differences relate to surface areas of 1,222 and 639 m^2/g, respectively. Similarly, AC from eucalyptus wood adsorbed 31.31 mmol/g, attributed to its high surface area of 2,097 m^2/g.

4. COMMERCIAL ACTIVATED CARBON TO CO_2 CAPTURE

Commercial activated carbons (ACs) are often chemically treated to enhance CO_2 adsorption. Several studies have used commercial ACs, or fiber modified chemically and physically, with their textural properties shown in Table 5. The highest CO_2 adsorption capacities were reported for non-modified ACs, though obtained at high pressures and room temperature, which implies high energy consumption and potential CO_2 emissions if powered by fossil fuels.

On the other hand, activated carbons chemically activated by KOH, O_2+HNO_3, and K_2CO_3 were studied at low-pressure values and room temperature; these studies reported CO_2 adsorption capacities ranged from 5.33 to 5.8 mmol/g; moreover, the lowest adsorption capacity was reported when activated carbon physically activated with steam was used. These studies demonstrate that modifying ACs can be helpful to improve the CO_2 adsorption capacity.

Some non-modified CACs show high CO_2 adsorption capacities. For instance, Jribi et al. (2017) used powdered commercial activated carbon and reported 25.68 mmol CO_2/g. In contrast, Plaza et al. (2010) reported the lowest capacity using steam-modified CAC. Using more resources does not guarantee better performance; Lee & Park used CAC fibers chemically activated with KOH at elevated temperature, yet CO_2 adsorption was low. Thus, resource-intensive methods are not justified.

Table 5. Commercial AC applied to CO_2 adsorption.

Activated carbon type	**Textural properties**			**Activation process and agents**	**Adsorption capacity (mmol/g)**	**Activation Temperature (°C)**	**Ref**
	S_A (m^2/g)	V_P (cm^3/g)	D_p (nm)				
CAC Fibers	900-2318	0.456-1.124	0.7-1	Chemical (KOH)	5.68	900	(Lee & Park, 2013)
CAC	1228	---	---	Chemical (O_2+HNO_3)	5.33	400 or 600	(Caglayan & Aksoylu, 2013)
CAC	3045	1.7	0.1-4.4	Non-modified	25.68	---	(Jribi et al., 2017)
CAC	---	---	---	Non-modified	10.99	---	(V. K. Singh & Anil Kumar, 2016)
CAC	2406	0.46	0.3-0.8	Chemical (KOH+K_2CO_3)	5.8	600-850	(Sreńscek-Nazzal & Kiełbasa, 2019a)
CAC	942	---	---	Physical (steam)	0.7	n.d.	(Plaza et al., 2010)

5. TEXTURAL MODIFICATIONS OF ACTIVATED CARBON

The adsorption occurs on the surface of adsorbent, therefore enhancing the surface area and the pore size distribution are essential. The fundamental textural properties for adsorbent materials to adsorb CO_2 are specific area, pore volume, and micropore diameter (Li et al., 2022; Rehman & Park, 2018; J. Singh et al., 2018; Sreńscek-Nazzal & Kiełbasa, 2019b; H. Yuan et al., 2019).

The surface-specific area of adsorbent material includes the internal pores and external surfaces. Activated carbon has a porous structure made up of macropores (diameter > 50 nm), mesopores (2 nm < diameter < 50 nm), and micropores (diameter < 2 nm); most of the specific area is found in the mesopores and micropores. Therefore, most of the studies on developing adsorbent materials to capture CO_2 are focused on developing micropores. Studies performed by Singh et al. (G. Singh, Kim, et al., 2017b) and Wang et al. (Z. Wang et al., 2011) showed that the adsorption capacity of CO_2 at pressure of 1 atm by activated carbons is directly proportional to the increase of surface area.

The textural properties of adsorbent materials, the diameter and volume of the micropores affect the adsorption capacity and its selectivity due to the exclusion principle. The ideal pore size in laminar-type structures was theatrically determined as 0.57 and 0.72 nm by Heuchel et al. (Heuchel et al., 1999), for CO_2 adsorption and storage, considering that the dynamic diameter of CO_2 is 0.33 nm (Z. Wang et al., 2011). For these systems, these pore sizes represent the minimum energy potential.

Micropores provide extensive space for CO_2 movement and interaction with surface functional groups, while offering higher resistance to diffusion of larger molecules like nitrogen. The optimal micropore range was empirically confirmed for activated and heteroatom-functionalized carbons (Rehman & Park, 2018) and non-modified types (W. Hao et al., 2013; Sevilla & Fuertes, 2011).

The size of micropores can be achieved within a range of 0.57 to 0.72 nm by controlling the activation conditions, introducing pore-forming agents, or using templating agents. Potassium hydroxide is an activating agent highly effective in developing micropores (Wu et al., 2005); additionally, KOH activation often results in AC with high adsorption capacities and good mechanical strength. Several authors have used KOH as activating agent for activated carbons used in CO_2 adsorption (Wickramaratne & Jaroniec, 2013), i.e., Sreńscek-Nazzal and Kiełbasa (Sreńscek-Nazzal & Kiełbasa, 2019b) used KOH as modification agents for commercial activated carbons, varying temperature between 600 and 850 °C and achieved a high volume of micropores (0.49 cm^3/g) which led to increase the CO_2 adsorption. Singh et al. used solid KOH to synthesize lignocellulosic activated microporous carbons with a high specific surface area (1,122 m^2/g) and high micropore volume (0.50 cm^3/g) by (G. Singh, Kim, et al., 2017b).

Moreover, Yuan et al. (Li et al., 2022) proposed a synthesis of AC using biomass as precursor, the adsorbent material was highly microporous, a surface area of 2,424 m^2/g was attained. When AC is doped with nitrogen, other authors suggested that K_2CO_3 favored the formation of micropores and degraded the total nitrogen content of the material, achieving higher CO_2 adsorption capacity (Rehman & Park, 2018).

6. SURFACE FUNCTIONALIZATION OF ACTIVATED CARBON

Heteroatomic doping is a good methodology to enhance the chemical characteristics of carbons. Saha et al. (Saha & Kienbaum, 2019) studied the inclusion of heteroatoms such as oxygen, sulfur, phosphorus, and nitrogen on activated carbons matrix, according to their results a higher adsorption capacity and selectivity were achieved in comparison with non-modified carbon. This increase is due to the increasing electrostatic interactions with acid-type gases such as CO_2. Oxygen functional groups have been the most studied due to their natural formation on the surface of carbon when exposed to the atmosphere. Oxidative treatments in the gas or liquid phase can modify their nature/amount.

Unlike other heteroatoms, sulfur atoms protrude from the graphene plane when used as dopants, resulting in a more irregular surface morphology due to their larger atomic radius compared to carbon and other dopants such as nitrogen. Additionally, sulfur exhibits high adsorption energies (approximately 59 kJ/mol), which may hinder regeneration and reduce the cyclability of the adsorbent. Furthermore, the synthesis of sulfur-doped materials often requires sulfur-containing polymers, some of which are high-value chemicals, posing significant challenges for large-scale production. (Saha, Orkoulas, et al., 2017).

Recently, activated carbons containing nitrogen on their surface have been studied to enhance CO_2 capture on non-polar or non-acidic gases, for example N_2 or CH_4 (Saha, Van Bramer, et al., 2017). The nitrogen on the surface of carbon generates greater basicity in the carbon matrix, improving the capture of CO_2 more efficiently. The primary interaction between CO_2 and the surface is a weak acid-Lewis base interaction involving electron donation and van der Waals interactions. However, according to specific authors, incorporating nitrogen onto a carbon surface enhances the hydrogen-bonding interactions between the carbon surface and CO_2 molecules (Xing et al., 2012).

Theoretical studies by Lim et al. (2016) demonstrated that pyridinic and pyrrolic nitrogen groups exhibit shorter bond junction distances and greater adsorption energy differences compared to N_2, resulting in high CO_2 selectivity. These functional groups have been identified as the most influential in CO_2 adsorption (Saha & Kienbaum, 2019). X-ray photoelectron spectroscopy (XPS) enables detailed qualitative and quantitative characterization of nitrogen functionalities.

Unlike sulfur doping, nitrogen-doped AC synthesis is flexible and uses abundant lignocellulosic precursors rich in carbon, oxygen, and nitrogen, which interact well with inexpensive compounds like urea (Yue et al., 2018), enabling scalable carbon matrices. Saha et al. (2017) produced nitrogen-doped hierarchical porous carbon from lignin via KOH and NH_3 activation, achieving a BET surface area of 2,922

m^2/g and 7.1% nitrogen content. The material showed pyridinic, amino, and pyrrolic/pyridone groups (Figure 6).

Likewise, theoretical and experimental studies carried out by Ma et al. (Ma et al., 2019) and others (Han et al., 2019; Rehman & Park, 2018, 2019; Rouzitalab et al., 2018; Yue et al., 2018) have demonstrated a synergy for CO_2 capture and selectivity by combining different oxygenated and nitrogenous functional groups that can produce this type of precursor, combined with a narrow porosity in the range of micropores.

Figure 6. Schematic of nitrogen functionalities on carbon surface, adapted from(Dipendu Saha, 2019).

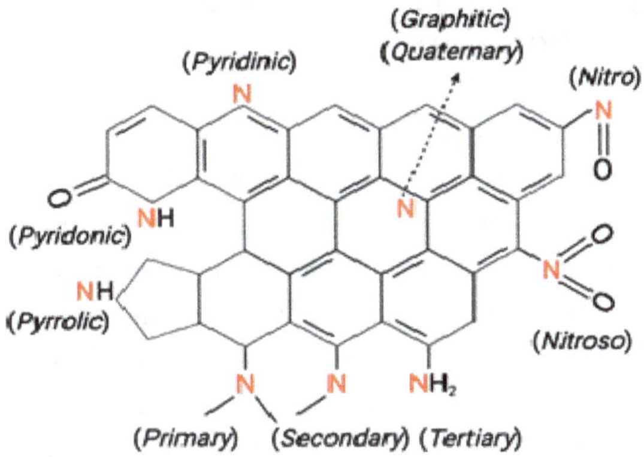

Various methods have been employed to obtain nitrogen-enriched carbons. These methods involved incorporating nitrogen into carbon precursors, utilizing activation agents containing nitrogen, and subjecting the carbon to elevated temperatures with nitrogen compounds.

The most employed agents for introducing nitrogen are ammonia (Saha, Van Bramer, et al., 2017), urea (Rangel-Sequeda et al., 2021), ammonium hydroxide (Aijaz et al., 2014), acetonitrile (L. Wang & Yang, 2012; Xia et al., 2011), sodium amide (Huang et al., 2015), and amines.

Horikawa et al. (Horikawa et al., 2012), studied the treatment of carbon with ammonia gas was highly effective for introducing nitrogen atoms into the carbon matrix. They also observed that ammonia gas acted as an activation agent for the carbon material. The most prominent nitrogen functionalities that emerged from this treatment were pyridine-like structures, pyrrolic and pyridone-N moieties, and quaternary nitrogen. When directly preparing nitrogen-containing porous carbons using nitrogen-rich precursors polyvinyl pyridine (Frackowiak & Béguin, 2001), polyacrylonitrile (Vagner et al., 2003), cyanamide (Goettmann et al., 2006), melamine

(Pevida et al., 2008), lysine (G.-P. Hao et al., 2010; W. Hao et al., 2013), glucosamine (Zhao et al., 2010), and microalgae (Saha et al., 2018; Sevilla et al., 2012; B. Yuan et al., 2016; Yue et al., 2018) have been used. Due to the exceptionally high protein content found in several types of algae (60 - 70%), algae have gained significant popularity as a preferred natural source for nitrogen-doped carbons (Saha & Kienbaum, 2019).

7. CO_2 ADSORPTION BY COMMERCIAL AC IN CONTINUOUS SYSTEMS.

7.1. Structuring of Powdered Activated Carbon for CO_2 Adsorption

Although activated carbons have demonstrated superior performance for CO_2 adsorption, these adsorbents present a shortcoming like most currently synthesized materials, for instance, a particle size of microns (95% smaller than 150 μm (Cooney, 1998)). Powdered adsorbents represent an operational problem when integrating these materials into a fixed bed to carry out the separation efficiently.

Material aggregation and structuring address specific issues. Structured adsorbents offer advantages in managing kinetic and engineering aspects of separation processes, comprising pressure drop, adsorption enthalpy, heat integration in TSA, performance, mechanical stability, gas flow optimization, and mass and energy transfer. (Stephen J A DeWitt, 2018).

Recent scholarly articles describe various material extrusion techniques, broadly categorized into two main groups(Yao Ma, 2020). The first method involves impregnating inert structures with materials through washing or hydrothermal treatment, though this technique often results in low adsorbent material content and issues with coating adhesion (Stephan Utgenannt, 2015). The second method, active structuring, addresses these challenges by incorporating a binder system, which may alter the adsorbent materials' textural properties and surface chemistry. Prominent active structuring techniques include mechanical extrusion, reverse replication, dry gel conversion, and additive manufacturing (J. Lefevere, 2017).

The optimal combination of structural and geometric parameters for active structuring remains under investigation. Rezaei et al. proposed a comparative methodology using analytical and numerical models to evaluate the performance of various adsorbent configurations—including monolithic, laminated, and foam structures—against conventional packed beds.

The authors demonstrate that laminated systems are potentially very promising structures, provided that these structures have small gaps (<0.2 mm) and sheet widths (0.2 mm). Monoliths can be excellent candidates if appropriate cell densities (>1000 psi) are used, and the foams exhibit comparable properties to compact bedding (Rezaei & Webley, 2009).

Monoliths are solid channel bundles resembling honeycombs, with geometries like hexagonal, circular, square, or triangular (Figure 7). Geometry critically affects performance. Studies comparing shapes with equivalent cylindrical bases showed hexagonal channels perform best, followed by square ones. Minimizing wall thickness is key for optimal design (Table 6) (Patton et al., 2004).

Figure 7. Schematic of channel geometries of monolithic structures.

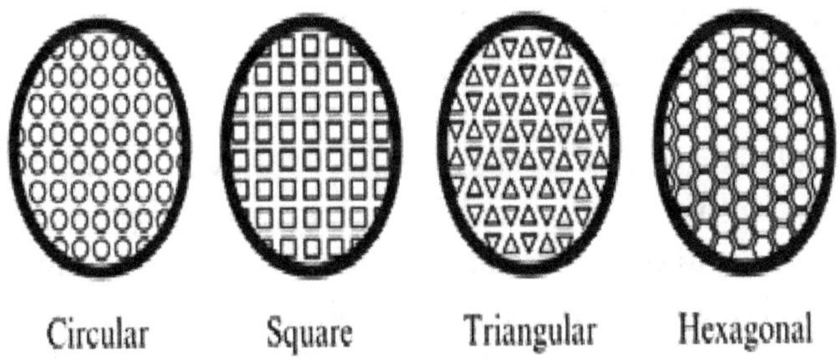

Table 6. Characteristics of monoliths with different channel geometries (Patton et al., 2004)

Channel geometry	Hydraulic diameter	Internal transfer coefficient	External transfer coefficient	Pressure drops
	d_h, mm	K_m, mm^{-2}	$\frac{K_d}{D}$, mm^{-1}	K_p, m^{-2}
Circle	2.00	41.51	1.83	2.55
Square	1.52	38.94	1.96	5.36
Equilateral triange	1.11	35.51	2.24	13.45
Regular hexagon	1.79	40.54	1.92	3.36
Rectangle α=0.5	1.35	38.94	2.51	8.36
Rectangle α=0.25	0.97	38.94	4.57	26.26

continued on following page

Table 6. Continued

Channel geometry	Hydraulic diameter	Internal transfer coefficient	External transfer coefficient	Pressure drops
	d_h, mm	K_m, mm^{-2}	$\frac{K_d}{D}, mm^{-1}$	K_p, m^{-2}
Hexagon θ=60° γ=0.333	1.31	40.54	3.23	10.00
Hexagon θ=75° γ=0.333	1.25	40.11	3.51	11.86
Hexagon θ=75° γ=1.5	1.60	40.11	2.18	4.90

7.2 Structuring of Materials Through Additive Manufacturing (3D Printing)

Additive manufacturing has gained significant relevance recently because this methodology is useful to create complex macroscopic morphologies and change designs immediately without incurring extra costs for manufacturing new molds (Thakkar et al., 2018). Although there are many 3D printing technologies for structuring materials, the most widely used is the so-called extrusion-free formation (EFF), which creates objects by deposition through a nozzle (Ruiz-Morales et al., 2017).

According to recent studies, additive manufacturing has potential benefits which are listed below (Tofail et al., 2018):

- Direct translation from design to component
- Generation of parts with greater customization without additional tools or manufacturing costs.
- Functional design that allows the fabrication of complex internal features
- Manufacture of flexible and light components with hollow or lattice structures.
- Ability to manufacture components directly to their final or near-final form with minimal or no additional processing.
- Potential to move closer to the "Zero waste" approach by maximizing material utilization.
- A vast reduction in overall product development and manufacturing time, leading to faster transfer to market.
- Smaller operational footprint toward manufacturing a wide variety of parts.
- Manufacture on demand
- Excellent scalability

Key factors for high CO_2 adsorption in extruded materials—reduced wall thickness and increased contact surface—are effectively controlled by 3D printing. Stable geometries via EFF require control of rheology, extruder type, and printing parameters. Paste rheology depends on adsorbent properties (density, particle size), binder nature, and solvent. A proper formulation allows smooth deposition and durable structures that retain the solid's characteristics (Sarah Couck, 2018). Table 7 summarizes binder characteristics and their effects on final paste properties in 3D printing.

Table 7. Summary of paste additives' physicochemical, rheological, and mechanical effects on 3D printed adsorbents (Lawson et al., 2021).

Additive	Concentration (%w)	Mechanical strength	Solvent behavior	Rheology	Adsorbent deactivation
Binding agent (silica, clays)	0	Low	Expulsion	Fluid	Low
	0-15	High	Retention	Compact	Variable
	> 15	Extremely high	Expulsion	Compact	High
Agent co-binder (polymers)	0	Variable	Retention	Compact	Low
	0-5	High	Variable	Slight hardening	Slightly high
	> 5	Extremely high	Expulsion	High hardening	High
Plasticizer (cellulose polymers)	0	High	Expulsion	Slight dispersion	Low
	0-3	High	Retention	Small dispersion	Low
	> 3	Cracking	Retention	Unpredictable	High
Water		High	Slow dry	Material dependent	Material dependent
Organic solvent		Low	Fast dry	Compact and hard	Low
Water-organic combination		Medium	Variable	Compact and lightweight	Low

Inorganic porous solids often use clay or silica-based binders, which require high-temperature treatment for solidification. Polymeric binders, however, form rigid structures without heat (Couck et al., 2018), ideal for heat-sensitive materials like nitrogen-doped carbon.

In 3D paste printing, fluid dosing methods include pneumatic, mechanical (piston or screw), and solenoid-based systems (Figure 8). Pneumatic extruders suit low viscosity pastes like hydrogels but lack uniformity for semi-solid materials and require a carrier. Piston systems offer direct flow control; screw systems suit high-viscosity bio-inks but may cause pressure drops. Mechanical systems enhance printability of semi-solid waste materials (Ibrahim T. Ozbolat, 2017).

Figure 8. Schematic illustration of extrusion-based printing methods.

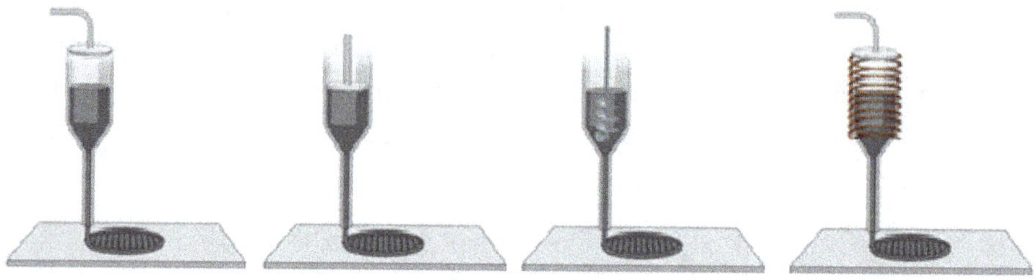

Solenoid-based microextrusion dispenses microliter-scale pastes and suits low-viscosity inks crosslinked by UV or ion irradiation. Despite high precision, accuracy and reproducibility are affected by coil activation delay, seal compression, and the high pressure required for viscous inks.

Table 8 summarizes studies on 3D-printed porous adsorbents. J. Lefevere (2017) analyzed how binders and printing conditions affect mechanical and physicochemical properties. Paste extrusion depends on rheology and particle size, which must be smaller than the nozzle. Shear-thinning enables smooth flow, while high viscosity prevents collapse.

Table 8. Structuring of porous solids using the 3D methodology.

Material	Binders	Binder mix	Printing parameters	Application	Ref.
Zeolite SAPO-34	Polyacrylic acid 0.6% Methylcellulose 2-3%	48%w/w 48% w/w +4%grafite	PS=4 mm/s ØN= 330 µm Td=25 °C td=48 h	Gas separation	(Couck et al., 2018)
ZSM-5	Bentonite 17.5% Ludox (coloidal silica) 17.5%	35%w/w	PS=NR ØN= 400-900 µm Td=20 °C 80%HR td=24 h	Gas separation	(Couck et al., 2017)
SiO_2	Starch 19% Agar- Agar 6% Melamine 19%.	90% w/w	PS=N.R. ØN= 400 µm Td=Freeze drying td=24 h	Carbon synthesis via hard template	(Liu et al., 2019)
Zeolite ZSM-5	Bentonite Silica Methylcellulose 0.5% Litopix P-1 (aluminophosphate solution) 1%	37-65% w/w	PS=1.58 mm/s ØN= 900 µm Td=20°C 80%HR td=N.R.	Gas separation	(Lefevere et al., 2017)
Zeolite 13X Activated carbon MCS-30	Carboxymethylcellulose (CMC) 3.5%	60% w/w	PS=N.R. ØN= 840 µm Td=20°C td=48h	CO_2 adsorption	(Regufe et al., 2019)
Commercial activated carbon	Carboxymethylcellulose (CMC) 3%	1:2 mass ratio	PS= 7 mm/s ØN= 420–820 µm Td= 25 oC td= 24 h	CO_2 adsorption	(Rangel-Sequeda et al., 2022)

PS: Printing speed, ØN: Nozzle diameter, Td: Drying temperature, td: Drying time, NR: Not reported

Regufe et al. developed a 13× slit-shaped zeolite monolith for CO_2 adsorption using CMC as binder. The 3D-printed monolith showed 3.49 mmol/g adsorption—about 15% less than its powdered form—due to surface area loss from particle compression during extrusion or binder blocking micropores. However, it tripled the capacity of a traditionally extruded monolith with the same material, due to binder amount and extrusion pressure (Maria João Regufe, 2019).

3D-printed monoliths have also been tested for multicomponent gas separation. Couck et al. (2017a) found that bentonite binder reduced N_2 and CH_4 adsorption by 35% compared to unstructured ZSM-5, linked to binder quantity. CO_2 removal dropped only 10%, as bentonite also adsorbs CO_2, enhancing selectivity.

8. CONCLUSIONS

In recent years, the increasing of CO_2 levels in atmospherics has become a global concern due to its contribution to global warming and its effect as greenhouse gas. By May of 2025, the atmospheric concentration of carbon dioxide had exceeded 430.55 ppm, without indication of a decline in this trend. Scientists have extensively investigated adsorption processes using activated carbon to capture carbon dioxide from both batch and continuous systems. Activated carbon, a material that can be produced from abundant agricultural waste, has been widely studied as a hopeful solution. In 2019, agricultural activities generated approximately 2.62×10^9 tons of agro-waste, making them a readily available resource to produce activated carbon.

The production of activated carbon from agro-wastes comprises various activation methods, such as hydrothermal, chemical, or physical processes, which enhance the textural properties of the resulting activated carbon. For example, the use of Arundo donax as a precursor and chemical activation with zinc chloride has yielded activated carbon with a surface area as high as 3,298 m^2/g and an impressive carbon dioxide adsorption capacity of 30.2 mmol/g. Similarly, activated carbon derived from corncobs using hydrothermal activation exhibited an adsorption capacity of 35.35 mmol/g.

To enhance adsorption capacity and selectivity, researchers have incorporated heteroatoms (O, S, P, N) into activated carbon, boosting electrostatic interactions with acid gases like CO_2. Although powdered activated carbons perform well in batch systems, their use in fixed beds is limited by small particle size (<150 μm), causing pressure drops and mechanical instability. To overcome this, material impregnation and structural modifications have been explored to improve stability. Additive manufacturing offers a promising route to fabricate structured activated carbons with complex geometries, enabling direct design translation, customizable architecture, and lightweight, flexible structures.

9. PERSPECTIVE AND FUTURE WORK

CO_2 adsorption by biomass activated carbons has been studied in batch systems, but their powdered form limits use in continuous systems due to pressure drop. Enhancing their properties is essential for large-scale CO_2 capture from combustion in boilers and furnaces, which emit greenhouse gases including SOx and NOx.

Simultaneous adsorption processes must consider CO_2 selectivity and the impact of these contaminants on capture efficiency.

Additive manufacturing (3D printing) has recently been studied for structuring granular activated carbons for continuous systems. Key operational aspects—rheology, extruder type, and printing parameters—must be optimized to ensure stable geometries and improve CO_2 capture performance.

Studies show that binders like bentonite reduce CO_2 adsorption capacity, highlighting the need for further research on binder types and mass ratios. Conventional CO_2 capture equipment is large and requires significant space and services, making the development of compact, efficient systems.

Finally, an important issue in adsorption processes is the reusing of the adsorbents, it is well known that the regeneration temperatures of the regeneration of adsorbent materials in carbon dioxide capture are up to 120°C, under these conditions the thermal stability of adsorbent materials is very important, therefore studies to improve thermal stability of these materials are essential to increase the number of adsorption-desorption cycles to be carried out in carbon dioxide capture.

REFERENCES

Acevedo, S., Giraldo, L., & Moreno-Piraján, J. C. (2020). Adsorption of CO2 on Activated Carbons Prepared by Chemical Activation with Cupric Nitrate. *ACS Omega*, *5*(18), 10423–10432. DOI: 10.1021/acsomega.0c00342

Ahmadpour, A., & Do, D. D. (1996). THE PREPARATION OF ACTIVE CARBONS FROM COAL BY CHEMICAL AND PHYSICAL ACTIVATION. *Carbon*, *34*(4), 471–479. DOI: 10.1016/0008-6223(95)00204-9

Aijaz, A., Fujiwara, N., & Xu, Q. (2014). From Metal–Organic Framework to Nitrogen-Decorated Nanoporous Carbons: High CO_2 Uptake and Efficient Catalytic Oxygen Reduction. *Journal of the American Chemical Society*, *136*(19), 6790–6793. DOI: 10.1021/ja5003907

Alvarez, J., Lopez, G., Amutio, M., Bilbao, J., & Olazar, M. (2015). Physical Activation of Rice Husk Pyrolysis Char for the Production of High Surface Area Activated Carbons. *Industrial & Engineering Chemistry Research*, *54*(29), 7241–7250. DOI: 10.1021/acs.iecr.5b01589

Amin, M., Chung, E., & Shah, H. H. (2023). Effect of different activation agents for activated carbon preparation through characterization and life cycle assessment. *International Journal of Environmental Science and Technology*, *20*(7), 7645–7656. DOI: 10.1007/s13762-022-04472-6

Anastopoulos, I., Lima, E., Meili, L., & Giannakoudakis, D. (2022). Biomass-Derived Materials for Environmental Applications. In *Biomass-Derived Materials for Environmental Applications*. Elsevier., DOI: 10.1016/C2021-0-00449-1

Brewer, C. E., & Brown, R. C. L. D. a. (2012). Biochar characterization and engineering. *Graduate Teses and Dissertations*, 12284. http://search.proquest.com/docview/1023114544?accountid=27932

Caglayan, B. S., & Aksoylu, A. E. (2013). CO2 adsorption on chemically modified activated carbon. *Journal of Hazardous Materials*, *252–253*, 19–28. DOI: 10.1016/j.jhazmat.2013.02.028

Christian Di Stasi, D. A. G. G. B. G. J. J. M. (2019). Physically activated wheat straw-derived biochar for biomass pyrolysis vapors upgrading with high resistance against coke deactivation. *Fuel*, *255*, 115807. DOI: 10.1016/j.fuel.2019.115807

Columbia Climate School, & Sarah Fecht. (2021, February 25). *State of the Planet*. Https://News.Climate.Columbia.Edu/2021/02/25/Carbon-Dioxide-Cause-Global-Warming/

Cooney, D. O. (1998). *Adsorption design for wastewater treatment.*

Couck, S., Cousin-Saint-Remi, J., Van der Perre, S., Baron, G. V., Minas, C., Ruch, P., & Denayer, J. F. M. (2018). 3D-printed SAPO-34 monoliths for gas separation. *Microporous and Mesoporous Materials, 255*, 185–191. DOI: 10.1016/j.micromeso.2017.07.014

Couck, S., Lefevere, J., Mullens, S., Protasova, L., Meynen, V., Desmet, G., Baron, G. V., & Denayer, J. F. M. (2017). CO2, CH4 and N2 separation with a 3DFD-printed ZSM-5 monolith. *Chemical Engineering Journal, 308*, 719–726. DOI: 10.1016/j.cej.2016.09.046

Dalai, A. K., & Azargohar, R. (2007). Production of activated carbon from biochar using chemical and physical activation: Mechanism and modeling. *ACS Symposium Series, 954*, 463–476. DOI: 10.1021/bk-2007-0954.ch029

Dipendu Saha, M. J. K. (2019). Role of oxygen, nitrogen and sulfur functionalities on the surface of nanoporous carbons in CO2 adsorption: A critical review. *Microporous and Mesoporous Materials, 287*, 29–55. DOI: 10.1016/j.micromeso.2019.05.051

Earth System Research Laboratory. (n.d.). *Global Monitoring Laboratory*. Https://Gml.Noaa.Gov/Ccgg/Trends/Mlo.Html

Eduardo, S. R., Dávila-Guzmán, N. E., & Margarita, L. C. (2022). Agro-wastes as precursors of biochar, a cleaner adsorbent to remove pollutants from aqueous solutions. In *Biomass-Derived Materials for Environmental Applications* (pp. 349–376). Elsevier., DOI: 10.1016/B978-0-323-91914-2.00001-5

Frackowiak, E., & Béguin, F. (2001). Carbon materials for the electrochemical storage of energy in capacitors. *Carbon, 39*(6), 937–950. DOI: 10.1016/S0008-6223(00)00183-4

Goettmann, F., Fischer, A., Antonietti, M., & Thomas, A. (2006). Chemical Synthesis of Mesoporous Carbon Nitrides Using Hard Templates and Their Use as a Metal-Free Catalyst for Friedel–Crafts Reaction of Benzene. *Angewandte Chemie International Edition, 45*(27), 4467–4471. DOI: 10.1002/anie.200600412

Guo, Y., Tan, C., Sun, J., Li, W., Zhang, J., & Zhao, C. (2020). Porous activated carbons derived from waste sugarcane bagasse for CO2 adsorption. *Chemical Engineering Journal, 381*, 122736. Advance online publication. DOI: 10.1016/j.cej.2019.122736

Han, J., Zhang, L., Zhao, B., Qin, L., Wang, Y., & Xing, F. (2019). The N-doped activated carbon derived from sugarcane bagasse for CO2 adsorption. *Industrial Crops and Products, 128*, 290–297. DOI: 10.1016/j.indcrop.2018.11.028

Hao, G.-P., Li, W.-C., Qian, D., & Lu, A.-H. (2010). Rapid Synthesis of Nitrogen-Doped Porous Carbon Monolith for CO_2 Capture. *Advanced Materials, 22*(7), 853–857. DOI: 10.1002/adma.200903765

Hao, W., Björkman, E., Lilliestråle, M., & Hedin, N. (2013). Activated carbons prepared from hydrothermally carbonized waste biomass used as adsorbents for CO2. *Applied Energy, 112*, 526–532. DOI: 10.1016/j.apenergy.2013.02.028

Heidari, A., Younesi, H., Rashidi, A., & Ghoreyshi, A. A. (2014). Evaluation of CO2 adsorption with eucalyptus wood based activated carbon modified by ammonia solution through heat treatment. *Chemical Engineering Journal, 254*, 503–513. DOI: 10.1016/j.cej.2014.06.004

Heuchel, M., Davies, G. M., Buss, E., & Seaton, N. A. (1999). Adsorption of Carbon Dioxide and Methane and Their Mixtures on an Activated Carbon: Simulation and Experiment. *Langmuir, 15*(25), 8695–8705. DOI: 10.1021/la9904298

Hidayu, A. R., & Muda, N. (2016). Preparation and Characterization of Impregnated Activated Carbon from Palm Kernel Shell and Coconut Shell for CO2 Capture. *Procedia Engineering, 148*, 106–113. DOI: 10.1016/j.proeng.2016.06.463

Horikawa, T., Sakao, N., Sekida, T., Hayashi, J., Do, D. D., & Katoh, M. (2012). Preparation of nitrogen-doped porous carbon by ammonia gas treatment and the effects of N-doping on water adsorption. *Carbon, 50*(5), 1833–1842. DOI: 10.1016/j.carbon.2011.12.033

Huang, K., Chai, S.-H., Mayes, R. T., Veith, G. M., Browning, K. L., Sakwa-Novak, M. A., Potter, M. E., Jones, C. W., Wu, Y.-T., & Dai, S. (2015). An efficient low-temperature route to nitrogen-doping and activation of mesoporous carbons for CO_2 capture. *Chemical Communications (Cambridge), 51*(97), 17261–17264. DOI: 10.1039/C5CC05619E

Ibrahim, T. Ozbolat. (2017). Extrusion-Based Bioprinting**With minor contributions by Monika Hospodiuk, The Pennsylvania State University. In *3D Bioprinting* (pp. 93–125). Academic Press.

Jiménez, V., Díaz, J. A., Sánchez, P., Valverde, J. L., & Romero, A. (2009). Influence of the activation conditions on the porosity development of herringbone carbon nanofibers. *Chemical Engineering Journal, 155*(3), 931–940. DOI: 10.1016/j.cej.2009.09.035

Jribi, S., Miyazaki, T., Saha, B. B., Pal, A., Younes, M. M., Koyama, S., & Maalej, A. (2017). Equilibrium and kinetics of CO2 adsorption onto activated carbon. *International Journal of Heat and Mass Transfer*, *108*, 1941–1946. DOI: 10.1016/j.ijheatmasstransfer.2016.12.114

Kazmi, S. M. S., Abbas, S., Saleem, M. A., Munir, M. J., & Khitab, A. (2016). Manufacturing of sustainable clay bricks: Utilization of waste sugarcane bagasse and rice husk ashes. *Construction & Building Materials*, *120*, 29–41. DOI: 10.1016/j.conbuildmat.2016.05.084

Khalili, S., Khoshandam, B., & Jahanshahi, M. (2015). Optimization of production conditions for synthesis of chemically activated carbon produced from pinecone using response surface methodology for CO2 adsorption. *RSC Advances*, *5*(114), 94115–94129. DOI: 10.1039/C5RA18986A

Lawson, S., Li, X., Thakkar, H., Rownaghi, A. A., & Rezaei, F. (2021). Recent Advances in 3D Printing of Structured Materials for Adsorption and Catalysis Applications. *Chemical Reviews*, *121*(10), 6246–6291. DOI: 10.1021/acs.chemrev.1c00060

Lee, S. Y., & Park, S. J. (2013). Determination of the optimal pore size for improved CO 2 adsorption in activated carbon fibers. *Journal of Colloid and Interface Science*, *389*(1), 230–235. DOI: 10.1016/j.jcis.2012.09.018

Lefevere, J., Protasova, L., Mullens, S., & Meynen, V. (2017). 3D-printing of hierarchical porous ZSM-5: The importance of the binder system. *Materials & Design*, *134*, 331–341. DOI: 10.1016/j.matdes.2017.08.044

Li, D., Yang, J., Zhao, Y., Yuan, H., & Chen, Y. (2022). Ultra-highly porous carbon from Wasted soybean residue with tailored porosity and doped structure as renewable multi-purpose absorbent for efficient CO2, toluene and water vapor capture. *Journal of Cleaner Production*, *337*, 130283. DOI: 10.1016/j.jclepro.2021.130283

Lim, G., Lee, K. B., & Ham, H. C. (2016). Effect of N-Containing Functional Groups on CO_2 Adsorption of Carbonaceous Materials: A Density Functional Theory Approach. *The Journal of Physical Chemistry. C, Nanomaterials and Interfaces*, *120*(15), 8087–8095. DOI: 10.1021/acs.jpcc.5b12090

Liu, Z., Zhou, X., & Liu, C. (2019). N-doped porous carbon material prepared via direct ink writing for the removal of methylene blue. *Diamond and Related Materials*, *95*(April), 121–126. DOI: 10.1016/j.diamond.2019.04.010

López-Cano, I., Cayuela, M. L., Mondini, C., Takaya, C. A., Ross, A. B., & Sánchez-Monedero, M. A. (2018). Suitability of different agricultural and urban organic wastes as feedstocks for the production of Biochar-Part 1: Physicochemical characterisation. *Sustainability (Basel), 10*(7), 2265. Advance online publication. DOI: 10.3390/su10072265

Loredo-Cancino, M., Soto-Regalado, E., Cerino-Córdova, F. J., García-Reyes, R. B., García-León, A. M., & Garza-González, M. T. (2013). Determining optimal conditions to produce activated carbon from barley husks using single or dual optimization. *Journal of Environmental Management, 125*, 117–125. DOI: 10.1016/j.jenvman.2013.03.028

Ma, J., Huang, D., Zou, J., Li, L., Kong, Y., & Komarneni, S. (2014). Adsorption of methylene blue and Orange II pollutants on activated carbon prepared from banana peel. *Journal of Porous Materials, 22*(2), 301–311. DOI: 10.1007/s10934-014-9896-2

Ma, X., Li, L., Zeng, Z., Chen, R., Wang, C., Zhou, K., & Li, H. (2019). Experimental and theoretical demonstration of the relative effects of O-doping and N-doping in porous carbons for CO2 capture. *Applied Surface Science, 481*, 1139–1147. DOI: 10.1016/j.apsusc.2019.03.162

Ma, Y. F. Z. R. P. L. (2020). Manufacturing Nanoporous Materials for Energy-Efficient Separations: Application and Challenges. In *Sustainable Nanoscale Engineering* (pp. 33–81). Elsevier.

Manyà, J. J., García-Morcate, D., & González, B. (2020). Adsorption performance of physically activated biochars for postcombustion Co2 capture from dry and humid flue gas. *Applied Sciences (Basel, Switzerland), 10*(1), 376. Advance online publication. DOI: 10.3390/app10010376

Marcos, J., & Prauchner, F. R.-R. (2012). Chemical versus physical activation of coconut shell: A comparative study. *Microporous and Mesoporous Materials, 152*, 163–171. DOI: 10.1016/j.micromeso.2011.11.040

Maria João Regufe, A. F. P. F. J. M. L. A. R. A. M. R. (2019). Electrical conductive 3D-printed monolith adsorbent for CO2 capture. *Microporous and Mesoporous Materials, 278*, 403–413. DOI: 10.1016/j.micromeso.2019.01.009

Mehfooz, T., Mohsin Ali, T., Arif, S., & Hasnain, A. (2018). Effect of barley husk addition on rheological, textural, thermal and sensory characteristics of traditional flat bread (chapatti). *Journal of Cereal Science, 79*, 376–382. DOI: 10.1016/j.jcs.2017.11.020

Nartey, O. D., & Zhao, B. (2014). Biochar preparation, characterization, and adsorptive capacity and its effect on bioavailability of contaminants: An overview. *Advances in Materials Science and Engineering, 2014*, 1–12. Advance online publication. DOI: 10.1155/2014/715398

NASA. (2022). *Global Climate Change*. Https://Climate.Nasa.Gov/Vital-Signs/Global-Temperature/

Ogungbenro, A. E., Quang, D. V., Al-Ali, K., & Abu-Zahra, M. R. M. (2017). Activated Carbon from Date Seeds for CO2 Capture Applications. *Energy Procedia, 114*, 2313–2321. DOI: 10.1016/j.egypro.2017.03.1370

Paraskeva, P., Kalderis, D., & Diamadopoulos, E. (2008). Production of activated carbon from agricultural by-products. In *Journal of Chemical Technology and Biotechnology* (Vol. 83, Issue 5, pp. 581–592). DOI: 10.1002/jctb.1847

Patton, A., Crittenden, B. D., & Perera, S. P. (2004). Use of the Linear Driving Force Approximation to Guide the Design of Monolithic Adsorbents. *Chemical Engineering Research & Design, 82*(8), 999–1009. DOI: 10.1205/0263876041580749

Pevida, C., Drage, T. C., & Snape, C. E. (2008). Silica-templated melamine–formaldehyde resin derived adsorbents for CO2 capture. *Carbon, 46*(11), 1464–1474. DOI: 10.1016/j.carbon.2008.06.026

Plaza, M. G., García, S., Rubiera, F., Pis, J. J., & Pevida, C. (2010). Post-combustion CO2 capture with a commercial activated carbon: Comparison of different regeneration strategies. *Chemical Engineering Journal, 163*(1–2), 41–47. DOI: 10.1016/j.cej.2010.07.030

Prauchner, M. J., & Rodríguez-reinoso, F. (2012). Microporous and Mesoporous Materials Chemical versus physical activation of coconut shell : A comparative study. *Microporous and Mesoporous Materials, 152*, 163–171. DOI: 10.1016/j.micromeso.2011.11.040

Rangel-Sequeda, J., Loredo-Cancino, M., Agueda Maté, V. I., Delgado Dobladez, J. A., De Haro-Del Rio, D. A., Salazar-Rábago, J. J., & Soto-Regalado, E. (2021). Synthesis of Zeolite-Templated Carbonaceous Materials Doped With Nitrogen for Carbon Dioxide Adsorption. *Water, Air, and Soil Pollution, 232*(7), 253. DOI: 10.1007/s11270-021-05184-2

Rangel-Sequeda, J. F., Loredo-Cancino, M., Águeda Maté, V. I., De Haro-Del Rio, D. A., & Dávila-Guzmán, N. E. (2022). 3D printing of powdered activated carbon monoliths: Effect of structuring on physicochemical and mechanical properties and its influence on the adsorption performance. *Materials Today. Communications, 33*, 104758. DOI: 10.1016/j.mtcomm.2022.104758

Regufe, M. J., Ferreira, A. F. P., Loureiro, J. M., Rodrigues, A., & Ribeiro, A. M. (2019). Electrical conductive 3D-printed monolith adsorbent for CO 2 capture. *Microporous and Mesoporous Materials, 278*(October 2018), 403–413. DOI: 10.1016/j.micromeso.2019.01.009

Rehman, A., & Park, S. J. (2018). Comparative study of activation methods to design nitrogen-doped ultra-microporous carbons as efficient contenders for CO2 capture. *Chemical Engineering Journal, 352*, 539–548. DOI: 10.1016/j.cej.2018.07.046

Rehman, A., & Park, S.-J. (2019). Tunable nitrogen-doped microporous carbons: Delineating the role of optimum pore size for enhanced CO2 adsorption. *Chemical Engineering Journal, 362*, 731–742. DOI: 10.1016/j.cej.2019.01.063

Rezaei, F., & Webley, P. (2009). Optimum structured adsorbents for gas separation processes. *Chemical Engineering Science, 64*(24), 5182–5191. DOI: 10.1016/j.ces.2009.08.029

Rouzitalab, Z., Mohammady Maklavany, D., Rashidi, A., & Jafarinejad, S. (2018). Synthesis of N-doped nanoporous carbon from walnut shell for enhancing CO2 adsorption capacity and separation. *Journal of Environmental Chemical Engineering, 6*(5), 6653–6663. DOI: 10.1016/j.jece.2018.10.035

Ruiz-Morales, J. C., Tarancón, A., Canales-Vázquez, J., Méndez-Ramos, J., Hernández-Afonso, L., Acosta-Mora, P., Marín Rueda, J. R., & Fernández-González, R. (2017). Three dimensional printing of components and functional devices for energy and environmental applications. *Energy & Environmental Science, 10*(4), 846–859. DOI: 10.1039/C6EE03526D

Sadh, P. K., Duhan, S., & Duhan, J. S. (2018). Agro-industrial wastes and their utilization using solid state fermentation: a review. In *Bioresources and Bioprocessing* (Vol. 5, Issue 1). Springer Science and Business Media Deutschland GmbH. DOI: 10.1186/s40643-017-0187-z

Saha, D., & Kienbaum, M. J. (2019). Role of oxygen, nitrogen and sulfur functionalities on the surface of nanoporous carbons in CO2 adsorption: A critical review. *Microporous and Mesoporous Materials, 287*, 29–55. DOI: 10.1016/j.micromeso.2019.05.051

Saha, D., Orkoulas, G., Chen, J., & Hensley, D. K. (2017). Adsorptive separation of CO2 in sulfur-doped nanoporous carbons: Selectivity and breakthrough simulation. *Microporous and Mesoporous Materials*, *241*, 226–237. DOI: 10.1016/j.micromeso.2016.12.015

Saha, D., Thorpe, R., Van Bramer, S. E., Alexander, N., Hensley, D. K., Orkoulas, G., & Chen, J. (2018). Synthesis of Nitrogen and Sulfur Codoped Nanoporous Carbons from Algae: Role in CO_2 Separation. *ACS Omega*, *3*(12), 18592–18602. DOI: 10.1021/acsomega.8b02892

Saha, D., Van Bramer, S. E., Orkoulas, G., Ho, H. C., Chen, J., & Henley, D. K. (2017). CO2 capture in lignin-derived and nitrogen-doped hierarchical porous carbons. *Carbon*, *121*, 257–266. DOI: 10.1016/j.carbon.2017.05.088

Sarah Couck, J. C.-S.-R. S. V. der P. G. V. B. C. M. P. R. J. F. M. D. (2018). 3D-printed SAPO-34 monoliths for gas separation. *Microporous and Mesoporous Materials, 255*, 185–191.

Sarwar, A., Ali, M., Khoja, A. H., Nawar, A., Waqas, A., Liaquat, R., Naqvi, S. R., & Asjid, M. (2021). Synthesis and characterization of biomass-derived surface-modified activated carbon for enhanced CO2adsorption. *Journal of CO2 Utilization, 46*. DOI: 10.1016/j.jcou.2021.101476

Sevilla, M., Falco, C., Titirici, M.-M., & Fuertes, A. B. (2012). High-performance CO2 sorbents from algae. *RSC Advances*, *2*(33), 12792. DOI: 10.1039/c2ra22552b

Sevilla, M., & Fuertes, A. B. (2011). Sustainable porous carbons with a superior performance for CO2 capture. *Energy & Environmental Science*, *4*(5), 1765. DOI: 10.1039/c0ee00784f

Shahkarami, S., Azargohar, R., Dalai, A. K., & Soltan, J. (2015). Breakthrough CO2 adsorption in bio-based activated carbons. *Journal of Environmental Sciences (China)*, *34*, 68–76. DOI: 10.1016/j.jes.2015.03.008

Singh, G., Kim, I. Y., Lakhi, K. S., Srivastava, P., Naidu, R., & Vinu, A. (2017a). Single step synthesis of activated bio-carbons with a high surface area and their excellent CO2 adsorption capacity. *Carbon*, *116*, 448–455. DOI: 10.1016/j.carbon.2017.02.015

Singh, G., Kim, I. Y., Lakhi, K. S., Srivastava, P., Naidu, R., & Vinu, A. (2017b). Single step synthesis of activated bio-carbons with a high surface area and their excellent CO2 adsorption capacity. *Carbon*, *116*, 448–455. DOI: 10.1016/j.carbon.2017.02.015

Singh, G., Lakhi, K. S., Kim, I. Y., Kim, S., Srivastava, P., Naidu, R., & Vinu, A. (2017). Highly Efficient Method for the Synthesis of Activated Mesoporous Biocarbons with Extremely High Surface Area for High-Pressure CO2 Adsorption. *ACS Applied Materials & Interfaces*, *9*(35), 29782–29793. DOI: 10.1021/acsami.7b08797

Singh, J., Bhunia, H., & Basu, S. (2018). Synthesis of porous carbon monolith adsorbents for carbon dioxide capture: Breakthrough adsorption study. *Journal of the Taiwan Institute of Chemical Engineers*, *89*, 140–150. DOI: 10.1016/j.jtice.2018.04.031

Singh, V. K., & Anil Kumar, E. (2016). Measurement and analysis of adsorption isotherms of CO2 on activated carbon. *Applied Thermal Engineering*, *97*, 77–86. DOI: 10.1016/j.applthermaleng.2015.10.052

Song, T., Liao, J. M., Xiao, J., & Shen, L. H. (2015). Effect of micropore and mesopore structure on CO2 adsorption by activated carbons from biomass. *Xinxing Tan Cailiao. New Carbon Materials*, *30*(2), 156–166. DOI: 10.1016/S1872-5805(15)60181-0

Sreńscek-Nazzal, J., & Kiełbasa, K. (2019a). Advances in modification of commercial activated carbon for enhancement of CO2 capture. *Applied Surface Science*, *494*, 137–151. DOI: 10.1016/j.apsusc.2019.07.108

Sreńscek-Nazzal, J., & Kiełbasa, K. (2019b). Advances in modification of commercial activated carbon for enhancement of CO2 capture. *Applied Surface Science*, *494*, 137–151. DOI: 10.1016/j.apsusc.2019.07.108

Stephan Utgenannt, F. H. O. K. S. J. A. W. P. K. (2015). Control of porosity and composition of carbon-based catalysts prepared by template assisted routes. *Catalysis Today*, *249*, 38–44. DOI: 10.1016/j.cattod.2014.10.036

Stephen, J. A., & DeWitt, A. S. J. K. F. Z. M. J. R. R. P. L. (2018). Critical Comparison of Structured Contactors for Adsorption-Based Gas Separations. *Annual Review of Chemical and Biomolecular Engineering*, *9*(1), 129–152. DOI: 10.1146/annurev-chembioeng-060817-084120

Thakkar, H., Lawson, S., Rownaghi, A. A., & Rezaei, F. (2018). Development of 3D-printed polymer-zeolite composite monoliths for gas separation. *Chemical Engineering Journal*, *348*, 109–116. DOI: 10.1016/j.cej.2018.04.178

Tofail, S. A. M., Koumoulos, E. P., Bandyopadhyay, A., Bose, S., O'Donoghue, L., & Charitidis, C. (2018). Additive manufacturing: Scientific and technological challenges, market uptake and opportunities. *Materials Today*, *21*(1), 22–37. DOI: 10.1016/j.mattod.2017.07.001

Vagner, C., Finqueneisel, G., Zimny, T., Burg, P., Grzyb, B., Machnikowski, J., & Weber, J. V. (2003). Characterization of the surface properties of nitrogen-enriched carbons by inverse gas chromatography methods. *Carbon*, *41*(14), 2847–2853. DOI: 10.1016/S0008-6223(03)00412-3

Vilén, A., Laurell, P., & Vahala, R. (2022). Comparative life cycle assessment of activated carbon production from various raw materials. *Journal of Environmental Management*, *324*, 116356. Advance online publication. DOI: 10.1016/j.jenvman.2022.116356

Wang, L., & Yang, R. T. (2012). Significantly Increased CO_2 Adsorption Performance of Nanostructured Templated Carbon by Tuning Surface Area and Nitrogen Doping. *The Journal of Physical Chemistry. C, Nanomaterials and Interfaces*, *116*(1), 1099–1106. DOI: 10.1021/jp2100446

Wang, Z., Zhan, L., Ge, M., Xie, F., Wang, Y., Qiao, W., Liang, X., & Ling, L. (2011). Pith based spherical activated carbon for CO2 removal from flue gases. *Chemical Engineering Science*, *66*(22), 5504–5511. DOI: 10.1016/j.ces.2011.06.070

Wei, H., Deng, S., Hu, B., Chen, Z., Wang, B., Huang, J., & Yu, G. (2012). Granular bamboo-derived activated carbon for high CO2 adsorption: The dominant role of narrow micropores. *ChemSusChem*, *5*(12), 2354–2360. DOI: 10.1002/cssc.201200570

Wickramaratne, N. P., & Jaroniec, M. (2013). Importance of small micropores in CO_2 capture by phenolic resin-based activated carbon spheres. *Journal of Materials Chemistry. A, Materials for Energy and Sustainability*, *1*(1), 112–116. DOI: 10.1039/C2TA00388K

Wu, F. C., Tseng, R. L., & Juang, R. S. (2005). Comparisons of porous and adsorption properties of carbons activated by steam and KOH. *Journal of Colloid and Interface Science*, *283*(1), 49–56. DOI: 10.1016/j.jcis.2004.08.037

Xia, Y., Mokaya, R., Walker, G. S., & Zhu, Y. (2011). Superior CO2 Adsorption Capacity on N-doped, High-Surface-Area, Microporous Carbons Templated from Zeolite. *Advanced Energy Materials*, *1*(4), 678–683. DOI: 10.1002/aenm.201100061

Xing, W., Liu, C., Zhou, Z., Zhang, L., Zhou, J., Zhuo, S., Yan, Z., Gao, H., Wang, G., & Qiao, S. Z. (2012). Superior CO2 uptake of N-doped activated carbon through hydrogen-bonding interaction. *Energy & Environmental Science*, *5*(6), 7323. DOI: 10.1039/c2ee21653a

Yuan, B., Wang, J., Chen, Y., Wu, X., Luo, H., & Deng, S. (2016). Unprecedented performance of N-doped activated hydrothermal carbon towards C_2H_6/CH_4, CO_2/CH_4, and CO_2/H_2 separation. *Journal of Materials Chemistry. A, Materials for Energy and Sustainability*, *4*(6), 2263–2276. DOI: 10.1039/C5TA08436A

Yuan, H., Chen, J., Li, D., Chen, H., & Chen, Y. (2019). 5 Ultramicropore-rich renewable porous carbon from biomass tar with excellent adsorption capacity and selectivity for CO2 capture. *Chemical Engineering Journal*, *373*, 171–178. DOI: 10.1016/j.cej.2019.04.206

Yue, L., Xia, Q., Wang, L., Wang, L., DaCosta, H., Yang, J., & Hu, X. (2018). CO2 adsorption at nitrogen-doped carbons prepared by K2CO3 activation of urea-modified coconut shell. *Journal of Colloid and Interface Science*, *511*, 259–267. DOI: 10.1016/j.jcis.2017.09.040

Zachary, D., & Herde, R. D. G. S. J. S. T. J. S. (2020). Impact of hydrolysis on surface area and energy storage applications of activated carbons produced from corn fiber and soy hulls. *Carbon Resources Conversion*, *3*, 19–28. DOI: 10.1016/j.crcon.2019.12.002

Zhao, L., Fan, L.-Z., Zhou, M.-Q., Guan, H., Qiao, S., Antonietti, M., & Titirici, M.-M. (2010). Nitrogen-Containing Hydrothermal Carbons with Superior Performance in Supercapacitors. *Advanced Materials*, *22*(45), 5202–5206. DOI: 10.1002/adma.201002647

Chapter 12 Innovative Pavement Materials and Sustainable Engineering Practices for Next-Generation Infrastructure

Bader Eddin Alasali
https://orcid.org/0009-0004-3587-4530
National University of Science and Technology, Muscat, Oman

Mohanad Talal Alfach
https://orcid.org/0000-0002-3448-423X
School of Mathematics, Computer Science, and Engineering, City St. George's, University of London, UK

ABSTRACT

Because people rely heavily on durable surface transportation, there is now a strong demand for new pavement solutions. For decades, conventional asphalt and concrete were reliable, but now climate change, more vehicles and new regulations often make them less effective. It gives a review of new pavement engineering techniques and materials using polymer- and nano-modified asphalts, advanced concrete and recycled industrial waste. Showing what improves a material's toughness, helps the planet and increases its usefulness, we present engineers with practical choices for building strong and useful products. These materials are shown and highlight how connected and responsive building materials can transform infrastructure

DOI: 10.4018/979-8-3373-3556-8.ch012

management. A lifecycle assessment acts as an approach designed to measure the sustainability of materials. Eventually, current problems and future research directions in building sustainable pavements are covered.

1. INTRODUCTION

Due to the rapid increase in global transportation and the emergence of more serious natural and monetary problems, new solutions are required in the field of pavement engineering. Resource heavy building materials and construction methodologies are safe but pose an environmental issue and should be avoided in building future infrastructure (Huang, 2007). Consequently, professionals and players in the industry come up with solutions for building and construction that are sustainable, affordable and friendly to the environment.

Asphalt pavement material suffers some distress that requires rehabilitation. The lack of skilled manpower, the absence of proper pavement management system, budget constraint, the absence of proper machineries, and the absence of monetary and quality control are just some of the possibilities (Kenea L. M., et al, 2021). Therefore, it should be enhanced with the help of different methods. The current advances in the pavement materials that are tough and those that conserve resources such as high-performance polymers, warm mix asphalts (WMA), reclaimed asphalt pavement (RAP) and bio-based binders allow us to construct the infrastructure that is both tough and resource-saving (Giani et al., 2015; Zalghout et al., 2022). When made and delivered, these materials can minimize emission of greenhouse gases and can also last longer with minimal maintenance and repairs (Yousefi et al., 2022). With the help of smart compaction, environmentally evaluated tools and advanced monitoring aides, the data-based pavement maintenance and design process can be improved (Fahad et al., 2022).

Sustainable engineering does not hold back on the selection of appropriate materials. They implement the mechanisms of circular economy, reduce construction impact, and reuse byproducts, fly ash and steel slag (Priya et al., 2020b). Today, sustainable practices have been incorporated in all aspects of the pavement engineering process and thus is weatherproof as well as sustainable in the long run.

The chapter takes a closer look at the contemporary materials and environmentally friendly methods that define the development of the pavement systems. It summarizes the latest studies, case studies and models employed in practice to give useful ideas to the engineers, planners and policy makers to make the infrastructure safer and greener.

2. ADVANCED PAVEMENT MATERIALS

With the increase in demand for transport systems because of the growth in traffic systems and being environmentally friendly, engineers must investigate and implement new materials in pavements systems. The chapter also discusses some of the new materials that have been recently developed which are promising to be alternatives or more advanced forms of traditional binder and nanomaterials in concrete also. They include polymer-modified asphalts, nano-modified binders, fiber-reinforced concrete, geopolymers and bio-based binders.

The geotechnical test disclosed the normal density, and fine-grained composition of silt and clay predominantly. A reduction in the pavement distresses such as rutting and cracking the results demonstrate the possibility of the coir geotextiles to make pavement construction tougher and longer lasting (Kumar et al., 2024).

The studies of the multiple innovative applications of nanomaterials to the construction industry are quite intensive. One of the studies deals with nanomaterial and its use in the construction industry, nanotechnology in concrete, smart concrete materials and smart concrete self-awareness, carbon sources of nanoparticles, nanomaterials are used to enhance the strength of composite concrete, durability and sustainability measures of concrete and the implementation of nanotechnology solutions to concrete degradation, and potential environmental benefits of nanomaterials and nanotechnology application (Utsev, T., et al., 2022).

2.1 Polymer-Modified Asphalts

Polymers on asphalt have significantly enhanced the roles of flexible pavements when incorporated. Though most roads are constructed using conventional asphalt, it cannot cope with more forces and high variations in temperature nowadays. PMAs solve these issues by incorporating polymers into asphalt, which reinforces bitumen against hot and cold weather damage.

2.1.1. Types of Polymers Used in Asphalt

In asphalt modification, two common polymers are mostly used, including SBS and EVA.

Styrene-Butadiene-Styrene (SBS) is a synthetically produced material that forms a network through the layer of bitumen. Due to such network, the flexibility and the ability to tear the asphalt is improved which assists in re-forming the road bends and preventing the cracks under the heat. SBS-modified asphalt has far superior fatigue and durability when subjected to repeated loading (Yan et al., 2021).

Ethylene-Vinyl Acetate (EVA) is random copolymer which improves the rigidity of the binder and heat performance. EVA gives superior materials for withstanding heat stress and does not rut easily due to heavy traffic loads (Airey, 2002).

EVA is used to make the modified binder more viscous and stiffer although it lacks elasticity as compared to SBS.

2.1.2. Performance Advantages

Ample field and laboratory investigations have established that PMAs can enhance asphalt pavement mechanical strength immensely.

Permanent deformation characteristics: the character of PMAs at high temperatures means they are tough in high service temperature and busy intersection climates (Hunter, Self, & Read, 2015).

On a par with their elastic structure, polymer-containing binders assist in better holding stress, thereby hindering the initiation and development of fatigue cracks (Yan et al., 2021).

They are used to prevent lower temperatures in the colder regions because of the immediate elasticity of SBS-modified PMAs (Lu and Isacsson, 1997).

The FHA states that PMAs are resilient to aging and hardening that contribute to increasing the lifespan of the pavement and reducing the number of maintenance procedures (Mazumder et al., 2016).

2.1.3. Suitability in Climatic Conditions

They are appealing since they may be applied in various climatic conditions.

SBS and EVA ensure that the mixture is strong in hot regions and it does not flow or rut.

Application of SBS in low temperature areas enables the binder to bend and withstand the cracking of the material by low temperatures.

Where the climatic conditions are unpredictable, such issues as cracking and rutting are addressed with the help of PMAs.

Due to their performance under numerous climatic conditions, asphalt binders are a significant factor in the pavement design systems including Superpave, which is based on climatic-based evaluations (Yan et al., 2021).

2.1.4. Practical Considerations

Despite the obvious benefits of PMAs, there are some practical aspects that have to be addressed.

Initially, polymers increase the cost of the binder materials, however they tend to have lower maintenance and a longer lasting binder.

Among the polymers, EVA would separate out of bitumen in storage unless it is frequently stirred necessary to prevent separation.

Polymer modification affects the viscosity and mixing temperature of asphalt binder and therefore, the designers must plan the recipe of HMA carefully (Airey, 2002).

2.2. Nano-Modified Binders

Nanotechnology gives new options to pavement engineering in respect to improving the performance of asphalt binders. Between the sizes of 1-100 nanometers, materials do not act the same in terms of strength, heat/chemical reactions as compared to the same materials in normal (bulk) forms. The said features imply that nano-modified binders have the potential of making asphalt pavements tougher, more resistant and stronger to the environment.

Nano-silica (SiO2), nano-titanium dioxide (TiO2) and graphene oxide (GO) have proved to be useful as binders to enhance stiffness, less oxidation, and enhanced water resistance. They are so small that they are uniformly distributed throughout the binder and this permits variations in the microstructure along with improved contacts between asphalt and the other constituents.

2.2.1. Types of Nanomaterials in Asphalt Binders

Nanomaterials and functionalized nanomaterials are very advantageous, and their properties are better than the traditional ones. The properties are enhanced physical, chemical, mechanical and other properties that allow functionalized nanomaterials to overtake the conventional materials (Tiza et al., 2023).

Figure 1. Types of nanomaterials in asphalt binders

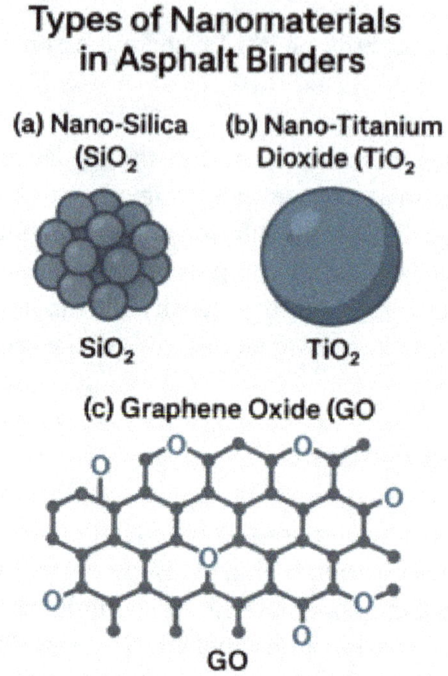

(a) Nano-Silica (SiO_2)

One of the primary nanomaterials under study in the asphalt modification area is nano-silica as depicted in figure 1. The high area of nano-silica and their powerful qualities make nano-silica to be a good additive in increasing the stiffness, viscosity of the binder and enhances its resistance to heat (Yao et al., 2012). Evenly dispersed, nano-silica enhances strength of the material to damage caused by rutting, breaking and aging.

Besides that, the application of nano-silica reduces the space between the binder matrix that closes most of the paths through which moisture and oxygen can access the binder and thereby deteriorates it, causing any form of damage (Nazari et al., 2018).

(b) Nano-Titanium Dioxide (TiO_2)

Nano-TiO2 in the form of figure 1 is a mechanical modifier, which is photocatalytic when compared to the traditional TiO2. Thus, it can be useful in minimizing surface pollution and heat capture. (H. The addition of TiO2 nanoparticles to asphalt pavement (Wang et al., 2011) serves to make the pavement less susceptible to sun exposure and slower ageing with an effect of increasing the working life of the pavement.

TiO2 can enhance its thermal stress resistance to enable the binder to enhance the ability of the road structures to withhold high-traffic pressure in hot weather (Shafabakhsh et al., 2014).

(c) Graphene Oxide (GO)

Another form of graphene bonded with oxygen, which is called graphene oxide as shown in figure 1 is receiving a lot of interest due to its robust characteristics, large surface area and solid heat conduction. With the addition of GO, the bitumen becomes much stronger, it can go back to its original form more easily and spend a lot of time in this condition (Li et al., 2022).

GO has functional groups, which makes it more compatible with bitumen and enhances the moisture resistance that is available at the contact point between binder and aggregate.

2.2.2. Performance Enhancements

It is proved by laboratory studies that nano-modified binders lead to improved performance:

- A layer of nanoparticles develops which retards the aging process of bitumen by oxidation (Yao et al., 2012).
- The growth in stiffening on use at high temperatures causes the reduction in rutting and the non-change of flexible performance at low temperatures (Nazari et al., 2018).
- The binder is resistant to moisture due to its nano properties and sticks to the aggregates while mixing very well thus it is minimally damaged during the rain.
- Due to nano-additives, the binder can recover and release pressure more effectively which allows the mix to last longer with the fatigue (H. Wang et al., 2011).

2.2.3. Dispersion and Compatibility Challenges

Even dispersion of nanoparticles in bitumen is crucial to the performance of bitumen. Combining nanomaterials in high concentrations will potentially diminish their benefits and deteriorate the performance of the binder. High shear mixing, ultrasonic and surfactants are normally used to improve dispersion.

There should be enhanced performance of bitumen through superior preservation and renovation methods which include recycled or substitute materials. It can be used in conjunction with sustainable approach using Recycled Concrete, Recycled Asphalt Pavement, Bio-binders of asphalt, asphalt cement, warm mix asphalt, tack coat, rubberized asphalt (RA) and polymer modified asphalt (PMA) (Tiza et al., 2022).

The interactions of nanomaterials with asphalt in the long run should also be taken into consideration to ensure there are high levels of benefits obtained. Surface functionalization of nanoparticles is a common strategy of improving chemical connection to the binder matrix.

2.2.4. Environmental and Economic Considerations

The technology behind nanotechnology is quite fast developing due to the excellent physio-chemical and biological properties of some nanoparticles in comparison with bulk analogs (Mondal et al., 2023). Although nanomaterials are more expensive to incorporate initially, results with respect to saving in maintenance and repair of the pavements prove to be worthwhile in the long-term. TiO2 are good nanomaterials in the environment as they serve to cool the city and degrade surface pollutants (Shafabakhsh et al., 2014). Human beings are yet to understand the design of environmentally friendly and cost-effective nanomaterials to use in huge road works.

2.3. Fiber-Reinforced Concrete

FRC is a significant addition that strengthens and enhances life cycle of traditional cementitious compositions. One of the greatest fears in using the traditional concrete is that it cannot retain tension and fractures once anything damages it. It incorporates discontinuous fibers into concrete that seal discontinuities, increase hardness and augment the capacity of the concrete to dissipate energy. When tensile or flexural stress is applied to FRC, it no longer acts like brittle concrete but becomes more ductile as the fibers prevent the development of cracks on small and large scales as the concrete reaches old age.

It looks at the various types of fiber utilized, how its quantity can be hedged, the outcome after numerous years in the industry and the instructions on applying it in both structural and non-structural applications.

2.3.1. Fiber Types and Characteristics

Figure 2. Fiber types

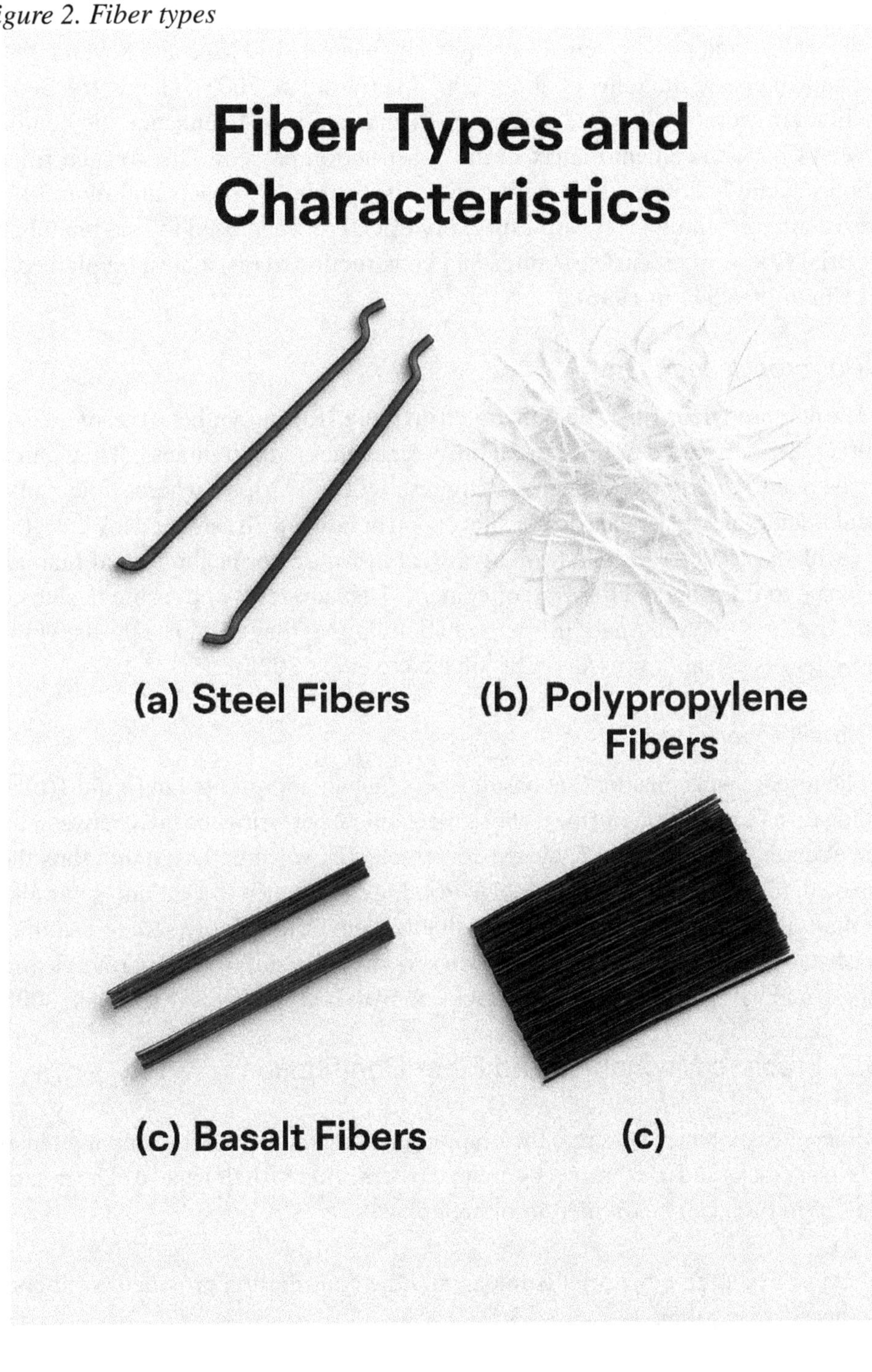

(a) Steel Fibers

The steel fibers as presented in figure 1(a) demonstrate better outcomes in the tests of strength. Typically, these materials are made using cold-drawn, cut or melt-extracted wire and possess tensile strengths between 1, 100 Mpa and beyond 2,800 Mpa and an elastic modulus of about 200 Gpa (Naaman, 2003). Geometry Geometry In their geometry, hooked and crimped ends are used to enhance the bonding between fiber and cement matrix in the construction process. Due to such fibers, composites are less prone to bending, less susceptible to impacts and more likely to resist after cracking. They are especially effective when used to line tunnels, in industrial floors, in precast structures and construction to resist blast (explained by ACI Committee 544 in 1996).

(b) Polypropylene Fibers

The polypropylene fibers demonstrated in figure 1(b) are synthetic thermoplastics and they are selected due to their stability, cheapness and lightness. They can be utilized due to their moderate tensile strength (400-700 Mpa), whereas their small modulus and water resistance do not allow good bonding with cement. However, they can regulate plastic cracking shrinkage, slice moisture permeability and maintain resistance to freezing and thawing of cement. They are most applicable in slab-on-grade, overlays, shotcrete and fire-resistant buildings as they eliminate the dangerous consequences of vapor pressure (Da Silva Neto et al., 2025).

(c) Basalt Fibers

The researchers consider that basalt fibers that are represented in figure 1(c) are coming up as a new form of fiber whose mechanical performance falls between that of steel and synthetic fiber. They are constructed of volcanic basalt and thus they possess high tensile strength (~4,800 MPa), high resistance to heat and wear alkaline materials well. They are non-corrosive and environmentally safe to use in the marine and offshore buildings. It has been shown that basalt fibers improve flexural strength and have greater control of cracks on hybrid composites (Sim et al., 2005).

2.3.2. Fracture Mechanics and Fiber Contribution

Fibers also enhance concrete by wrapping across cracks and transferring tensile loads over cracks and use the energy created by cracking to their benefit. The fracture behavior in FRC can be divided into three phases.

- At this point, the material is linear, and the contribution provided by fibers is nearly negligible.

- Fibers prevent crack development, which causes kinking.
- When cracking is achieved, toughness, ductility and load-carrying capacities are enhanced since fibers do not pull out and rupture easily.

Primarily the function of steel fibers is to restrain crack opening by mechanical bonding and fuel, but primarily polymeric fibers defend against crack expansion and enhance strength by ductile mechanisms. The big tensile strength as well as bondage to other materials within basalt fibers enables the fibers to absorb more energy (Neville, 1992).

2.3.3. Dosage Optimization

When establishing the optimal dosage of fiber, engineers must consider the intended strength, easiness of use, the cost and the requirements of the intended use of the fiber. Increase in level of fiber in self-consolidating concrete does not translate to equal increase in performance and these levels may result in concrete being harder to place and harder to segregate.

- The most common application of steel fibers is at a rate range of 20-80kg/m3, or 0.5 to 2.0 percent by volume. The mix is more flexible and stronger impact-wise at higher doses but must be flowable through the addition of superplasticizers.
- Polypropylene fibers can manage the shrinkage using low dosage (0.1-0.3 by volume). Microfibers are used in early cracks, whereas the macrofibers offer strength in crack development and healing.
- The level of basalt fiber to be utilized must be between 0.25 and 1.0 percent by volume depending on the length of the fiber and its interaction with the matrix. Dosage optimization has yet to be perfected by many people.

Hybrid fiber systems are on the increase in the use of different types of fiber within one system due to producing positive outcomes. The composite made of steel and polypropylene fibers tends to enhance the strength and restrict the amount of shrinking of the concrete, when combined (Afthroughsabet and Ozbakkaloglu, 2015).

2.3.4. Long-Term Performance and Durability

The environment directs what FRC does with time as it weakens at the fiber-matrix boundaries and loading cycles. The essential elements of working place include:

- There are chances of rusting steel fibres in Findings Stainless steel or epoxy coating prevents rust on materials (Neville, 1992).
- Polypropylene and basalt are equal in terms of their resistance to alkaline-induced changes. Scientists have found out that basalt fibres perform better than glass fibres in the presence of alkalinity (Sim et al., 2005).
- Cyclic loading and fatigue lead to reduction in strength of fiber-matrix bond. Nevertheless, the presence of fibres that are well distributed slows down the crack propagation that causes the material to be less susceptible to stress.
- Freeze-thaw tests established that polypropylene fibers are tough, whilst basalt fibers maintain their shape even in hot temperatures, which is necessary in fire-proof buildings.

According to the experience on real projects, fiber-reinforced surfaces in industries, tunnels (e.g. in Barcelona and Singapore) and airports are resistant to adverse utilization and environmental risks (Banthia & Gupta, 2006).

2.3.5. Practical Considerations in Mixing and Placing

When you use FRC then you must use standards to mix fiber scattering and fiber locations of installation. The clumping of fibers is likely to occur with poor blending, and the poor-compaction fiber bridging is not as helpful.

- Silica fume, fly ash or slag as a blend is used to enable FRC have a stronger and better-quality matrix.
- As the fibers are added, the slump or consistency of concrete decreases. The common way of enhancing the workability of a concrete mixture is by adding superplasticizers or viscosity-modifying agents.

FRC and 3D-printed fiber-reinforced concrete can be used as a self-compacting substance and enable smoother construction as well as more creativity in the design of structures.

2.3.6. Environmental and Economic Considerations

Sustainable fiber choice should look at its impact via the life of the product and its cost-efficiency of the choice. Amongst the three, polypropylene and basalt require least input energy as compared to the steel fibers. Basalt is a natural product, and it is abundant and can be recycled, which appeals to those buildings in an environmentally friendly way.

Despite the high initial costs of FRC compared to regular concrete, in the long-term, it is ultimately more economical by reducing the cracking, repairs and maintenance costs.

2.4. Geopolymers and Bio-based Binders

Due to the demand to have sustainable construction, various new binders are being developed. Geopolymers and bio-based binders are the main examples of the low-carbon and environmentally friendly options of construction materials. They last longer and perform equally as efficiently as systems that use Portland cement and provide valuable environmental and practical advantages. This chapter discusses these materials in detail, including their material background, their mechanism of operation, their disadvantages and the current use in construction.

2.4.1. Geopolymer Binders

2.4.1.1. Advanced Chemistry and Reaction Mechanisms

In geopolymer technology, alkaline activation of material that contains aluminosilicates produces an amorphous N-A-S-H or semi-crystalline C-A-S-H gel when calcium is also present. A section of the response involves:

- Silicate and aluminate are decomposed in high pH conditions.
- Formation of a three-dimensional firm structure through the unification of dissolved molecules.
- Aluminosilicate gel is polymerized with prolonged heating that ensures that it becomes the bonding material in the system (Provis, 2017).

Compared to geopolymerization, where it refers to the creation of CO2, OPC hydration relies on the formation of calcium silicate hydrate (C-S-H) gel and hence OPC hydration has lower CO2 emissions and better chemistry.

2.4.1.2. Raw Material Sources and Global Availability

The availability of source material that is easy to use and find is often a determining factor in how successful and workable geopolymers are. Most of these occurrences begin as illustrated in figure 3 and they are as follows:

Figure 3. Raw material sources and global availability

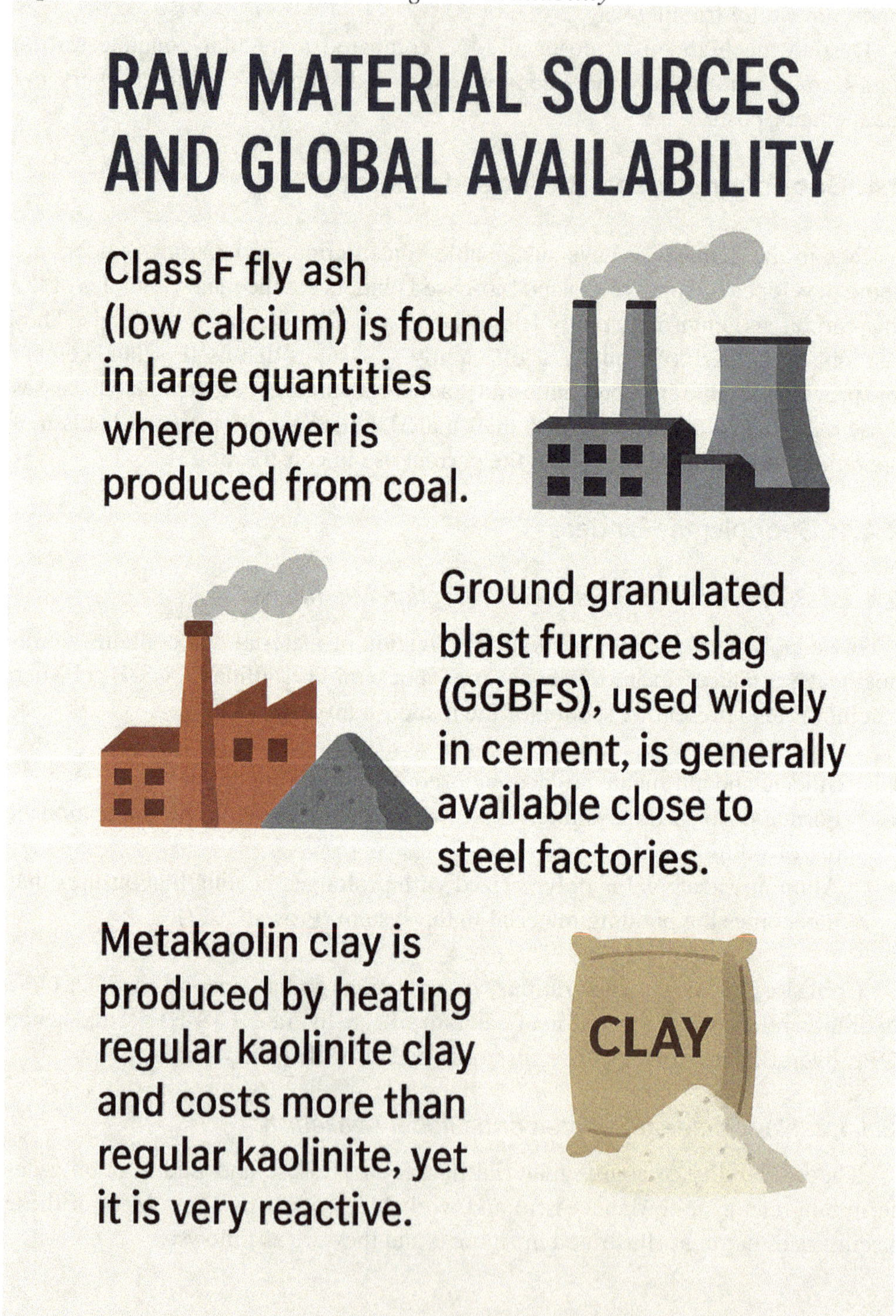

- Fly ash of Class F (low calcium) is abundant where power generation is done using coal.
- Ground granulated blast furnace slag (GGBFS) is one of the most common types of cement additives; it is typically located near steel factories.
- Metakaolin clay is prepared through heating of regular kaolinite clay and is more expensive than regular kaolinite, but it is highly reactive.

To counteract the scarcity of Chinese fly ash in Chinese economies, rice husk ash, mine tailings and red mud are being offered as alternatives in the region.

2.4.1.3. Structural and Functional Performance

It has been shown that geopolymers are significantly stronger (40-100 Mpa) than OPC, are more resistant to acid and sulfate as well as shrink less. They cannot be destroyed by high temperatures since they can survive temperatures above 800degC, which is why they are ideal in jobs that demand high heat.

- Walls or shells of the tunnel
- Industrial type flooring
- Nuclear waste is buried in seals.

Research in geopolymer concrete leads to the development of reinforced geopolymer concrete since it can bond effectively and can be used with steel and basalt reinforcement (Hardjito & Rangan, 2005). Table 1 below is an illustration of some performance aspects of structural and functional performance of geopolymer binders.

Table 1. Structural and functional performance of geopolymer binders

Performance Aspect	Description	Key Benefits	References
Compressive Strength	High early and long-term strength due to polymeric gel matrix (N-A-S-H or C-A-S-H).	Suitable for high-load applications like pavements, precast elements, and structural members.	Provis & van Deventer, 2014
Flexural Strength	Moderate to high, depending on mix design and reinforcement.	Enhance resistance to bending and cracking under dynamic loads.	Davidovits, 2015
Durability	High resistance to acid, sulfate, and chloride environments.	Ideal for marine, industrial, and wastewater environments.	Wiley, 2015
Thermal Resistance	Excellent resistance up to 1000°C without significant loss of strength.	Preferred in fire-prone structures and high-temperature zones.	Duxson et al., 2007

continued on following page

Table 1. Continued

Performance Aspect	Description	Key Benefits	References
Shrinkage and Creep	Lower shrinkage and creep compared to OPC when properly cured.	Maintains dimensional stability in structural elements.	Rangan, 2008
Setting Time	Variable—can be adjusted with activator concentration and curing conditions.	Flexibility in casting and placement processes.	Hardjito & Rangan, 2005
Workability	Dependent on mix design; improved with superplasticizers.	Enables ease of casting in complex formwork and precast operations.	Nath & Sarker, 2015
Carbon Footprint	Significantly lower than OPC due to use of industrial by-products (e.g., fly ash, slag).	Supports sustainable construction and reduction of greenhouse gas emissions.	van Deventer et al., 2012
Water Permeability	Low permeability when well-cured and compacted.	Improved durability against freeze-thaw cycles and corrosion.	Mehta & Monteiro, 2014
Adhesion to Aggregates	Strong interfacial bonding with aggregates.	Enhances composite action and overall strength of the concrete.	Temuujin et al., 2010
Alkali Silica Reaction (ASR)	Minimal risk due to absence of Portland cement.	Ideal for use with reactive aggregates in concrete mixtures.	Shi et al., 2017

2.4.1.4. Case Studies

- Infrastructure projects in Western Australia were undertaken using precast geopolymer concrete panel retaining walls and footpaths. These panels realized more than 30 percent of embodied carbon reduction over the conventional use of OPC-based concrete. The project proved to have not only positive environmental results but also high-performing results in the conditions of harsh environment, which corresponds to the sustainability requirements in construction aimed at the population.
- Pilot road projects in India using geopolymer concrete with Class F fly ash was applied on performance in terms of load bearing and cyclic loading behaviour. Findings showed that the pavement could handle severe traffic loads, which meant that it was structurally sound (to 50 MPa) and with thermal resistance in hot environments. The project presented geopolymer concrete as a resource of sustainable and long-lasting road infrastructure in coal-intensive areas.

2.4.2. Bio-based Binders

2.4.2.1 Biochemical Origins and Processing

The bio-based construction binders are based on natural polymers and resins, which are modified where necessary.

- Lignin is a by-product of pulp and paper industry; as it is constructed, lignin assists in the incorporation of mineral particles, it decreases the quantity of water required as well as an agent that enables the mixture to set.
- The process of tall oil estrification enables increased adhesion of tall oil to metal and oil particles in organic-inorganic composites.
- Microalgae like Spirulina and Chlorella contain polysaccharides and proteins that assist in the formation of a gel by the mixtures of mineral powder.

The concept of using waste materials and natural fibers to enhance the quality of soil is advantageous, as they are cheap, locally available as well as Ecological. Natural Fiber (coconut coir) was found to induce a stabilizing effect on the properties of the soil (Kaushik and Singh, 2021). It is also found that bamboo natural fibre and sheet wool has the potency to be used as a binder in a sustainable manner (Tiza et al., 2021).

2.4.2.2. Hybridization with Mineral Binders

Bio-based binders are not very frequent in load-bearing structures since they lack good compressive characteristics (less than 10 MPa). However, hybrid systems prove to be quite effective in case of the introduction of geopolymers or other pozzolanic materials.

For instance:

- Ding et al. (2016) confirm that the strength of lignin slag blends may reach 25-30 Mpa at a 28 day.
- Biochar-cement composites are associated with the trapping of carbon and enhancing thermal resistance.
- The process of carbonate hardening of ecological bricks is made possible using microalgae-lime systems.

Biodegradable Properties and Special Uses:

These systems are implemented in situations when the ecological safety or when composability is needed.

- Parking area and eating spots and shaded parking spots are short-term.
- Stopping soil erosion on the land that is susceptible to landslides.
- Structures with low impact on safeguarded ecosystems.

LEED, BREEAM and Living Building Challenge begin to consider them as a potential green building since their environmental impact is less than zero.

2.4.3. Performance, Durability, and Aging

Due to the increased attention given to sustainable construction materials, researchers are exploring the various binders which will not have a significant effect on the environment compared to ordinary cement. Improved performance, enhanced durability and a more friendly environmental effect has led to the rise of geopolymers and bio-based binders as a good alternative to Ordinary Portland Cement as replacements. To make materials last long and perform efficiently over the years it is vital to comprehend their behavior as they age, respond to stress within the environment and perform in real life situations.

2.4.3.1. Performance and Aging Characteristics (as presented in figure 4)

Figure 4. Performance and aging characteristics of bio-based binders

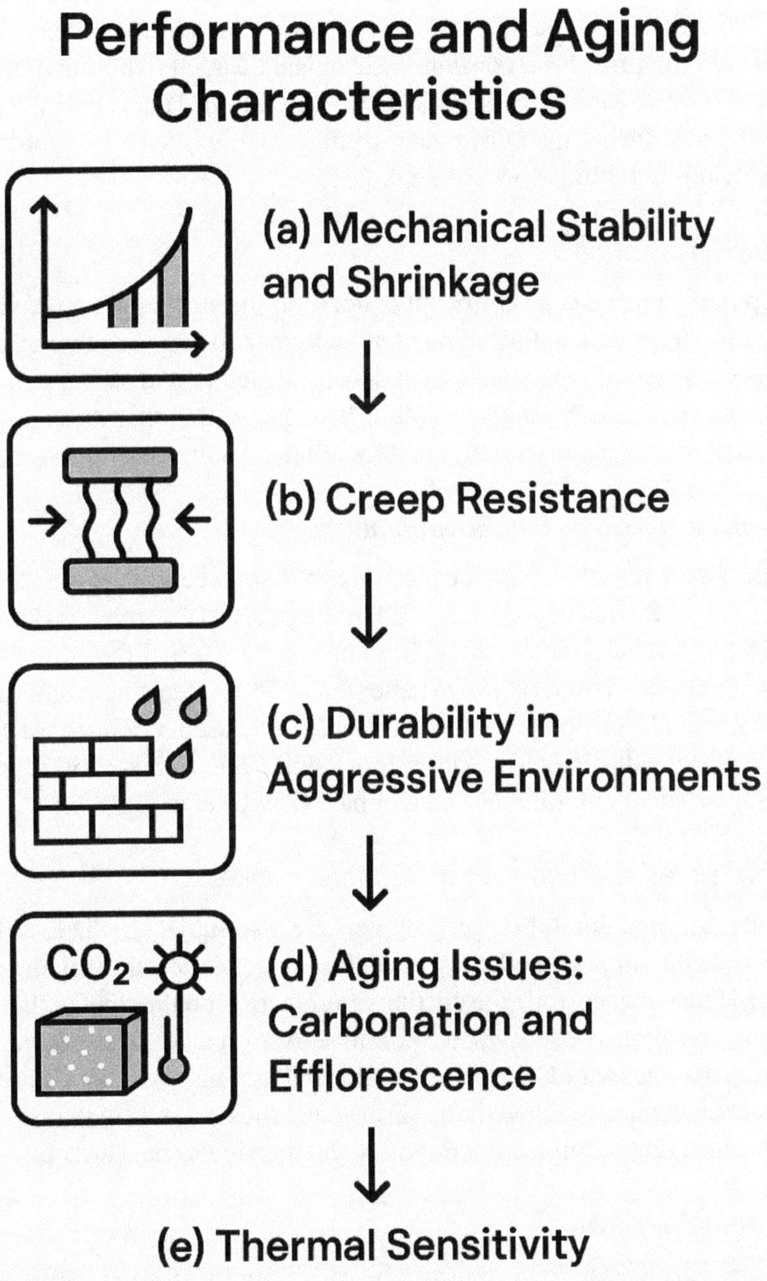

(a) Mechanical Stability and Shrinkage

With the help of alkaline treatment, fly ash, metakaolin and GGBFS can also create geopolymers based on the three-dimensional network of the Si-O-Al bonds. The network provides the cement with superior compressive strength at a tender age and superior thermal qualities. Long-term experiments have indicated that properly prepared mixtures have constant mechanical characteristics, and compressive strength of over 60 Mpa (Provis & van Deventer, 2014). This is why the movement of geopolymers during the curing and drying process is virtually zero relative to that of calcium rich binders (Wiley, n.d. 2015).

(b) Creep Resistance

Creep performance is a significant concern in the structural application of materials. Time-dependent deformation of geopolymers is less than that of OPC based materials since geopolymers have high density structure and strong gels that resist longer loads (Hardjito & Rangan, 2005). It is due to this that these polymers are usually used in pavements and bridge decks that are subject to long-term loading.

(c) Durability in Aggressive Environments

Geopolymers have been appreciated due to their excellent resistance to chemical degradation. Such cement lacks Portlandite (Ca(OH)2) which makes them less susceptible to sulfates and leaching. Temuujin et al. (2010) have found that their donor nanoparticle systems perform better than OPC systems in strong sulfuric and hydrochloric acids. Geopolymers are favorable to be used in reinforced concrete in marine water or salt-removal solution-prone areas due to their ability to generate low passages where chloride ions cannot pass (Zivica et al., 2014).

(d) Aging Issues: Carbonation and Efflorescence

Despite the good durability, geopolymers are also subject to the loss of strength because of carbonation and efflorescence. Filling concrete carbonated cans can reduce the pH that may corrode the coating of protection on the steel within it. This is more pronounced on systems where sodium activation takes place as they are more sensitive to the release of CO2 (Wang & Ma, 2018). This is due to efflorescence as the unreacted alkali is dragged to the surface and thus it reacts with carbon dioxide or moisture leading to unpleasant deposits and damage to the surface.

(e) Thermal Sensitivity

Thermal curing results in improved early strengths of geopolymers produced using low-calcium fly ash faster. However, mass or ambient-cured concrete is difficult to use with such groups of peroxides. Geopolymers based on slag or high calcium content that are hybrid have a high promise of providing desirable strength at ambient conditions (Wallah and Rangan, 2006).

2.4.4. Life Cycle, Standards, and Market Integration

The shift to sustainable materials in building has prompted people to seek some new solutions, and both geopolymers and bio-based binders are currently seen as alternatives to OPC. Being produced using industrial or natural resources, these materials are a potentially good option in terms of infrastructure that will allow reducing the impact of the sector on the environment and maintaining a high level of performance. Meanwhile, an increased uptake depends on the technical characteristics, any environmental consideration, application of engineering specifications and a successful market introduction of building materials.

(a) Life Cycle Assessment (LCA)

LCA encompasses the entire environmental life of a material including the source and the end of life when the material is scrapped up. LCA is significant in the determination of the sustainability of geopolymers and bio-based binders.

(b) Geopolymer Binders

With geopolymers, a greater benefit is greenhouse gas emissions; this has been largely since limestone calcination is not involved in the procedure, and the process is carried out at lower temperatures. Geopolymers made using fly ash or slag conserve up to 80 percent of carbon emission when the alkali activators are made using low carbon emission activators (Turner and Collins, 2013). Moreover, geopolymers have a long-life cycle as a value to builders, which is in part since they are resistant to chemical as well as other harmful agents.

Nonetheless, changing the kind and quantity of activators may have an impact on the environment. Given that the two substances are likely to be used as activators: sodium silicate and sodium hydroxide, the energy represented by the two can influence the overall environmental impact of the net life cycle. Continuous sustainability outcomes will be attained by enhancing the supply chain of both precursors and activators.

(c) Bio-based Binders

We can produce materials with no or reduced carbon content using agricultural or forestry waste by making binders to substitute wood-based materials. Techniques used to produce polymer, or cementitious materials are generally less energy-consuming and less temperature-demanding when compared to other methods. However, bio-based binders can only be sustainable when the source of the biomass is sustainable, the use of biomass does not cause harm to the land, and the process of processing is not harmful to the environment (Elahi et al., 2020).

Nevertheless, there are instances of added chemical processes to ensure that the particles are not destabilised thus the environmental advantage can be compromised. Bio-based binders help a lot in improving sustainability by incorporating the fundamentals of a circular economy like regenerative recycling.

(d) Standards and Regulatory Framework

Standardization must occur before several construction companies move towards the use of new materials. Although the much more popular systems are covered with the help of developed codes, other binders are not yet completely integrated into the rules.

(e) Geopolymer Standards

The world is yet to reach a consensus on geopolymers performance criteria. Even though initial recommendations are given in RILEM TC 247-DTA and ASTM C1709, the current codes do not typically acknowledge alkali-activated binders. Currently, performance-based standards should be employed since materials and activators are highly different (Provis, 2018).

In places like Australia and parts of Europe, there are project-specific standards which have resulted in the application of geopolymer concrete to infrastructure projects like the newest rules of VicRoads on geopolymer on precast bridges. However, it should have a globally accepted standard to assist in trading and global regulatory acceptance.

(f) Bio-based Binder Standards

Converting bio-based binders to a standard has not been as successful as it is in the case of bio-based fuels. The current standards including ISO 16620 and ASTM D6866 tend to consider bio-based content and biodegradability rather than the behavior of a material with stress. We need to establish criteria on the way mechanical, thermal and chemical characteristics are to be tested on materials applied in construction.

Testing methods of advanced concrete should include biodegradation kinetics, hydrophobicity, and UV radiation resistance and fire behavior reaction because they do not exist in more basic testing standards (Arias et al., 2021). A connection between the work of material scientists and environmental organizations and authorities will be required to translate such requirements into guidelines on engineering industries.

(g) Applications

Instead, there is an introduction of geopolymers and bio-based binders into construction using what they are good at:

- Geopolymer technology is used to make precast bridge components, runways, sewer lining materials and concrete which has enhanced chemical resistance.
- Research wall panels, acoustic tiles, formwork and interior layers that are not only healthy for the health of people but also biodegradable.
- Hybrid systems are combinations of geopolymer matrices and bio-based fillers which enhance sustainability and introduce a wide range of functions.

3. SUSTAINABLE INTEGRATION OF MATERIALS

Better utilization of materials on pavements will aid in saving on environment, consumption of fewer resources and production of stronger roads that can be in use over many years. This chapter discusses three key elements of sustainable material integration, which are re-using Reclaimed Asphalt Pavement (RAP) and ensuring that this is effective, waste-based components in the production of asphalt and concrete, and circular economy approaches to make these pavements more recycleable, reusable, and durable. The chapter provides examples of material science and mix design techniques along with sustainable building methods to explain how they find application to create future-oriented sustainable infrastructure.

3.1. RAP and Rejuvenation

RAP is currently being implemented as an environmentalist method of ensuring that resources are not manufactured new and that the overall cost of the roadway costs is reduced. RAP is created by recycling old binder and aggregate that was used in the previous pavement and then it is reintroduced to asphalt.

The process of mixing and selecting an appropriate H.E.L meeting Rejuvenator:

YOLOv3 can be applied to detect and localize the examples of concrete cracks and fatigue indicators in the image and video frames taken in the different urban environment (U. Archana, et al., 2023). The principal problem with RAP integration is restoring the properties of the aging binder to maintain its performance in the mixture. When RAP increases, there is often the result of the concrete being stiffer and of lower fatigue-resistance strength. Rejuvenators chemicals which soften the olden binder are added in place of this. Using these substances, such as bio-oils, tall oils and other engineered petroleum products, the missing maltene fractions of oxidation are compensated (Xu et al., 2014).

Each batch should be tested to ensure the compatibility between virgin and RAP bitumen to mix the bitumen properly. Performance-grade (PG) blending charts, Fourier-transform infrared spectroscopy (FTIR) and a dynamic shear rheometer (DSR) may be used as the best evaluation methods to determine rejuvenator efficacy. Recent research is beginning to employ nano-rejuvenators and reactive polymers, which fundamentally chemically revive aging of bitumen (Arafat et al., 2023).

Emulsification is crucial as well as the In-Place Recycling of potable water.

With the help of hot in-place recycling (HIPR) and cold central plant recycling (CCPR), transportation pollution will decrease, and the resources will be less. Technologies have been implemented to help in making asphalt easier to handle and make the rejuvenators more uniform in the jobsites. The emulsions prepared to certain viscosity and cure rates allow improved compaction on the field and improved quality in the finished pavement (Z. Liu et al., 2022).

3.2. Waste-Based Additives

Consumption of wastage-based resources assists with constructing more sustainable pavements and lessens the ecological issues relating to landfill waste and the utilization of natural materials.

Ground Tire Rubber (GTR):

Most asphalt manufacturers apply GSR to enhance elasticity, reduce the noise of the extra road and reinforce their pavements against cracks and grooves. The bonding of the asphalt is altered by the crumb rubber, which swells the asphalt and bonds with it forming a network within the asphalt. Consequently, the viscoelastic range of asphalt is extended (Jones et al., 2017). Wet and dry method of GTR lead to diverse varying mixing temperature and operation.

Post-Consumer Polymers and R-Cycle Polymers are made of plastic Pellets. Polyethylene (PE) and polypropylene (PP) are post-consumer plastic materials under evaluation to determine their ability to make asphalt mixtures stronger and tougher. To substitute part of the binder employed, pelletized formulas may be added to your mix. Environmental impacts need to be screened since the issue of microplastic leakages and emissions should be researched during the processing as a part of the life-cycle assessment (Awad, 2017).

Construction and Demolition Waste: Processed Collection:

Other pavements like the crushed concrete and masonry are constructed using C&D waste that has undergone processing. Well assessed materials in terms of size, strength, and contamination can be utilized to mitigate fresh aggregate consumption without affecting the structure (Toka & Olgun, 2023).

3.3. Circular Economy Principles

With the application of a circular economy, the usage of resources is repeated numerous times, and it leads to a reduced level of waste and a lot of extended product life. Due to this reason, the choice of pavement materials and planning of the system alters significantly.

(i) Design for Recycling and Reusability

All decisions concerning the building materials must be taken keeping in mind the possibility of recycling at the end of their life. This will entail the use of friendly binders and aggregates which are recyclable, minimization of use of complex components which complicate separation of the products and making the products modular so that they can easily be disassembled (Kanoungo et al., 2020). Design-for-recycling that can be broadly utilized can be seen by geopolymer treatment of the base layer or interlocking concrete blocks.

(ii) Modular Road Systems

Precast slabs and interlocking pavers reduce the amount of time to install, reduce the disturbance of the site and it is much easier to repair or replace. They are most extensively used in cities and temporary building works. The fact that they are designed to be dismantled means that they can be reused and suited well to the current environmental objectives.

(iii) Waste Minimization During Construction

When digital material tracking tools are implemented on the lean construction strategies, additional materials are less frequently utilized, and procurement becomes increasingly precise. The introduction of BIM technology to material passports creates opportunities of improved inventory of stores, re-using what they have and limiting the quantity of waste materials generated in the process of building.

4. INNOVATIVE CONSTRUCTION PROCESS AND SELF-HEALING PAVEMENTS

The construction and materials are evolving leading to new changes in the way pavement engineering is performed. The three developments as discussed in this chapter are to enhance the lifespan, environmental friendliness and performance of transportation infrastructure like dicing, electrically active pavements to be used in cold conditions, self-healing systems in asphalt and MEMS used to monitor critical road structures. The technological aid in making equipment lasts longer, less maintenance checkups and increase safety. The examples and scientific works show how these technologies may be implemented and what issues they face and how to make them successful.

4.1 Deicing and Electrically Active Pavements

Using chemicals to thaw frozen roads is being eliminated in cold areas by laying pavement heating systems. The pavement type of solar energy system incorporates such materials as carbon fillers, steel fibers and conductive plastics that enable heat generation through Joule heating any time an electric current is transferred into the system. Some of the processes are as follows:

(a) Conductive Fillers and Heating Mechanism

Under the functional principle, carbon nanotubes, graphite or carbon black are incorporated into the mixture of asphalt or concrete. These fillers get electrified and produce sufficient heat to melt snow and inhibit the formation of ice. The main advantage is that it would decrease the usage of salt-based deicers, as they are damaging the pavement and they are damaging the environment (Garcia et al., 2009).

There is also the use of fiber-reinforced conductive composite that have made the materials to be stronger as well as more effective in heat dissipation. The addition of steel or carbon fibers leads to the enhanced electric conductivity and strength of the material without cracking (Xie et al., 2020).

(b) Benefits and Implementation

The roads are also made safe during winter due to these pavements, the number of closures is also minimized, and the issue of rusting vehicles and other infrastructures is also minimized. Although they are still more expensive, electric vehicles are more economical in terms of maintenance and lead to less environmental issues in the long term.

4.2 Self-Healing Asphalt

The issue of resistance to cracking due to fatigue, heating and cooling and corrosion is a significant one in the planning of flexible pavements. Self-healing asphalt technologies are designed to improve their integrity by encouraging repairs on a small-scale structure. The following are some types:

(a) Microcapsules of Rejuvenators

One of the approaches is to add microcapsules containing rejuvenating agents typically oils or polymers to asphalt binder. The coating cracks lead to the bursting of the capsules releasing the rejuvenator to the crack location. This renders the material more flexible and slows down the additional cracking (Su et al., 2016).

(b) Induction Heating and Steel Fibers

Today, infrastructure monitoring involves a significant amount of data, and embedded sensors provide fundamental facts regarding the pavement condition and functionality. These sensors, in particular MEMS-based ones, are tiny, tough and can detect stress, strain, temperature and moisture simultaneously.

Numerous varieties of sensors exist including:

- Strain gauges are used to measure mechanical deformation that occurs on roads because of vehicles.
- Pavement internal temperatures are monitored using thermistors.
- In the capacitive moisture sensors, water content in the soil could be determined under the surface.
- The accelerometers detect abrupt acceleration brought about by the weight of the vehicle.
- Such sensors can be embedded in pavement as new roads are being laid or fitted to existing roads later and they comprise part of an automated pavement monitoring system (Wang et al., 2021).

4.3.1. Implementation Challenges

Even with their usefulness in positioning, practical problems for sensors include keeping them reliable in rough conditions, the tendency for their alignment to change as time goes on and analyzing extensive datasets. Recently, improvements in how data is transmitted, collected using harvested energy and analyzed using advanced algorithms are solving these issues (Zhu et al., 2020).

5. LIFE CYCLE ASSESSMENT AND ENVIRONMENTAL IMPACT

Planning sustainable infrastructure requires creating detailed systems to evaluate how various materials, building methods and maintenance solutions will influence the environment years into the future. The quantitative basis for sustainable pavement engineering comes from Life Cycle Assessment (LCA) and Life Cycle Cost Analysis (LCCA). The chapter investigates standards in LCA such as ISO 14040/44 and EN 15804, talks about setting system boundaries and deciding on a functional unit and analyzes the balance between spending resources and lasting cost savings. How aligning environmental impact with economic sustainability can be seen in actual cases and applied in practice is discussed.

5.1 Frameworks and Metrics

ISO 14040 and ISO 14044 Standards

No LCIA process can exist without following ISO 14040 and ISO 14044. It is defined by these frameworks that LCA is a method for checking environmental effects and problems related to a product, process or service throughout its whole life cycle (ISO, 2006a; ISO, 2006b). Four stages make up the main methods in this approach.

- In this step, the purpose, systems boundaries and main unit (for example, a five-decade lane-kilometer of road) are defined.
- LCI records and measures all the raw materials, energy used, emissions produced, and waste created.
- LCIA converts inventory information into environmental consequences such as GWP, acidification and depletion of resources.
- The results are interpreted by considering potential uncertainties and assumptions that affect decisions.
    - This way of working guarantees that assessments are fair, repeatable and consistent with others.

EN 15804 Standard

EN 15804 is a European standard for assessing the environmental performance of construction products by making sure rules are the same for every Environmental Product Declaration (EPD). Pavement materials are required by EN 15804 to consider various impacts such as ozone depletion, eutrophication and resource depletion, using modules for assessment during manufacturing (A1–A3), installation (A4–A5), use (B1–B7) and disposal/recycling (C1–C4) (CEN, 2012).

Because of this, practitioners can separately look at each phase or the overall impact of a product or service.

Functional Unit and System Boundaries

The functional unit tells us what service the system provides in measurable terms. Researchers in pavement LCA use "1 lane-kilometer of roadway over 50 years of service life" to compare different pavement designs and materials (Santero et al., 2011).

What processes are included in the analysis is decided by the location of the system boundaries. A cradle-to-grave boundary is defined by the activities of extracting raw materials, moving them, manufacturing, constructing, emitting them in use, keeping them running and demolition. Some companies use a cradle-to-gate scope for product-level checks but, since the assessment stops at the factory gate, it does not fully show how infrastructure operates.

5.2 Economic Trade-Offs and Life Cycle Cost Analysis (LCCA)

Project bids usually look at first expenses, though this focus often does not include how well and long something will function or last. LCCA helps review whether an economic system is sustainable by including all the expenses related to a product, throughout its life span.

The first phase of construction.
Taking care of and repairing roads
The cost caused by slow users
What the salvage option would give

(a) LCCA Methodology

LCCA typically uses Net Present Value (NPV) calculations to discount future costs to present-day equivalents. A typical formula is:

Where:

- NPV: Net Present Value
- Ct: Cost in year t
- r: Discount rate
- n: Analysis period

The model allows for the comparison of traditional pavements with other types such as asphalt with a high proportion of recycled materials, geopolymer concrete or fiber-reinforced composite pavements (Walls & Smith, 1998).

(b) Trade-Off Examples

- Even though polymer-modified and self-healing asphalts are more expensive upfront, they save maintenance cost over time because they last much longer than standard asphalt (based on Koch et al., 2019).
- Even so, Reclaimed Asphalt Pavement (RAP) helps cut the costs and can ease environmental pressure, but inappropriate rejuvenator uses without careful consideration through LCCA and LCA can create premature problems.

(c) Integration of Environmental and Economic Assessment

Today, experts suggest assessing environmental performance through combinations of LCA and LCCA. The FHWA has developed the INVEST framework which considers cost and the environmental impact to direct sustainable highway project development. MCDA tools can consider both technical, economic and environmental factors.

5.3 Applications in Pavement Engineering

Many LCA tools can be found for pavement professionals to use these days.

- UC Berkeley created PaLATE which measures the environmental and financial effects of several pavement designs.
- The ATHENA Pavement LCA Tool caters to Canadian situations and breaks down emissions into greater detail.
- GREET Model: Formulated by Argonne National Laboratory for the purpose of transport emissions.

The case studies are as follows:

- A Texas DOT study reports that WMA produces 15–20% less GHG emissions than HMA and costs 10% less (Aurangzeb et al., 2014).
- An EU-funded investigation found that geopolymer concrete roads produced a 30% lower carbon footprint than OPC-based roads on a 50-year basis (Habert et al., 2011).

6. CHALLENGES, OPPORTUNITIES AND THE FUTURE PATH

When sustainability, circular economy solutions and digitalization are introduced in pavement engineering, the field must handle pressing challenges and good chances. In this chapter, the main obstacles to wider use of innovative materials and technologies in pavement construction, including problems related to technology, rules and money, are explored. In addition, it discusses new research areas in artificial intelligence (AI), bio-based materials and digital twin systems and finishes by suggesting a plan for working together between academia, industry and government. Some technical challenges are presented in figure 5.

Figure 5. Understanding challenges in adopting sustainable pavement materials for construction

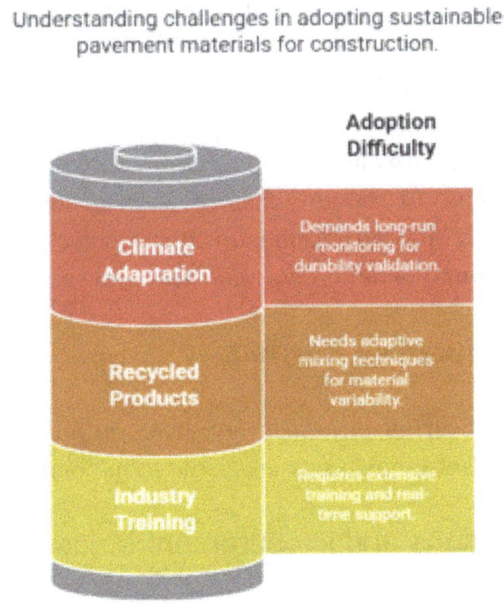

6.1 Technical Challenges

Scaling up and ensuring long-term performance for using geopolymer, recycled asphalt and bio-based additives in pavement is not straightforward because of several technological problems.

(a) Climate Adaptation and Durability Validation

Such materials must resist being affected by changes in temperature and sunlight in both frigid and sun-blasted regions. The early understanding we get from lab simulations is limited by the fact that validation over large areas is still rare (Chen et al., 2020). We need well-defined, long-run monitoring systems to determine properties, aging trends and degradation mechanisms in genuine working environments.

(b) Quality Assurance of Recycled Products

Asphalt made with Reclaimed Asphalt Pavement (RAP) or Ground Tire Rubber (GTR) is not uniform, so it can affect the composition and results of the mix (Zaumanis et al., 2015). Contractors should design mixing techniques that adjust to the kind of material used in different areas and the expected level of traffic on the road.

(c) Industry Familiarity and Training

Difficulty facing contractors and field engineers comes from not understanding the properties and methods of applying these new materials and methods. We then need to organize specially tailored training, update construction guides and give real-time technical help to boost the industry's confidence and skills (Huang et al., 2007).

6.2 Policy and Regulatory Barriers

Usually, the rules covering pavement construction are not updated with new technologies, so important innovations take time to happen.

Outdated Specifications and Procurement Models

There are still many transportation agencies working with specifications that restrict what can be used from recycling or that do not accept certain types of binders (Van Dam et al., 2015). In addition, using lowest-bid models reduces the use of sustainable materials that, while costly at first, are beneficial over time.

Pathways to Reform

1. By using performance specifications instead of prescriptive ones, we allow for innovation and safety.

2. Pilot projects are more achievable when funding models exist to fund research on big, test highways and pilot programs which deliver field results, lower perceived risk and encourage more general acceptance.
3. The European Commission identified (2021) that lifecycle-based public procurement methods with attention to environmental impacts are vital for meeting climate goals.

6.3 Research Needs and Future Directions

Using new technologies and combining different fields offers a lot of potential, but these efforts require strong research backing, joint growth and practical application.

Figure 6. Research needs and future directions

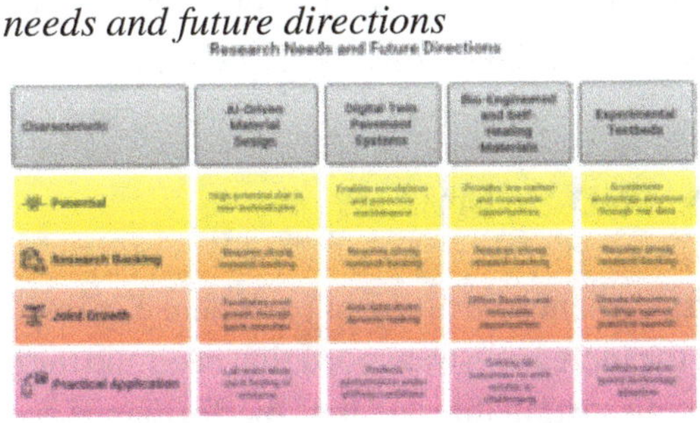

(a) AI-Driven Material Design

AI and machine learning now allow scientists to predict mechanical properties, age of materials and environmental impact of materials depending on their basic ingredients. Lab tests involving AI allow for quick search and finding of mixtures with increased durability, reduced expense and more sustainable properties (Zhang et al., 2021).

(b) Digital Twin Pavement Systems

Thanks to digital twins, it's possible to run simulations, watch for issues and carry out predictive maintenance for pavements. By bringing together IoT sensors, finite element models and machine learning, digital twins aid in using data to make decisions and predict performance when roads face shifting traffic and climate.

(c) Bio-Engineered and Self-Healing Materials

More and more, bio-based polymers, calcite precipitation guided by bacteria and microcapsule-based self-healing are being used in materials science. They provide low-carbon, flexible and possibly renewable opportunities. Even so, getting outcomes found in the lab to work reliably in the field is still a major challenge (De Belie et al., 2018).

(d) Experimental Testbeds

Larger experimental highways with built-in sensors which experience real traffic flow, help gather important data to speed technology adoption. These testbeds make it possible to check laboratory findings against practical construction, upkeep costs and what maintenance is needed.

7. CONCLUSION AND SUMMARY

The current chapter presents the future of pavement engineering in terms of adaptation to the reality of the increasing transportation need and the concerns towards the environment by adopting sustainable and high-performance construction materials. The problem of existing conventional materials is that they have low durability and cause high environmental expenditure, which led to the introduction of new and advanced alternatives (polymer-modified asphalt: PMAs; nano-modified binders; fiber-reinforced concrete: FRC). The PMAs are used in enhancing cracking and deformation resistance and on the other hand, nano/additives are used in enhancing stiffness, delaying the aging process, as well as their pollutant reduction properties. Although more expensive to install, these materials earned more economically and ecologically beneficial long-term benefits since they prolong the pavement life and decrease maintenance.

FRC is important in concrete innovation, increasing crack resistance, ductility, and energy absorption. Choice of fibers to use steel, polypropylene or basalt is based on the depending on application requirements and environment. The hybrid systems and accurate doses maximize functionality and workability. Although FRC appears to be highly expensive at first glance, it is more economical in the long term with increased lifespan and decreased repair expenses. Geopolymers and bio-based binders fall under complementary innovations, and these are eco-friendly substitutes for cement, and they use the waste and biomass and reduce carbon emissions.

Another circular economy option highlighted in the chapter is the reuse of materials to reuse them and cut down on emissions by adding reclaimed asphalt pavement, recycled plastics, and tire rubber. On the one hand, smarter, low-maintenance infrastructure assists with technological innovations such as self-healing asphalt, embedded sensors and electrically activating pavements. Such frameworks as Life Cycle Assessment (LCA) and Life Cycle Cost Analysis (LCCA) are used in making sustainable decisions.

There is still some way to go like inadequate policy support, technological barriers, and the reluctance of the industry. The challenge of overcoming these obstacles involves regulating changes, developing a workforce and pilot projects. In the future, the next-generation infrastructure may be possible due to AI-based designing, digital twins, and bioengineered materials. The chapter culminates with the thought that all stakeholders need to have a collective strategy to achieve a resilient, environmentally conscious, and a ready transportation infrastructure to face the 21 st century.

REFERENCES

Airey, G. D. (2002). Rheological evaluation of ethylene vinyl acetate polymer modified bitumens. *Construction & Building Materials, 16*(8), 473–487. DOI: 10.1016/S0950-0618(02)00103-4

Arafat, S., Wasiuddin, N. M., & Mohammad, L. N. (2023). Evaluation of bio-based and petroleum-based rejuvenator based on cracking susceptibility of hot mix asphalt with high RAP content. *Construction & Building Materials, 371*, 130725. DOI: 10.1016/j.conbuildmat.2023.130725

Archana, U., Sharma, S., & Singh, S. K. S. R, S. P and T. H. Babu, "Analysis Of Concrete Cracks And Fatigue In Smart Cities Using Yolov3," *2023 International Conference on Research Methodologies in Knowledge Management, Artificial Intelligence and Telecommunication Engineering (RMKMATE)*, Chennai, India, 2023, pp. 1-6, DOI: 10.1109/RMKMATE59243.2023.10368774

Arias, A., González-Rodríguez, S., Barros, M. V., Salvador, R., De Francisco, A. C., Piekarski, C. M., & Moreira, M. T. (2021). Recent developments in bio-based adhesives from renewable natural resources. *Journal of Cleaner Production, 314*, 127892. DOI: 10.1016/j.jclepro.2021.127892

Awad, A. (2017). UTILIZATION OF WASTE PLASTICS TO ENHANCE THE PERFORMANCE OF MODIFIED HOT MIX ASPHALT. *International Journal of Geomate, 13*(40). Advance online publication. DOI: 10.21660/2017.40.170603

Banthia, N., & Gupta, R. (2006). Influence of polypropylene fiber geometry on plastic shrinkage cracking in concrete. *Cement and Concrete Research, 36*(7), 1263–1267. DOI: 10.1016/j.cemconres.2006.01.010

Da Silva Neto, J. T., & Soares, P. R. R. (2025). Fiber-reinforced cementitious composites: Recent advances and future perspectives on key properties for high-performance design. *Deleted Journal, 2*(1), 65. Advance online publication. DOI: 10.1007/s44290-025-00209-9

De Belie, N., Gruyaert, E., Al-Tabbaa, A., Antonaci, P., Baera, C., Bajare, D., & Jonkers, H. M. (2018). A review of self-healing concrete for damage management of structures. *Advanced Materials Interfaces, 5*(17), 1800074.

Elahi, M. M. A., Hossain, M. M., Karim, M. R., Zain, M. F. M., & Shearer, C. (2020). A review on alkali-activated binders: Materials composition and fresh properties of concrete. *Construction & Building Materials, 260*, 119788. DOI: 10.1016/j.conbuildmat.2020.119788

Fahad, M., Nagy, R., & Gosztola, D. (2022). Pavement Sensing Systems: Literature Review. *Civil and Environmental Engineering*, *18*(2), 603–630. DOI: 10.2478/cee-2022-0057

García, Á., Schlangen, E., Van De Ven, M., & Liu, Q. (2009). Electrical conductivity of asphalt mortar containing conductive fibers and fillers. *Construction & Building Materials*, *23*(10), 3175–3181. DOI: 10.1016/j.conbuildmat.2009.06.014

Giani, M. I., Dotelli, G., Brandini, N., & Zampori, L. (2015). Comparative life cycle assessment of asphalt pavements using reclaimed asphalt, warm mix technology and cold in-place recycling. *Resources, Conservation and Recycling*, *104*, 224–238. DOI: 10.1016/j.resconrec.2015.08.006

Giani, M. I., Dotelli, G., Brandini, N., & Zampori, L. (2015). Comparative life cycle assessment of asphalt pavements using reclaimed asphalt, warm mix technology and cold in-place recycling. *104*, 224–238. DOI: 10.1016/j.resconrec.2015.08.006

Habert, G., De Lacaillerie, J. D., & Roussel, N. (2011). An environmental evaluation of geopolymer based concrete production: Reviewing current research trends. *Journal of Cleaner Production*, *19*(11), 1229–1238. DOI: 10.1016/j.jclepro.2011.03.012

Hardjito, D., & Rangan, B. V. (2005). Development and properties of low-calcium fly ash-based geopolymer concrete (Research Report GC1). Faculty of Engineering, Curtin University of Technology.

Huang, Y., Bird, R. N., & Heidrich, O. (2007). A review of the use of recycled solid waste materials in asphalt pavements. *Resources, Conservation and Recycling*, *52*(1), 58–73. DOI: 10.1016/j.resconrec.2007.02.002

Hunter, R. N., Self, A., & Read, J. (2015). *The Shell Bitumen Handbook* (6th ed.). ICE Publishing.

Jones, D., Liang, Y., & Harvey, J. (2017, September 1). *Performance based specifications: Literature review on increasing crumb rubber usage by adding small amounts of crumb rubber modifier in hot mix asphalt*. https://escholarship.org/uc/item/9cw18024

Kanoungo, S., Sharma, U., & Singh, S. (2020). Assessment of treatment methods of recycled aggregates for utilization in flexible pavements. *Materials Today: Proceedings*, *43*, 1320–1324. DOI: 10.1016/j.matpr.2020.09.119

Kaushik, D., & Singh, S. K. (2021). Use of coir fiber and analysis of geotechnical properties of soil. *Materials Today: Proceedings*, *47*, 4418–4422. DOI: 10.1016/j.matpr.2021.05.255

Kenea, L. M., Kumar, L., Singh, S. K., Chala, G., & Chimdi, J. (2021). Assessment of asphalt road pavement rehabilitation and maintenance practices: The case of West Showa zone in Oromia, Ethiopia. *IOP Conference Series. Materials Science and Engineering, 1187*(1), 012033. DOI: 10.1088/1757-899X/1187/1/012033

Kumar, S., Singh, S. K., & Shree, D. (2024). A comprehensive investigation of Coir geotextiles in sustainable pavement engineering. *International Journal of Civil Engineering, 11*(7), 73–87. DOI: 10.14445/23488352/IJCE-V11I7P107

Li, S., Xu, W., Zhang, F., Wu, H., & Ge, Q. (2022). Effect of graphene oxide on aging properties of Polyurethane-SBS modified asphalt and asphalt mixture. *Polymers, 14*(17), 3496. DOI: 10.3390/polym14173496 PMID: 36080571

Liu, Z., Sun, L., Zhai, J., & Huang, W. (2022). A review of design methods for cold in-place recycling asphalt mixtures: Design processes, key parameters, and evaluation. *Journal of Cleaner Production, 370*, 133530. DOI: 10.1016/j.jclepro.2022.133530

Lu, X., & Isacsson, U. (1997). Rheological characterization of styrene-butadiene-styrene copolymer modified bitumens. *Construction & Building Materials, 11*(1), 23–32. DOI: 10.1016/S0950-0618(96)00033-5

Mazumder, M., Kim, H., & Lee, S. (2016). Performance properties of polymer modified asphalt binders containing wax additives. *International Journal of Pavement Research and Technology, 9*(2), 128–139. DOI: 10.1016/j.ijprt.2016.03.004

Mondal, P., Nandan, A., Ajithkumar, S., Siddiqui, N. A., Raja, S., Kola, A. K., & Balakrishnan, D. (2023). Sustainable application of nanoparticles in wastewater treatment: Fate, current trend & paradigm shift. *Environmental Research, 232*, 116071. DOI: 10.1016/j.envres.2023.116071 PMID: 37209979

Naaman, A. E. (2003). Engineered Steel Fibers with Optimal Properties for Reinforcement of Cement Composites. *Journal of Advanced Concrete Technology, 1*(3), 241–252. DOI: 10.3151/jact.1.241

Nath, P., & Sarker, P. K. (2015). Use of OPC to improve setting and early strength properties of low calcium fly ash geopolymer concrete cured at room temperature. *Cement and Concrete Composites, 55*, 205–214.

Nazari, H., Naderi, K., & Nejad, F. M. (2018). Improving aging resistance and fatigue performance of asphalt binders using inorganic nanoparticles. *Construction & Building Materials, 170*, 591–602. DOI: 10.1016/j.conbuildmat.2018.03.107

Neville, A. M. (1992). Fibre reinforced cement and concrete. In *CRC Press eBooks*. DOI: 10.1201/9781482271225

Priya, P. K., Vanitha, S., & Meyyappan, P. (2020b). Utilization of waste for building materials – a review. *IOP Conference Series. Materials Science and Engineering*, *955*(1), 012048. DOI: 10.1088/1757-899X/955/1/012048

Provis, J. L. (2017). Alkali-activated materials. *Cement and Concrete Research*, *114*, 40–48. DOI: 10.1016/j.cemconres.2017.02.009

Shafabakhsh, G., Mirabdolazimi, S., & Sadeghnejad, M. (2014). Evaluation of the effect of nano-TiO2 on the rutting and fatigue behavior of asphalt mixtures. *Construction & Building Materials*, *54*, 566–571. DOI: 10.1016/j.conbuildmat.2013.12.064

Su, J., Yang, P., Wang, Y., Han, S., Han, N., & Li, W. (2016). Investigation of the Self-Healing behaviors of Microcapsules/Bitumen composites by a repetitive direct tension test. *Materials (Basel)*, *9*(7), 600. DOI: 10.3390/ma9070600 PMID: 28773722

Temuujin, J., Minjigmaa, A., Rickard, W., Lee, M., Williams, I., & Van Riessen, A. (2010). Fly ash based geopolymer thin coatings on metal substrates and its thermal evaluation. *Journal of Hazardous Materials*, *180*(1–3), 748–752. DOI: 10.1016/j.jhazmat.2010.04.121 PMID: 20488615

Tiza, T. M., Kpur, G., Ogunleye, E., Sharma, S., Singh, S. K., & Likassa, D. M. (2023). The potency of functionalized nanomaterials for industrial applications. *Materials Today: Proceedings*. Advance online publication. DOI: 10.1016/j.matpr.2023.03.212

Tiza, T. M., Mogbo, O., Singh, S. K., Shaik, N., & Shettar, M. P. (2022). Bituminous pavement sustainability improvement strategies. *Energy Nexus*, *6*, 100065. DOI: 10.1016/j.nexus.2022.100065

Tiza, T. M., Singh, S. K., Kumar, L., Shettar, M. P., & Singh, S. P. (2021). Assessing the potentials of Bamboo and sheep wool fiber as sustainable construction materials: A review. *Materials Today: Proceedings*, *47*, 4484–4489. DOI: 10.1016/j.matpr.2021.05.322

Toka, E. B., & Olgun, M. (2023). Mechanical and economic evaluation of recycled concrete aggregate as granular road base and subbase material. *Journal of Materials in Civil Engineering*, *35*(5), 04023097. Advance online publication. DOI: 10.1061/JMCEE7.MTENG-14388

Turner, L. K., & Collins, F. G. (2013). Carbon dioxide equivalent (CO_2-e) emissions: A comparison between geopolymer and OPC cement concrete. *Construction & Building Materials*, *43*, 125–130. DOI: 10.1016/j.conbuildmat.2013.01.023

Utsev, T., Tiza, T. M., Mogbo, O., Singh, S. K., Chakravarti, A., Shaik, N., & Singh, S. P. (2022). Application of nanomaterials in civil engineering. *Materials Today: Proceedings*, *62*, 5140–5146. DOI: 10.1016/j.matpr.2022.02.480

Wallah, S. E., & Rangan, B. V. (2006). *Low-Calcium fly ash-based geopolymer concrete: Long-term properties*. https://www.geopolymer.org/dl/?get=curtin_flyash_GC-2.pdf

Walls, J., & Smith, M. R. (1998). *Life-cycle cost analysis in pavement design: Interim technical bulletin (No. FHWA-SA-98-079)*. Federal Highway Administration.

Wang, G., & Ma, Y. (2018). Drying shrinkage of alkali-activated fly ash/slag blended system. *Journal of Sustainable Cement-Based Materials*, *7*(4), 203–213. DOI: 10.1080/21650373.2018.1471424

Wang, H., You, Z., Mills-Beale, J., & Hao, P. (2011). Laboratory evaluation on high temperature viscosity and low temperature stiffness of asphalt binder with high percent scrap tire rubber. *Construction & Building Materials*, *26*(1), 583–590. DOI: 10.1016/j.conbuildmat.2011.06.061

Wiley. (2015). High-Resolution X-ray diffraction and fluorescence microscopy characterization of Alkali-Activated Slag-Metakaolin binders - White Rose Research online. https://eprints.whiterose.ac.uk/id/eprint/86418/

Xu, S., Wu, Z., Xu, Y., & Chen, S. (2014). Performance Evaluation of High RAP Asphalt Mixture with Fiber and Warm Mixing Agent. *Conference: 10th Asia Pacific Transportation Development Conference*, 343–351. DOI: 10.1061/9780784413364.042

Yan, Y., Hernando, D., Park, B., Allen, C., & Roque, R. (2021). Understanding asphalt binder cracking characterization at intermediate temperatures: Review and evaluation of two approaches. *Construction & Building Materials*, *312*, 125163. DOI: 10.1016/j.conbuildmat.2021.125163

Yao, H., You, Z., Li, L., Goh, S. W., Lee, C. H., Yap, Y. K., & Shi, X. (2012). Rheological properties and chemical analysis of nanoclay and carbon microfiber modified asphalt with Fourier transform infrared spectroscopy. *Construction & Building Materials*, *38*, 327–337. DOI: 10.1016/j.conbuildmat.2012.08.004

Yousefi, A. A., Haghshenas, H. F., Underwood, B. S., Harvey, J., & Blankenship, P. (2022). Performance of warm asphalt mixtures containing reclaimed asphalt pavement, an anti-stripping agent, and recycling agents: A study using a balanced mix design approach. *Construction & Building Materials*, *363*, 129633. DOI: 10.1016/j.conbuildmat.2022.129633

Zalghout, A., Castro, S., Karam, J., & Kaloush, K. (2022). Laboratory and field evaluation of plant produced asphalt mixtures containing RAP in hot climate: A case study from Phoenix, Arizona. *Construction & Building Materials*, *351*, 128322. DOI: 10.1016/j.conbuildmat.2022.128322

Compilation of References

. Coutinho Lemos, M. (2021). The conception of a Coca-Cola Bottle.

.Chourpagar, A., Chandak, P. & Chavan, G. Technological Innovations and Alternatives: Advances in Materials Science and Sustainable Practices. Microplastics: Impacts and Implications: p. 76.

Aabid, A., Raheman, M. A., Ibrahim, Y. E., Anjum, A., Hrairi, M., Parveez, B., & Mohammed Zayan, J. (2021). A systematic review of piezoelectric materials and energy harvesters for industrial applications. *Sensors (Basel)*, *21*(12), 414. DOI: 10.3390/s21124145

Ab Alim, N. N. N., Mohamed Saheed, M. S., Mohamed, N. M., & Mohamed Saheed, M. S. (2019). Highly flexible and stretchable 3D graphene/MXene composite thin film. *Materials Today: Proceedings*, *7*, 738–743. DOI: 10.1016/j.matpr.2018.12.069

Abbas, A., Mariana, L. T., & Phan, A. N. (2018). Biomass-waste derived graphene quantum dots and their applications. *Carbon*, *140*, 77–99.

Abbasi, K. R., Abbas, J., & Tufail, M. (2021). Revisiting electricity consumption, price, and real GDP: A modified sectoral level analysis from Pakistan. *Energy Policy*, *149*, 112087. DOI: 10.1016/j.enpol.2020.112087

Abdelfatah, M., A., Hosny, M., S. Elbay, A., El-Maghrabi, N., & Fawzy, M. (2024). From Waste to Worth: Upcycling Plastic into High-Value Carbon-Based Nanomaterials. *Polymers*, *17*(1), 63.

Abdulkadir, B. A., Teh, L. P., Rahman Khan, M. M., & Setiabudi, H. D. (2024). Potential Applications of Metal-Doped Nanomaterials for Enhancing Solid-State Hydrogen Storage. *ChemistrySelect*, *9*(14), e202400051. DOI: 10.1002/slct.202400051

Abdullah, T. A. A., Mohd, S. M. Z., & Ali, W. (2021). A Review of Interpretable Ml in Healthcare: Taxonomy, Applications, Challenges, and Future Directions. *Symmetry*, *13*(12), 2439. Advance online publication. DOI: 10.3390/sym13122439

Abel, D., Arumugam, D., Lehnert, L., & Littman, M. (2018, July). State abstractions for lifelong reinforcement learning. In *International Conference on Machine Learning* (pp. 10-19). PMLR.

Abid, N., Khan, A. M., Shujait, S., Chaudhary, K., Ikram, M., Imran, M., Haider, J., Khan, M., Khan, Q., & Maqbool, M. (2022). Synthesis of nanomaterials using various top-down and bottom-up approaches, influencing factors, advantages, and disadvantages: A review. *Advances in Colloid and Interface Science, 300*, 102597. DOI: 10.1016/j.cis.2021.102597

Abolhasani, M., & Kumacheva, E. (2023). The Rise of Self-Driving Labs in Chemical and Materials Sciences. *Nature Synthesis, 2*(6), 483–492. Advance online publication. DOI: 10.1038/s44160-022-00231-0

Acevedo, S., Giraldo, L., & Moreno-Piraján, J. C. (2020). Adsorption of CO2 on Activated Carbons Prepared by Chemical Activation with Cupric Nitrate. *ACS Omega, 5*(18), 10423–10432. DOI: 10.1021/acsomega.0c00342

Adamiak, K., & Sionkowska, A. (2023). State of innovation in alginate-based materials. *Marine Drugs, 21*(6), 353.

Addin, A. H. S., Ramli, R. M., Sufian, S., Al-Mahbashi, N., Idris, S. S., Farooqi, A., & Bakri, S. H. S. (2025). Sustainable synthesis of carbon nanomaterials from waste tire-derived char using microwave pyrolysis. *Journal of Ecological Engineering, 26*(6), 189–201.

Adeoye, A. E., Adeaga, O. A., Oloore, L. E., & Ogunmola, E. D. (2023). Review of Peculiarities of Multifunctional Nano Materials for Harvesting and Storing Light Energy. *Adeleke University Journal of Engineering and Technology, 6*(2), 329–349.

Afif, K., Rebolledo, C., & Roy, J. (2022). Drivers, barriers and performance outcomes of sustainable packaging: A systematic literature review. *British Food Journal, 124*(3), 915–935.

Agrawal, P., Thethi, H. P., Mohammad, Q., Gupta, N., Asha, V., & Reddy, K. J. (2025). Integrating AI and Multi-objective Optimization for Enhanced Microgrid Energy Management Using Quadratic Programming. *E3S Web of Conferences, 616*, 02007. DOI: 10.1051/e3sconf/202561602007

Ahmadi, A. A., Arabbeiki, M., Ali, H. M., Goodarzi, M., & Safaei, M. R. (2020). Configuration and Optimization of a Minichannel Using Water–Alumina Nanofluid by Non-Dominated Sorting Genetic Algorithm and Response Surface Method. *Nanomaterials (Basel, Switzerland), 10*, 901.

Ahmadi, M. H., Ahmadi, M. A., Nazari, M. A., Mahian, O., & Ghasempour, R. (2019). A Proposed Model to Predict Thermal Conductivity Ratio of Al2O3/EG Nanofluid by Applying Least Squares Support Vector Machine (LSSVM) and Genetic Algorithm as a Connectionist Approach. *Journal of Thermal Analysis and Calorimetry*, *135*(1), 271–281. DOI: 10.1007/s10973-018-7035-z

Ahmadpour, A., & Do, D. D. (1996). THE PREPARATION OF ACTIVE CARBONS FROM COAL BY CHEMICAL AND PHYSICAL ACTIVATION. *Carbon*, *34*(4), 471–479. DOI: 10.1016/0008-6223(95)00204-9

Ahmad, Z., Kumar, S., Trinh, C. K., Shim, J. J., & Lee, J. S. (2023). Decoupling electrochemical parameters of molecular-level-controlled polypyrrole and graphene oxide nanocomposite. *Applied Surface Science*, *610*, 155464.

Ahmed, I., Jeon, G., & Piccialli, F. (2022). From Artificial Intelligence to Explainable Artificial Intelligence in Industry 4.0: A Survey on What, How, and Where. *IEEE Transactions on Industrial Informatics*, *18*(8), 5031–5042. DOI: 10.1109/TII.2022.3146552

Ahmed, S. F., Mofijur, M., Ahmed, B., Mehnaz, T., Mehejabin, F., Maliat, D., Hoang, A. T., & Shafiullah, G. M. (2022). Nanomaterials as a sustainable choice for treating wastewater. *Environmental Research*, *214*, 113807. DOI: 10.1016/j.envres.2022.113807

Aijaz, A., Fujiwara, N., & Xu, Q. (2014). From Metal–Organic Framework to Nitrogen-Decorated Nanoporous Carbons: High CO_2 Uptake and Efficient Catalytic Oxygen Reduction. *Journal of the American Chemical Society*, *136*(19), 6790–6793. DOI: 10.1021/ja5003907

Airey, G. D. (2002). Rheological evaluation of ethylene vinyl acetate polymer modified bitumens. *Construction & Building Materials*, *16*(8), 473–487. DOI: 10.1016/S0950-0618(02)00103-4

Ajala, A., Kashim, I., Akinbogun, T., & Aramide, F. (2019). Sustainable materials and role of professionals in built environment sustainability. In *Procs West Africa Built Environment Research (WABER)*. Conference.

Akbarian, M., & Chen, S. H. (2022). Instability challenges and stabilization strategies of pharmaceutical proteins. *Pharmaceutics*, *14*(11), 2533.

Akhai, S., Abbass, M., Kaur, P., & Kaur, T. (2025). Digital Transformation Across Generations: Robotics and AI in Action. In Impacts of Digital Technologies Across Generations (pp. 23-40). IGI Global Scientific Publishing

Akhai, S., Arif, S. M., & Dadhwal, R. (2025). High-Performance Aluminium Alloys for Aircraft Structures: Advances in Properties, Processing, and Manufacturing. In Innovative Materials for Next-Generation Defense Applications: Cost, Performance, and Mass Production (pp. 61-106). IGI Global Scientific Publishing.

Akhtar, K., Khan, S. B., Asiri, A. M., & Khan, S. A. (2018). Scanning Electron Microscopy: Principle and Applications in Nanomaterials Characterization (pp. 113–145). springer. DOI: 10.1007/978-3-319-92955-2_4

Akin, D. E. (2013). Linen most useful: Perspectives on structure, chemistry, and enzymes for retting flax. *International Scholarly Research Notices, 2013*(1), 186534.

Akter, M., Uddin, M. H., & Tania, I. S. (2022). Biocomposites based on natural fibers and polymers: A review on properties and potential applications. *Journal of Reinforced Plastics and Composites, 41*(17-18), 705–742.

Al Alim, M., Hore, S., Zohora, F., Ghosh, D., & Hore, R. (2025). Advances in Smart Polymeric Materials for Sustainable and Next-generation Applications. *Civil Engineering & Building Science, 2*(1).

Al Atrach, J., Aitblal, A., Amedlous, A., Xiong, Y., Desmurs, M., Ruaux, V., Guillet-Nicolas, R., & Valtchev, V. (2024). Nanosized zeolite P for enhanced CO2 adsorption kinetics. *ACS Applied Materials & Interfaces, 16*(29), 38006–38016. DOI: 10.1021/acsami.4c05988

Al Ghamdi, H. S., & Al-Ghamdi, A. A. (2023). Opening band gap of multi-color graphene quantum dots using D-fructose as a green precursor. *Alexandria Engineering Journal, 79*, 155–163. DOI: 10.1016/j.aej.2023.08.019

Al Miad, A., Saikat, S. P., Alam, M. K., Hossain, M. S., Bahadur, N. M., & Ahmed, S. (2024). Metal oxide-based photocatalysts for the efficient degradation of organic pollutants for a sustainable environment: A review. *Nanoscale Advances, 6*(19), 4781–4803. DOI: 10.1039/D4NA00517A

Al-Akayleh, F., Ali Agha, A. S. A., Al-Remawi, M., & Abdel Rahem, R. A. (2024). A mini review on the applications of artificial intelligence (AI) in surface chemistry and catalysis. *Tenside, Surfactants, Detergents, 61*(4), 285–296. DOI: 10.1515/tsd-2024-2580

Alayan, H. M., Aljumaily, M. M., Alsaadi, M. A., Mjalli, F. S., & Hashim, M. A. (2021). A review exploring the adsorptive removal of organic micropollutants on tailored hierarchical carbon nanotubes. *Toxicological and Environmental Chemistry, 103*(4), 282–325. DOI: 10.1080/02772248.2021.1932888

Albalawi, F., Hussein, M. Z., Fakurazi, S., & Masarudin, M. J. (2021). Engineered Nanomaterials: The Challenges and Opportunities for Nanomedicines. *International Journal of Nanomedicine*, *16*, 161–184. DOI: 10.2147/IJN.S288236

Alemu, D., Tafesse, M., & Mondal, A. K. (2022). Mycelium-based composite: The future sustainable biomaterial. *International Journal of Biomaterials*, *2022*(1), 8401528.

Al-Harbi, N., & Abd-Elrahman, N. K. (2024). Physical methods for preparation of nanomaterials, their characterization and applications: A review. *Journal of Umm Al-Qura University for Applied Sciences*, *11*(2), 356–377. DOI: 10.1007/s43994-024-00165-7

Alharkan, H., Habib, S., & Islam, M. (2023). Solar Power Prediction Using Dual Stream CNN-LSTM Architecture. *Sensors (Basel)*, *23*(2), 945. Advance online publication. DOI: 10.3390/s23020945

Ali, A., & Mashwani, W. (2023). A Supervised Machine Learning Algorithms: Applications, Challenges, and Recommendations. Proceedings of the Pakistan Academy of Sciences: A. Physical and Computational Sciences, 60(4). DOI: 10.53560/PPASA(60-4)831

Ali, A., & Shaikh, M. N. (2024). Recent developments in catalyst design for liquid organic hydrogen carriers: Bridging the gap to affordable hydrogen storage. *International Journal of Hydrogen Energy*, *78*, 1–21.

Ali, M. L., Thakur, K., Schmeelk, S., Debello, J., & Dragos, D. (2025). Deep Learning vs. Machine Learning for Intrusion Detection in Computer Networks: A Comparative Study. *Applied Sciences (Basel, Switzerland)*, *15*(4), 1903. DOI: 10.3390/app15041903

Almansoori, A., Balázsi, K., & Balázsi, C. (2024). Advances, Challenges, and Applications of Graphene and Carbon Nanotube-Reinforced Engineering Ceramics. *Nanomaterials (Basel, Switzerland)*, *14*(23), 1881. Advance online publication. DOI: 10.3390/nano14231881 PMID: 39683269

Alnuaimi, A. F. A. H., & Albaldawi, T. H. K. (2024). An overview of machine learning classification techniques. BIO Web of Conferences, 97, 00133. DOI: 10.1051/bioconf/20249700133

Al-Oqla, F. M., & Sapuan, S. (2020). *Advanced processing, properties, and applications of starch and other bio-based polymers*. Elsevier.

Al-Qaseemi, S. A., Almulhim, M. F., Chaudhry, S. R., & Almulhim, H. A. (2016, December 1). IoT architecture challenges and issues: Lack of standardization. DOI: 10.1109/FTC.2016.7821686

Alsabery, A. I., Ghalambaz, M., Armaghani, T., Chamkha, A., Hashim, I., & Pour, M. S. (2020). Role of Rotating Cylinder toward Mixed Convection inside a Wavy Heated Cavity via Two-Phase Nanofluid Concept. *Nanomaterials (Basel, Switzerland)*, *10*, 1138.

Alshammari, B. H., Lashin, M. M. A., Mahmood, M. A., Al-Mubaddel, F. S., Ilyas, N., Rahman, N., Sohail, M., Khan, A., Abdullaev, S. S., & Khan, R. (2023). Organic and inorganic nanomaterials: Fabrication, properties and applications. *RSC Advances*, *13*(20), 13735–13785. DOI: 10.1039/D3RA01421E

Alufaisan, Y., Bakdash, J. Z., Zhou, Y., Marusich, L. R., & Kantarcioglu, M. (2021). Does Explainable Artificial Intelligence Improve Human Decision-Making? *Proceedings of the AAAI Conference on Artificial Intelligence*, *35*(8), 6618–6626. DOI: 10.1609/aaai.v35i8.16819

Alvarez, J., Lopez, G., Amutio, M., Bilbao, J., & Olazar, M. (2015). Physical Activation of Rice Husk Pyrolysis Char for the Production of High Surface Area Activated Carbons. *Industrial & Engineering Chemistry Research*, *54*(29), 7241–7250. DOI: 10.1021/acs.iecr.5b01589

Alzubaidi, L., Bai, J., Al-Sabaawi, A., Santamaría, J., Albahri, A. S., Al-dabbagh, B. S. N., Fadhel, M. A., Manoufali, M., Zhang, J., Al-Timemy, A. H., Duan, Y., Abdullah, A., Farhan, L., Lu, Y., Gupta, A., Albu, F., Abbosh, A., & Gu, Y. (2023). A Survey on Deep Learning Tools Dealing with Data Scarcity: Definitions, Challenges, Solutions, Tips, and Applications. *Journal of Big Data*, *10*(1), 46. Advance online publication. DOI: 10.1186/s40537-023-00727-2

Amburi, P. K., Senthilkumar, G., & Nithya, A. (2023). Novel use of CuO nanoparticles additive for improving thermal conductivity of MgO/water and MWCNT/water nanofluids. *Journal of Thermal Analysis and Calorimetry*, *148*(19), 10389–10398.

Amici, J., Asinari, P., Ayerbe, E., Barboux, P., Bayle-Guillemaud, P., Behm, R. J., & Edström, K. (2022). A roadmap for transforming research to invent the batteries of the future designed within the European large scale research initiative battery. *Advanced Energy Materials*, *12*(17), 2102785.

Amin Mir, M. (2025). Bimal Krishna Banik, Synthesis of Schiff base ligands under environment friendly conditions: A systematic review. *Inorganic Chemistry Communications*, *174*, 1.

Amin, M., Chung, E., & Shah, H. H. (2023). Effect of different activation agents for activated carbon preparation through characterization and life cycle assessment. *International Journal of Environmental Science and Technology, 20*(7), 7645–7656. DOI: 10.1007/s13762-022-04472-6

Amin, M., Shah, H. H., Bashir, B., Iqbal, M. A., Shah, U. H., & Ali, M. U. (2023). Environmental assessment of hydrogen utilization in various applications and alternative renewable sources for hydrogen production: A review. *Energies, 16*(11), 4348. DOI: 10.3390/en16114348

Amutha, C., Gopan, A., Pushbalatatha, I., Ragavi, M., & Reneese, J. A. (2024). Nanotechnology and Governance: Regulatory Framework for Responsible Innovation. Nanotechnology in Societal Development, 481–503. DOI: 10.1007/978-981-97-6184-5_14

Anastopoulos, I., Lima, E., Meili, L., & Giannakoudakis, D. (2022). Biomass-Derived Materials for Environmental Applications. In *Biomass-Derived Materials for Environmental Applications*. Elsevier., DOI: 10.1016/C2021-0-00449-1

Andersen, S. C., Larsen, H., Raffnsøe, L., & Melvang, C. (2019). Environmental product declarations (EPDs) as a competitive parameter within sustainable buildings and building materials. In *IOP Conference Series: Earth and Environmental Science*. IOP Publishing.

Anwar, A., Liu, Y., Dong, R., Bai, L., Yu, X., & Li, Y. (2018). The physiological and molecular mechanism of brassinosteroid in response to stress: A review. *Biological Research, 51*(1), 51. DOI: 10.1186/s40659-018-0195-2

Arafat, S., Wasiuddin, N. M., & Mohammad, L. N. (2023). Evaluation of bio-based and petroleum-based rejuvenator based on cracking susceptibility of hot mix asphalt with high RAP content. *Construction & Building Materials, 371*, 130725. DOI: 10.1016/j.conbuildmat.2023.130725

Archana, U., Sharma, S., & Singh, S. K. S. R, S. P and T. H. Babu, "Analysis Of Concrete Cracks And Fatigue In Smart Cities Using Yolov3," *2023 International Conference on Research Methodologies in Knowledge Management, Artificial Intelligence and Telecommunication Engineering (RMKMATE)*, Chennai, India, 2023, pp. 1-6, DOI: 10.1109/RMKMATE59243.2023.10368774

Arias, A., González-Rodríguez, S., Barros, M. V., Salvador, R., De Francisco, A. C., Piekarski, C. M., & Moreira, M. T. (2021). Recent developments in bio-based adhesives from renewable natural resources. *Journal of Cleaner Production, 314*, 127892. DOI: 10.1016/j.jclepro.2021.127892

Arif, M. "Neutron imaging for fuel cell research," in *Proceedings of the Imaging and Neutron Workshop*, Oak Ridge, Tenn, USA, October 2006.

Arif, Z. U., Khalid, M. Y., Sheikh, M. F., Zolfagharian, A., & Bodaghi, M. (2022). Biopolymeric sustainable materials and their emerging applications. *Journal of Environmental Chemical Engineering*, *10*(4), 108159.

Aryanfar, Y., Wu, W., Alcaraz, J. L. G., Fernandez, J. B., Sadabad, A. N., & Keçebaş, A. (2024). Production of Biodiesel from Industrial Sludge: Recent Progress, Challenges, Perspective. *Processes (Basel, Switzerland)*, *12*(11), 2517. DOI: 10.3390/pr12112517

Asadi, A., Molana, M., & Ghasemiasl, R. (2020). A New Thermal Conductivity Model and Two-Phase Mixed Convection of CuO–Water Nanofluids in a Novel I-Shaped Porous Cavity Heated by Oriented Triangular Hot Block. *Nanomaterials (Basel, Switzerland)*, *10*, 2219.

Asamoah, G. A., Korsah, M., Jeyasundar, P. G. S. A., Ahmed, M., Lau, S. Y., & Danquah, M. K. (2024). Nanotechnology-based lithium-ion battery energy storage systems. *Sustainability (Basel)*, *16*(21), 9231. DOI: 10.3390/su16219231

Asdrubali, F., Berardi, U., & Stasi, R. (2025). The impact of the building sector. In *Sustainability Certifications, Labels and Tools in the Built Environment* (pp. 4-21). Routledge. DOI: 10.1201/9781032705149-2

Aslam, A., Kanwal, F., Javied, S., Nisar, N., & Torriero, A. A. J. (2025). Microbial biosorption: A sustainable approach for metal removal and environmental remediation. *International Journal of Environmental Science and Technology*, *22*(13), 1–32. DOI: 10.1007/s13762-025-06611-1

Asperti, A., Evangelista, D., & Piccolomini, E. L. (2021). A Survey on Variational Autoencoders from a Green AI Perspective. *SN Computer Science*, *2*(4), 301. Advance online publication. DOI: 10.1007/s42979-021-00702-9

Asrafali, S. P., Periyasamy, T., & Bari, G. A. R. (2025). Advances in Carbon-Based Aerogels for CO2 Capture: Fundamental Design Strategies and Technological Progress. *Gels (Basel, Switzerland)*, *11*(5), 361. DOI: 10.3390/gels11050361

Assyl, S., Botakoz, S., & Saule, Z. (2025). Dimensions, structure, and morphology variations of carbon-based materials for hydrogen storage: A review. *Discover Nano*, *20*(1), 1–41. DOI: 10.1186/s11671-025-04229-3

Auteri, N. (2023). Characterization of modified cactus pear (Opuntia ficus indica (L.) Mill) pruning waste as a sustainable and innovative solution to recycle phosphorus from agricultural streams.

Avila, F., Puertas, E., & Gallego, R. (2021). Characterization of the mechanical and physical properties of unstabilized rammed earth: A review. *Construction & Building Materials*, *270*, 121435.

Awad, A. (2017). UTILIZATION OF WASTE PLASTICS TO ENHANCE THE PERFORMANCE OF MODIFIED HOT MIX ASPHALT. *International Journal of Geomate*, *13*(40). Advance online publication. DOI: 10.21660/2017.40.170603

Ayanda, O. S., Mmuoegbulam, A. O., Okezie, O., Durumin Iya, N. I., Mohammed, S. E., James, P. H., Muhammad, A. B., Unimke, A. A., Alim, S. A., Yahaya, S. M., Ojo, A., Adaramoye, T. O., Ekundayo, S. K., Abdullahi, A., & Badamasi, H. (2024). Recent progress in carbon-based nanomaterials: Critical review. *Journal of Nanoparticle Research*, *26*(5), 106. Advance online publication. DOI: 10.1007/s11051-024-06006-2

Ayanleye, S., Udele, K., Nasir, V., Zhang, X., & Militz, H. (2022). Durability and protection of mass timber structures: A review. *Journal of Building Engineering*, *46*, 103731.

Baburaj, A., Banerjee, S., Aliyana, A. K., Shee, C., Banakar, M., Bairagi, S., & Ali, S. W. (2024). Biodegradable based TENGs for self-sustaining implantable medical devices. *Nano Energy*, •••, 109785.

Back, G.-S., Park, H. S., Seo, S. M., & Jung, W.-G. (2015). Exploring high-strength glass-ceramic materials for upcycling of industrial wastes. *Metals and Materials International*, *21*, 1061–1067.

Badini, S., Regondi, S., & Pugliese, R. (2023). Unleashing the Power of Artificial Intelligence in Materials Design. *Materials (Basel)*, *16*(17), 5927. DOI: 10.3390/ma16175927 PMID: 37687620

Bae, G., Han, S., Oh, H. S., & Choi, C. H. (2023). Operando stability of single-atom electrocatalysts. *Angewandte Chemie*, *135*(19), e202219227.

Baghel, N., & Kumar, A. (2025). Sustainable thermoelectric materials for solar energy applications: A review. *Solid State Sciences*, *160*, 107784. DOI: 10.1016/j.solidstatesciences.2024.107784

Bagheri, S., TermehYousefi, A., & Do, T.-O. (2017). Photocatalytic pathway toward degradation of environmental pharmaceutical pollutants: Structure, kinetics and mechanism approach. *Catalysis Science & Technology*, *7*(20), 4548–4569. DOI: 10.1039/C7CY00468K

Baig, N., Kammakakam, I., & Falath, W. (2021). Nanomaterials: A review of synthesis methods, properties, recent progress, and challenges. *Materials Advances*, *2*(6), 1821–1871. DOI: 10.1039/D0MA00807A

Bai, J., Yang, Y., Wen, C., Chen, J., Zhou, G., Jiang, B., Peng, X., & Pan, F. (2023). Applications of magnesium alloys for aerospace: A review. *Journal of Magnesium and Alloys*, *11*(10), 3609–3619. DOI: 10.1016/j.jma.2023.09.015

Bai, X., & Zhang, X. (2025). Artificial intelligence-powered materials science. *Nano-Micro Letters*, *17*(1), 1–30.

Balakrishnan, D. (2024). Environmental sustainability through the utilization of seashell wastes in the development of MWCNT-reinforced nylon 66 composites. *Carbon Letters*, *34*(2), 731–742.

Balea, A., Fuente, E., Monte, M. C., Blanco, A., & Negro, C. (2021). Recycled fibers for sustainable hybrid fiber cement based material: A review. *Materials (Basel)*, *14*(9), 2408.

Bandpey, M., & Barz, D. P. J. (2024). Effects of interlayer space engineering and surface modification on the charge storage mechanisms of MXene nanomaterials: A review on recent developments. *Nanoscale*, *16*(32), 15078–15093. DOI: 10.1039/D4NR01317D

Bansal, S. A., Kumar, V., Karimi, J., Singh, A. P., & Kumar, S. (2020). Role of gold nanoparticles in advanced biomedical applications. *Nanoscale Advances*, *2*(9), 3764–3787. DOI: 10.1039/D0NA00472C

Bansal, S., Chaudhary, P., Sharma, B. B., Saini, S., & Joshi, A. (2024). Review of MXenes and their composites for energy storage applications. *Journal of Energy Storage*, *87*, 111420. DOI: 10.1016/j.est.2024.111420

Banthia, N., & Gupta, R. (2006). Influence of polypropylene fiber geometry on plastic shrinkage cracking in concrete. *Cement and Concrete Research*, *36*(7), 1263–1267. DOI: 10.1016/j.cemconres.2006.01.010

Bappy, M. A. (2024). Exploring the Integration of Informed Machine Learning in Engineering Applications: A Comprehensive Review. *American Journal of Science and Learning for Development*, *3*(2), 11–21. DOI: 10.51699/ajsld.v3i2.3459

Barakat, N. A. M., Khil, M. S., Omran, A. M., Sheikh, F. A., & Kim, H. Y. (2009). Extraction of pure natural hydroxyapatite from the bovine bones bio waste by three different methods. *Journal of Materials Processing Technology*, *209*(7), 3408–3415. DOI: 10.1016/j.jmatprotec.2008.07.040

Baranwal, J., Barse, B., Fais, A., Delogu, G. L., & Kumar, A. (2022). Biopolymer: A sustainable material for food and medical applications. *Polymers*, *14*(5), 983.

Barbhuiya, S., Das, B. B., Adak, D., Kapoor, K., & Tabish, M. (2025). Low carbon concrete: Advancements, challenges and future directions in sustainable construction. *Discover Concrete and Cement*, *1*(1), 1–24.

Barelli, L., Bidini, G., Pelosi, D., & Sisani, E. (2021). Enzymatic biofuel cells: A review on flow designs. *Energies*, *14*(4), 910.

Barhoum, A., García-Betancourt, M. L., Jeevanandam, J., Hussien, E. A., Mekkawy, S. A., Mostafa, M., Omran, M. M. S., Abdalla, M., & Bechelany, M. (2022). Review on Natural, Incidental, Bioinspired, and Engineered Nanomaterials: History, Definitions, Classifications, Synthesis, Properties, Market, Toxicities, Risks, and Regulations. *Nanomaterials (Basel, Switzerland)*, *12*(2), 177. DOI: 10.3390/nano12020177

Barsce, J. C., Martinez, E. C., & Palombarini, J. A. (2017). Towards autonomous reinforcement learning: Automatic setting of hyper-parameters using Bayesian optimization. 36, 1–9. DOI: 10.1109/CLEI.2017.8226439

Basiago, A. D. (1995). Methods of defining 'sustainability'. *Sustainable Development*, *3*(3), 109–119.

Bassyouni, M., & Javaid, U. (2017). S.W. ul Hasan, Bio-based hybrid polymer composites: a sustainable high performance material. In *Hybrid polymer composite materials* (pp. 23–70). Elsevier.

Batra, R., Song, L., & Ramprasad, R. (2020). Emerging materials intelligence ecosystems propelled by machine learning. *Nature Reviews. Materials*, *6*(8), 655–678. DOI: 10.1038/s41578-020-00255-y

Baumbauer, C. L. (2022). *Printed Biodegradable Wireless Soil Nitrate Sensor Nodes*. University of California.

Behera, M., Philip, A., & Kumar, A. R. (2025). Optimized Hydroxyapatite–Carbon Nanotube Composite Electrodes for Enhanced Supercapacitor Performance. ChemistrySelect, 10(29). Portico. DOI: 10.1002/slct.202501845

Ben Dassi, R., & Chamam, B. (2025). Nanomaterial-enhanced membranes for advanced water and wastewater treatment: A comprehensive review. *Reviews in Environmental Science and Biotechnology*, *24*(2), 1–35. DOI: 10.1007/s11157-025-09723-9

Bennett, J. A., & Abolhasani, M. (2022). Autonomous Chemical Science and Engineering Enabled by Self-Driving Laboratories. *Current Opinion in Chemical Engineering*, *36*, 100831. Advance online publication. DOI: 10.1016/j.coche.2022.100831

Bennett, J. A., Orouji, N., Khan, M., Sadeghi, S., Rodgers, J., & Abolhasani, M. (2024). Autonomous Reaction Pareto-Front Mapping with a Self-Driving Catalysis Laboratory. *Nature Chemical Engineering*, *1*(3), 240–250. Advance online publication. DOI: 10.1038/s44286-024-00033-5

Bespalov, V. A., Dyuzhev, N. A., & Kireev, V. Yu. (2022). Possibilities and Limitations of CMOS Technology for the Production of Various Microelectronic Systems and Devices. *Nanobiotechnology Reports*, *17*(1), 24–38. DOI: 10.1134/S2635167622010037

Bhanvase, B. A., Shende, T. P., & Sonawane, S. H. (2017). A review on graphene–TiO2 and doped graphene–TiO2 nanocomposite photocatalyst for water and wastewater treatment. *Environmental Technology Reviews*, *6*(1), 1–14. DOI: 10.1080/21622515.2016.1264489

Bhat, A., Anwer, S., Bhat, K. S., Mohideen, M. I. H., Liao, K., & Qurashi, A. (2021). Prospects challenges and stability of 2D MXenes for clean energy conversion and storage applications. *Npj 2D Materials and Applications*, *5*(1), 61. DOI: 10.1038/s41699-021-00239-8

Bhat, M. A., Rather, R. A., & Shalla, A. H. (2021). PEDOT and PEDOT: PSS conducting polymeric hydrogels: A report on their emerging applications. *Synthetic Metals*, *273*, 116709.

Bhatt, M. D., Kim, H., & Kim, G. (2022). Various defects in graphene: A review. *RSC Advances*, *12*(33), 21520–21547. DOI: 10.1039/D2RA01436J

Bianco, A., Kostarelos, K., & Prato, M. (2005). Applications of carbon nanotubes in drug delivery. *Current Opinion in Chemical Biology*, *9*(6), 674–679.

Bilisik, K., & Akter, M. (2022). Graphene nanocomposites: A review on processes, properties, and applications. *Journal of Industrial Textiles*, *51*(3, suppl), 3718S–3766S. DOI: 10.1177/15280837211024252

Biswal, M. (2025). Advances in nanotechnology for heavy metal removal from contaminated water sources. Journal of Nanomaterials and Devices. DOI: 10.61577/jnd.2025.100003

Blaiszik, B., Kramer, S. L. B., Grady, M. E., McIlroy, D. A., Moore, J. S., White, S. R., & Sottos, N. R. (2015). Autonomic restoration of electrical conductivity. *Advanced Materials*, *27*(40), 5962–5967.

Bogmans, C., Gomez-Gonzalez, P., Ganpurev, G., Melina, G., Pescatori, A., & Thube, S. (2025). Power hungry: How ai will drive energy demand. *IMF Working Paper*, *81*(4), 2025.

Boiko, D. A., Pentsak, E. O., Cherepanova, V. A., Gordeev, E. G., & Ananikov, V. P. (2021). Deep neural network analysis of nanoparticle ordering to identify defects in layered carbon materials. *Chemical Science (Cambridge)*, *12*(21), 7428–7441. DOI: 10.1039/D0SC05696K

Borah, R., AG, K. R., Minja, A. C., & Verbruggen, S. W. (2023). A Review on Self-Assembly of Colloidal Nanoparticles into Clusters, Patterns, and Films: Emerging Synthesis Techniques and Applications. Small Methods, 7(6). Portico. DOI: 10.1002/smtd.202201536

Boretti, A. (2025). A narrative review of breakthroughs and future horizons in zero-dimensional nanomaterials. *Optical and Quantum Electronics*, *57*(6), 370. Advance online publication. DOI: 10.1007/s11082-025-08288-9

Bostwick, J. E., Zanelotti, C. J., Yu, D., Pietra, N. F., Williams, T. A., Madsen, L. A., & Colby, R. H. (2022). Ionic interactions control the modulus and mechanical properties of molecular ionic composite electrolytes. *Journal of Materials Chemistry. C, Materials for Optical and Electronic Devices*, *10*(3), 947–957.

Boungiorno, J., Hu, L.-W., Kim, S. J., Hannink, R., Truong, B., & Forrest, E. (2008). Nanofluids for enhanced economics and safety of nuclear reactors: An evaluation of the potential features issues, and research gaps. *Nuclear Technology*, *162*(1), 80–91.

Boz, Z., Korhonen, V., & Koelsch Sand, C. (2020). Consumer considerations for the implementation of sustainable packaging: A review. *Sustainability*, *12*(6), 2192.

Brandić, I., Voća, N., Gunjača, J., Lončar, B., Bilandžija, N., Peter, A., Šurić, J., & Pezo, L. (2024). Biomass higher heating value prediction machine learning insights into ultimate, proximate, and structural analysis datasets. *Energy Sources. Part A, Recovery, Utilization, and Environmental Effects*, *46*(1), 2842–2854. DOI: 10.1080/15567036.2024.2309303

Brewer, C. E., & Brown, R. C. L. D. a. (2012). Biochar characterization and engineering. *Graduate Teses and Dissertations*, 12284. http://search.proquest.com/docview/1023114544?accountid=27932

Bridgens, B., Powell, M., Farmer, G., Walsh, C., Reed, E., Royapoor, M., Gosling, P., Hall, J., & Heidrich, O. (2018). Creative upcycling: Reconnecting people, materials and place through making. *Journal of Cleaner Production*, *189*, 145–154.

Bringezu, S., & Moriguchi, Y. (2018). Material flow analysis. In *Green accounting* (pp. 149–166). Routledge.

Budiutama, G., Li, R., Manzhos, S., & Ihara, M. (2023). Hybrid Density Functional Tight Binding (DFTB)—Molecular Mechanics Approach for a Low-Cost Expansion of DFTB Applicability. *Journal of Chemical Theory and Computation*, *19*(15), 5189–5198. Advance online publication. DOI: 10.1021/acs.jctc.3c00310

Bugnicourt, E., Cinelli, P., Lazzeri, A., & Alvarez, V. A. (2014). Polyhydroxy-alkanoate (PHA): Review of synthesis, characteristics, processing and potential applications in packaging.

Buongiorno, J., Hu, L. W., Apostolakis, G., Hannink, R., Lucas, T., & Chupin, A. (2009). A feasibility assessment of the use of nanofluids to enhance the in-vessel retention capability in light-water reactors. *Nuclear Engineering and Design*, *239*(5), 941–948.

Busarac, N., Adamovic, D., Grujovic, N., & Zivic, F. (2022, December). Lightweight materials for automobiles. [). IOP Publishing.]. *IOP Conference Series. Materials Science and Engineering*, *1271*(1), 012010.

Butakova, M. A., Chernov, A. V., Kartashov, O. O., & Soldatov, A. V. (2022). Data-Centric Architecture for Self-Driving Laboratories with Autonomous Discovery of New Nanomaterials. *Nanomaterials (Basel, Switzerland)*, *12*(1), 12. Advance online publication. DOI: 10.3390/nano12010012

Caglayan, B. S., & Aksoylu, A. E. (2013). CO2 adsorption on chemically modified activated carbon. *Journal of Hazardous Materials*, *252–253*, 19–28. DOI: 10.1016/j.jhazmat.2013.02.028

Calderon, C. E., Plata, J. J., Toher, C., Oses, C., Levy, O., Fornari, M., Natan, A., Mehl, M. J., Hart, G., Buongiorno Nardelli, M., & Curtarolo, S. (2015). The AFLOW Standard for High-Throughput Materials Science Calculations. *Computational Materials Science*, *108*, 233–238. Advance online publication. DOI: 10.1016/j.commatsci.2015.07.019

Calkins, M. (2008). *Materials for sustainable sites: a complete guide to the evaluation, selection, and use of sustainable construction materials*. John wiley & sons.

Capone, M., Narzi, D., & Guidoni, L. (2021). Mechanism of oxygen evolution and Mn_4CaO_5 cluster restoration in the natural water-oxidizing catalyst. *Biochemistry*, *60*(30), 2341–2348.

Carvalho, D., Ferreira, N., França, B., Marques, R., Silva, M., Silva, S., & Oliveira, C. (2024). Advancing sustainability in the automotive industry: Bioprepregs and fully bio-based composites. *Composites Part C: Open Access, 14*, 100459.

Chakraborty, A., Islam, M., Shahriyar, F., Uddin, M. M., Uddin, M. H., & Minar, M. R. (2023, July). Economic and Environmental Feasibility Study of Fossil Fuel, Electric and Hydrogen Fuel Cell Vehicles in Perspective of Bangladesh. In *2023 14th International Conference on Computing Communication and Networking Technologies (ICCCNT)* (pp. 1-6). IEEE.

Chan, Q.-H., Alias, S. A., Quek, S. W., Ng, C. Y., & Ku Marsilla, K. I. (2023). A review of the preparations, properties, and applications of smart biodegradable polymers. Polymer-Plastics Technology and Materials, ahead-of-print(ahead-of-print), 1273–1289. DOI: 10.1080/25740881.2023.2204954

Chatziparaskeva, G., Papamichael, I., Voukkali, I., Loizia, P., Sourkouni, G., Argirusis, C., & Zorpas, A. A. (2022). End-of-life of composite materials in the framework of the circular economy. *Microplastics, 1*(3), 377–392.

Chávez-Angel, E., Eriksen, M. B., Castro-Alvarez, A., Garcia, J. H., Botifoll, M., Avalos-Ovando, O., Arbiol, J., & Mugarza, A. (2025). Applied Artificial Intelligence in Materials Science and Material Design. *Advanced Intelligent Systems, 7*(8), 2400986. Advance online publication. DOI: 10.1002/aisy.202400986

Che, L., Li, N., Wei, W., Li, J., Ji, J., Zhao, X., Yu, B., & Wang, Q. (2025). Day-Night energy harvesting: Photovoltaics-driven moisture evaporation and absorption for simultaneous 24-hour power and dehumidification. *The Innovation Energy, 2*(2), 100078–1. DOI: 10.59717/j.xinn-energy.2025.100078

Chen, B. B., Liu, M. L., & Huang, C. Z. (Eds.). (2023). *Green Development of Photoluminescent Carbon Dots: For Analytical, Biomedical and Environmental Applications* (Vol. 80). Royal Society of Chemistry.

Chen, C. T., & Gu, G. X. (2020). Generative Deep Neural Networks for Inverse Materials Design Using Backpropagation and Active Learning. *Advanced Science (Weinheim, Baden-Wurttemberg, Germany), 7*(5), 1902607. Advance online publication. DOI: 10.1002/advs.201902607

Chen, F. D., Xie, Z. Y., Li, M. T., Chen, S. G., Ding, W., Li, L., Li, J., & Wei, Z. D. (2024). Series Reports from Professor Wei's Group of Chongqing University: Advancements in Electrochemical Energy Conversions (1/4): Report 1: High-Performance Oxygen Reduction Catalysts for Fuel Cells. *Dian Hua Xue, 30*(7), 1. DOI: 10.61558/2993-074X.3459

Chen, G., & Tang, D.-M. (2024). Machine Learning as a 'Catalyst' for Advancements in Carbon Nanotube Research. *Nanomaterials (Basel, Switzerland)*, *14*(21), 1688. DOI: 10.3390/nano14211688

Cheng, D., Zhao, Y., An, T., Wang, X., Zhou, H., & Fan, T. (2019). 3D interconnected crumpled porous carbon sheets modified with high-level nitrogen doping for high performance lithium sulfur batteries. *Carbon*, *154*, 58–66. DOI: 10.1016/j.carbon.2019.07.094

Cheng, H., Chen, L., McClements, D. J., Yang, T., Zhang, Z., Ren, F., Miao, M., Tian, Y., & Jin, Z. (2021). Starch-based biodegradable packaging materials: A review of their preparation, characterization and diverse applications in the food industry. *Trends in Food Science & Technology*, *114*, 70–82.

Cheng, X., Sun, Z., Yang, X., Li, Z., Zhang, Y., Wang, P., & Shao, L. (2020). Construction of superhydrophilic hierarchical polyacrylonitrile nanofiber membranes by in situ asymmetry engineering for unprecedently ultrafast oil–water emulsion separation. *Journal of Materials Chemistry. A, Materials for Energy and Sustainability*, *8*(33), 16933–16942. DOI: 10.1039/D0TA03011B

Chen, H. L., Liu, Y., Gong, X. Z., Hao, L. W., Sun, B. X., & Li, X. Q. (2020). A review on water footprint research of materials industry. []. Trans Tech Publications Ltd.]. *Materials Science Forum*, *993*, 404–413.

Chen, Q., He, Z., Zhao, Y., Liu, X., Wang, D., Zhong, Y., Hu, C., Hao, C., Lu, K., & Wang, Z. (2024). Stacking Ensemble Learning Assisted Design of Al-Nb-Ti-V-Zr Lightweight High-Entropy Alloys with High Hardness. *Materials & Design*, *246*(October), 113363. DOI: 10.1016/j.matdes.2024.113363

Chen, S., Zhang, C., & Lu, X. (2023). Energy conversion from Fossil fuel to renewable energy. In *Handbook of Air Quality and Climate Change* (pp. 1–44). Springer Nature Singapore.

Chen, X., Sierens, Z., Gruyaert, E., & Li, J. (2023). Precast concrete wall panels incorporating mixed recycled aggregates. *ACI Materials Journal*, *120*(1), 75–88.

Chiou, C. S., Chen, H. W., & Chen, Z. T. (2018). Application of magnetic adsorbent with silicate and phenyl polymers to adsorb Rhodamine B. *International Journal of Environmental Science and Technology*, *15*(9), 1879–1886. DOI: 10.1007/s13762-017-1553-2

Choi, J. S., Fortunato, G. V., Jung, D. C., Lourenço, J. C., Lanza, M. R., & Ledendecker, M. (2024). Catalyst durability in electrocatalytic H_2O_2 production: Key factors and challenges. *Nanoscale Horizons*, *9*(8), 1250–1261.

Chong, S. S., Ng, Y. S., Hui, Q. W., & Jin, C. Z. (2024). Advances of Machine Learning in Materials Science: Ideas and Techniques. *Frontiers of Physics*, *19*(1), 13501. Advance online publication. DOI: 10.1007/s11467-023-1325-z

Chopkar, M., Das, P. K., & Manna, I. (2006). Synthesis and characterization of nanofluid for advanced heat transfer applications. *Scripta Materialia*, *55*(6), 549–552.

Choubisa, H., Askerka, M., Ryczko, K., Voznyy, O., Mills, K., Tamblyn, I., & Sargent, E. H. (2020). Crystal Site Feature Embedding Enables Exploration of Large Chemical Spaces. *Matter*, *3*(2), 433–448. Advance online publication. DOI: 10.1016/j.matt.2020.04.016

Choudhury, I. A., & Hashmi, M. (2020). *Encyclopedia of renewable and sustainable materials*. Elsevier.

Choy, L. H. T., & Ho, W. K. O. (2023). The Use of Machine Learning in Real Estate Research. *Land (Basel)*, *12*(4), 740. DOI: 10.3390/land12040740

Christian Di Stasi, D. A. G. G. B. G. J. J. M. (2019). Physically activated wheat straw-derived biochar for biomass pyrolysis vapors upgrading with high resistance against coke deactivation. *Fuel*, *255*, 115807. DOI: 10.1016/j.fuel.2019.115807

Chung, J. H., Hasyimah, N., & Hussein, N. (2022). Application of carbon nanotubes (CNTs) for remediation of emerging pollutants-a review. *Tropical Aquatic and Soil Pollution*, *2*(1), 13–26. DOI: 10.53623/tasp.v2i1.27

Chung, Y. H., Beiss, V., Fiering, S. N., & Steinmetz, N. F. (2020). COVID-19 Vaccine Frontrunners and Their Nanotechnology Design. *ACS Nano*, *14*(10), 12522–12537. DOI: 10.1021/acsnano.0c07197

Cichy, C., & Rass, S. (2019). An Overview of Data Quality Frameworks. *IEEE Access : Practical Innovations, Open Solutions*, *7*, 24634–24648. Advance online publication. DOI: 10.1109/ACCESS.2019.2899751

Coelho, P. M., Corona, B., ten Klooster, R., & Worrell, E. (2020). Sustainability of reusable packaging–Current situation and trends. *Resources, Conservation and Recycling. X*, *6*, 100037.

Columbia Climate School, & Sarah Fecht. (2021, February 25). *State of the Planet*. Https://News.Climate.Columbia.Edu/2021/02/25/Carbon-Dioxide-Cause-Global-Warming/

Compagnucci, L., & Spigarelli, F. (2018). Fostering cross-sector collaboration to promote innovation in the water sector. *Sustainability (Basel)*, *10*(11), 4154. DOI: 10.3390/su10114154

Cong, H., & Zhang, N. (2022). Perspectives in translating microfluidic devices from laboratory prototyping into scale-up production. *Biomicrofluidics, 16*(2).

Cooney, D. O. (1998). *Adsorption design for wastewater treatment.*

Couck, S., Cousin-Saint-Remi, J., Van der Perre, S., Baron, G. V., Minas, C., Ruch, P., & Denayer, J. F. M. (2018). 3D-printed SAPO-34 monoliths for gas separation. *Microporous and Mesoporous Materials, 255*, 185–191. DOI: 10.1016/j.micromeso.2017.07.014

Couck, S., Lefevere, J., Mullens, S., Protasova, L., Meynen, V., Desmet, G., Baron, G. V., & Denayer, J. F. M. (2017). CO2, CH4 and N2 separation with a 3DFD-printed ZSM-5 monolith. *Chemical Engineering Journal, 308*, 719–726. DOI: 10.1016/j.cej.2016.09.046

Crespilho, F. N., Costa, C. M., & Lanceros-Méndez, S. (2025). Sustainable Battery Biomaterials. *ChemElectroChem, 12*(3), e202400530.

Cui, J., Li, F., Wang, Y., Zhang, Q., Ma, W., & Huang, C. (2020). Electrospun nanofiber membranes for wastewater treatment applications. *Separation and Purification Technology, 250*, 117116. DOI: 10.1016/j.seppur.2020.117116

Cui, Y., Zhu, J., Tong, H., & Zou, R. (2023). Advanced perspectives on MXene composite nanomaterials: Types synthetic methods, thermal energy utilization and 3D-printed techniques. *iScience, 26*(1), 105824. DOI: 10.1016/j.isci.2022.105824

Curran, M. A. (2000). Biobased materials. Kirk-Othmer encyclopedia of chemical technology, 1-19.

Cywar, R. M., Rorrer, N. A., Hoyt, C. B., Beckham, G. T., & Chen, E. Y.-X. (2022). Bio-based polymers with performance-advantaged properties. *Nature Reviews. Materials, 7*(2), 83–103.

Da Silva Neto, J. T., & Soares, P. R. R. (2025). Fiber-reinforced cementitious composites: Recent advances and future perspectives on key properties for high-performance design. *Deleted Journal, 2*(1), 65. Advance online publication. DOI: 10.1007/s44290-025-00209-9

Dadda, K., Alleg, S., Rabhi, S., Abadlia, L., Elmohri, A., Suňol, J. J., & Hlil, E. K. (2025). Sensing the future: Smart nanomaterials revolutionizing environmental monitoring. In *Breaking Boundaries: Pioneering Sustainable Solutions Through Materials and Technology* (pp. 237–255). Springer Nature Singapore. DOI: 10.1007/978-981-97-9827-8_11

Dahal, S., Yilma, W., Sui, Y., Atreya, M., Bryan, S., Davis, V., Whiting, G. L., & Khosla, R. (2020). Degradability of biodegradable soil moisture sensor components and their effect on maize (Zea mays L.) growth. *Sensors (Basel)*, *20*(21), 6154.

Dai, X., & Chen, Y. (2022). Computational Biomaterials: Computational Simulations for Biomedicine. Advanced Materials, 35(7). Portico. DOI: 10.1002/adma.202204798

Dalai, A. K., & Azargohar, R. (2007). Production of activated carbon from biochar using chemical and physical activation: Mechanism and modeling. *ACS Symposium Series, 954*, 463–476. DOI: 10.1021/bk-2007-0954.ch029

Dash, R. K., Borca-Tasciuc, T., Purkayastha, A., & Ramanath, G. "Electrowetting on dielectric-actuation of microdroplets of aqueous bismuth telluride nanoparticle suspensions," Nanotechnology, vol. 18, no. 47, Article ID 475711, 6 pages, 2007.

Das, N., Panda, S., Das, D. K., Nayak, S. K., Parthasarathy, S., Kirgiz, M. S., & Patnaik, P. (2025). Polymer Nanocomposites: Innovations in Material Design and Applications. In *Exploring Nanomaterial Synthesis, Characterization, and Applications* (pp. 169–196). IGI Global.

Das, O., Restás, Á., Shanmugam, V., Sas, G., Försth, M., Xu, Q., Jiang, L., Hedenqvist, M. S., & Ramakrishna, S. (2021). Demystifying low-carbon materials. *Materials Circular Economy*, *3*, 1–11.

Das, R., Lindstrom, T., Sharma, P. R., Chi, K., & Hsiao, B. S. (2022). Nanocellulose for sustainable water purification. *Chemical Reviews*, *122*(9), 8936–9031.

David, M. E., Ion, R.-M., Grigorescu, R. M., Iancu, L., Holban, A. M., Nicoara, A. I., Alexandrescu, E., Somoghi, R., Ganciarov, M., Vasilievici, G., & Gheboianu, A. I. (2021). Hybrid Materials Based on Multi-Walled Carbon Nanotubes and Nanoparticles with Antimicrobial Properties. *Nanomaterials (Basel, Switzerland)*, *11*(6), 1415. DOI: 10.3390/nano11061415

Davies, I. (2016). Sustainable construction timber: sourcing and specifying local timber (pp. 54-pp).

De Araujo, V. (2023). Timber construction as a multiple valuable sustainable alternative: Main characteristics, challenge remarks and affirmative actions. *International Journal of Construction Management*, *23*(8), 1334–1343.

De Belie, N., Gruyaert, E., Al-Tabbaa, A., Antonaci, P., Baera, C., Bajare, D., & Jonkers, H. M. (2018). A review of self-healing concrete for damage management of structures. *Advanced Materials Interfaces*, *5*(17), 1800074.

De Blasio, C., Salierno, G., & Magnano, A. (2021). Implications on feedstock processing and safety issues for semi-batch operations in supercritical water gasification of biomass. *Energies*, *14*(10), 2863.

de Carvalho, A. B. G., Rahimnejad, M., Oliveira, R. L., Sikder, P., Saavedra, G. S., Bhaduri, S. B., & Bottino, M. C. (2024). Personalized bioceramic grafts for craniomaxillofacial bone regeneration. *International Journal of Oral Science*, *16*(1), 62. DOI: 10.1038/s41368-024-00327-7

de Mattos Nascimento, D. L., Mury Nepomuceno, R., Caiado, R. G. G., Maqueira, J. M., Moyano-Fuentes, J., & Garza-Reyes, J. A. (2022). A sustainable circular 3D printing model for recycling metal scrap in the automotive industry. *Journal of Manufacturing Technology Management*, *33*(5), 876–892.

Deepanraj, B., Saravanan, A. M., Senthilkumar, N., Afzal, A. A., & Afzal, A. R. (2025). Drilling Characteristics Optimization of Polymer Composite Fortified with Eggshells using Box-Behnken Design and Zebra Optimization Algorithm. *Results in Engineering*, •••, 104102.

Deepthi Jayan, K., & Ragin Ramdas, M. (2022). Exploring the Photovoltaic Capabilities of Sc4C3 MXene Using Density Functional Theory. *Materials Letters*, *318*, 132217. Advance online publication. DOI: 10.1016/j.matlet.2022.132217

Deng, T., Ran, Y., Yin, Y., & Liu, P. (2020). Multi-Objective Optimization Design of Thermal Management System for Lithium-Ion Battery Pack Based on Non-Dominated Sorting Genetic Algorithm II. *Applied Thermal Engineering*, *164*, 114394. Advance online publication. DOI: 10.1016/j.applthermaleng.2019.114394

Devadas, V. V., Khoo, K. S., Chia, W. Y., Chew, K. W., Munawaroh, H. S. H., Lam, M.-K., Lim, J.-W., Ho, Y.-C., Lee, K. T., & Show, P. L. (2021). Algae biopolymer towards sustainable circular economy. *Bioresource Technology*, *325*, 124702.

Dewangan, S. K., Mohan, M., Kumar, V., Sharma, A., & Ahn, B. (2022). A comprehensive review of the prospects for future hydrogen storage in materials-application and outstanding issues. *International Journal of Energy Research*, *46*(12), 16150–16177. DOI: 10.1002/er.8322

Dhaliwal, H., Browne, M., Flanagan, W., Laurin, L., & Hamilton, M. (2014). A life cycle assessment of packaging options for contrast media delivery: Comparing polymer bottle vs. glass bottle. *The International Journal of Life Cycle Assessment*, *19*, 1965–1973.

Dhal, S. B., & Kar, D. (2025). Leveraging artificial intelligence and advanced food processing techniques for enhanced food safety, quality, and security: A comprehensive review. *Discover Applied Sciences*, *7*(1), 75. Advance online publication. DOI: 10.1007/s42452-025-06472-w

Ding, G. K. (2014). Life cycle assessment (LCA) of sustainable building materials: an overview. Eco-efficient construction and building materials, 38-62.

Dipendu Saha, M. J. K. (2019). Role of oxygen, nitrogen and sulfur functionalities on the surface of nanoporous carbons in CO2 adsorption: A critical review. *Microporous and Mesoporous Materials*, *287*, 29–55. DOI: 10.1016/j.micromeso.2019.05.051

Donzelli, G., Cerbino, R., & Vailati, A. "Bistable heat transfer in a nanofluid," Physical Review Letters, vol. 102, no. 10, Article ID 104503, 4 pages, 2009.

Dorozhkin, S. V. (2017). Calcium Orthophosphate-Based Bioceramics and Its Clinical Applications. In Kaur, G. (Ed.), *Clinical Applications of Biomaterials*. Springer., DOI: 10.1007/978-3-319-56059-5_5

Drzymała, T., Jackiewicz-Rek, W., Tomaszewski, M., Kuś, A., Gałaj, J., & Šukys, R. (2017). Effects of high temperature on the properties of high performance concrete (HPC). *Procedia Engineering*, *172*, 256–263.

Du, A., Lu, H., Liu, S., Chen, S., Chen, Z., Li, W., & Yang, C. (2024). Breaking the Trade-Off between Ionic Conductivity and Mechanical Strength in Solid Polymer Electrolytes for High-Performance Solid Lithium Batteries. *Advanced Energy Materials*, *14*(31), 2400808.

Duan, Z., Deng, Q., Liang, C., Ma, Z., & Wu, H. (2023). Upcycling of recycled plastic fiber for sustainable cementitious composites: A critical review and new perspective. *Cement and Concrete Composites*, *142*, 105192.

Dusastre, V. (2011). *Materials for sustainable energy: a collection of peer-reviewed research and review articles from Nature Publishing Group*. World Scientific.

Dutta, A., Mitra, S., Basak, M., & Banerjee, T. (2023). A comprehensive review on batteries and supercapacitors: Development and challenges since their inception. *Energy Storage*, *5*(1), e339. DOI: 10.1002/est2.339

Dwivedi, C., Manjare, S., & Rajan, S. K. (2020). Recycling of waste tire by pyrolysis to recover carbon black: Alternative & environment-friendly reinforcing filler for natural rubber compounds. *Composites. Part B, Engineering*, *200*, 108346.

Earley, R. (2022). Sustainable Textiles for Fashion. *JPhys Materials*, *5*(032001), 81–84.

Earth System Research Laboratory. (n.d.). *Global Monitoring Laboratory*. Https://Gml.Noaa.Gov/Ccgg/Trends/Mlo.Html

Eckhardt, C. M., Ollivier, M., Madjarova, S. J., Williams, R. J., Pareek, A., Karlsson, J., & Nwachukwu, B. U. (2022). Unsupervised machine learning methods and emerging applications in healthcare. *Knee Surgery, Sports Traumatology, Arthroscopy : Official Journal of the ESSKA*, *31*(2), 376–381. DOI: 10.1007/s00167-022-07233-7 PMID: 36378293

Egwu, C. N., Babalola, R., Udoh, T. H., & Esio, O. O. (2022). Nanotechnology: Applications, Challenges, and Prospects. Advanced Manufacturing in Biological, Petroleum, and Nanotechnology Processing, 3–15. DOI: 10.1007/978-3-030-95820-6_1

El Abbassi, M., Perrin, M. L., Calame, M., Braun, O., Overbeck, J., & van der Zant, H. S. J. (2021). Benchmark and application of unsupervised classification approaches for univariate data. *Communications on Physics*, *4*(1), 50. Advance online publication. DOI: 10.1038/s42005-021-00549-9

El Majd, A., Younsi, Z., Youssef, N., Belouaggadia, N., & El Bouari, A. (2023). Experimental study of thermal characteristics of bio-based textiles integrating microencapsulated phase change materials. *Energy and Building*, *297*, 113465.

Elahi, M. M. A., Hossain, M. M., Karim, M. R., Zain, M. F. M., & Shearer, C. (2020). A review on alkali-activated binders: Materials composition and fresh properties of concrete. *Construction & Building Materials*, *260*, 119788. DOI: 10.1016/j.conbuildmat.2020.119788

El-Azazy, M., AlReyashi, A., Al-Saad, K., Al-Hashimi, N., Al-Ghouti, M. A., Shibl, M. F., & El-Shafie, A. S. (2024). Mandarin Peels-Derived Carbon Dots: A Multifaceted Fluorescent Probe for Cu (II) Detection in Tap and Drinking Water Samples. *Nanomaterials (Basel, Switzerland)*, *14*(20), 1666.

Elton, D. C., Boukouvalas, Z., Butrico, M. S., Fuge, M. D., & Chung, P. W. (2018). Applying Machine Learning Techniques to Predict the Properties of Energetic Materials. *Scientific Reports*, *8*(1), 9059. Advance online publication. DOI: 10.1038/s41598-018-27344-x

Emin, A. (2021). How composite materials influence sustainable development. *Euro Economica*, *40*(2), 202–210.

Enache, A. C., Grecu, I., & Samoila, P. (2024). Polyethylene terephthalate (PET) recycled by catalytic glycolysis: A bridge toward circular economy principles. *Materials (Basel)*, *17*(12), 2991.

Eneh, A. E. (2015). Application of recycled plastics and its composites in the built environment. *Int. J. Manag. Inf, 3*, 9–16.

Epps, R. W., Volk, A. A., Reyes, K. G., & Abolhasani, M. (2021). Accelerated AI development for autonomous materials synthesis in flow. *Chemical Science (Cambridge), 12*(17), 6025–6036. DOI: 10.1039/D0SC06463G PMID: 34976336

Erzen, B., Karataş, M., Orhan, R., & Aydoğmuş, E. (2025). Advancements and Challenges in Biomaterials: Innovations, Sustainability, and Future Prospects. *Journal of Macromolecular Science, Part B: Physics*, •••, 1–27. DOI: 10.1080/00222348.2025.2476852

Esiri, A. E., Kwakye, J. M., Ekechukwu, D. E., Ogundipe, O. B., & Ikevuje, A. H.Andrew Emuobosa EsiriJephta Mensah KwakyeDarlington Eze EkechukwuOlorunshogo Benjamin OgundipeAugusta Heavens Ikevuje. (2024). Leveraging regional resources to address regional energy challenges in the transition to a low-carbon future. *Open Access Research Journal of Multidisciplinary Studies, 8*(1), 105–114. DOI: 10.53022/oarjms.2024.8.1.0052

Espasa-Valdepeñas, A., Vega, J. F., Cruz, V., Ramos, J., Müller, A. J., & Martínez-Salazar, J. (2021). Revisiting polymer–particle interaction in PEO solutions. *Langmuir, 37*(13), 3808–3816.

Eyube, M., Enuesueke, C., & Alimikhena, M. (2025). The Future of Nanomaterials in Manufacturing.

Fahad, M., Nagy, R., & Gosztola, D. (2022). Pavement Sensing Systems: Literature Review. *Civil and Environmental Engineering, 18*(2), 603–630. DOI: 10.2478/cee-2022-0057

Fahim, M., Haris, M., Khan, W., & Zaman, S. (2022). Bamboo as a construction material: prospects and challenges. Advances in Science and Technology. Research Journal, 16(3).

Fang, J., Liu, L., Yang, H., & Du, H. (2024). Applications of Fenton/Fenton-like photocatalytic degradation in g-C3N4 based composite materials. *Journal of Environmental Chemical Engineering, 12*(6), 114153. DOI: 10.1016/j.jece.2024.114153

Fang, Y., Yang, X., Lin, Y., Shi, J., Prominski, A., Clayton, C., & Tian, B. (2021). Dissecting biological and synthetic soft–hard interfaces for tissue-like systems. *Chemical Reviews, 122*(5), 5233–5276.

Fernandes, C., Jathar, M., Sawant, B. K. S., & Warde, T. (2023). Scale-up of nanoparticle manufacturing process. In *Pharmaceutical Process Engineering and Scale-up Principles* (pp. 173–203). Springer Nature Switzerland.

Ferraz-Caetano, J., Teixeira, F., & Cordeiro, M. N. D. S. (2023). Explainable Supervised Machine Learning Model To Predict Solvation Gibbs Energy. *Journal of Chemical Information and Modeling*, *64*(7), 2250–2262. DOI: 10.1021/acs.jcim.3c00544 PMID: 37603608

Fiedler, L., Shah, K., Bussmann, M., & Cangi, A. (2022). Deep dive into machine learning density functional theory for materials science and chemistry. *Physical Review Materials*, *6*(4), 040301.

Finkbeiner, M. (2014). The international standards as the constitution of life cycle assessment: the ISO 14040 series and its offspring. In *Background and future prospects in life cycle assessment* (pp. 85–106). Springer Netherlands.

Flaherty, D. W., & Bhan, A. (2024). Improving the rigor and reproducibility of catalyst testing and evaluation in the laboratory. *Journal of Catalysis*, *431*, 115408.

Formalik, F., Shi, K., Snurr, R. Q., Wang, X., & Joodaki, F. (2023). Exploring the Structural, Dynamic, and Functional Properties of Metal-Organic Frameworks through Molecular Modeling. *Advanced Functional Materials*. Advance online publication. DOI: 10.1002/adfm.202308130

Frackowiak, E., & Béguin, F. (2001). Carbon materials for the electrochemical storage of energy in capacitors. *Carbon*, *39*(6), 937–950. DOI: 10.1016/S0008-6223(00)00183-4

Fry, C., & Cox, D. (2020). Crossing the Valley of Death: Solutions to the Challenges of Translating Academic Research into Industrial Innovation. *Nature Reviews. Materials*, *5*(9), 637–643.

Frydrych, K., Karimi, K., Pecelerowicz, M., Alvarez, R., Dominguez-Gutiérrez, F. J., Rovaris, F., & Papanikolaou, S. (2021). Materials Informatics for Mechanical Deformation: A Review of Applications and Challenges. *Materials (Basel)*, *14*(19), 5764. Advance online publication. DOI: 10.3390/ma14195764

Fuhg, J. N., Hamel, C. M., Johnson, K., Jones, R., & Bouklas, N. (2023). Modular Machine Learning-Based Elastoplasticity: Generalization in the Context of Limited Data. *Computer Methods in Applied Mechanics and Engineering*, *407*, 115930. Advance online publication. DOI: 10.1016/j.cma.2023.115930

Fuhr, A. S., & Sumpter, B. G. (2022). Deep Generative Models for Materials Discovery and Machine Learning-Accelerated Innovation. *Frontiers in Materials*, *9*, 865270. Advance online publication. DOI: 10.3389/fmats.2022.865270

Gajjar, P., Pettee, B., Britt, D. W., Huang, W., Johnson, W. P., & Anderson, A. J. (2009). Antimicrobial activities of commercial nanoparticles against an environmental soil microbe, pseudomonas putida KT2440. *Journal of Biological Engineering*, *3*(1), 1–13.

Gao, Q., Dukker, T., Schweidtmann, A. M., & Weber, J. M. (2024). Self-Supervised Graph Neural Networks for Polymer Property Prediction. *Molecular Systems Design & Engineering*, *9*(11), 1130–1143. DOI: 10.1039/D4ME00088A

Gao, X., Jamil, N., Ramli, M. I., & Syed, M. Z. S. Z. A. (2024). A Comparative Analysis of Combination of CNN-Based Models with Ensemble Learning on Imbalanced Data. *JOIV: International Journal on Informatics Visualization*, *8*(1), 456. Advance online publication. DOI: 10.62527/joiv.8.1.2194

Gao, X., Wu, J., Yu, Y., Chu, Z., Shi, H., & Dong, S. (2018). Giant piezoelectric coefficients in relaxor piezoelectric ceramic PNN-PZT for vibration energy harvesting. *Advanced Functional Materials*, *28*(30), 1706895. DOI: 10.1002/adfm.201706895

Gao, Y., Li, Y., Niu, Q., & Zhang, P. (2025). Transforming petroleum asphalt into carbon fibers and related metal/oxide composites by electrospinning synthesis. *Green Chemistry*, *27*(26), 8007–8017. DOI: 10.1039/D5GC01775K

Garavand, F., Jafarzadeh, S., Cacciotti, I., Vahedikia, N., Sarlak, Z., Tarhan, Ö., Yousefi, S., Rouhi, M., Castro-Muñoz, R., & Jafari, S. M. (2022). Different strategies to reinforce the milk protein-based packaging composites. *Trends in Food Science & Technology*, *123*, 1–14.

García, Á., Schlangen, E., Van De Ven, M., & Liu, Q. (2009). Electrical conductivity of asphalt mortar containing conductive fibers and fillers. *Construction & Building Materials*, *23*(10), 3175–3181. DOI: 10.1016/j.conbuildmat.2009.06.014

García-Beltrán, G., & Mercado-Zuniga, C. (2020). Navigation of Silver/Carbon Nano-antennas in Organic Fluids Explored by a Two-Wave Mixing. *Nanomaterials (Basel, Switzerland)*, *10*, 1886.

García-Salaberri, P. A. (2024) Battery manufacturing—from laboratory to industry challenges, Elsevier, 465-489.

Gardetti, M. Á., & Muthu, S. S. (2019). *Organic Cotton*. Textile Science and Clothing Technology.

Gautam, M. K., Mondal, T., Nath, R., Mahajon, B., Chincholikar, M., Bose, A., & Mondal, S. (2024). Harnessing activated hydrochars: A novel approach for pharmaceutical contaminant removal. *C*, *10*(1), 8.

Gautam, R., Marriwala, N., & Devi, R. (2023). A review: Study of Mxene and graphene together. *Measurement. Sensors*, *25*, 100592. DOI: 10.1016/j.measen.2022.100592

Geng, X., Wang, S., Mao, X., Wang, F., Wu, G., Zhao, H., Zhang, C., Gao, J., & Wu, H. (2023). Data-driven and artificial intelligence accelerated steel material research and intelligent manufacturing technology. *Materials Genome Engineering Advances*, *1*(1), e10. Advance online publication. DOI: 10.1002/mgea.10

Ghimire, G., Noormohammadi, F., Kohl, J., Devenport, H., & Johnson, P. A. (2025). Machine learning optimization of laser-induced graphene parameters for surface-enhanced Raman spectroscopic detection of glucose. *Sensors and Actuators. B, Chemical*, *432*, 137466. DOI: 10.1016/j.snb.2025.137466

Gholampour, A., & Ozbakkaloglu, T. (2020). A review of natural fiber composites: Properties, modification and processing techniques, characterization, applications. *Journal of Materials Science*, *55*(3), 829–892.

Ghosh, M., Ghosh, A., & Roy, A. (2020). Renewable and sustainable materials in automotive industry. Encyclopedia of Renewable and Sustainable Materials: p. 162-179.

Ghosh, A., & Kiran, B. (2017). Carbon concentration in algae: Reducing CO2 from exhaust gas. *Trends in Biotechnology*, *35*(9), 806–808. DOI: 10.1016/j.tibtech.2017.05.003

Giani, M. I., Dotelli, G., Brandini, N., & Zampori, L. (2015). Comparative life cycle assessment of asphalt pavements using reclaimed asphalt, warm mix technology and cold in-place recycling. *Resources, Conservation and Recycling*, *104*, 224–238. DOI: 10.1016/j.resconrec.2015.08.006

Giglio, E., Pirone, R., & Bensaid, S. (2021). Dynamic modelling of methanation reactors during start-up and regulation in intermittent power-to-gas applications. *Renewable Energy*, *170*, 1040–1051.

Giulimondi, V., Mitchell, S., & Pérez-Ramírez, J. (2023). Challenges and opportunities in engineering the electronic structure of single-atom catalysts. *ACS Catalysis*, *13*(5), 2981–2997.

Godja, N.-C., & Munteanu, F.-D. (2024). Hybrid Nanomaterials: A Brief Overview of Versatile Solutions for Sensor Technology in Healthcare and Environmental Applications. *Biosensors (Basel)*, *14*(2), 67. DOI: 10.3390/bios14020067

Goettmann, F., Fischer, A., Antonietti, M., & Thomas, A. (2006). Chemical Synthesis of Mesoporous Carbon Nitrides Using Hard Templates and Their Use as a Metal-Free Catalyst for Friedel–Crafts Reaction of Benzene. *Angewandte Chemie International Edition*, *45*(27), 4467–4471. DOI: 10.1002/anie.200600412

Goh, Y., Yap, S. P., & Tong, T. Y. (2020). Bamboo: the emerging renewable material for sustainable construction. Encyclopedia of Renewable and Sustainable Materials. 2(3): p. 365-376.

Goloshchapov, D. L., Kashkarov, V. M., Rumyantseva, N. A., Seredin, P. V., Lenshin, A. S., Agapov, B. L., & Domashevskaya, E. P. (2013). Synthesis of nanocrystalline hydroxyapatite by precipitation using hen's eggshell. *Ceramics International*, *39*(4), 4539–4549. DOI: 10.1016/j.ceramint.2012.11.050

Gonçalves, R., Martinho, A., & Oliveira, J. (2022). Evaluating the potential use of recycled glass fibers for the development of gypsum-based composites. *Construction & Building Materials*, *321*, 126320.

Gounder, S., Hasan, A., Shrestha, A., & Elmualim, A. (2023). Barriers to the use of sustainable materials in Australian building projects. *Engineering, Construction, and Architectural Management*, *30*(1), 189–209.

Gouveia, J. D., Galvão, T. L. P., Iben Nassar, K., & Gomes, J. R. B. (2025). First-principles and machine-learning approaches for interpreting and predicting the properties of MXenes. *Npj 2D Materials and Applications, 9*(1), 8. DOI: 10.1038/s41699-025-00529-5

Govindaraj, D., & Rajan, M. (2016). Synthesis and spectral characterization of novel nanohydroxyapatite from Moringaoleifera leaves. *Materials Today: Proceedings*, *3*(6), 2394–2398. DOI: 10.1016/j.matpr.2016.04.153

Gui, R., & Jin, H. (2025). Synthesis, characterizations, properties, modifications and applications of zero-dimensional monoelemental selenium nanomaterials. *Coordination Chemistry Reviews*, *535*, 216626. DOI: 10.1016/j.ccr.2025.216626

Guo, P., Song, H., & Chen, X. (2009). Electrochemical performance of graphene nanosheets as anode material for lithium-ion batteries. *Electrochemistry Communications*, *11*(6), 1320–1324. DOI: 10.1016/j.elecom.2009.04.036

Guo, Q., He, Y., Chen, Y., & Qiao, C. (2025). Design and application of computational modeling in science education research: A systematic review. *Interactive Learning Environments*, *33*(2), 969–992.

Guo, Y., Tan, C., Sun, J., Li, W., Zhang, J., & Zhao, C. (2020). Porous activated carbons derived from waste sugarcane bagasse for CO2 adsorption. *Chemical Engineering Journal, 381*, 122736. Advance online publication. DOI: 10.1016/j.cej.2019.122736

Guo, Y., Wei, Y., Li, H., & Zhai, T. (2017). Layer Structured Materials for Advanced Energy Storage and Conversion. *Small, 13*(45), 1701649. Advance online publication. DOI: 10.1002/smll.201701649

Gupta, D., Varghese, B. S., Suresh, M., Panwar, C., & Gupta, T. K. (2022). Nanoarchitectonics: Functional nanomaterials and nanostructures—A review. *Journal of Nanoparticle Research, 24*(10), 196. DOI: 10.1007/s11051-022-05577-2

Gupta, P. K. (2023). Introduction and historical background. In *Nanotoxicology in Nanobiomedicine* (pp. 1–22). Springer International Publishing., DOI: 10.1007/978-3-031-24287-8_1

Güvensoy-Morkoyun, A., Baysal, T., Tantekin-Ersolmaz, Ş. B., & Velioğlu, S. (2024). Guide for Nonequilibrium Molecular Dynamics Simulations of Organic Solvent Transport in Nanopores: The Case of 2D MXene Membranes. *Journal of Chemical Theory and Computation, 20*(21), 9642–9654. DOI: 10.1021/acs.jctc.4c00693

Habert, G., De Lacaillerie, J. D., & Roussel, N. (2011). An environmental evaluation of geopolymer based concrete production: Reviewing current research trends. *Journal of Cleaner Production, 19*(11), 1229–1238. DOI: 10.1016/j.jclepro.2011.03.012

Hachem, K., Ansari, M. J., Saleh, R. O., Kzar, H. H., Al-Gazally, M. E., Altimari, U. S., Hussein, S. A., Mohammed, H. T., Hammid, A. T., & Kianfar, E. (2022). Methods of Chemical Synthesis in the Synthesis of Nanomaterial and Nanoparticles by the Chemical Deposition Method: A Review. *BioNanoScience, 12*(3), 1032–1057. DOI: 10.1007/s12668-022-00996-w

Hafiz, W., Meng, Y., Qi, D., & Zhu, F. (2025). *Nis/ Hollow Carbon Sphere Composites for High-Performance and Long-Life Lithium-Ion Battery Anodes*. DOI: 10.1016/j.jelechem.2025.119237

Haider, J., Shahzadi, A., Akbar, M. U., Hafeez, I., Shahzadi, I., Khalid, A., Ashfaq, A., Ahmad, S. O. A., Dilpazir, S., Imran, M., Ikram, M., Ali, G., Khan, M., Khan, Q., & Maqbool, M. (2022). A review of synthesis, fabrication, and emerging biomedical applications of metal-organic frameworks. *Biomaterials Advances, 140*, 213049. DOI: 10.1016/j.bioadv.2022.213049

Han, Z. H., Cao, F. Y., & Yang, B. "Synthesis and thermal characterization of phase-changeable indium/poly alpha olefin nanofluids," Applied Physics Letters, vol. 92, no. 24, Article ID 243104, 3 pages, 2008.

Han, G., Sun, Y., Feng, Y., Lin, G., & Lu, N. (2021). *Machine Learning Regression Guided Thermoelectric Materials Discovery – A Review*. ES Materials & Manufacturing., DOI: 10.30919/esmm5f451

Hangarter, C. M., Dyatkin, B., Laskoski, M., Palenik, M. C., Miller, J. B., Tyagi, M., & Klug, C. A. (2024). A cobalt-based layered MOF material for supercapacitor applications. *Journal of Energy Storage*, *89*, 111476.

Han, J., Zhang, L., Zhao, B., Qin, L., Wang, Y., & Xing, F. (2019). The N-doped activated carbon derived from sugarcane bagasse for CO2 adsorption. *Industrial Crops and Products*, *128*, 290–297. DOI: 10.1016/j.indcrop.2018.11.028

Hao, G.-P., Li, W.-C., Qian, D., & Lu, A.-H. (2010). Rapid Synthesis of Nitrogen-Doped Porous Carbon Monolith for CO_2 Capture. *Advanced Materials*, *22*(7), 853–857. DOI: 10.1002/adma.200903765

Hao, W., Björkman, E., Lilliestråle, M., & Hedin, N. (2013). Activated carbons prepared from hydrothermally carbonized waste biomass used as adsorbents for CO2. *Applied Energy*, *112*, 526–532. DOI: 10.1016/j.apenergy.2013.02.028

Hardjito, D., & Rangan, B. V. (2005). Development and properties of low-calcium fly ash-based geopolymer concrete (Research Report GC1). Faculty of Engineering, Curtin University of Technology.

Hargittai, I. (2022). Robert F. Curl, Jr. (1933–2022) structural chemist, Nobel laureate co-discoverer of buckminsterfullerene. *Structural Chemistry*, *33*(6), 2265–2266. DOI: 10.1007/s11224-022-02041-8

Hariram, N. P., Mekha, K. B., Suganthan, V., & Sudhakar, K. (2023). Sustainalism: An integrated socio-economic-environmental model to address sustainable development and sustainability. *Sustainability (Basel)*, *15*(13), 10682. DOI: 10.3390/su151310682

Hariram, V., John, J. G., Sangeethkumar, E., Gajalakshmi, B., Ramanathan, V., Vinothkumar, M., & Balachandar, M. (2022). Scenedesmus obliquus and Chlorella vulgaris–A Prospective Algal Fuel Source. *Nature Environment and Pollution Technology*, *21*(5), 2129–2139. DOI: 10.46488/NEPT.2022.v21i05.009

Hariram, V., Sathishbabu, R., John, J. G., Kailiappan, N., Vijayakumar, K., Kumar, E. S., & Priya, K. K. (2025). Maximizing the combustion phenomenon with reduced emission of a compression ignition engine fulled with Dunaliella salina biodiesel and hydrogen blends. *Scientific Reports*, *15*(1), 30372. DOI: 10.1038/s41598-025-16285-x

Harren, T., Rarey, M., Hessler, G., Grebner, C., & Gutermuth, T. (2024). Modern machine-learning for binding affinity estimation of protein–ligand complexes: Progress, opportunities, and challenges. *Wiley Interdisciplinary Reviews. Computational Molecular Science*, *14*(3), e1716. Advance online publication. DOI: 10.1002/wcms.1716

Harte, A. M. (2017). Mass timber–the emergence of a modern construction material. *Journal of Structural Integrity and Maintenance*, *2*(3), 121–132.

Hasan, M. N., Sahlan, S., Osman, K., & Mohamed Ali, M. S. (2021). Energy harvesters for wearable electronics and biomedical devices. *Advanced Materials Technologies*, *6*(3), 2000771. DOI: 10.1002/admt.202000771

Hashemi, S., & Akhavan-Behabadi, M. (2012). An empirical study on heat transfer and pressure drop characteristics of CuO–base oil nanofluid flow in a horizontal helically coiled tube under constant heat flux. *International Communications in Heat and Mass Transfer*, *39*(1), 144–151.

Hassan, Q., Azzawi, I. D., Sameen, A. Z., & Salman, H. M. (2023). Hydrogen fuel cell vehicles: Opportunities and challenges. *Sustainability (Basel)*, *15*(15), 11501. DOI: 10.3390/su151511501

Hegazy, H. H., Khan, J., Shakeel, N., Alabdullkarem, E. A., Saleem, M. I., Alrobei, H., & Yahia, I. S. (2024). 2D-based electrode materials for supercapacitors – status, challenges, and prospects. *RSC Advances*, *14*(45), 32958–32977. DOI: 10.1039/D4RA05473C

Heidari, A., Younesi, H., Rashidi, A., & Ghoreyshi, A. A. (2014). Evaluation of CO2 adsorption with eucalyptus wood based activated carbon modified by ammonia solution through heat treatment. *Chemical Engineering Journal*, *254*, 503–513. DOI: 10.1016/j.cej.2014.06.004

Heidari, M., Zad, S., & Rafatirad, S. (2021).. . *Ensemble of Supervised and Unsupervised Learning Models to Predict a Profitable Business Decision.*, *7*, 1–6. DOI: 10.1109/IEMTRONICS52119.2021.9422649

Heng, T., Zhang, L., Wang, R., Lu, Y., Yang, D., & Du, G. (2021). Progress in Research on Artificial Intelligence Applied to Polymorphism and Cocrystal Prediction. *ACS Omega*, *6*(24), 15543–15550. DOI: 10.1021/acsomega.1c01330 PMID: 34179597

Hepel, M. (2023). Advances in micro-supercapacitors (MSCs) with high energy density and fast charge-discharge capabilities for flexible bioelectronic devices—A review. *Electrochemical Science Advances*, *3*(3), e2100222. DOI: 10.1002/elsa.202100222

Herdiana, Y., Wathoni, N., Shamsuddin, S., & Muchtaridi, M. (2022). Scale-up polymeric-based nanoparticles drug delivery systems: Development and challenges. *OpenNano*, *7*, 100048.

He, S., Hu, Z., He, C., Tian, S., Ouyang, G., Lin, Z., & Fang, J. (2025). Nano-confinement-Enabled Resistance to Water Matrix Components for Selective Degradation of Micropollutants. *Angewandte Chemie*, *137*(14), e202424827. DOI: 10.1002/ange.202424827

Heuchel, M., Davies, G. M., Buss, E., & Seaton, N. A. (1999). Adsorption of Carbon Dioxide and Methane and Their Mixtures on an Activated Carbon: Simulation and Experiment. *Langmuir*, *15*(25), 8695–8705. DOI: 10.1021/la9904298

He, Y., Ye, H., Li, H., Miao, G., Hu, Y., Zeng, X., & Xu, F. (2025). Fabrication of lignin nanoparticles with adjustable size, antioxidant, antibacterial, and hydrophobic properties by a two-step fractionation. *International Journal of Biological Macromolecules*, *297*, 139618.

Hidayu, A. R., & Muda, N. (2016). Preparation and Characterization of Impregnated Activated Carbon from Palm Kernel Shell and Coconut Shell for CO2 Capture. *Procedia Engineering*, *148*, 106–113. DOI: 10.1016/j.proeng.2016.06.463

Himanen, L., Jäger, M. O. J., Morooka, E. V., Canova, F. F., Ranawat, Y. S., Gao, D. Z., Rinke, P., & Foster, A. S. (2020). DScribe: Library of Descriptors for Machine Learning in Materials Science. *Computer Physics Communications*, *247*, 106949. Advance online publication. DOI: 10.1016/j.cpc.2019.106949

Hinton, E., & Owen, D. R. J. (1986). *Finite elements in plasticity: Theory and practice*. Pineridge Press Limited.

Hiremath, A., Murthy, A. A., & Thipperudrappa, S., & KN, B. (2021). Nanoparticles filled polymer nanocomposites: A technological review. *Cogent Engineering*, *8*(1), 1991229.

Holloway, M. D. (2025). *Synthesizing Materials in Microgravity: Unlocking Novel Materials Beyond Earth's Limits*. CRC Press. DOI: 10.1201/9781003656081

Hopewell, J., Dvorak, R., & Kosior, E. (2009). Plastics recycling: Challenges and opportunities. *Philosophical Transactions of the Royal Society of London. Series B, Biological Sciences*, *364*(1526), 2115–2126.

Horikawa, T., Sakao, N., Sekida, T., Hayashi, J., Do, D. D., & Katoh, M. (2012). Preparation of nitrogen-doped porous carbon by ammonia gas treatment and the effects of N-doping on water adsorption. *Carbon*, *50*(5), 1833–1842. DOI: 10.1016/j.carbon.2011.12.033

Hospital, A., Goñi, J. R., Orozco, M., & Gelpí, J. L. (2015). Molecular Dynamics Simulations: Advances and Applications. *Advances and Applications in Bioinformatics and Chemistry : AABC*, 37. Advance online publication. DOI: 10.2147/AABC.S70333

Hossain, M. K., Islam, S., Sakib, M. N., Uddin, M. S., Toki, G. F., Rubel, M. H., & Bencherif, H. (2025). Exploring the optoelectronic and photovoltaic characteristics of lead-free Cs2TiBr6 double perovskite solar cells: A DFT and SCAPS-1D investigations. *Advanced Electronic Materials*, *11*(2), 2400348.

Hosseinzadeh, E., Davarpanah, M., Nemati, N. H., & Tavakoli, S. A. (2014). Fabrication of a hard tissue replacement using natural hydroxyapatite derived from bovine bones by thermal decomposition method. *International Journal of Organ Transplantation Medicine*, *5*, 23–31.

Ho, W. K. O., Tang, B.-S., & Wong, S. W. (2020). Predicting property prices with machine learning algorithms. *Journal of Property Research*, *38*(1), 48–70. DOI: 10.1080/09599916.2020.1832558

Hu, X., Dai, Y., & Liu, A. (2024). Combining ChatGPT and knowledge graph for explainable machine learning-driven design: a case study. Journal of Engineering Design, ahead-of-print(ahead-of-print), 1–23. DOI: 10.1080/09544828.2024.2355758

Huang, G., Huang, F., & Dong, W. (2024). Machine learning in energy storage material discovery and performance prediction. *Chemical Engineering Journal*, *492*, 152294. DOI: 10.1016/j.cej.2024.152294

Huang, K., Chai, S.-H., Mayes, R. T., Veith, G. M., Browning, K. L., Sakwa-Novak, M. A., Potter, M. E., Jones, C. W., Wu, Y.-T., & Dai, S. (2015). An efficient low-temperature route to nitrogen-doping and activation of mesoporous carbons for CO_2 capture. *Chemical Communications (Cambridge)*, *51*(97), 17261–17264. DOI: 10.1039/C5CC05619E

Huang, M., Li, Z., & Zhu, H. (2022). Recent Advances of Graphene and Related Materials in Artificial Intelligence. *Advanced Intelligent Systems*, *4*(10), 2200077. Advance online publication. DOI: 10.1002/aisy.202200077

Huang, Y., Bird, R. N., & Heidrich, O. (2007). A review of the use of recycled solid waste materials in asphalt pavements. *Resources, Conservation and Recycling*, *52*(1), 58–73. DOI: 10.1016/j.resconrec.2007.02.002

Huang, Y., Shafiee, M., Charnley, F., & Encinas-Oropesa, A. (2022). Designing a framework for materials flow by integrating circular economy principles with end-of-life management strategies. *Sustainability*, *14*(7), 4244.

Huang, Z.-M., Zhang, Y.-Z., Kotaki, M., & Ramakrishna, S. (2003). A review on polymer nanofibers by electrospinning and their applications in nanocomposites. *Composites Science and Technology*, *63*(15), 2223–2253. DOI: 10.1016/S0266-3538(03)00178-7

Hunter, R. N., Self, A., & Read, J. (2015). *The Shell Bitumen Handbook* (6th ed.). ICE Publishing.

Hu, Q., Albolkany, M. K., Liu, C., Zhang, T., Cui, S., Ye, F., & Liu, B. (2024). Photothermal graphite phase carbon nitride membrane intercalated with polyoxovanadate for efficient solar steam generation. *Desalination*, *583*, 117650. DOI: 10.1016/j.desal.2024.117650

Hurtado, A., Aljabali, A. A., Mishra, V., Tambuwala, M. M., & Serrano-Aroca, Á. (2022). Alginate: Enhancement strategies for advanced applications. *International Journal of Molecular Sciences*, *23*(9), 4486.

Hussain, A., Ghaffar, I., Sattar, S., Muneeb, M., Hasan, A., & Deepanraj, B. (2025). Eco-friendly catalysts revolutionizing energy and environmental applications: An overview. *Topics in Catalysis*, *68*(5), 487–509.

Hussain, T., Chandio, I., Ali, A., Hyder, A., Memon, A. A., Yang, J., & Thebo, K. H. (2024). Recent developments of artificial intelligence in MXene-based devices: From synthesis to applications. *Nanoscale*, *16*(38), 17723–17760. DOI: 10.1039/D4NR03050H

Ibrahim, T. Ozbolat. (2017). Extrusion-Based Bioprinting**With minor contributions by Monika Hospodiuk, The Pennsylvania State University. In *3D Bioprinting* (pp. 93–125). Academic Press.

Ibrahim, I. D., Hamam, Y., Sadiku, E. R., Ndambuki, J. M., Kupolati, W. K., Jamiru, T., Eze, A. A., & Snyman, J. (2022). Need for sustainable packaging: An overview. *Polymers*, *14*(20), 4430.

Ichou, H., Arrousse, N., Berdimurodov, E., & Aliev, N. (2023). Exploring the Advancements in Physical Vapor Deposition Coating: A Review. *Journal of Bio- and Tribo-Corrosion*, *10*(1), 3. Advance online publication. DOI: 10.1007/s40735-023-00806-0

Imoisili, P. E., Ukoba, K. O., & Jen, T. C. (2020a). Synthesis and characterization of amorphous mesoporous silica from palm kernel shell ash. boletín de la sociedad española de cerámica y vidrio, 59(4), 159-164. DOI: 10.1016/j.bsecv.2019.09.006

Imoisili, P. E., & Jen, T. C. (2024). [a]). Microwave-assisted sol–gel synthesis of V-TiO2 nanocatalyst for wastewater treatment. *Materials Today: Proceedings*, *105*, 247–252. Advance online publication. DOI: 10.1016/j.matpr.2023.09.021

Imoisili, P. E., & Jen, T.-C. (2021). Facile Extraction and Characterization of Silica Nanoparticles from Corn Stalk by Sol-Gel Hydrothermal Methods. *IOP Conference Series. Materials Science and Engineering*, *1107*(1), 012030. DOI: 10.1088/1757-899X/1107/1/012030

Imoisili, P. E., & Jen, T.-C. (2024). [b]). Synthesis and characterization of amorphous nano silica from South African coal fly ash. *Materials Today: Proceedings*, *105*, 21–26. DOI: 10.1016/j.matpr.2023.06.077

Imoisili, P. E., Makhatha, M. E., & Jen, T.-C. (2024). Artificial Intelligence prediction and optimization of the mechanical strength of modified Natural Fibre/MWCNT polymer nanocomposite. *Journal of Science. Advanced Materials and Devices*, *9*(2), 100705. DOI: 10.1016/j.jsamd.2024.100705

Imoisili, P. E., Ukoba, K., & Jen, T.-C. (2020b). Physical, mechanical and thermal properties of high frequency microwave treated plantain (Musa Paradisiaca) fibre/MWCNT hybrid epoxy nanocomposites. *Journal of Materials Research and Technology*, *9*(3), 4933–4939. DOI: 10.1016/j.jmrt.2020.03.012

Indira, V., & Abhitha, K. (2023). Mesoporogen-free synthesis of hierarchical zeolite A for CO2 capture: Effect of freeze drying on surface structure, porosity and particle size. *Results in Engineering*, *17*, 100886. DOI: 10.1016/j.rineng.2023.100886

Innes, J. L. (2016). Sustainable forest management: From concept to practice. In *Sustainable Forest Management* (pp. 21–52). Routledge.

Iravani, S., Khosravi, A., Nazarzadeh Zare, E., Varma, R. S., Zarrabi, A., & Makvandi, P. (2024). MXenes and artificial intelligence: Fostering advancements in synthesis techniques and breakthroughs in applications. *RSC Advances*, *14*(49), 36835–36851. DOI: 10.1039/D4RA06384H

Iravani, S., Zarepour, A., Nazarzadeh Zare, E., Makvandi, P., Khosravi, A., & Zarrabi, A. (2024). Synergistic advancements: Exploring MXene/graphene oxide and MXene/reduced graphene oxide composites for next-generation applications. *FlatChem*, *48*, 100759. DOI: 10.1016/j.flatc.2024.100759

Irzhak, V. I., Uflyand, I. E., & Dzhardimalieva, G. I. (2022). Self-healing of polymers and polymer composites. *Polymers*, *14*(24), 5404.

Islam, M. M. U., Li, J., & Roychand, R., Saberia.n, M., & Chen, F. (2022). A comprehensive review on the application of renewable waste tire rubbers and fibers in sustainable concrete. *Journal of Cleaner Production*, *374*, 133998.

Islam, M. M., Yu, T., Giannoccaro, G., Mi, Y., La Scala, M., Rajabi, M. N., & Wang, J. (2024). Improving reliability and stability of the power systems: A comprehensive review on the role of energy storage systems to enhance flexibility. *IEEE Access : Practical Innovations, Open Solutions*.

Islam, S. (2024). FUTURE TRENDS IN SQL DATABASES AND BIG DATA ANALYTICS: IMPACT OF MACHINE LEARNING AND ARTIFICIAL INTELLIGENCE. *International Journal of Science and Engineering*, *1*(4), 47–62. DOI: 10.62304/ijse.v1i04.188

Istrate, O. M., & Chen, B. (2018). Structure and properties of clay/recycled plastic composites. *Applied Clay Science*, *156*, 144–151.

Jabeen, S., Khan, T., Jaiswal, A., & Bala, S. (2024). Green Nanotechnology for Clean Energy and Environmental Sustainability. Sustainable Nanomaterials, 1–20. DOI: 10.1007/978-981-97-2761-2_1

Jaber, H. L., Hammood, A. S., & Parvin, N. (2018). Synthesis and characterization of hydroxyapatite powder from natural Camelus bone. *J. Aust. Ceram. Soc.*, *54*(1), 1–10. DOI: 10.1007/s41779-017-0120-0

Jafarizadeh, H., Yamini, E., Zolfaghari, S. M., Esmaeilion, F., Assad, M. E. H., & Soltani, M. (2024). Navigating challenges in large-scale renewable energy storage: Barriers, solutions, and innovations. *Energy Reports*, *12*, 2179–2192.

Jalal, R., Goharshadi, E. K., Abareshi, M., Moosavi, M., Yousefi, A., & Nancarrow, P. (2010). ZnO nanofluids: Green synthesis, characterization, and antibacterial activity. *Materials Chemistry and Physics*, *121*(1–2), 198–201.

Jamal-Abad, M. T., Zamzamian, A., & Dehghan, M. (2013). Experimental studies on the heat transfer and pressure drop characteristics of Cu–water and Al–water nanofluids in a spiral coil. *Experimental Thermal and Fluid Science*, *47*, 206–212.

Janasik, M., Jagusiak-Kocik, M., & Ulewicz, R. (2023). Barriers to technology transfer and commercialization of research findings: Case study. *Organization and Management Series*, (179), 164–180.

Jayaramulu, K., Mukherjee, S., Morales, D. M., Dubal, D. P., Nanjundan, A. K., Schneemann, A., Masa, J., Kment, S., Schuhmann, W., Otyepka, M., Zbořil, R., & Fischer, R. A. (2022). Graphene-Based Metal–Organic Framework Hybrids for Applications in Catalysis, Environmental, and Energy Technologies. *Chemical Reviews*, *122*(24), 17241–17338. DOI: 10.1021/acs.chemrev.2c00270

Jia, H., & Sundar, S. S. (2024). Vivid and engaging: Effects of interactive data visualization on perceptions and attitudes about social issues. *Digital Journalism (Abingdon, England)*, *12*(8), 1205–1229. DOI: 10.1080/21670811.2023.2250815

Jiang, Z., Li, L., Huang, H., He, W., & Ming, W. (2022). Progress in Laser Ablation and Biological Synthesis Processes: "Top-Down" and "Bottom-Up" Approaches for the Green Synthesis of Au/Ag Nanoparticles. *International Journal of Molecular Sciences*, *23*(23), 14658. DOI: 10.3390/ijms232314658

Jiménez, V., Díaz, J. A., Sánchez, P., Valverde, J. L., & Romero, A. (2009). Influence of the activation conditions on the porosity development of herringbone carbon nanofibers. *Chemical Engineering Journal*, *155*(3), 931–940. DOI: 10.1016/j.cej.2009.09.035

Jin, J., Song, M., & Hourston, D. J. (2004). Novel chitosan-based films cross-linked by genipin with improved physical properties. *Biomacromolecules*, *5*(1), 162–168.

Joarder, M. S. A., Islam, M. S., Hasan, M. H., Kutub, A., Kabir, M. F., Rashid, F., & Joarder, T. A. (2025). A comprehensive review of carbon dioxide capture, transportation, utilization, and storage: A source of future energy. *Environmental Science and Pollution Research International*, *32*(15), 1–34. DOI: 10.1007/s11356-025-36284-9

Jolliffe, I. (2022). A 50-Year Personal Journey through Time with Principal Component Analysis. *Journal of Multivariate Analysis*, *188*, 104820. Advance online publication. DOI: 10.1016/j.jmva.2021.104820

Jones, D., Liang, Y., & Harvey, J. (2017, September 1). *Performance based specifications: Literature review on increasing crumb rubber usage by adding small amounts of crumb rubber modifier in hot mix asphalt.* https://escholarship.org/uc/item/9cw18024

Jordan, A., Hall, C. G., Thorp, L. R., & Sneddon, H. F. (2022). Replacement of less-preferred dipolar aprotic and ethereal solvents in synthetic organic chemistry with more sustainable alternatives. *Chemical Reviews*, *122*(6), 6749–6794.

Joshi, J., Homburg, S. V., & Ehrmann, A. (2022). Atomic Force Microscopy (AFM) on Biopolymers and Hydrogels for Biotechnological Applications—Possibilities and Limits. *Polymers, 14*(6), 1267. DOI: 10.3390/polym14061267

Jribi, S., Miyazaki, T., Saha, B. B., Pal, A., Younes, M. M., Koyama, S., & Maalej, A. (2017). Equilibrium and kinetics of CO2 adsorption onto activated carbon. *International Journal of Heat and Mass Transfer, 108*, 1941–1946. DOI: 10.1016/j.ijheatmasstransfer.2016.12.114

Jung, Y. J., Han, S. H., & Choi, H. J. (2021). Explaining CNN and RNN Using Selective Layer-Wise Relevance Propagation. *IEEE Access : Practical Innovations, Open Solutions, 9*, 18670–18681. Advance online publication. DOI: 10.1109/ACCESS.2021.3051171

Jussambayev, M., Shakenov, K., Sultakhan, S., Zhantikeyev, U., Askaruly, K., Toshtay, K., & Azat, S. (2025). MXenes for sustainable energy: A comprehensive review on conservation and storage applications. *Carbon Trends, 19*, 100471. DOI: 10.1016/j.cartre.2025.100471

Kabakçi, G. C., Aslan, O., & Bayraktar, E. (2022). A review on analysis of reinforced recycled rubber composites. *Journal of Composites Science, 6*(8), 225.

Kadulkar, S., Sherman, Z. M., Ganesan, V., & Truskett, T. M. (2022). Machine Learning-Assisted Design of Material Properties. *Annual Review of Chemical and Biomolecular Engineering, 13*(1), 235–254. Advance online publication. DOI: 10.1146/annurev-chembioeng-092220-024340

Kadupitiya, J., Fox, G. C., & Jadhao, V. (2020). Machine learning for parameter auto-tuning in molecular dynamics simulations: Efficient dynamics of ions near polarizable nanoparticles. *International Journal of High Performance Computing Applications, 34*(3), 357–374. DOI: 10.1177/1094342019899457

Kai, D., Tan, M. J., Chee, P. L., Chua, Y. K., Yap, Y. L., & Loh, X. J. (2016). Towards lignin-based functional materials in a sustainable world. *Green Chemistry, 18*(5), 1175–1200.

Kalair, A., Abas, N., Saleem, M. S., Kalair, A. R., & Khan, N. (2021). Role of energy storage systems in energy transition from fossil fuels to renewables. *Energy Storage, 3*(1), e135. DOI: 10.1002/est2.135

Kamyab, H., Chelliapan, S., Hayder, G., Yusuf, M., Taheri, M. M., Rezania, S., Hasan, M., Yadav, K. K., Khorami, M., Farajnezhad, M., & Nouri, J. (2023). Exploring the potential of metal and metal oxide nanomaterials for sustainable water and wastewater treatment: A review of their antimicrobial properties. *Chemosphere*, *335*, 139103. DOI: 10.1016/j.chemosphere.2023.139103

Kang, H. Y., & Schoenung, J. M. (2004, May). Used consumer electronics: A comparative analysis of materials recycling technologies. In IEEE International Symposium on Electronics and the Environment, 2004. Conference Record. 2004 (pp. 226-230). IEEE.

Kang, D., & Wolverton, C. (2022). Data-driven materials discovery and manufacturing: Challenges and future directions. *MRS Bulletin*, *47*(3), 227–233.

Kanoungo, S., Sharma, U., & Singh, S. (2020). Assessment of treatment methods of recycled aggregates for utilization in flexible pavements. *Materials Today: Proceedings*, *43*, 1320–1324. DOI: 10.1016/j.matpr.2020.09.119

Kanti, P. K., Sharma, P., Maiya, M. P., & Sharma, K. V. (2023). The stability and thermophysical properties of Al2O3-graphene oxide hybrid nanofluids for solar energy applications: Application of robust autoregressive modern machine learning technique. *Solar Energy Materials and Solar Cells*, *253*, 112207.

Kanti, P. K., Shrivastav, A. P., Sharma, P., & Maiya, M. P. (2024). Thermal performance enhancement of metal hydride reactor for hydrogen storage with graphene oxide nanofluid: Model prediction with machine learning. *International Journal of Hydrogen Energy*, *52*, 470–484.

Kara, S., Hauschild, M., Sutherland, J., & McAloone, T. (2022). Closed-loop systems to circular economy: A pathway to environmental sustainability? *CIRP Annals*, *71*(2), 505–528.

Karnwal, A., Kumar Sachan, R. S., Devgon, I., Devgon, J., Pant, G., Panchpuri, M., Ahmad, A., Alshammari, M. B., Hossain, K., & Kumar, G. (2024). Gold Nanoparticles in Nanobiotechnology: From Synthesis to Biosensing Applications. *ACS Omega*, *9*(28), 29966–29982. DOI: 10.1021/acsomega.3c10352

Kathpalia, R., & Verma, A. K. (2021). Bio-inspired nanoparticles for artificial photosynthesis. *Materials Today: Proceedings*, *45*, 3825–3832.

Kaur, J., Sengupta, P., & Mukhopadhyay, S. (2022). Critical review of bioadsorption on modified cellulose and removal of divalent heavy metals (Cd, Pb, and Cu). *Industrial & Engineering Chemistry Research*, *61*(5), 1921–1954.

Kaushik, D., & Singh, S. K. (2021). Use of coir fiber and analysis of geotechnical properties of soil. *Materials Today: Proceedings*, *47*, 4418–4422. DOI: 10.1016/j.matpr.2021.05.255

Kazmi, S. M. S., Abbas, S., Saleem, M. A., Munir, M. J., & Khitab, A. (2016). Manufacturing of sustainable clay bricks: Utilization of waste sugarcane bagasse and rice husk ashes. *Construction & Building Materials*, *120*, 29–41. DOI: 10.1016/j.conbuildmat.2016.05.084

Ke, G., Meng, Q., Finley, T., Wang, T., Chen, W., Ma, W., Ye, Q., & Tie, Y. L. 2017. "LightGBM: A Highly Efficient Gradient Boosting Decision Tree." In *Advances in Neural Information Processing Systems*. Vol. 2017-December.

Kee, C. W. (2023). Molecular Understanding and Practical In Silico Catalyst Design in Computational Organocatalysis and Phase Transfer Catalysis—Challenges and Opportunities. *Molecules (Basel, Switzerland)*, *28*(4), 1715.

Kenea, L. M., Kumar, L., Singh, S. K., Chala, G., & Chimdi, J. (2021). Assessment of asphalt road pavement rehabilitation and maintenance practices: The case of West Showa zone in Oromia, Ethiopia. *IOP Conference Series. Materials Science and Engineering*, *1187*(1), 012033. DOI: 10.1088/1757-899X/1187/1/012033

Kesidou, S., & Sovacool, B. K. (2019). Supply chain integration for low-carbon buildings: A critical interdisciplinary review. *Renewable & Sustainable Energy Reviews*, *113*, 109274.

Ke, W., & Kanatzidis, M. G. (2019). Prospects for low-toxicity lead-free perovskite solar cells. *Nature Communications*, *10*(1), 965. DOI: 10.1038/s41467-019-08918-3

Khairani, I. Y., Mínguez-Vega, G., Doñate-Buendía, C., & Gökce, B. (2023). Green nanoparticle synthesis at scale: A perspective on overcoming the limits of pulsed laser ablation in liquids for high-throughput production. *Physical Chemistry Chemical Physics*, *25*(29), 19380–19408. DOI: 10.1039/D3CP01214J

Khalid, N., Othman, H., & Hassabo, A. G. (2024). Manufacturing and Usage of Lyocell Fabric. *Egyptian Journal of Chemistry*, *67*(13), 151–157.

Khalid, R., Shah, A., Javed, M., & Hussain, H. (2025). Progress and obstacles in electrode materials for lithium-ion batteries: A journey towards enhanced energy storage efficiency. *RSC Advances*, *15*(20), 15951–15998. DOI: 10.1039/D5RA02042E

Khalili, S., Khoshandam, B., & Jahanshahi, M. (2015). Optimization of production conditions for synthesis of chemically activated carbon produced from pinecone using response surface methodology for CO2 adsorption. *RSC Advances*, *5*(114), 94115–94129. DOI: 10.1039/C5RA18986A

Khan, A., Al Rashid, H., Roy, P. K., Chowdhury, S. I., & Sathi, S. A. (2025). Challenges and the Way to Improve Lithium-Ion Battery Technology for Next-Generation Energy Storage. *Energy & Environmental Materials*, *8*(6), 70088. DOI: 10.1002/eem2.70088

Khanam, Z., Gogoi, N., & Srivastava, D. N. (2022). Challenges, Commercial Aspect. *2D Nanomaterials for Energy and Environmental Sustainability*, 267.

Khandan, A., Khosravi, M., Roustazadeh, D., & Aghadavoudi, F. (2024). Impact of alumina and carbon nanotubes on mechanical properties of a composite: Molecular dynamic (MD) simulation. *Iranian Journal of Chemistry and Chemical Engineering*, *43*(8), 2866–2877.

Khan, M. Z. U., Uddin, E., Akbar, B., Akram, N., Naqvi, A. A., Sajid, M., & Ali, Z. (2020). Investigation of Heat Transfer and Pressure Drop in Microchannel Heat Sink Using Al2O3 and ZrO2 Nanofluids. *Nanomaterials (Basel, Switzerland)*, *10*, 1796.

Khan, M., & McNally, C. (2024). *Recent developments on low carbon 3D printing concrete: Revolutionizing construction through innovative technology*. Cleaner Materials.

Khan, S., Naushad, M., Al-Gheethi, A., & Iqbal, J. (2021). Engineered nanoparticles for removal of pollutants from wastewater: Current status and future prospects of nanotechnology for remediation strategies. *Journal of Environmental Chemical Engineering*, *9*(5), 106160. DOI: 10.1016/j.jece.2021.106160

Khatoon, U. T., & Velidandi, A. (2025). An Overview on the Role of Government Initiatives in Nanotechnology Innovation for Sustainable Economic Development and Research Progress. *Sustainability*, *17*(3), 1250.

Khorsand Zak, A., Yazdi, S. T., Abrishami, M. E., & Hashim, A. M. (2024). A review on piezoelectric ceramics and nanostructures: Fundamentals and fabrications. *Journal of the Australian Ceramic Society*, *60*(3), 723–753. DOI: 10.1007/s41779-024-00990-3

Khurana, K., & Jaggi, N. (2021). Localized surface plasmonic properties of Au and Ag nanoparticles for sensors: A review. *Plasmonics*, *16*(4), 981–999. DOI: 10.1007/s11468-021-01381-1

Khutso Ntobeng, M., Ehi Imoisili, P., & Jen, T.-C. (2020). Synthesis and photocatalytic performance of Ag-TiVOX nanocomposite. *Journal of King Saud University. Science*, *32*(7), 3103–3110. DOI: 10.1016/j.jksus.2020.08.021

Kiani, M. (2025). Nanomaterials Revolution: Paving the Way for Next-Gen Applications. *Available at SSRN* 5086617. DOI: 10.2139/ssrn.5086617

Kim, H., Lee, K., Kim, C., Lim, J., & Kim, W. Y. (2023). DFRscore: Deep Learning-Based Scoring of Synthetic Complexity with Drug-Focused Retrosynthetic Analysis for High-Throughput Virtual Screening. *Journal of Chemical Information and Modeling, 64*(7), 2432–2444. DOI: 10.1021/acs.jcim.3c01134 PMID: 37651152

Kim, S., Jeon, H., Koo, J. M., Oh, D. X., & Park, J. (2024). Practical Applications of Self-Healing Polymers Beyond Mechanical and Electrical Recovery. *Advanced Science (Weinheim, Baden-Wurttemberg, Germany), 11*(16), 2302463. Advance online publication. DOI: 10.1002/advs.202302463

Kim, S., Kim, J., Lee, Y., Lee, W. B., Lee, C. W., Jin, K., & Na, J. (2024). Economic and environmental insights into the hybrid ethylene oxide production processes. *Chemical Engineering Journal, 494*, 152879.

Kirklin, S., Saal, J. E., Meredig, B., Thompson, A., Doak, J. W., Aykol, M., Rühl, S., & Wolverton, C. (2015). The Open Quantum Materials Database (OQMD): Assessing the Accuracy of DFT Formation Energies. *npj Computational Materials, 1*(1), 15010. Advance online publication. DOI: 10.1038/npjcompumats.2015.10

Kirubakaran, D., Wahid, J. B. A., Karmegam, N., Jeevika, R., Sellapillai, L., & Rajkumar, M., & SenthilKumar, K. J. (2025). A comprehensive review on the green synthesis of nanoparticles: Advancements in biomedical and environmental applications. *Biomedical Materials & Devices, 4*, 1–26.

Kleimann, D., Poitiers, N., Sapir, A., Tagliapietra, S., Véron, N., Veugelers, R., & Zettelmeyer, J. (2023). *How Europe should answer the US inflation reduction act.* Bruegel Policy Contribution.

Koller, M., & Mukherjee, A. (2022). A new wave of industrialization of PHA biopolyesters. *Bioengineering (Basel, Switzerland), 9*(2), 74.

Konstantopoulos, G., Koumoulos, E. P., & Charitidis, C. A. (2022). Digital Innovation Enabled Nanomaterial Manufacturing; Machine Learning Strategies and Green Perspectives. *Nanomaterials (Basel, Switzerland), 12*(15), 2646. Advance online publication. DOI: 10.3390/nano12152646

Kostarelos, K., Lacerda, L., Pastorin, G., Wu, W., Wieckowski, S., Luangsivilay, J., Godefroy, S., Pantarotto, D., Briand, J.-P., & Muller, S.. (2007). Cellular uptake of functionalized carbon nanotubes is independent of functional group and cell type. *Nature Nanotechnology, 2*(2), 108–113.

Kraus, O. Z., & Frey, B. J. (2016). Computer vision for high content screening. *Critical Reviews in Biochemistry and Molecular Biology, 51*(2), 102–109. DOI: 10.3109/10409238.2015.1135868 PMID: 26806341

Krishani, M., Shin, W. Y., Suhaimi, H., & Sambudi, N. S. (2023). Development of scaffolds from bio-based natural materials for tissue regeneration applications: A review. *Gels (Basel, Switzerland)*, *9*(2), 100.

Krzywanski, J., Sosnowski, M., Grabowska, K., Zylka, A., Lasek, L., & Kijo-Kleczkowska, A. (2024). Advanced Computational Methods for Modeling, Prediction and Optimization-A Review. *Materials (Basel)*, *17*(14), 3521. DOI: 10.3390/ma17143521 PMID: 39063813

Kuczynski, J., & Boday, D. (2012). Bio-based materials for high-end electronics applications. *International Journal of Sustainable Development and World Ecology*, *19*(6), 557–563.

Kumar, A., Dixit, S., Singh, S., Sreenivasa, S., Bains, P. S., & Sharma, R. (2025). Recent developments in the mechanical properties and recycling of fiber-reinforced polymer composites. *Polymer Composites*, *46*(5), 3883–3908.

Kumar, A., Tyagi, P. K., Tyagi, S., & Ghorbanpour, M. (2024). Integrating green nanotechnology with sustainable development goals: A pathway to sustainable innovation. *Discover Sustainability*, *5*(1), 364. Advance online publication. DOI: 10.1007/s43621-024-00610-x

Kumar, J. S., Senthilkumar, G., & Ramachandran, S. (2023). Novel strategy of mixing MgO in CuO/water nanofluid for thermal conductivity improvement: Experimental study. *Case Studies in Thermal Engineering*, *52*, 103723.

Kumar, S., Singh, S. K., & Shree, D. (2024). A comprehensive investigation of Coir geotextiles in sustainable pavement engineering. *International Journal of Civil Engineering*, *11*(7), 73–87. DOI: 10.14445/23488352/IJCE-V11I7P107

Kunde, G. B., & Sehgal, B. (2021). Industrial innovation through sustainable materials. In *Handbook of Smart Materials, Technologies, and Devices: Applications of Industry 4.0* (pp. 1–42). Springer.

Kundu, S., Khandaker, T., Anik, M. A. A. M., Hasan, M. K., Dhar, P. K., Dutta, S. K., & Hossain, M. S. (2024). A comprehensive review of enhanced CO 2 capture using activated carbon derived from biomass feedstock. *RSC Advances*, *14*(40), 29693–29736.

Kurc, B., Pigłowska, M., Rymaniak, Ł., & Fuć, P. (2021). Modern Nanocomposites and Hybrids as Electrode Materials Used in Energy Carriers. *Nanomaterials (Basel, Switzerland)*, *11*(2), 538. DOI: 10.3390/nano11020538

Kurzydlowski, K. J. (2010). Modelling of the Microstructure and Properties in the Length Scales Varying from Nanoto Macroscopic. *Bulletin of the Polish Academy of Sciences. Technical Sciences, 58*(2). Advance online publication. DOI: 10.2478/v10175-010-0021-7

La Rosa, A. D., Cozzo, G., Latteri, A., Mancini, G., Recca, A., & Cicala, G. (2013). A comparative life cycle assessment of a composite component for automotive. *Chemical Engineering (Albany, N.Y.)*, •••, 32.

Larsson, E. (2017). *The Corrosive Effect of Chlorine Containing Species on Waterwalls and Superheater Materials in Waste and Biomass-Fired Power Plants*. Chalmers Tekniska Hogskola.

Larsson, M. (2018). *Recycling and upcycling*. Circular Business Models.

Latif, A. (2024). RECYCLING AND UPCYCLING METAL SCRAPS BY SOLID STATE PROCESSING.

Lau, M. L., Terry, J., Long, M., & Burleigh, A. (2023). Materials characterization: Can artificial intelligence be used to address reproducibility challenges? *Journal of Vacuum Science & Technology. A, Vacuum, Surfaces, and Films, 41*(6), 060801. Advance online publication. DOI: 10.1116/6.0002809

Lawson, S., Li, X., Thakkar, H., Rownaghi, A. A., & Rezaei, F. (2021). Recent Advances in 3D Printing of Structured Materials for Adsorption and Catalysis Applications. *Chemical Reviews, 121*(10), 6246–6291. DOI: 10.1021/acs.chemrev.1c00060

Le Ferrand, H., Chabi, S., & Agarwala, S. (2020). 3D Assembly of Graphene Nanomaterials for Advanced Electronics. *Advanced Intelligent Systems, 2*(5), 1900151. Advance online publication. DOI: 10.1002/aisy.201900151

Le, A. T., Pandey, A., Sirohi, R., Sharma, P., Chen, W.-H., Nguyen, D. K. P., Tran, V. D., Nguyen, X. P., & Hoang, A. T. (2023). Precise Prediction of Biochar Yield and Proximate Analysis by Modern Machine Learning and SHapley Additive ExPlanations. *Energy & Fuels, 37*(22), 17310–17327. DOI: 10.1021/acs.energyfuels.3c02868

Lee, H. J., Yeo, S. Y., & Jeong, S. H. (2003). Antibacterial effect of nanosized silver colloidal solution on textile fabrics. *Journal of Materials Science, 38*, 2199–2204.

Lee, H., You, S., Pikhitsa, P. V., Kim, J., Kwon, S., Woo, C. G., & Choi, M. (2010). Three-Dimensional Assembly of Nanoparticles from Charged Aerosols. *Nano Letters, 11*(1), 119–124. DOI: 10.1021/nl103787k

Lee, S. Y., & Park, S. J. (2013). Determination of the optimal pore size for improved CO 2 adsorption in activated carbon fibers. *Journal of Colloid and Interface Science*, *389*(1), 230–235. DOI: 10.1016/j.jcis.2012.09.018

Lee, S., & Popovics, J. S. (2022). Applications of Physics-informed Neural Networks for Property Characterization of Complex Materials. *RILEM Technical Letters*, *7*, 178–188. Advance online publication. DOI: 10.21809/rilemtechlett.2022.174

Lefevere, J., Protasova, L., Mullens, S., & Meynen, V. (2017). 3D-printing of hierarchical porous ZSM-5: The importance of the binder system. *Materials & Design*, *134*, 331–341. DOI: 10.1016/j.matdes.2017.08.044

Lefèvre Martín, N. (2024). Carmen Hijosa. Piñatex, el sustituto sostenible del cuero. GDI. Revista de investigación de Género. *Diseño e Innovación*, *1*(1), 1.

Lelivelt, R. J. J., Lindner, G., Teuffel, P., & Lamers, H. (2015). The production process and compressive strength of mycelium-based materials. In *First International Conference on Bio-based Building Materials*. 22-25 June 2015, Clermont-Ferrand, France (pp. 1-6).

Leonas, K. K. (2016). The use of recycled fibers in fashion and home products. Textiles and clothing sustainability: Recycled and Upcycled textiles and fashion, 55-77.

León, G., Fernanda, M., & Dewulf, J. (2020). Data Quality Assessment Framework for Critical Raw Materials. The Case of Cobalt. *Resources, Conservation and Recycling*, *157*, 104564. Advance online publication. DOI: 10.1016/j.resconrec.2019.104564

Liang, X., Rangom, Y., Kwok, C. Y., Pang, Q., & Nazar, L. F. (2017). Interwoven MXene Nanosheet/Carbon-Nanotube Composites as Li–S Cathode Hosts. *Advanced Materials*, *29*(3), 1603040. Advance online publication. DOI: 10.1002/adma.201603040

Li, D., Yang, J., Zhao, Y., Yuan, H., & Chen, Y. (2022). Ultra-highly porous carbon from Wasted soybean residue with tailored porosity and doped structure as renewable multi-purpose absorbent for efficient CO2, toluene and water vapor capture. *Journal of Cleaner Production*, *337*, 130283. DOI: 10.1016/j.jclepro.2021.130283

Liew, K., & Akbar, A. (2020). The recent progress of recycled steel fiber reinforced concrete. *Construction & Building Materials*, *232*, 117232.

Li, G., Lian, S., Wang, J., Xie, G., Zhang, N., & Xie, X. (2023). Surface chemistry engineering and the applications of MXenes. *Journal of Materiomics*, *9*(6), 1160–1184. DOI: 10.1016/j.jmat.2023.08.003

Li, K., DeCost, B., Choudhary, K., Greenwood, M., & Hattrick-Simpers, J. (2023). A Critical Examination of Robustness and Generalizability of Machine Learning Prediction of Materials Properties. *npj Computational Materials*, *9*(1), 55. Advance online publication. DOI: 10.1038/s41524-023-01012-9

Lim, C. T. (2017). Nanofiber technology: Current status and emerging developments. *Progress in Polymer Science*, *70*, 1–17.

Lim, G., Lee, K. B., & Ham, H. C. (2016). Effect of N-Containing Functional Groups on CO_2 Adsorption of Carbonaceous Materials: A Density Functional Theory Approach. *The Journal of Physical Chemistry. C, Nanomaterials and Interfaces*, *120*(15), 8087–8095. DOI: 10.1021/acs.jpcc.5b12090

Lin, S. P., Khumsupan, D., Chou, Y. J., Hsieh, K. C., Hsu, H. Y., Ting, Y., & Cheng, K. C. (2022). Applications of atmospheric cold plasma in agricultural, medical, and bioprocessing industries. *Applied Microbiology and Biotechnology*, *106*(23), 7737–7750. DOI: 10.1007/s00253-022-12252-y

Lin, Y.-H., Kang, S.-W., & Chen, H.-L. (2008). Effect of silver nanofluid on pulsating heat pipe thermal performance. *Applied Thermal Engineering*, *28*(11-12), 1312–1317.

Li, P., Zhao, X., Ding, Y., Chen, L., Wang, X., & Xie, H. (2024). Application studies on MXene-based flexible composites. *Frontiers in Thermal Engineering*, *4*, 1440165. Advance online publication. DOI: 10.3389/fther.2024.1440165

Li, R., Zhou, Y., Li, W., Zhu, J., & Huang, W. (2020). Structure Engineering in Biomass-Derived Carbon Materials for Electrochemical Energy Storage. *Research*, *2020*, 2020/8685436. Advance online publication. DOI: 10.34133/2020/8685436

Li, S., Xu, W., Zhang, F., Wu, H., & Ge, Q. (2022). Effect of graphene oxide on aging properties of Polyurethane-SBS modified asphalt and asphalt mixture. *Polymers*, *14*(17), 3496. DOI: 10.3390/polym14173496 PMID: 36080571

Liu, B., Wang, Y., Rabczuk, T., Olofsson, T., & Lu, W. (2024). Multi-Scale Modeling in Thermal Conductivity of Polyurethane Incorporated with Phase Change Materials Using Physics-Informed Neural Networks. *Renewable Energy*, *220*, 119565. Advance online publication. DOI: 10.1016/j.renene.2023.119565

Liu, C., Li, F., Ma, L., & Cheng, H. (2010). Advanced Materials for Energy Storage. *Advanced Materials*, *22*(8). Advance online publication. DOI: 10.1002/adma.200903328

Liu, C., Ma, H., Yuan, S., Jin, Y., & Tian, W. (2025). Living Cell-Mediated Self-Assembly: From Monomer Design and Morphology Regulation to Biomedical Applications. *ACS Nano*, *19*(2), 2047–2069. DOI: 10.1021/acsnano.4c16669

Liu, C., Xu, M., Wang, Y., Yin, Q., Hu, J., Chen, H., Sun, Z., Liu, C., Li, X., Zhou, W., & Liu, H. (2024). Exploring the potential of hydroxyapatite-based materials in biomedicine: A comprehensive review. *Materials Science and Engineering R Reports*, *161*, 100870. DOI: 10.1016/j.mser.2024.100870

Liu, D., Chen, S., Li, R., & Peng, T. (2021). Review of Z-scheme heterojunctions for photocatalytic energy conversion. *Wuli Huaxue Xuebao*, *37*(6).

Liu, D., Ge, H., Song, M., Jiang, Y., Gong, X., You, T., Fu, L., Fu, Z., & Zhang, Y. (2025). Machine Learning Driven Optimization of Electrolyte for Highly Reversible Zn-Air Batteries with Superior Long-Term Cycling Performance. *Advanced Materials*, *37*(7), 2417161. Advance online publication. DOI: 10.1002/adma.202417161

Liu, D., Gu, W., Zhou, L., Wang, L., Zhang, J., Liu, Y., & Lei, J. (2022). Recent advances in MOF-derived carbon-based nanomaterials for environmental applications in adsorption and catalytic degradation. *Chemical Engineering Journal*, *427*, 131503.

Liu, H., Liu, H., Liu, H., Chen, P., Cheng, M., Wang, H., Wang, X., Wang, H., Hu, B., Wang, X., Cheng, M., Wang, S., Chen, P., Chen, P., Wang, J., Wang, J., Wang, H., Lin, B., Wang, S., & Hu, B. (2024). Docking Score ML: Target-Specific Machine Learning Models Improving Docking-Based Virtual Screening in 155 Targets. *Journal of Chemical Information and Modeling*, *64*(14), 5413–5426. DOI: 10.1021/acs.jcim.4c00072 PMID: 38958413

Liu, K., Niri, M. F., Apachitei, G., Lain, M., Greenwood, D., & Marco, J. (2022). Interpretable machine learning for battery capacities prediction and coating parameters analysis. *Control Engineering Practice*, *124*, 105202. DOI: 10.1016/j.conengprac.2022.105202

Liu, L., Chen, T., & Chen, Z. (2024). Understanding the Dynamic Aggregation in Single-Atom Catalysis. *Advancement of Science*, *11*(13), 2308046.

Liu, M., Qin, Z., Yang, X., Lin, Z., & Guo, T. (2019). Fabricating controllable hierarchical pores on smooth carbon sheet for synthesis of supercapacitor materials. *Vacuum*, *168*, 108806. DOI: 10.1016/j.vacuum.2019.108806

Liu, M., Song, A., Zhang, X., Wang, J., Fan, Y., Wang, G., Tian, H., Ma, Z., & Shao, G. (2025). Interfacial lithium-ion transportation in solid-state batteries: Challenges and prospects. *Nano Energy*, *136*, 110749. DOI: 10.1016/j.nanoen.2025.110749

Liu, Q., Deng, M., Shi, Y., & Wang, J. (2012). A Density-Based Spatial Clustering Algorithm Considering Both Spatial Proximity and Attribute Similarity. *Computers & Geosciences, 46*, 296–309. Advance online publication. DOI: 10.1016/j.cageo.2011.12.017

Liu, R., Dai, L., Xu, C., Wang, K., Zheng, C., & Si, C. (2020). Lignin-based micro- and nanomaterials and their composites in biomedical applications. *ChemSusChem, 13*(17), 4266–4283.

Liu, S., Zhang, H., Chen, J., Peng, X., Chai, Y., Shao, X., He, Y., Wang, X., & Ding, B. (2025). Functionalization Strategies of MXene Architectures for Electrochemical Energy Storage Applications. *Energies, 18*(5), 1223. DOI: 10.3390/en18051223

Liu, X., Zhang, J., & Pei, Z. (2023). Machine Learning for High-Entropy Alloys: Progress, Challenges and Opportunities. *Progress in Materials Science, 131*, 101018. Advance online publication. DOI: 10.1016/j.pmatsci.2022.101018

Liu, Y. X., Dong, Y. W., Jiang, Z. H., Li, Y. S., Zha, W., Du, Y. X., & Du, S. Y. (2023). XGBoost-based model for predicting hydrogen content in electroslag remelting. *Journal of Iron and Steel Research International, 30*(5), 887–896.

Liu, Y., Liu, D., Avdeev, M., Shi, S., Zou, X., Ma, S., & Yang, Z. (2023). Data quantity governance for machine learning in materials science. *National Science Review, 10*(7), nwad125. Advance online publication. DOI: 10.1093/nsr/nwad125 PMID: 37323811

Liu, Y., Niu, C., Wang, Z., Gan, Y., Zhu, Y., Sun, S., & Shen, T. (2020). Machine Learning in Materials Genome Initiative: A Review. *Journal of Materials Science and Technology, 57*, 113–122. Advance online publication. DOI: 10.1016/j.jmst.2020.01.067

Liu, Y., Zhao, T., Ju, W., & Shi, S. (2017). Materials Discovery and Design Using Machine Learning. *Journal of Materiomics, 3*(3), 159–177. Advance online publication. DOI: 10.1016/j.jmat.2017.08.002

Liu, Z., Shi, Y., Chen, H., Qin, T., Zhou, X., Huo, J., Dong, H., Yang, X., Zhu, X., Chen, X., Zhang, L., Yang, M., Gao, Y., & Ma, J. (2021). Machine Learning on Properties of Multiscale Multisource Hydroxyapatite Nanoparticles Datasets with Different Morphologies and Sizes. *npj Computational Materials, 7*(1), 142. Advance online publication. DOI: 10.1038/s41524-021-00618-1

Liu, Z., Sun, L., Zhai, J., & Huang, W. (2022). A review of design methods for cold in-place recycling asphalt mixtures: Design processes, key parameters, and evaluation. *Journal of Cleaner Production, 370*, 133530. DOI: 10.1016/j.jclepro.2022.133530

Liu, Z., Zhou, X., & Liu, C. (2019). N-doped porous carbon material prepared via direct ink writing for the removal of methylene blue. *Diamond and Related Materials*, *95*(April), 121–126. DOI: 10.1016/j.diamond.2019.04.010

Li, X., Mo, Y., Qing, W., Shao, S., Tang, C. Y., & Li, J. (2019). Membrane-based technologies for lithium recovery from water lithium resources: A review. *Journal of Membrane Science*, *591*, 117317. DOI: 10.1016/j.memsci.2019.117317

Li, Y. (2021). Carbon Nanotube Research in Its 30th Year. *ACS Nano*, *15*(6), 9197–9200. DOI: 10.1021/acsnano.1c04972

Ljungberg, L. Y. (2007). Materials selection and design for development of sustainable products. *Materials & Design*, *28*(2), 466–479.

Lo, C.-H., Tsung, T.-T., Chen, L.-C., Su, C.-H., & Lin, H.-M. (2005). Fabrication of copper oxide nanofluid using submerged arc nanoparticle synthesis system (SANSS). *Journal of Nanoparticle Research*, *7*, 313–320.

López-Cano, I., Cayuela, M. L., Mondini, C., Takaya, C. A., Ross, A. B., & Sánchez-Monedero, M. A. (2018). Suitability of different agricultural and urban organic wastes as feedstocks for the production of Biochar-Part 1: Physicochemical characterisation. *Sustainability (Basel)*, *10*(7), 2265. Advance online publication. DOI: 10.3390/su10072265

Loredo-Cancino, M., Soto-Regalado, E., Cerino-Córdova, F. J., García-Reyes, R. B., García-León, A. M., & Garza-González, M. T. (2013). Determining optimal conditions to produce activated carbon from barley husks using single or dual optimization. *Journal of Environmental Management*, *125*, 117–125. DOI: 10.1016/j.jenvman.2013.03.028

Lovrenčić Butković, L., Mihić, M., & Sigmund, Z. (2023). Assessment methods for evaluating circular economy projects in construction: A review of available tools. *International Journal of Construction Management*, *23*(5), 877–886.

Lu, X., Wang, C., & Wei, Y. (2009). One-Dimensional Composite Nanomaterials: Synthesis by Electrospinning and Their Applications. Small, 5(21), 2349–2370. Portico. DOI: 10.1002/smll.200900445

Lu, B., Xia, Y., Ren, Y., Xie, M., Zhou, L., Vinai, G., Morton, S. A., Wee, A. T. S., van der Wiel, W. G., Zhang, W., & Wong, P. K. J. (2024). When Machine Learning Meets 2D Materials: A Review. *Advanced Science (Weinheim, Baden-Wurttemberg, Germany)*, *11*(13), 2305277. Advance online publication. DOI: 10.1002/advs.202305277

Lu, D., Gangrade, S., Kao, S.-C., Konapala, G., & Painter, S. L. (2021). Streamflow simulation in data-scarce basins using Bayesian and physics-informed machine learning models. *Journal of Hydrometeorology*, *1*(6). Advance online publication. DOI: 10.1175/JHM-D-20-0082.1

Lu, G., Zhang, Y., Zhang, J., Du, X., Lv, Z., Du, J., & Cui, G. (2023). Trade-offs between ion-conducting and mechanical properties: The case of polyacrylate electrolytes. *Carbon Energy*, *5*(2), e287.

Lu, T. D., Chen, B. Z., Wang, J., Jia, T. Z., Cao, X. L., Wang, Y., Xing, W., Lau, C. H., & Sun, S. P. (2018). Electrospun nanofiber substrates that enhance polar solvent separation from organic compounds in thin-film composites. *Journal of Materials Chemistry. A, Materials for Energy and Sustainability*, *6*(31), 15047–15056. DOI: 10.1039/C8TA04504F

Lützkendorf, T. (2022). LCA of building materials within the framework of the Construction Products Regulation (CPR) in Europe. ce/papers, 5(5), 43-47.

Lu, X., & Isacsson, U. (1997). Rheological characterization of styrene-butadiene-styrene copolymer modified bitumens. *Construction & Building Materials*, *11*(1), 23–32. DOI: 10.1016/S0950-0618(96)00033-5

Lu, Y., Lou, X., Jiang, J., Wang, J., Peng, X., Yao, H., & Wu, J. (2024). Antioxidative, Anti-Inflammatory, Antibacterial, Photo-Cross-Linkable Hydrogel of Gallic Acid–Chitosan Methacrylate: Synthesis, In Vitro, and In Vivo Assessments. *Biomacromolecules*, *25*(7), 4358–4373.

Lv, K., Wan, H., Qi, C., Wang, W., Liu, X., Zhang, L., Wei, H., Zhu, H., Wang, Y., Yu, J., & Du, S. (2025). *Integrating AI and material science: MXene synthesis, preparation, and applications*. AI & Materials., DOI: 10.55092/aimat20250004

Lv, Q., Li, Q., Cao, P., Wei, C., Li, Y., Wang, Z., & Wang, L. (2025). Designing Silk Biomaterials toward Better Future Healthcare: The Development and Application of Silk-Based Implantable Electronic Devices in Clinical Diagnosis and Therapy. *Advanced Materials*, *37*(8), 2411946.

Ma, Y. F. Z. R. P. L. (2020). Manufacturing Nanoporous Materials for Energy-Efficient Separations: Application and Challenges. In *Sustainable Nanoscale Engineering* (pp. 33–81). Elsevier.

MacCarley, C. A. (2025). Hydrogen. In *Non-Petroleum Automotive Transportation* (pp. 199–333). Springer Nature Switzerland. DOI: 10.1007/978-3-031-85028-8_9

Machín, A., Fontánez, K., Arango, J. C., Ortiz, D., De León, J., Pinilla, S., Nicolosi, V., Petrescu, F. I., Morant, C., & Márquez, F. (2021). One-Dimensional (1D) Nanostructured Materials for Energy Applications. *Materials (Basel), 14*(10), 2609. DOI: 10.3390/ma14102609

Madadkhani, S., Nandy, S., Chae, K. H., Aleshkevych, P., & Najafpour, M. M. (2024). Advances in Understanding Tungsten Disulfide Dynamics during the Hydrogen-Evolution Reaction: An Initial Step in Elucidating the Mechanism. *The Journal of Physical Chemistry Letters, 15*(19), 5112–5119. DOI: 10.1021/acs.jpclett.4c00924 PMID: 38709179

Madkour, L. H. (2019). Introduction to Nanotechnology (NT) and Nanomaterials (NMs). Nanoelectronic Materials, 1–47. DOI: 10.1007/978-3-030-21621-4_1

Ma, H. B., Wilson, C., Yu, Q., Park, K., Choi, U. S., & Tirumala, M. (2006). An experimental investigation of heat transport capability in a nanofluid oscillating heat pipe. *Journal of Heat Transfer, 128*(11), 1213–1216.

Maham, M., Rabia, C., Anum, I., & Ahad, I. U. (2025). Critical and Strategic Raw Materials for Energy Storage Devices. *Batteries, 11*(4), 163.

Mahapatra, D. M., Kumar, A., Kumar, R., Gupta, N. K., Ethiraj, B., & Singh, L. (2025). Artificial intelligence interventions in 2D MXenes-based photocatalytic applications. *Coordination Chemistry Reviews, 529*, 216460. DOI: 10.1016/j.ccr.2025.216460

Mahapatra, O., Bhagat, M., Gopalakrishnan, C., & Arunachalam, K. D. (2008). Ultrafine dispersed CuO nanoparticles and their antibacterial activity. *Journal of Experimental Nanoscience, 3*(3), 185–193.

Mahmoud, M. A., Alsehli, B. R., Alotaibi, M. T., Hosni, M., & Shahat, A. (2024). A comprehensive review on the application of semiconducting materials in the degradation of effluents and water splitting. *Environmental Science and Pollution Research International, 31*(3), 3466–3494. DOI: 10.1007/s11356-023-31353-3

Mahmud, M. Z. A., Rabbi, S. F., Islam, M. D., & Hossain, N. (2025). Synthesis and applications of natural fiber-reinforced epoxy composites: A comprehensive review. *SPE Polymers, 6*(1), e10161.

Ma, J., Huang, D., Zou, J., Li, L., Kong, Y., & Komarneni, S. (2014). Adsorption of methylene blue and Orange II pollutants on activated carbon prepared from banana peel. *Journal of Porous Materials, 22*(2), 301–311. DOI: 10.1007/s10934-014-9896-2

Makul, N., Fediuk, R., Amran, M., Zeyad, A. M., Klyuev, S., Chulkova, I., & Azevedo, A. (2021). Design strategy for recycled aggregate concrete: A review of status and future perspectives. *Crystals, 11*(6), 695.

Malavolta, L., Bracaglia, I., Cazzador, G., Kovtun, A., Tomasi, L., Zanardi, C., & Palermo, V. (2025). Multivariate sensing of ions using machine learning and composite 2D-3D graphene oxide-hexacyanoferrate electrodes. *Sensors and Actuators. B, Chemical, 427*, 137194. DOI: 10.1016/j.snb.2024.137194

Maleki, A., Haghighi, A., & Mahariq, I. (2021). Machine Learning-Based Approaches for Modeling Thermophysical Properties of Hybrid Nanofluids: A Comprehensive Review. *Journal of Molecular Liquids, 322*(January), 114843. Advance online publication. DOI: 10.1016/j.molliq.2020.114843

Malik, S., Muhammad, K., & Waheed, Y. (2023). Emerging Applications of Nanotechnology in Healthcare and Medicine. *Molecules (Basel, Switzerland), 28*(18), 6624. DOI: 10.3390/molecules28186624

Malik, S., Muhammad, K., & Waheed, Y. (2023). Nanotechnology: A revolution in modern industry. *Molecules (Basel, Switzerland), 28*(2), 661. DOI: 10.3390/molecules28020661

Ma, M.-Y., Han, D., Chen, N.-K., Wang, D., & Li, X.-B. (2022). Recent Progress in Double-Layer Honeycomb Structure: A New Type of Two-Dimensional Material. *Materials (Basel), 15*(21), 7715. DOI: 10.3390/ma15217715

Mangaiyarkkarasi, J. (2025). Nanostructural Innovations: Powering the Future of Energy Storage. In *Design, Fabrication, and Significance of Advanced Nanostructured Materials* (pp. 51-74). IGI Global.

Manhart, A., Vogt, R., Priester, M., Dehoust, G., Auberger, A., Blepp, M., Dolega, P., Kämper, C., Giegrich, J., & Schmidt, G. (2019). The environmental criticality of primary raw materials–A new methodology to assess global environmental hazard potentials of minerals and metals from mining. *Mineral Economics, 32*, 91–107.

Manica, M., Das, P., Wehden, K., Hoffman, S. C., Padhi, I., Mchugh, L., Chenthamarakshan, V., Born, J., Christofidellis, D., Dave, A., Cadow, J., Teukam, Y., Schilter, O., Kishimoto, A., Donovan, T., Khrabrov, A., Hamada, L., Zipoli, F., Giannone, G., & Smith, J. (2023). Accelerating material design with the generative toolkit for scientific discovery. *npj Computational Materials, 9*(1), 69. Advance online publication. DOI: 10.1038/s41524-023-01028-1

Manjunatheshwara, K., & Vinodh, S. (2021). Sustainable electronics product design and manufacturing: State of art review. *International Journal of Sustainable Engineering*, *14*(4), 541–551.

Man, X., & Chan, E. P. (2021). The Best Way to Select Features? Comparing MDA, LIME, and SHAP. *The Journal of Financial Data Science*, *3*(1), 127–139. Advance online publication. DOI: 10.3905/jfds.2020.1.047

Manyà, J. J., García-Morcate, D., & González, B. (2020). Adsorption performance of physically activated biochars for postcombustion Co2 capture from dry and humid flue gas. *Applied Sciences (Basel, Switzerland)*, *10*(1), 376. Advance online publication. DOI: 10.3390/app10010376

Manyepedza, T., & Makuve, N. (2024). Catalysts for biomass upgrading: Transforming 5-hydroxymethyl furfural into 2, 5-furan dicarboxylic acid for fine chemicals, featuring mil-100 (Fe) Mof as a novel catalyst. *ChemistrySelect*, *9*(28), e202401412.

Manzo, D., Thai, R., Le, H. T., & Venayagamoorthy, G. K. (2025). Fuel cell technology review: Types, economy, applications, and vehicle-to-grid scheme. *Sustainable Energy Technologies and Assessments*, *75*, 104229. DOI: 10.1016/j.seta.2025.104229

Marcos, J., & Prauchner, F. R.-R. (2012). Chemical versus physical activation of coconut shell: A comparative study. *Microporous and Mesoporous Materials*, *152*, 163–171. DOI: 10.1016/j.micromeso.2011.11.040

Maria João Regufe, A. F. P. F. J. M. L. A. R. A. M. R. (2019). Electrical conductive 3D-printed monolith adsorbent for CO2 capture. *Microporous and Mesoporous Materials*, *278*, 403–413. DOI: 10.1016/j.micromeso.2019.01.009

Marijan, D., & Gotlieb, A. (2020). Industry-Academia research collaboration in software engineering: The Certus model. *Information and Software Technology*, *132*, 106473. DOI: 10.1016/j.infsof.2020.106473

Marino, A., Genchi, G. G., Mattoli, V., & Ciofani, G. (2017). Piezoelectric nanotransducers: The future of neural stimulation. *Nano Today*, *14*, 9–12. DOI: 10.1016/j.nantod.2016.12.005

Martínez-Hincapié, R., Wegner, J., Anwar, M. U., Raza-Khan, A., Franzka, S., Kleszczynski, S., & Čolić, V. (2024). The determination of the electrochemically active surface area and its effects on the electrocatalytic properties of structured nickel electrodes produced by additive manufacturing. *Electrochimica Acta*, *476*, 143663. DOI: 10.1016/j.electacta.2023.143663

Martin, N., Madrid-López, C., Villalba-Méndez, G., & Talens-Peiró, L. (2022). New techniques for assessing critical raw material aspects in energy and other technologies. *Environmental Science & Technology*, *56*(23), 17236–17245.

Ma, S., Wang, S., Du, Y., & Su, Y. (2023). Application of machine learning in material corrosion research. *Corrosion Reviews*, *41*(4), 417–426. DOI: 10.1515/corrrev-2022-0089

Ma, S., Yuan, G., Zhang, Y., Yang, N., Li, Y., & Chen, Q. (2022). Development of encapsulation strategies towards the commercialization of perovskite solar cells. *Energy & Environmental Science*, *15*(1), 13–55. DOI: 10.1039/D1EE02882K

Mateus, P., Moonen, J., Beran, M., Bron, E., Jansen, W., Heuvelink, J., Van Der Landen, S. M., Birhanu, M., Dekker, A., Oomens, J., Cats, D., Harms, A. G. J., Wolters, F. J., Bermejo, I., Mei, H., Schram, M. T., & Jaarsma, E. (2024). Data harmonization and federated learning for multi-cohort dementia research using the OMOP common data model: A Netherlands consortium of dementia cohorts case study. *Journal of Biomedical Informatics*, *155*, 104661. DOI: 10.1016/j.jbi.2024.104661 PMID: 38806105

Mathotaarachchi, K. V., Mahmood, S., & Hasan, R. (2024). Advanced Machine Learning Techniques for Predictive Modeling of Property Prices. *Information (Basel)*, *15*(6), 295. DOI: 10.3390/info15060295

Mattur, M. N., Nagappan, N., Rath, S., & Thomas, T. (2022). Prediction of nature of band gap of perovskite oxides (ABO3) using a machine learning approach. *Journal of Materiomics*, *8*(5), 937–948.

Maughan, P. A., Seymour, V. R., Bernardo-Gavito, R., Kelly, D. J., Shao, S., Tantisriyanurak, S., Dawson, R., Haigh, S. J., Young, R. J., Tapia-Ruiz, N., & Bimbo, N. (2020). Porous Silica-Pillared MXenes with Controllable Interlayer Distances for Long-Life Na-Ion Batteries. *Langmuir*, *36*(16), 4370–4382. DOI: 10.1021/acs.langmuir.0c00462

Ma, X., Li, L., Zeng, Z., Chen, R., Wang, C., Zhou, K., & Li, H. (2019). Experimental and theoretical demonstration of the relative effects of O-doping and N-doping in porous carbons for CO2 capture. *Applied Surface Science*, *481*, 1139–1147. DOI: 10.1016/j.apsusc.2019.03.162

Ma, Z., Zhou, X., Deng, W., Lei, D., & Liu, Z. (2018). 3D Porous MXene (Ti 3 C 2)/Reduced Graphene Oxide Hybrid Films for Advanced Lithium Storage. *ACS Applied Materials & Interfaces*, *10*(4), 3634–3643. DOI: 10.1021/acsami.7b17386

Mazumder, M., Kim, H., & Lee, S. (2016). Performance properties of polymer modified asphalt binders containing wax additives. *International Journal of Pavement Research and Technology*, *9*(2), 128–139. DOI: 10.1016/j.ijprt.2016.03.004

Meereboer, K. W., Misra, M., & Mohanty, A. K. (2020). Review of recent advances in the biodegradability of polyhydroxyalkanoate (PHA) bioplastics and their composites. *Green Chemistry*, *22*(17), 5519–5558.

Megahed, S. H., Abdel-Halim, M., El-shabrawy, Y. I., Saad, E. M., Hefnawy, A., Handoussa, H., Mizaikoff, B., & El Gohary, N. A. (2025). Design, synthesis and medical prospects of electrospun molecularly imprinted fibers. *Scientific Reports*, *15*(1), 26082. Advance online publication. DOI: 10.1038/s41598-025-11114-7

Mehfooz, T., Mohsin Ali, T., Arif, S., & Hasnain, A. (2018). Effect of barley husk addition on rheological, textural, thermal and sensory characteristics of traditional flat bread (chapatti). *Journal of Cereal Science*, *79*, 376–382. DOI: 10.1016/j.jcs.2017.11.020

Mehmood, S., Wang, X., Ahmed, W., Imtiaz, M., Ditta, A., Rizwan, M., & Li, W. (2021). Removal mechanisms of slag against potentially toxic elements in soil and plants for sustainable agriculture development: A critical review. *Sustainability (Basel)*, *13*(9), 5255. DOI: 10.3390/su13095255

Mei, J., Wang, J., Gu, H., Du, Y., Wang, H., Yamauchi, Y., Liao, T., Sun, Z., & Yin, Z. (2020). Nano Polymorphism-Enabled Redox Electrodes for Rechargeable Batteries. Advanced Materials, 33(8). Portico. DOI: 10.1002/adma.202004920

Mekonnin, A. S., Wacławiak, K., Humayun, M., Zhang, S., & Ullah, H. (2025). Hydrogen storage technology, and its challenges: A review. *Catalysts*, *15*(3), 260. DOI: 10.3390/catal15030260

Mekuye, B., & Abera, B. (2023). Nanomaterials: An overview of synthesis, classification, characterization, and applications. Nano Select, 4(8), 486–501. Portico. DOI: 10.1002/nano.202300038

Menachery, N., Deepanraj, B., & Thomas, S. (2025). Characterization and analysis of carbon fiber and nano hBN reinforced hybrid Aluminium Metal Matrix Composites by conventional sintering. *Turkish Journal of Engineering*, *9*(2), 378–384.

Mendrela, P., Stanek, W., & Simla, T. (2024). Thermo-ecological cost–System evaluation of energy-ecological efficiency of hydrogen production from renewable and non-renewable energy resources. *International Journal of Hydrogen Energy*, *50*, 1–14. DOI: 10.1016/j.ijhydene.2023.06.150

Meng, Q., Chen, W., Wu, L., Lei, J., Liu, X., Zhu, W., & Duan, T. (2019). A strategy of making waste profitable: Nitrogen doped cigarette butt derived carbon for high performance supercapacitors. *Energy*, *189*, 116241.

Meng, Y., Feng, Y., Bai, X., Yu, Q., Zhou, J., & Wang, J. (2025). Application of nanotechnology in agricultural sustainability: Absorption, translocation, and challenges of nanoparticles. *Current Plant Biology*, *42*, 100492. DOI: 10.1016/j.cpb.2025.100492

Meramo-Hurtado, S. I., & González-Delgado, A. D. (2020). Application of techno-economic and sensitivity analyses as decision-making tools for assessing emerging large-scale technologies for production of chitosan-based adsorbents. *ACS Omega*, *5*(28), 17601–17610.

Miah, M. H., Singh Chand, D., & Malhi, G. S. (2024). Accuracy compensation method with the combined dissimilar measurement devices for enhanced measurement quality: A digital aircraft assembly technology. *Aircraft Engineering and Aerospace Technology*, *96*(6), 780–797. DOI: 10.1108/AEAT-09-2022-0253

Miłek, D., Nowak, P., & Latosińska, J. (2022). The development of renewable energy sources in the European Union in the light of the European Green Deal. *Energies*, *15*(15), 5576.

Minello, L. V. P., Ahmad, I., Aguzzoli, C., & Sperotto, R. A. (2025). Surface engineering of nanoparticles via physical vapor deposition (PVD): Opportunities and challenges for sustainable agriculture. *Plant Nano Biology*, *13*, 100187. DOI: 10.1016/j.plana.2025.100187

Miri, Z., Haugen, H. J., Loca, D., Rossi, F., Perale, G., Moghanian, A., & Ma, Q. (2024). Review on the strategies to improve the mechanical strength of highly porous bone bioceramic scaffolds. *Journal of the European Ceramic Society*, *44*(1), 23–42. DOI: 10.1016/j.jeurceramsoc.2023.09.003

Mir, M. A., Chang, S. K., & Hefni, D. (2024). A comprehensive review on challenges and choices of food waste in Saudi Arabia: Exploring environmental and economic impacts. *Environmental Systems Research*, *13*, 40.

Mir, M. A., Mahmoud, M., Andrews, K., Nayak, A., Kukretee, N., & Kotnala, S. (2025). Experimental, Spectroscopic, Computational and Molecular Studies of Plant Growth Promotor 2-Naphthylacetic Acid. *Physics and Chemistry of Solid State*, *26*(1), 158–165.

Mitchell, S., Qin, R., Zheng, N., & Pérez-Ramírez, J. (2021). Nanoscale engineering of catalytic materials for sustainable technologies. *Nature Nanotechnology*, *16*(2), 129–139.

Mohamad Razali, N. A. I., Pramanik, S., Abu Osman, N. A., & Radzi, Z., Pingguan-Murphy, B. (2016). Conversion of calcite from cockle shells to bioactive nanorod hydroxyapatite for biomedical applications. *Journal of Ceramic Processing Research*, *17*, 699–706.

Mohammed, H., Mia, M. F., Wiggins, J., & Desai, S. (2025). Nanomaterials for Energy Storage Systems—A Review. *Molecules (Basel, Switzerland)*, *30*(4), 883. DOI: 10.3390/molecules30040883

Mohanty, A. K., Vivekanandhan, S., Pin, J.-M., & Misra, M. (2018). Composites from renewable and sustainable resources: Challenges and innovations. *Science*, *362*(6414), 536–542.

Mok, D. H., & Back, S. (2024). Generative Pretrained Transformer for Heterogeneous Catalysts. *Journal of the American Chemical Society*, *146*(49), 33712–33722. DOI: 10.1021/jacs.4c11504 PMID: 39576215

Mondal, A., Ha, J., Jo, V. Y., Wu, F. Y., Kaza, A. K., & Dupont, P. E. (2021). Preclinical evaluation of a pediatric airway stent for tracheobronchomalacia. *The Journal of Thoracic and Cardiovascular Surgery*, *161*(1), e51–e60.

Mondal, P., Nandan, A., Ajithkumar, S., Siddiqui, N. A., Raja, S., Kola, A. K., & Balakrishnan, D. (2023). Sustainable application of nanoparticles in wastewater treatment: Fate, current trend & paradigm shift. *Environmental Research*, *232*, 116071. DOI: 10.1016/j.envres.2023.116071 PMID: 37209979

Mondal, S., & Shubhra Goswami, S. (2024). Machine learning techniques for quality assurance in additive manufacturing processes. *International Journal of AI for Materials and Design*, *0*(0), 3455. DOI: 10.36922/ijamd.3455

Mo, R., Li, F., Tan, X., Xu, P., Tao, R., Shen, G., Lu, X., Liu, F., Shen, L., Xu, B., Xiao, Q., Wang, X., Wang, C., Li, J., Wang, G., & Lu, Y. (2019). High-quality mesoporous graphene particles as high-energy and fast-charging anodes for lithium-ion batteries. *Nature Communications*, *10*(1), 1474. DOI: 10.1038/s41467-019-09274-y

Morini, A. A., Ribeiro, M. J., & Hotza, D. (2019). Early-stage materials selection based on embodied energy and carbon footprint. *Materials & Design*, *178*, 107861.

Mortazavi, B. (2025). Recent Advances in Machine Learning-Assisted Multiscale Design of Energy Materials. *Advanced Energy Materials*, *15*(9), 2403876.

Moses, O. A., Chen, W., Adam, M. L., Wang, Z., Liu, K., Shao, J., & Yu, X. (2021). Integration of data-intensive, machine learning and robotic experimental approaches for accelerated discovery of catalysts in renewable energy-related reactions. *Materials Reports: Energy*, *1*(3), 100049.

Moshood, T. D., Nawanir, G., Mahmud, F., Mohamad, F., & Ahmad, M. H., & AbdulGhani, A. (2022). Sustainability of biodegradable plastics: New problem or solution to solve the global plastic pollution? *Current Research in Green and Sustainable Chemistry*, *5*, 100273.

Motia, K., Kumar, R., & Akhai, S. (2024). AI and Smart Manufacturing: Building Industry 4.0. In Modern Management Science Practices in the Age of AI (pp. 1-28). IGI Global.

Mounsaveng, S., Vazquez, D., Laradji, I., Ayed, I. B., & Pedersoli, M. (2021).. . *Learning Data Augmentation with Online Bilevel Optimization for Image Classification.*, *13*, 1690–1699. DOI: 10.1109/WACV48630.2021.00173

Mu, B., Hao, X., Luo, X., Yang, Z., Lu, H., & Tian, W. (2024). Bioinspired polymeric supramolecular columns as efficient yet controllable artificial light-harvesting platform. *Nature Communications*, *15*(1), 903.

Muhammad, A., Ishaq, A. A., B, I. O., & Idris, M. B. (2023). Artificial Intelligence and Machine Learning for Real-time Energy Demand Response and Load Management. *Journal of Technology Innovations and Energy*, *2*(2), 20–29. DOI: 10.56556/jtie.v2i2.537

Mullani, S., Kim, C., Lokhande, V., & Ji, T. (2025). MXene structural and surface modifications for enhanced Li-ion diffusion in lithium-ion capacitors: A critical mini review of recent advances. *Chemical Engineering Journal*, *510*, 161565. DOI: 10.1016/j.cej.2025.161565

Müller, S., Sauter, C., Shunmugasundaram, R., Wenzler, N., De Andrade, V., De Carlo, F., Konukoglu, E., & Wood, V. (2021). Deep learning-based segmentation of lithium-ion battery microstructures enhanced by artificially generated electrodes. *Nature Communications*, *12*(1), 6205. DOI: 10.1038/s41467-021-26480-9

Murawski, A., Diaz, R., Inglesby, S., Delabar, K., & Quirino, R. L. (2019). Synthesis of bio-based polymer composites: fabrication, fillers, properties, and challenges. In *Polymer nanocomposites in biomedical engineering* (pp. 29–55). Springer International Publishing.

Murtagh, N., Scott, L., & Fan, J. (2020). Sustainable and resilient construction: Current status and future challenges. *Journal of Cleaner Production, 268*, 122264. DOI: 10.1016/j.jclepro.2020.122264

Mustaffa, R., Mohd Yusof, M. R., & Abdullah, Y. (2015). A novelty of synthetic hydroxyapatite from cockle shell and characterization. *Advanced Materials Research, 1087*, 429–433. DOI: 10.4028/www.scientific.net/AMR.1087.429

Muthu, S. S., & Gardetti, M. A. (2021). *Sustainable Design in Textiles and Fashion*. Springer.

Naaman, A. E. (2003). Engineered Steel Fibers with Optimal Properties for Reinforcement of Cement Composites. *Journal of Advanced Concrete Technology, 1*(3), 241–252. DOI: 10.3151/jact.1.241

Nagarajan, V., Solaiyappan, A., Mahalingam, S. K., Nagarajan, L., Salunkhe, S., Nasr, E. A., Shanmugam, R., & Hussein, H. M. A. M. (2022). Meta-Heuristic Technique-Based Parametric Optimization for Electrochemical Machining of Monel 400 Alloys to Investigate the Material Removal Rate and the Sludge. *Applied Sciences (Basel, Switzerland), 12*(6), 2793. Advance online publication. DOI: 10.3390/app12062793

Nagaswarupa, H. P., Sillanpää, M. E. T., Ananda Murthy, H. C., & Naik, R. (2024). *Multifunctional Inorganic Nanomaterials for Energy Applications*. CRC Press., DOI: 10.1201/9781003479239

Nahirniak, S., Ray, A., & Saruhan, B. (2023). Challenges and Future Prospects of the MXene-Based Materials for Energy Storage Applications. *Batteries, 9*(2), 126. DOI: 10.3390/batteries9020126

Nahshon, Y., Kumaraswamy, K., Zierep, P., Weber, M., Helm, D., Morand, L., De Andrés, P., & Büschelberger, M. (2024). Semantic Orchestration and Exploitation of Material Data: A Dataspace Solution Demonstrated on Steel and Copper Applications. *Advanced Engineering Materials*. Advance online publication. DOI: 10.1002/adem.202401448

Nakano, M., Matsuura, H., Ju, D.-Y., Kumazawa, T., Kimura, S., Uozumi, Y., Tonohata, N., Koide, K., Noda, N., Bian, P., ... Drug delivery system using nanomagnetic fluid, in: 2008 3rd International Conference on Innovative Computing Information and Control, IEEE, 2008, p. 338.

Nakka, R., Harursampath, D., & Ponnusami, S. A. (2023). A generalised deep learning-based surrogate model for homogenisation utilising material property encoding and physics-based bounds. *Scientific Reports, 13*(1), 9079. Advance online publication. DOI: 10.1038/s41598-023-34823-3 PMID: 37277405

Namdeti, R., Al-Kathiri, D. S. M. S., Rao, G. B., Khan, A. H., & Alshehri, H. A. M. (2025). A critical review of smart materials for efficient water purification: Towards sustainable clean water solutions. *Journal of Membrane Science and Research, 11*(1).

Nandy, A., Duan, C., & Kulik, H. J. (2022). Audacity of Huge: Overcoming Challenges of Data Scarcity and Data Quality for Machine Learning in Computational Materials Discovery. *Current Opinion in Chemical Engineering, 36*, 100778. Advance online publication. DOI: 10.1016/j.coche.2021.100778

Nanta, P., Kasemwong, K., & Grisdanurak, N. (2025). Production and characterization of starch microspheres using supercritical carbon dioxide assisted spray drying. *International Journal of Biological Macromolecules, 305*, 141197.

Nartey, O. D., & Zhao, B. (2014). Biochar preparation, characterization, and adsorptive capacity and its effect on bioavailability of contaminants: An overview. *Advances in Materials Science and Engineering, 2014*, 1–12. Advance online publication. DOI: 10.1155/2014/715398

NASA. (2022). *Global Climate Change*. Https://Climate.Nasa.Gov/Vital-Signs/Global-Temperature/

Nath, P., & Sarker, P. K. (2015). Use of OPC to improve setting and early strength properties of low calcium fly ash geopolymer concrete cured at room temperature. *Cement and Concrete Composites, 55*, 205–214.

Natrayan, L., Janardhan, G., Paramasivam, P., & Dhanasekaran, S. (2023). Enhancing Mechanical Performance of TiO2 Filler with Kevlar/Epoxy-Based Hybrid Composites in a Cryogenic Environment: A Statistical Optimization Study Using RSM and ANN Methods. *Frontiers in Materials, 10*, 1267514. Advance online publication. DOI: 10.3389/fmats.2023.1267514

Navarro, D., Fullana-I-Palmer, P., Lin, W., Wu, J., & Puig, R. (2020). Life cycle assessment and leather production. *Journal of Leather Science and Engineering, 2*(1), 26. Advance online publication. DOI: 10.1186/s42825-020-00035-y

Nayak, A. K., Maity, M., Barik, H., Behera, S. S., Dhara, A. K., & Hasnain, M. S. (2024). Bioceramic materials in bone-implantable drug delivery systems: A review. *Journal of Drug Delivery Science and Technology, 95*, 105524. DOI: 10.1016/j.jddst.2024.105524

Nazari, H., Naderi, K., & Nejad, F. M. (2018). Improving aging resistance and fatigue performance of asphalt binders using inorganic nanoparticles. *Construction & Building Materials, 170*, 591–602. DOI: 10.1016/j.conbuildmat.2018.03.107

Neelakandeswari, N., Sangami, G., Dharmaraj, N., (2011). Preparation and characterization of nanostructured hydroxyapatite using a biomaterial. Synthesis and Reactivity in inorganic, metal-organic, Nano Metal. Chem. ume 41, 513–516.

Neeleshwar, S., Saini, A., Bairwa, M. K., Bisht, N., Katre, A., & Rao, G. N. (2022). Major Challenges Toward. *Nanomaterials for Innovative Energy Systems and Devices*, 103.

Nematollahi, P., & Neyts, E. C. (2024). Enhancement of Mn-N-C single atom catalysts via sulfur and/or oxygen co-doping for oxygen reduction in acidic conditions: Unveiling the catalyst durability in fuel cells. *Molecular Catalysis*, *553*, 113745.

Nepal, D., Kang, S., Adstedt, K. M., Kanhaiya, K., Bockstaller, M. R., Brinson, L. C., & Heinz, H. (2023). Hierarchically structured bioinspired nanocomposites. *Nature Materials*, *22*(1), 18–35.

Neville, A. M. (1992). Fibre reinforced cement and concrete. In *CRC Press eBooks*. DOI: 10.1201/9781482271225

Nguyen, C. T., Roy, G., Gauthier, C., & Galanis, N. (2007). Heat transfer enhancement using Al2O3-water nanofluid for an electronic liquid cooling system. *Applied Thermal Engineering*, *27*(8-9), 1501–1506.

Nguyen, Q., Rizvandi, R., Karimipour, A., Malekahmadi, O., & Bach, Q. V. (2020). A Novel Correlation to Calculate Thermal Conductivity of Aqueous Hybrid Graphene Oxide/Silicon Dioxide Nanofluid: Synthesis, Characterizations, Preparation, and Artificial Neural Network Modeling. *Arabian Journal for Science and Engineering*, *45*(11), 9747–9758. DOI: 10.1007/s13369-020-04885-w

Nguyen, V. G., Sharma, P., Ağbulut, Ü., Le, H. S., Truong, T. H., Dzida, M., Tran, M. H., Le, H. C., & Tran, V. D. (2024). Machine Learning for the Management of Biochar Yield and Properties of Biomass Sources for Sustainable Energy. *Biofuels, Bioproducts & Biorefining*, *18*(February), 567–593. Advance online publication. DOI: 10.1002/bbb.2596

Nguyen, V. N., Tarełko, W., Sharma, P., El-Shafay, A. S., Chen, W.-H., Phuoc, Q. P. N., Nguyen, X. P., & Hoang, A. T. (2024). Potential of Explainable Artificial Intelligence in Advancing Renewable Energy: Challenges and Prospects. *Energy & Fuels*, *38*(3), 1692–1712. DOI: 10.1021/acs.energyfuels.3c04343

Nieder, R., & Benbi, D. K. (2024). Potentially toxic elements in the environment–a review of sources, sinks, pathways and mitigation measures. *Reviews on Environmental Health*, *39*(3), 561–575. DOI: 10.1515/reveh-2022-0161

Ninduwezuor-Ehiobu, N., Tula, O., Daraojimba, C., Ogunjobi, O., Egbokhaebho, B., Ofonagoro, K., Gidiagba, J., & Banso, A. (2023). Tracing The Evolution Of Ai And Machine Learning Applications In Advancing Materials Discovery And Production Processes. *Engineering Science & Technology Journal*, *4*(3), 66–83. DOI: 10.51594/estj.v4i3.552

Niu, G., Guo, X., & Wang, L. (2015). Review of recent progress in chemical stability of perovskite solar cells. *Journal of Materials Chemistry. A, Materials for Energy and Sustainability*, *3*(17), 8970–8980. DOI: 10.1039/C4TA04994B

Noh, J., Kim, J., Stein, H. S., Sanchez-Lengeling, B., Gregoire, J. M., Aspuru-Guzik, A., & Jung, Y. (2019). Inverse Design of Solid-State Materials via a Continuous Representation. *Matter*, *1*(5), 1370–1384. Advance online publication. DOI: 10.1016/j.matt.2019.08.017

Nolan, M., Iwaszuk, A., Lucid, A. K., Carey, J. J., & Fronzi, M. (2016). Design of novel visible light active photocatalyst materials: Surface modified TiO2. *Advanced Materials*, *28*(27), 5425–5446. DOI: 10.1002/adma.201504894

Noor, B., Fahel, M., Yasmin, N., & Besara, T. (2024). Machine Learning in High-Entropy Alloys: Phase Formation Predictions with Artificial Neural Networks. *Future Sustainability*, *2*(1), 47–58. Advance online publication. DOI: 10.55670/fpll.fusus.2.1.5

Norton, M. G. (2023). There's Still Plenty of Room at the Bottom. In *A Modern History of Materials: From Stability to Sustainability* (pp. 81–100). Springer International Publishing., DOI: 10.1007/978-3-031-23990-8_5

Noureen, L., Wang, Q., Humayun, M., Shah, W. A., Xu, Q., & Wang, X. (2023). Recent advances in structural engineering of photocatalysts for environmental remediation. *Environmental Research*, *219*, 115084. DOI: 10.1016/j.envres.2022.115084

Nunes Vidal, I. (2021). Autocratization outbreaks: The Polish and Hungarian responses to the Covid-19 pandemic and their early effects on democracy in comparison to Italy.

Nwanna, E. C., Imoisili, P. E., & Jen, T.-C. (2020). Fabrication and synthesis of SnOX thin films: A review. *International Journal of Advanced Manufacturing Technology*, *111*(9–10), 2809–2831. DOI: 10.1007/s00170-020-06223-8

Nwanna, E. C., Imoisili, P. E., & Jen, T.-C. (2022). Synthesis and characterization of SnO2 thin films using metalorganic precursors. *Journal of King Saud University. Science*, *34*(5), 102123. DOI: 10.1016/j.jksus.2022.102123

O'Hegarty, R., Kinnane, O., Newell, J., & West, R. (2021). High performance, low carbon concrete for building cladding applications. *Journal of Building Engineering*, *43*, 102566.

Obradovic, D., & Mishra, L. N. (2020). Mechanical properties of recycled paper and cardboard. The Journal of Engineering and Exact Sciences, 6(3), 0429-0434.

Ofuani, N. (2024). Carbon Dioxide Gasification of Biochar: A Study on Utilizing Captured CO2 to Mitigate Greenhouse gas Emission and Improving Carbon Conversion Efficiency.

Ogungbenro, A. E., Quang, D. V., Al-Ali, K., & Abu-Zahra, M. R. M. (2017). Activated Carbon from Date Seeds for CO2 Capture Applications. *Energy Procedia*, *114*, 2313–2321. DOI: 10.1016/j.egypro.2017.03.1370

Ojodun, O. E., Imoisili, P. E., & Jen, T.-C. (2025[b]). Chemical Solution–Based Deposition Techniques for Asymmetric Supercapacitors: Progress in Electrode Materials for High-Efficiency Energy Storage. Journal of Nanotechnology, 2025(1). Portico. DOI: 10.1155/jnt/5551493

Ojodun, O. E., Imoisili, P. E., & Jen, T.-C. (2025). [a]). Thermal annealing-induced structural modifications and electrochemical enhancement of NiO/CNT electrodes synthesized by spray pyrolysis for high-performance supercapacitors. *Electrochemistry Communications*, *178*, 108010. DOI: 10.1016/j.elecom.2025.108010

Ojo, G. G., Lottu, O. A., Ndiwe, T. C., Izuka, U., & Ehiobu, N. N. (2023). Solar energy adaptation and efficiency across diverse Nigerian and global climates: A review of technological advancement. [GWK]. *Engineering Heritage Journal*, *7*(1), 99–107. DOI: 10.26480/gwk.01.2023.99.107

Oliva, D., Houssein, E. H., Abualigah, L., & Elaziz, M. A. (2022). *Integrating Meta-Heuristics and Machine Learning for Real-World Optimization Problems*. Springer International Publishing.

Olivetti, E. A., & Cullen, J. M. (2018). Toward a sustainable materials system. *Science*, *360*(6396), 1396–1398.

Onyiriuka, E. (2023). Predictive modelling of thermal conductivity in single-material nanofluids: A novel approach. *Bulletin of the National Research Center*, *47*(1), 140. Advance online publication. DOI: 10.1186/s42269-023-01115-9

Opetubo, O. R., Kitalu, R., Oviroh, P. O., Oyinbo, S. T., Imoisili, P. E., & Jen, T. C. (2023). A mini-review on MoS2 membrane for water desalination: Recent development and challenges. *Nanotechnology Reviews*, *12*(1), 20220563. DOI: 10.1515/ntrev-2022-0563

Osman, A. I., Ayati, A., Krivoshapkin, P., Tanhaei, B., Farghali, M., Yap, P. S., & Abdelhaleem, A. (2024). Coordination-driven innovations in low-energy catalytic processes: Advancing sustainability in chemical production. *Coordination Chemistry Reviews*, *514*, 215900.

Othman, M., Abd Halimee, N. F., Osmi, S. K. C., Nordin, N. Z., & Sabri, A. (2024, October). The Effect of Immersion Weathering on the Tensile Properties of rHDPE Green Composite Material. *International Exchange and Innovation Conference on Engineering & Sciences*.

Ozin, G. A., & Loh, J. Y. (2022). *Energy Materials Discovery: Enabling a Sustainable Future*. Royal Society of Chemistry.

Page, S. B., Bryson, J. M., Stone, M. M., Crosby, B. C., & Seo, D. (2021). Ambidexterity in Cross-Sector Collaborations Involving Public Organizations. *Public Performance & Management Review*, *44*(6), 1161–1190. DOI: 10.1080/15309576.2021.1937243

Pai, S. M., Shah, K. A., Sunder, S., Albuquerque, R. Q., Brütting, C., & Ruckdäschel, H. (2025). Machine learning applied to the design and optimization of polymeric materials: A review. *Next Materials*, *7*, 100449. DOI: 10.1016/j.nxmate.2024.100449

Pal, A., Paul, S., Choudhury, A. R., Balla, V. K., Das, M., & Sinha, A. (2017). Synthesis of hydroxyapatite from Lates calcarifer fish bone for biomedical applications. *Materials Letters*, *203*, 89–92. DOI: 10.1016/j.matlet.2017.05.103

Panagiotopoulou, V. C., Stavropoulos, P., & Chryssolouris, G. (2022). A critical review on the environmental impact of manufacturing: A holistic perspective. *International Journal of Advanced Manufacturing Technology*, *118*(1), 603–625. DOI: 10.1007/s00170-021-07980-w

Panda, N. N., Pramanik, K., & Sukla, L. B. (2014). Extraction and characterization of biocompatible hydroxyapatite from fresh water fish scales for tissue engineering scaffold. *Bioprocess and Biosystems Engineering*, *37*(3), 433–440. DOI: 10.1007/s00449-013-1009-0

Panda, P. K., & Sahoo, B. (2020). PZT and lead-free piezo ceramics for aerospace and energy applications. In *Handbook of Advanced Ceramics and Composites* (pp. 1–23). Springer. DOI: 10.1007/978-3-319-73255-8_32-1

Pandey, P., Khan, F., Agarwal, S., & Singh, M. (2022). Nano adsorbents in wastewater treatment: A new paradigm in wastewater management. *Lett. Appl. Nanobiosci*, *12*(4), 125. DOI: 10.33263/LIANBS124.125

Pandit, P., Nadathur, G. T., Maiti, S., & Regubalan, B. (2018). Functionality and properties of bio-based materials. Bio-based materials for food packaging: Green and sustainable advanced packaging materials, 81-103.

Pandit, P., Nadathur, G. T., & Jose, S. (2019). Upcycled and low-cost sustainable business for value-added textiles and fashion. In *Circular economy in textiles and apparel* (pp. 95–122). Elsevier.

Panghal, H., & Kumar, A. (2025). Eco-Friendly Concrete: A Dual Surface Modification Approach for Enhanced Mechanical Performance of Recycled Coarse Aggregates. *Journal of Materials in Civil Engineering*, *37*(4), 04025052. DOI: 10.1061/JMCEE7.MTENG-19054

Pan, Y., Farmahini-Farahani, M., O'Hearn, P., Xiao, H., & Ocampo, H. (2016). An overview of bio-based polymers for packaging materials. *J. Bioresour. Bioprod*, *1*(3), 106–113.

Paraskeva, P., Kalderis, D., & Diamadopoulos, E. (2008). Production of activated carbon from agricultural by-products. In *Journal of Chemical Technology and Biotechnology* (Vol. 83, Issue 5, pp. 581–592). DOI: 10.1002/jctb.1847

Paraye, P., & Sarviya, R. (2024). Advances in polymer composites, manufacturing, recycling, and sustainable practices. *Polymer-Plastics Technology and Materials*, *63*(11), 1474–1497.

Park, D., Park, K., Ryu, S., & Lee, J. (2023). Hierarchical Generative Network: A Hierarchical Multitask Learning Approach for Accelerated Composite Material Design and Discovery. *Advanced Engineering Materials*, *25*(21), 2300867. Advance online publication. DOI: 10.1002/adem.202300867

Park, H., Shin, D. J., & Yu, J. (2021). Categorization of quantum dots, clusters, nanoclusters, and nanodots. *Journal of Chemical Education*, *98*(3), 703–709. DOI: 10.1021/acs.jchemed.0c01403

Patel, S., & Nanda, R. (2015). Nanotechnology in Healthcare: Applications and Challenges. *Medicinal Chemistry (Sunnyvale, Calif.)*, *05*(12). Advance online publication. DOI: 10.4172/2161-0444.1000312

Patras, V. (2021). Sustainability in the shoe market: Adidas on a move to sustainability.

Pattanayak, S. R., Thakur, P., & Narayanan, T. N. (2025). The oxygen paradox in alkaline zinc-air batteries creates radical-induced anodic corrosion. *Cell Reports. Physical Science*.

Patton, A., Crittenden, B. D., & Perera, S. P. (2004). Use of the Linear Driving Force Approximation to Guide the Design of Monolithic Adsorbents. *Chemical Engineering Research & Design*, *82*(8), 999–1009. DOI: 10.1205/0263876041580749

Paul, S., Pal, A., Choudhury, A. R., Bodhak, S., Balla, V. K., Sinha, A., & Das, M. (2017). Effect of trace elements on the sintering effect of fish scale derived hydroxyapatite and its bioactivity. *Ceramics International*, *43*(17), 15678–15684. DOI: 10.1016/j.ceramint.2017.08.127

Paunovska, K., Loughrey, D., Sago, C. D., Langer, R., & Dahlman, J. E. (2019). Using Large Datasets to Understand Nanotechnology. *Advanced Materials*, *31*(43), 1902798. DOI: 10.1002/adma.201902798

Pedersen, E. R. G., Lüdeke-Freund, F., Henriques, I., & Seitanidi, M. M. (2021). Toward collaborative cross-sector business models for sustainability. *Business & Society*, *60*(5), 1039–1058. DOI: 10.1177/0007650320959027

Peixoto, D. A., Silva, S. C., Borges, P. H. S., Lima, R. C., & Nossol, E. (2023). Hydrothermal synthesis as a versatile tool for the preparation of metal hexacyanoferrates: A review. *Journal of Materials Science*, *58*(7), 2993–3024. DOI: 10.1007/s10853-023-08190-3

Penazzato, L., Illampas, R., & Oliveira, D. V. (2024). The challenge of integrating seismic and energy retrofitting of buildings: An opportunity for sustainable materials? *Sustainability (Basel)*, *16*(8), 3465. DOI: 10.3390/su16083465

Peng, W., & Sadaghiani, O. K. (2024). Machine Learning for Sustainable Reutilization of Waste Materials as Energy Sources–a Comprehensive Review. *International Journal of Green Energy*, *21*(7), 1641–1666. Advance online publication. DOI: 10.1080/15435075.2023.2255647

Pereira, L. F., de Azevedo-Silva, L. J., Minim, P. R., Lisboa-Filho, P. N., Fortulan, C. A., Griggs, J. A., & Borges, A. F. S. (2024). Are we approaching the development of a novel calcium phosphate-based bioceramic dental material? *Dental Materials*, *40*(11), 1843–1853. DOI: 10.1016/j.dental.2024.07.019

Peretz, R., Mamane, H., Wissotzky, E., Sterenzon, E., & Gerchman, Y. (2021). Making cardboard and paper recycling more sustainable: Recycled paper sludge for energy production and water-treatment applications. *Waste and Biomass Valorization*, *12*, 1599–1608.

Perin, D., Rigotti, D., Fredi, G., Papageorgiou, G. Z., Bikiaris, D. N., & Dorigato, A. (2021). Innovative bio-based poly (lactic acid)/poly (alkylene furanoate) s fiber blends for sustainable textile applications. *Journal of Polymers and the Environment*, *29*(12), 3948–3963.

Pevida, C., Drage, T. C., & Snape, C. E. (2008). Silica-templated melamine–formaldehyde resin derived adsorbents for CO2 capture. *Carbon*, *46*(11), 1464–1474. DOI: 10.1016/j.carbon.2008.06.026

Pham, H. T., Awange, J., & Kuhn, M. (2022). Evaluation of Three Feature Dimension Reduction Techniques for Machine Learning-Based Crop Yield Prediction Models. *Sensors (Basel)*, *22*(17), 6609. DOI: 10.3390/s22176609

Phogat, P., Thakur, J., & Wan, M. (2025). Advancements in energy storage: A review of batteries and capacitors—properties, materials, and emerging applications. *Ionics*, *31*(10), 1–25. DOI: 10.1007/s11581-025-06588-4

Ping, R., Nie, Y., Ji, C., Hao, Z., Yang, S., Li, L., Peng, J., & Li, X. (2023). Enhanced proton pseudocapacitive of Ti3C2Tx in neutral electrolyte activated by acid regulation dynamic pillars. *Chemical Engineering Journal*, *455*, 140650. DOI: 10.1016/j.cej.2022.140650

Pinlova, B., Sudheshwar, A., Vogel, K., Malinverno, N., Hischier, R., & Som, C. (2024). What can we learn about the climate change impacts of polylactic acid from a review and meta-analysis of lifecycle assessment studies? *Sustainable Production and Consumption*.

Pisano, R., & Durlo, A. (2025). Nanosciences and Nanotechnologies: A Scientific–Historical Introductory Review. Nanoscience & Nanotechnologies: Critical Problems, Science in Society, Historical Perspectives, 3-90. DOI: 10.1007/978-3-031-85122-3_1

Plaza, M. G., García, S., Rubiera, F., Pis, J. J., & Pevida, C. (2010). Post-combustion CO2 capture with a commercial activated carbon: Comparison of different regeneration strategies. *Chemical Engineering Journal*, *163*(1–2), 41–47. DOI: 10.1016/j.cej.2010.07.030

Podara, C., Termine, S., Modestou, M., Semitekolos, D., Tsirogiannis, C., Karamitrou, M., & Charitidis, C. (2024). Recent trends of recycling and upcycling of polymers and composites: A comprehensive review. *Recycling*, *9*(3), 37.

Podgornik, B. (2023). Advanced materials and research for the green future. *Materials Technology*, *57*(1), 71–83.

Pogue, E. A., New, A., McElroy, K., Le, N. Q., Pekala, M. J., McCue, I., Gienger, E., Domenico, J., Hedrick, E., McQueen, T. M., Wilfong, B., Piatko, C. D., Ratto, C. R., Lennon, A., Chung, C., Montalbano, T., Bassen, G., & Stiles, C. D. (2023). Closed-Loop Superconducting Materials Discovery. *npj Computational Materials*, *9*(1), 181. Advance online publication. DOI: 10.1038/s41524-023-01131-3

Pohjakallio, M. (2020). Secondary plastic products examples and market trends. In *Plastic waste and recycling* (pp. 467–479). Academic Press.

Polizos, G., Goswami, M., Keum, J. K., He, L., Jafta, C. J., Sharma, J., & Li, J. (2024). Nanoscale ion transport enhances conductivity in solid polymer-ceramic lithium electrolytes. *ACS Nano*, *18*(4), 2750–2762.

Polman, A., Knight, M., Garnett, E. C., Ehrler, B., & Sinke, W. C. (2016). Photovoltaic materials: Present efficiencies and future challenges. *Science*, *352*(6283), aad4424. Advance online publication. DOI: 10.1126/science.aad4424

Pomerantseva, E., Bonaccorso, F., Feng, X., Cui, Y., & Gogotsi, Y. (2019). Energy storage: The future enabled by nanomaterials. *Science*, *366*(6468), eaan8285. DOI: 10.1126/science.aan8285

Ponce-Bobadilla, A. V., Schmitt, V., Maier, C. S., Mensing, S., & Stodtmann, S. (2024). Practical guide to SHAP analysis: Explaining supervised machine learning model predictions in drug development. *Clinical and Translational Science*, *17*(11), e70056. Advance online publication. DOI: 10.1111/cts.70056 PMID: 39463176

Pon-On, W., Suntornsaratoon, P., Charoenphandhu, N., Thongbunchoo, J., Krishnamra, N., & Tang, I. M. (2016). Hydroxyapatite from fish scale for potential use as bone scaffold or regenerative material. *Materials Science and Engineering C*, *62*, 183–189. DOI: 10.1016/j.msec.2016.01.051

Powell, P. C., & Housz, A. I. (2023). *Engineering with polymers*. CRC Press.

Prasad Kumara, P. A. A. S., Cooper, P. R., Cathro, P., Gould, M., Dias, G., & Ratnayake, J. (2025). Bioceramics in Endodontics: Limitations and Future Innovations—A Review. *Dentistry Journal*, *13*(4), 157. DOI: 10.3390/dj13040157

Prateek, S., Garg, R., & Saxena, K. K. (2024). Data-Driven Materials Science: Application of ML for Predicting Band Gap. *Advances in Materials and Processing Technologies*, *10*(2), 708–717. Advance online publication. DOI: 10.1080/2374068X.2023.2171666

Priyantha, N., & Karunathilake, M. (2025). Introduction to Carbon-Based Materials for Pollution Control. In *Carbon: Bulk-to-Nano Forms for Detection and Remediation of Environmental Contaminants* (pp. 1–25). Springer Nature Switzerland. DOI: 10.1007/978-3-031-90613-8_1

Priya, P. K., Vanitha, S., & Meyyappan, P. (2020b). Utilization of waste for building materials – a review. *IOP Conference Series. Materials Science and Engineering*, *955*(1), 012048. DOI: 10.1088/1757-899X/955/1/012048

Provis, J. L. (2017). Alkali-activated materials. *Cement and Concrete Research*, *114*, 40–48. DOI: 10.1016/j.cemconres.2017.02.009

Purabgola, A., Mayilswamy, N., & Kandasubramanian, B. (2022). Graphene-based TiO2 composites for photocatalysis & environmental remediation: Synthesis and progress. *Environmental Science and Pollution Research International*, *29*(22), 32305–32325. DOI: 10.1007/s11356-022-18983-9

Qadir, A., Le, T. K., Malik, M., Min-Dianey, K. A. A., Saeed, I., Yu, Y., & Pham, P. V. (2021). Representative 2D-material-based nanocomposites and their emerging applications: A review. *RSC Advances*, *11*(39), 23860–23880.

Quijano, L., Speight, R., & Payne, A. (2021). Future fashion, biotechnology and the living world: Microbial cell factories and forming new 'oddkins'. *Continuum*, *35*(6), 897–913.

Radi, A., Khalifa, A., Fattah, A. A., & Jalal, A. (2020). Study of Physical, Mechanical Properties and Sewability of Modal and Micromodal Fabrics and Comparing it with Cotton. [JAAS]. *Journal of Arts & Applied Sciences*, *7*(4), 101–124.

Rahavi, S. S., Ghaderi, O., Monshi, A., & Fathi, M. H. (2017). A comparative study on physicochemical properties of hydroxyapatite powders derived from natural and synthetic sources. *Russian Journal of Non-Ferrous Metals*, *58*(3), 276–286. DOI: 10.3103/S1067821217030178

Rajak, P., Krishnamoorthy, A., Mishra, A., Kalia, R., Nakano, A., & Vashishta, P. (2021). Autonomous Reinforcement Learning Agent for Chemical Vapor Deposition Synthesis of Quantum Materials. *npj Computational Materials*, *7*(1), 108. Advance online publication. DOI: 10.1038/s41524-021-00535-3

Rajak, P., Wang, B., Nomura, K., Luo, Y., Nakano, A., Kalia, R., & Vashishta, P. (2021). Autonomous Reinforcement Learning Agent for Stretchable Kirigami Design of 2D Materials. *npj Computational Materials*, *7*(1), 102. Advance online publication. DOI: 10.1038/s41524-021-00572-y

Rajendran, S., Scelsi, L., Hodzic, A., Soutis, C., & Al-Maadeed, M. A. (2012). Environmental impact assessment of composites containing recycled plastics. *Resources, Conservation and Recycling*, *60*, 131–139.

Ramakrishna, S., & Jose, R. (2022). Principles of materials circular economy. *Matter*, *5*(12), 4097–4099.

Ramesh, N., Ratnayake, J. T., & Dias, G. J. (2021). Calcium-based ceramic biomaterials. In *Structural biomaterials* (pp. 333–394). Woodhead Publishing. DOI: 10.1016/B978-0-12-818831-6.00011-2

Rana, B., & Rathore, S. S. (2023). Industry 4.0 – Applications, challenges and opportunities in industries and academia: A review. *Materials Today: Proceedings*, *79*, 389–394. DOI: 10.1016/j.matpr.2022.12.162

Rana, M. S., AlHumaidan, F. S., Bouresli, R., & Navvamani, R. (2021). Guard-bed catalyst: Impact of textural properties on catalyst stability and deactivation rate. *Molecular Catalysis*, *502*, 111375.

Rangel-Sequeda, J. F., Loredo-Cancino, M., Águeda Maté, V. I., De Haro-Del Rio, D. A., & Dávila-Guzmán, N. E. (2022). 3D printing of powdered activated carbon monoliths: Effect of structuring on physicochemical and mechanical properties and its influence on the adsorption performance. *Materials Today. Communications*, *33*, 104758. DOI: 10.1016/j.mtcomm.2022.104758

Rangel-Sequeda, J., Loredo-Cancino, M., Agueda Maté, V. I., Delgado Dobladez, J. A., De Haro-Del Rio, D. A., Salazar-Rábago, J. J., & Soto-Regalado, E. (2021). Synthesis of Zeolite-Templated Carbonaceous Materials Doped With Nitrogen for Carbon Dioxide Adsorption. *Water, Air, and Soil Pollution*, *232*(7), 253. DOI: 10.1007/s11270-021-05184-2

Rani, M., Yadav, J., Shanker, U., & Wang, C. (2024). Recent updates on remediation approaches of environmentally occurring pollutants using visible light-active nano-photocatalysts. *Environmental Science and Pollution Research International*, *31*(15), 22258–22283. DOI: 10.1007/s11356-024-32455-2

Ratnesh, R. K., Kumar, R., Singh, S., Chandra, R., & Singh, J. (2025). Recent advances in solar cell technology: Addressing technological challenges, scenarios, and environmental implications in the development of sustainable energy solutions. *New Journal of Chemistry*, *49*(17), 6861–6887. DOI: 10.1039/D5NJ00719D

Rausa, G., Calabrese, M., Velazquez, R., Del-Valle-Soto, C., Fazio, R. D., & Visconti, P. (2025). Mechanical, Thermal, and Environmental Energy Harvesting Solutions in Fully Electric and Hybrid Vehicles: Innovative Approaches and Commercial Systems. *Energies, 18*(8), 1970. DOI: 10.3390/en18081970

Raval, B., Joshi, N., Kathrotiya, R. M., Raval, S. Y., & Raval, V. H. (2025). Microbes and Nanoconjugate-Assisted Removal of Heavy Metals from Water Resources. Nano-Bioremediation for Wastewater Treatment, 107-145.

Recatala-Gomez, J., Suwardi, A., Nandhakumar, I., Abutaha, A., & Hippalgaonkar, K. (2020). Toward accelerated thermoelectric materials and process discovery. *ACS Applied Energy Materials, 3*(3), 2240–2257. DOI: 10.1021/acsaem.9b02222

Rees, W. (2018). Ecological footprint. In *Companion to environmental studies* (pp. 43–48). Routledge.

Rehman, A., & Park, S. J. (2018). Comparative study of activation methods to design nitrogen-doped ultra-microporous carbons as efficient contenders for CO2 capture. *Chemical Engineering Journal, 352*, 539–548. DOI: 10.1016/j.cej.2018.07.046

Rehman, A., & Park, S.-J. (2019). Tunable nitrogen-doped microporous carbons: Delineating the role of optimum pore size for enhanced CO2 adsorption. *Chemical Engineering Journal, 362*, 731–742. DOI: 10.1016/j.cej.2019.01.063

Rehman, K. U., Shatanawi, W., & Çolak, A. B. (2023). Levenberg–Marquardt Training Technique Analysis of Thermally Radiative and Chemically Reactive Stagnation Point Flow of Non-Newtonian Fluid with Temperature Dependent Thermal Conductivity. *Mathematics, 11*(3), 753. Advance online publication. DOI: 10.3390/math11030753

Reinoso, D. M., & Frechero, M. A. (2022). Strategies for rational design of polymer-based solid electrolytes for advanced lithium energy storage applications. *Energy Storage Materials, 52*, 430–464.

Rekhviashvili, S. S. (2023). Effect of Polarization on Exciton Properties in a Semiconductor Quantum Dot. *Russian Physics Journal, 65*(12), 2058–2067. DOI: 10.1007/s11182-023-02871-x

Rezaei, F., & Webley, P. (2009). Optimum structured adsorbents for gas separation processes. *Chemical Engineering Science, 64*(24), 5182–5191. DOI: 10.1016/j.ces.2009.08.029

Rezania, S., Darajeh, N., Rupani, P. F., Mojiri, A., Kamyab, H., & Taghavijeloudar, M. (2024). Recent advances in the adsorption of different pollutants from waste-water using carbon-based and metal-oxide nanoparticles. *Applied Sciences (Basel, Switzerland), 14*(24), 11492. DOI: 10.3390/app142411492

Rezvani Ghomi, E., Khosravi, F., Saedi Ardahaei, A., Dai, Y., Neisiany, R. E., Foroughi, F., Wu, M., Das, O., & Ramakrishna, S. (2021). The life cycle assessment for polylactic acid (PLA) to make it a low-carbon material. *Polymers*, *13*(11), 1854.

Rheima, A. M., Abdul-Rasool, A. A., Al-Sharify, Z. T., Zaidan, H. K., Athair, D. M., Mohammed, S. H., & kianfar, E. (2024). Nano bioceramics: Properties, applications, Hydroxyapatite, Nanohydroxyapatite and drug delivery. *Case Studies in Chemical and Environmental Engineering*, *10*, 100869. DOI: 10.1016/j.cscee.2024.100869

Rizvi, S. (2023). Unveiling the Potential of Artificial Intelligence and Machine Learning in the 5G Network Landscape: A Comprehensive Review. *Asian Journal of Research in Computer Science*, *16*(4), 23–31. DOI: 10.9734/ajrcos/2023/v16i4367

Roco, M. C. (2023). National Nanotechnology Initiative at 20 years: Enabling new horizons. *Journal of Nanoparticle Research*, *25*(10), 197. Advance online publication. DOI: 10.1007/s11051-023-05829-9

Rodríguez-Galván, A., & Contreras-Torres, F. F. (2022). Scanning tunneling microscopy of biological structures: An elusive goal for many years. *Nanomaterials (Basel, Switzerland)*, *12*(17), 3013. DOI: 10.3390/nano12173013

Rodriguez-Palomo, A., Østergaard, M., & Birkedal, H. (2024). Bone hierarchical structure: Heterogeneity and uniformity. *Advanced Functional Materials*, *34*(35), 2307026. DOI: 10.1002/adfm.202307026

Roggia, M., Natale, B., Amendola, G., Di Maro, S., & Cosconati, S. (2023). Streamlining Large Chemical Library Docking with Artificial Intelligence: The PyRMD2Dock Approach. *Journal of Chemical Information and Modeling*, *64*(7), 2143–2149. DOI: 10.1021/acs.jcim.3c00647 PMID: 37552222

Rolo, I., Costa, V. A., & Brito, F. P. (2023). Hydrogen-based energy systems: Current technology development status, opportunities and challenges. *Energies*, *17*(1), 180. DOI: 10.3390/en17010180

Romero-Castro, N., López-Cabarcos, M. Á., & Piñeiro-Chousa, J. (2022). Uncovering complexity in the economic assessment of derogations from the European industrial emissions directive. *Journal of Innovation & Knowledge*, *7*(1), 100159.

Rong, C., Zhou, L., Zhang, B., & Xuan, F.-Z. (2023). Machine learning for mechanics prediction of 2D MXene-based aerogels. *Composites Communications*, *38*, 101474. DOI: 10.1016/j.coco.2022.101474

Rouzitalab, Z., Mohammady Maklavany, D., Rashidi, A., & Jafarinejad, S. (2018). Synthesis of N-doped nanoporous carbon from walnut shell for enhancing CO2 adsorption capacity and separation. *Journal of Environmental Chemical Engineering, 6*(5), 6653–6663. DOI: 10.1016/j.jece.2018.10.035

Roy, D. B., & Das, S. (2025). Advanced Nanostructured Materials for Energy Storage Devices. *Design, Fabrication, and Significance of Advanced Nanostructured Materials*, 187-206.

Roy, S., & Ragunath, S. (2018). Emerging membrane technologies for water and energy sustainability: Future prospects, constraints and challenges. *Energies, 11*(11), 2997. DOI: 10.3390/en11112997

Ruiz-Morales, J. C., Tarancón, A., Canales-Vázquez, J., Méndez-Ramos, J., Hernández-Afonso, L., Acosta-Mora, P., Marín Rueda, J. R., & Fernández-González, R. (2017). Three dimensional printing of components and functional devices for energy and environmental applications. *Energy & Environmental Science, 10*(4), 846–859. DOI: 10.1039/C6EE03526D

Rzaij, J. M., & Abass, A. M. (2020). Review on: TiO2 thin film as a metal oxide gas sensor. *J. Chem. Rev, 2*(2), 114–121. DOI: 10.33945/SAMI/JCR.2020.2.4

Saal, J. E., Kirklin, S., Aykol, M., Meredig, B., & Wolverton, C. (2013). Materials Design and Discovery with High-Throughput Density Functional Theory: The Open Quantum Materials Database (OQMD). *Journal of the Minerals Metals & Materials Society, 65*(11), 1501–1509. Advance online publication. DOI: 10.1007/s11837-013-0755-4

Sadeghi, G., Mehrali, M., Shahi, M., & Brem, G. (2022). Experimental analysis of Shape-Stabilized PCM applied to a Direct-Absorption evacuated tube solar collector exploiting sodium acetate trihydrate and graphite. *Energy Conversion and Management, 269*, 116176.

Sadeghzadeh, M., Maddah, H., & Ahmadi, M. H. (2020). Prediction of Thermo-Physical Properties of TiO2-Al2O3/Water Nanoparticles by Using Artificial Neural Network. *Nanomaterials (Basel, Switzerland), 10*, 697.

Sadh, P. K., Duhan, S., & Duhan, J. S. (2018). Agro-industrial wastes and their utilization using solid state fermentation: a review. In *Bioresources and Bioprocessing* (Vol. 5, Issue 1). Springer Science and Business Media Deutschland GmbH. DOI: 10.1186/s40643-017-0187-z

Saeloo, B., Saisopa, T., Chavalekvirat, P., Iamprasertkun, P., Jitapunkul, K., Sirisaksoontorn, W., Lee, T. R., & Hirunpinyopas, W. (2024). Role of Transition Metal Dichalcogenides as a Catalyst Support for Decorating Gold Nanoparticles for Enhanced Hydrogen Evolution Reaction. *Inorganic Chemistry*, *63*(40), 18750–18762. DOI: 10.1021/acs.inorgchem.4c02668

Saha, D., Orkoulas, G., Chen, J., & Hensley, D. K. (2017). Adsorptive separation of CO2 in sulfur-doped nanoporous carbons: Selectivity and breakthrough simulation. *Microporous and Mesoporous Materials*, *241*, 226–237. DOI: 10.1016/j.micromeso.2016.12.015

Saha, D., Thorpe, R., Van Bramer, S. E., Alexander, N., Hensley, D. K., Orkoulas, G., & Chen, J. (2018). Synthesis of Nitrogen and Sulfur Codoped Nanoporous Carbons from Algae: Role in CO_2 Separation. *ACS Omega*, *3*(12), 18592–18602. DOI: 10.1021/acsomega.8b02892

Saha, D., Van Bramer, S. E., Orkoulas, G., Ho, H. C., Chen, J., & Henley, D. K. (2017). CO2 capture in lignin-derived and nitrogen-doped hierarchical porous carbons. *Carbon*, *121*, 257–266. DOI: 10.1016/j.carbon.2017.05.088

Saharudin, M., Ayub, A., Hasbi, S., Muhammad-Sukki, F., Shyha, I., & Inam, F. (2023). Recent advances in MXene composites research, applications and opportunities. *Materials Today: Proceedings*. Advance online publication. DOI: 10.1016/j.matpr.2023.02.435

Sahoo, K., Bergman, R., & Runge, T. (2021). Life-cycle assessment of redwood lumber products in the US. *The International Journal of Life Cycle Assessment*, *26*, 1702–1720.

Said, Z., Sharma, P., Bora, B. J., & Pandey, A. K. (2023). Sonication impact on thermal conductivity of f-MWCNT nanofluids using XGBoost and Gaussian process regression. *Journal of the Taiwan Institute of Chemical Engineers*, *145*, 104818.

Said, Z., Sharma, P., Sundar, L. S., Nguyen, V. G., Tran, V. D., & Le, V. V. (2022). Using Bayesian Optimization and Ensemble Boosted Regression Trees for Optimizing Thermal Performance of Solar Flat Plate Collector under Thermosyphon Condition Employing MWCNT-Fe3O4/Water Hybrid Nanofluids. *Sustainable Energy Technologies and Assessments*, *53*(October), 102708. DOI: 10.1016/j.seta.2022.102708

Saini, R. K., Saini, D. K., Verma, P., Thakur, R., Gandotra, N., Kumar, R., & Verma, A. (2025). Climate change and energy transition-a critical review. *International Journal of Green Energy*, *23*, 1–26.

Salami, R., Kordi, M., Bolouri, P., Delangiz, N., & Asgari Lajayer, B. (2021). Algae-based biorefinery as a sustainable renewable resource. *Circular Economy and Sustainability*, •••, 1–17.

Saleh, T. A. (2022). Nanomaterials and hybrid nanocomposites for CO 2 capture and utilization: Environmental and energy sustainability. *RSC Advances*, *12*(37), 23869–23888. DOI: 10.1039/D2RA03242B

Sani, A. R., Kouzani, A. Z., & Zolfagharian, A. (2024). Artificial Intelligence-Augmented Additive Manufacturing: Insights on Closed-Loop 3D Printing. *Advanced Intelligent Systems*, *6*(10), 2400102. Advance online publication. DOI: 10.1002/aisy.202400102

Santos, A. J., Martin, N., Outon, J., Blanco, E., Garcia, R., & Morales, F. M. (2023). A simple two-step approach to the fabrication of VO2-based coatings with unique thermochromic features for energy-efficient smart glazing. *Energy and Building*, *285*, 112892.

Sarah Couck, J. C.-S.-R. S. V. der P. G. V. B. C. M. P. R. J. F. M. D. (2018). 3D-printed SAPO-34 monoliths for gas separation. *Microporous and Mesoporous Materials, 255*, 185–191.

Sarker, I. H. (2022). AI-Based Modeling: Techniques, Applications and Research Issues Towards Automation, Intelligent and Smart Systems. *SN Computer Science*, *3*(2), 158. DOI: 10.1007/s42979-022-01043-x

Sarwar, A., Ali, M., Khoja, A. H., Nawar, A., Waqas, A., Liaquat, R., Naqvi, S. R., & Asjid, M. (2021). Synthesis and characterization of biomass-derived surface-modified activated carbon for enhanced CO2adsorption. *Journal of CO2 Utilization, 46*. DOI: 10.1016/j.jcou.2021.101476

Savari, M., Moghaddam, A. H., Amiri, A., Shanbedi, M., & Ayub, M. N. B. (2017). Comprehensive heat transfer correlation for water/ethylene glycol-based graphene (nitrogen-doped graphene) nanofluids derived by artificial neural network (ANN) and adaptive neuro-fuzzy inference system (ANFIS). *Heat and Mass Transfer*, *53*(10), 3073–3083.

Scenari, G. L. I. (2020). Making the renovation wave work: a Marshall plan for EU buildings in three simple steps. Una sfida per tutti noi, 44.

Schaarschmidt, J., Kahle, L., Wenzel, W., Hickel, T., Yuan, J., Bölle, F. T., Kondov, I., Huber, S. P., Castelli, I. E., Hanke, F., Neugebauer, J., Vegge, T., Pizzi, G., Rêgo, C. R. C., & Strunk, T. (2021). Workflow Engineering in Materials Design within the BATTERY 2030+ Project. *Advanced Energy Materials*, *12*(17), 2102638. DOI: 10.1002/aenm.202102638

Scheffran, J., Felkers, M., & Froese, R. (2020). Economic growth and the global energy demand. *Green energy to sustainability: strategies for global industries*, 1-44.

Schileo, G., & Grancini, G. (2021). Lead or no lead? Availability, toxicity, sustainability and environmental impact of lead-free perovskite solar cells. *Journal of Materials Chemistry. C, Materials for Optical and Electronic Devices*, *9*(1), 67–76.

Schwarzkopf, M. J., & Burnard, M. D. (2016). Wood-plastic composites—Performance and environmental impacts. Environmental impacts of traditional and innovative forest-based bioproducts, 19-43.

Sekhar, M. C., Veena, E., Kumar, N. S., Naidu, K. C. B., Mallikarjuna, A., & Basha, D. B. (2023). A review on piezoelectric materials and their applications. *Crystal Research and Technology*, *58*(2), 2200130. DOI: 10.1002/crat.202200130

Selvarani, K., Theivasanthi, T., & Ajitha, M. (2025). Nanotechnology in Agriculture: Enhancing Seed Quality, Yield and Contribution to Sustainable Development Goals (SDGs). Journal of Advances in Biology &amp. *Journal of Advances in Biology & Biotechnology*, *28*(1), 22–30. DOI: 10.9734/jabb/2025/v28i11857

Semelsberger, T., Graetz, J., Sutton, A., & Rönnebro, E. C. (2021). Engineering challenges of solution and slurry-phase chemical hydrogen storage materials for automotive fuel cell applications. *Molecules (Basel, Switzerland)*, *26*(6), 1722.

Semieniuk, G., Taylor, L., Rezai, A., & Foley, D. K. (2021). Plausible energy demand patterns in a growing global economy with climate policy. *Nature Climate Change*, *11*(4), 313–318. DOI: 10.1038/s41558-020-00975-7

Sengottaiyan, S., Shaisundaram, V. S., Shameer Basha, M., Deepanraj, B., & Senthilkumar, N. (2025). Effect of Duplex and Austenite Stainless Steel Fillers on Mechanical, Microstructural, and Corrosion Characteristics of Dissimilar Welds Using Cold Metal Transfer Welding. *Results in Engineering*, *26*, 104818. DOI: 10.1016/j.rineng.2025.104818

Senthilkumar, N., Deepanraj, B., Shaik, F., & Elsehly, E. M. (2024, March). Artificial Intelligence Based Investigation on Drilling of MWCNT-Seashell-Nylon66 Hybrid Polymer Composites by Fuzzy Inference System. In *2024 3rd International Conference for Innovation in Technology (INOCON)* (pp. 1-6). IEEE.

Senthilkumar, N., Deepanraj, B., Shaik, F., & Elsehly, E. M. "Artificial Intelligence Based Investigation on Drilling of MWCNT-Seashell-Nylon66 Hybrid Polymer Composites by Fuzzy Inference System," 2024 3rd International Conference for Innovation in Technology (INOCON), Bangalore, India, 2024, pp. 1-6, DOI: 10.1109/INOCON60754.2024.10511569

Senthilkumar, N., Ramu, S., & Deepanraj, B. (2024). Heat Conduction through Natural Fiber/Carbon Nanotubes Filler Matrix Polymer Composite Slabs: An Experimental and Analytical Comparisons (No. 2023-01-5107). SAE Technical Paper. DOI: 10.4271/2023-01-5107

Senturk Parreidt, T., Müller, K., & Schmid, M. (2018). Alginate-based edible films and coatings for food packaging applications. *Foods*, *7*(10), 170.

Sevilla, M., Falco, C., Titirici, M.-M., & Fuertes, A. B. (2012). High-performance CO2 sorbents from algae. *RSC Advances*, *2*(33), 12792. DOI: 10.1039/c2ra22552b

Sevilla, M., & Fuertes, A. B. (2011). Sustainable porous carbons with a superior performance for CO2 capture. *Energy & Environmental Science*, *4*(5), 1765. DOI: 10.1039/c0ee00784f

Shabbirahmed, A. M., Jacob, A., Dey, P., Somu, P., & Haldar, D. (2025). Biomass as eco-friendly adsorbents for the removal of emerging pollutants from wastewater: A review. *Discover Applied Sciences*, *7*(7), 771. DOI: 10.1007/s42452-025-07463-7

Shabeer, Y., Madani, S. S., Panchal, S., Mousavi, M., & Fowler, M. (2025). Different Metal–Air Batteries as Range Extenders for the Electric Vehicle Market: A Comparative Study. *Batteries*, *11*(1), 35.

Shafabakhsh, G., Mirabdolazimi, S., & Sadeghnejad, M. (2014). Evaluation of the effect of nano-TiO2 on the rutting and fatigue behavior of asphalt mixtures. *Construction & Building Materials*, *54*, 566–571. DOI: 10.1016/j.conbuildmat.2013.12.064

Shahab, S. A., Khan, S. J. M., & Nayan, S. (2024). Financial Development and Economic Growth: A Bibliometric Analysis. *Journal of Business & Finance Librarianship*, *29*(4), 243–273. DOI: 10.1080/08963568.2024.2362096

Shahkarami, S., Azargohar, R., Dalai, A. K., & Soltan, J. (2015). Breakthrough CO2 adsorption in bio-based activated carbons. *Journal of Environmental Sciences (China)*, *34*, 68–76. DOI: 10.1016/j.jes.2015.03.008

Shahzad, F., Iqbal, A., Kim, H., & Koo, C. M. (2020). 2D Transition Metal Carbides (MXenes): Applications as an Electrically Conducting Material. *Advanced Materials*, *32*(51), 2002159. Advance online publication. DOI: 10.1002/adma.202002159

Sharma, A., Sunny, S., Arulraj, J., & Hegde, G. (2024). Exploring the efficiency of green synthesized silver nanoparticles as photocatalysts for organic dye degradation: Unveiling key insights. *Nano Express*, *5*(2), 022002. DOI: 10.1088/2632-959X/ad4d09

Sharma, K. V., & Sesha Talpa Sai, P. H. V. (2023). Prognostic Modeling of Polydisperse SiO2/Aqueous Glycerol Nanofluids' Thermophysical Profile Using an Explainable Artificial Intelligence (XAI) Approach. *Engineering Applications of Artificial Intelligence*, *126*(November), 106967. DOI: 10.1016/j.engappai.2023.106967

Sharma, P., & Bora, B. J. (2022). A Review of Modern Machine Learning Techniques in the Prediction of Remaining Useful Life of Lithium-Ion Batteries. *Batteries*, *9*(1), 13. DOI: 10.3390/batteries9010013

Sharma, P., Said, Z., Kumar, A., Nižetić, S., Pandey, A., Hoang, A. T., Huang, Z., Afzal, A., Li, C., Le, A. T., Nguyen, X. P., & Tran, V. D. (2022). Recent Advances in Machine Learning Research for Nanofluid-Based Heat Transfer in Renewable Energy System. *Energy & Fuels*, *36*(13), 6626–6658. DOI: 10.1021/acs.energyfuels.2c01006

Sha, W., Lu, S., Cao, Y.-C., Yuan, Q., Zhang, X., Guo, X., Guo, Y., Cheng, S., & Tang, S. (2020). Artificial Intelligence to Power the Future of Materials Science and Engineering. *Advanced Intelligent Systems*, *2*(4), 1900143. DOI: 10.1002/aisy.201900143

Sheerazi, Z., & Ahmed, M. (2022). Nano-adsorbents and nano-catalysts for wastewater treatment. In *Emerging nanomaterials for advanced technologies* (pp. 517–539). Springer International Publishing. DOI: 10.1007/978-3-030-80371-1_17

Shen, C., Jiang, S. B., Chen, L., Jia, X., & Gonzalez, Y. (2018). Intelligent Parameter Tuning in Optimization-Based Iterative CT Reconstruction via Deep Reinforcement Learning. *IEEE Transactions on Medical Imaging*, *37*(6), 1430–1439. DOI: 10.1109/TMI.2018.2823679 PMID: 29870371

Shen, C., Shen, C., Song, J., Hsieh, C.-Y., Hsieh, C.-Y., Cao, D., Kang, Y., Ye, W., Wu, Z., Wang, J., Zhang, O., Zhang, X., Zeng, H., Cai, H., Chen, Y., Chen, L., Luo, H., Zhao, X., Jian, T., & Hou, T. (2024). DrugFlow: An AI-Driven One-Stop Platform for Innovative Drug Discovery. *Journal of Chemical Information and Modeling*, *64*(14), 5381–5391. DOI: 10.1021/acs.jcim.4c00621 PMID: 38920405

Shen, J., Griesemer, S. D., Gopakumar, A., Baldassarri, B., Saal, J. E., Aykol, M., Hegde, V. I., & Wolverton, C. (2022). Reflections on One Million Compounds in the Open Quantum Materials Database (OQMD). *JPhys Materials*, *5*(3), 031001. Advance online publication. DOI: 10.1088/2515-7639/ac7ba9

Shen, L., & Worrell, E. (2024). Plastic recycling. In *Handbook of recycling* (pp. 497–510). Elsevier.

Shen, T., Dong, H., & Wang, P. (2024). Research progress on nanocellulose and its composite materials as orthopedic implant biomaterials. *Alexandria Engineering Journal, 87*, 575–590. DOI: 10.1016/j.aej.2024.01.003

Shen, W., Cao, L., Li, Q., Wen, Z., Wang, J., Liu, Y., Dong, R., Tan, Y., & Chen, R. (2016). Is magnesia cement low carbon? Life cycle carbon footprint comparing with Portland cement. *Journal of Cleaner Production, 131*, 20–27.

Sheppard, J. P., Chamberlain, J., Agúndez, D., Bhattacharya, P., Chirwa, P. W., Gontcharov, A., Sagona, W. C. J., Shen, H.-l., Tadesse, W., & Mutke, S. (2020). Sustainable forest management beyond the timber-oriented status quo: Transitioning to co-production of timber and non-wood forest products—a global perspective. *Current Forestry Reports, 6*(1), 26–40.

Shin, Y., Kim, S., & Lee, Y. H. (2021). AI musculoskeletal clinical applications: How can AI increase my day-to-day efficiency? *Skeletal Radiology, 51*(2), 293–304. DOI: 10.1007/s00256-021-03876-8 PMID: 34341865

Shi, X., Lin, X., Luo, R., Wu, S., Li, L., Zhao, Z. J., & Gong, J. (2021). Dynamics of heterogeneous catalytic processes at operando conditions. *JACS, 1*(12), 2100–2120.

Shi, X., Mukhopadhyay, A., Zollinger, D., & Grasley, Z. (2019). Economic input-output life cycle assessment of concrete pavement containing recycled concrete aggregate. *Journal of Cleaner Production, 225*, 414–425.

Shi, X., Si, W., Zhu, J., & Zhang, S. (2024). Boosting the Electrical Performance of PLA-Based Triboelectric Nanogenerators for Sustainable Power Sources and Self-Powered Sensing. *Small, 20*(15), 2307620.

Shrestha, S., Barvenik, K. J., Chen, T., Yang, H., Li, Y., Kesavan, M. M., Little, J. M., Whitley, H. C., Teng, Z., Luo, Y., Tubaldi, E., & Chen, P.-Y. (2024). Machine intelligence accelerated design of conductive MXene aerogels with programmable properties. *Nature Communications, 15*(1), 4685. DOI: 10.1038/s41467-024-49011-8

Sikdar, S., Menezes, P. V., Maccione, R., Jacob, T., & Menezes, P. L. (2021). Plasma electrolytic oxidation (PEO) process—Processing, properties, and applications. *Nanomaterials (Basel, Switzerland), 11*(6), 1375.

Simões, S. (2024). High-performance advanced composites in multifunctional material design: State of the art, challenges, and future directions. *Materials (Basel), 17*(23), 5997. DOI: 10.3390/ma17235997

Singh, R., Dutt, S., Sharma, P., Sundramoorthy, A. K., Dubey, A., Singh, A., & Arya, S. (2023). Future of Nanotechnology in Food Industry: Challenges in Processing, Packaging, and Food Safety. Global Challenges, 7(4). Portico. DOI: 10.1002/gch2.202200209

Singh, A., Chaturvedi, K. R., & Sharma, T. (2025). Recent advances in carbon-based materials for carbon dioxide capture and separation applications. *Petroleum Science and Technology*, •••, 1–26. DOI: 10.1080/10916466.2025.2538735

Singh, D., Toutbort, J., & Chen, G.. (2006). Heavy vehicle systems optimization merit review and peer evaluation. In *Annual Report*. Argonne National Laboratory.

Singh, G., Kim, I. Y., Lakhi, K. S., Srivastava, P., Naidu, R., & Vinu, A. (2017a). Single step synthesis of activated bio-carbons with a high surface area and their excellent CO2 adsorption capacity. *Carbon*, *116*, 448–455. DOI: 10.1016/j.carbon.2017.02.015

Singh, G., Lakhi, K. S., Kim, I. Y., Kim, S., Srivastava, P., Naidu, R., & Vinu, A. (2017). Highly Efficient Method for the Synthesis of Activated Mesoporous Biocarbons with Extremely High Surface Area for High-Pressure CO2 Adsorption. *ACS Applied Materials & Interfaces*, *9*(35), 29782–29793. DOI: 10.1021/acsami.7b08797

Singh, J., Bhunia, H., & Basu, S. (2018). Synthesis of porous carbon monolith adsorbents for carbon dioxide capture: Breakthrough adsorption study. *Journal of the Taiwan Institute of Chemical Engineers*, *89*, 140–150. DOI: 10.1016/j.jtice.2018.04.031

Singh, J., & Nayak, P. (2023). pH-responsive polymers for drug delivery: Trends and opportunities. *Journal of polymer science*, *61*(22), 2828–2850.

Singh, M., Sharma, H. M., Gupta, R. K., & Kumar, A. (2024). Recent advancements and prospects in noble and non-noble electrocatalysts for materials methanol oxidation reactions. *Discover Nano*, *19*(1), 128.

Singh, R., & Lillard, J. W.Jr. (2009). Nanoparticle-based targeted drug delivery. *Experimental and Molecular Pathology*, *86*(3), 215–223.

Singh, S. K., Tiwari, A. K., & Paliwal, H. K. (2023). A state-of-the-art review on the utilization of machine learning in nanofluids, solar energy generation, and the prognosis of solar power. *Engineering Analysis with Boundary Elements*, *155*, 62–86.

Singh, V. K., & Anil Kumar, E. (2016). Measurement and analysis of adsorption isotherms of CO2 on activated carbon. *Applied Thermal Engineering*, *97*, 77–86. DOI: 10.1016/j.applthermaleng.2015.10.052

Skiba, R. (2025). *Advanced Materials Engineering Fundamentals*. After Midnight Publishing.

Smitthipong, W., Chollakup, R., & Nardin, M. (2014). *Bio-based composites for high-performance materials: from strategy to industrial application*. CRC Press.

Snellings, R., Suraneni, P., & Skibsted, J. (2023). Future and emerging supplementary cementitious materials. *Cement and Concrete Research, 171*, 107199.

Solangi, K., Kazi, S., Luhur, M., Badarudin, A., Amiri, A., Sadri, R., Zubir, M., Gharehkhani, S., & Teng, K. H. (2015). A comprehensive review of thermo-physical properties and convective heat transfer to nanofluids. *Energy, 89*, 1065–1086.

Song, T., Liao, J. M., Xiao, J., & Shen, L. H. (2015). Effect of micropore and mesopore structure on CO2 adsorption by activated carbons from biomass. *Xinxing Tan Cailiao. New Carbon Materials, 30*(2), 156–166. DOI: 10.1016/S1872-5805(15)60181-0

Soori, M., Dastres, R., & Arezoo, B. (2023). AI-powered blockchain technology in industry 4.0, a review. *Journal of Economy and Technology, 1*, 222–241. DOI: 10.1016/j.ject.2024.01.001

Souto, E. B., Blanco-Llamero, C., Blanco-Llamero, C., Krambeck, K., Kiran, N. S., Yashaswini, C., Postwala, H., Severino, P., Severino, P., Priefer, R., Prajapati, B. G., & Maheshwari, R. (2024). Regulatory insights into nanomedicine and gene vaccine innovation: Safety assessment, challenges, and regulatory perspectives. *Acta Biomaterialia, 180*, 1–17. DOI: 10.1016/j.actbio.2024.04.010 PMID: 38604468

Spiwok, V., & Kříž, P. (2020). Time-Lagged t-Distributed Stochastic Neighbor Embedding (t-SNE) of Molecular Simulation Trajectories. *Frontiers in Molecular Biosciences, 7*, 132. Advance online publication. DOI: 10.3389/fmolb.2020.00132

Sreńscek-Nazzal, J., & Kiełbasa, K. (2019a). Advances in modification of commercial activated carbon for enhancement of CO2 capture. *Applied Surface Science, 494*, 137–151. DOI: 10.1016/j.apsusc.2019.07.108

Srivastava, N., Srivastava, M., Mishra, P. K., & Gupta, V. K. (Eds.). (2020). *Green Synthesis of Nanomaterials for Bioenergy Applications*., DOI: 10.1002/9781119576785

Stephan Utgenannt, F. H. O. K. S. J. A. W. P. K. (2015). Control of porosity and composition of carbon-based catalysts prepared by template assisted routes. *Catalysis Today, 249*, 38–44. DOI: 10.1016/j.cattod.2014.10.036

Stephen, J. A., & DeWitt, A. S. J. K. F. Z. M. J. R. R. P. L. (2018). Critical Comparison of Structured Contactors for Adsorption-Based Gas Separations. *Annual Review of Chemical and Biomolecular Engineering*, *9*(1), 129–152. DOI: 10.1146/annurev-chembioeng-060817-084120

Subrahmanian, S. K., & Narayanan, B. N. (2025). Three-dimensional array of holey graphene as a high-performance anode material for supercapacitors. *Journal of Power Sources*, *629*, 235952. DOI: 10.1016/j.jpowsour.2024.235952

Su, H., Zheng, J., Tang, L., Liu, F., Lu, M., Wang, F., Wang, L., Zheng, H., Cai, B., & Yu, W. W. (2025). One-step electrospinning synthesis of lead-free perovskite nanofiber-based triboelectric nanogenerator for wearable intelligent fall alarm. *Journal of Colloid and Interface Science*, *696*, 137836. DOI: 10.1016/j.jcis.2025.137836

Sui, F., Guo, R., Zhang, Z., Gu, G. X., & Lin, L. (2021). Deep Reinforcement Learning for Digital Materials Design. *ACS Materials Letters*, *3*(10), 1433–1439. Advance online publication. DOI: 10.1021/acsmaterialslett.1c00390

Su, J., Chen, L., Tan, C., Okulov, I., Zhu, H., Jiang, F., Zhang, L., Requena, G., Teng, J., Yan, M., & Wang, Y. M. (2024). Recent innovations in laser additive manufacturing of titanium alloys. *International Journal of Extreme Manufacturing*, *6*(3), 032001. DOI: 10.1088/2631-7990/ad2545

Su, J., Yang, P., Wang, Y., Han, S., Han, N., & Li, W. (2016). Investigation of the Self-Healing behaviors of Microcapsules/Bitumen composites by a repetitive direct tension test. *Materials (Basel)*, *9*(7), 600. DOI: 10.3390/ma9070600 PMID: 28773722

Sulzer, V., Mohtat, P., Aitio, A., Lee, S., Yeh, Y. T., Steinbacher, F., Khan, M. U., Lee, J. W., Siegel, J. B., Stefanopoulou, A. G., & Howey, D. A. (2021). The Challenge and Opportunity of Battery Lifetime Prediction from Field Data. *Joule*, *5*(8), 1934–1955. Advance online publication. DOI: 10.1016/j.joule.2021.06.005

Sunil, B. R., & Jagannatham, M. (2016). Producing hydroxyapatite from fish bones by heat treatment. *Materials Letters*, *185*, 411–414. DOI: 10.1016/j.matlet.2016.09.039

Sun, L., Dong, Y., Li, H., Meng, H., Liu, J., Cao, Q., & Pan, C. (2024). Research Progress and Challenges of Carbon/MXene Composites for Supercapacitors. *Batteries*, *10*(11), 395. DOI: 10.3390/batteries10110395

Sun, T., Wang, Z., Zeng, L., & Feng, G. (2025). Identifying MOFs for electrochemical energy storage via density functional theory and machine learning. *npj Computational Materials*, *11*(1), 90. DOI: 10.1038/s41524-025-01590-w

Surat'man, N. E. B., Quek, X. L., Wang, N., Ye, E., Xu, J., Li, Z., & Li, B. (2025). Sustainable nanofibrous membranes for air filtration, water purification and oil removal. *Nanoscale*, *17*(11), 6427–6447. DOI: 10.1039/D4NR04673K

Suresh, S., Chandrasekar, M., Selvakumar, P., & Page, T. (2012). Experimental studies on heat transfer and friction factor characteristics of Al2O3/water nanofluid under laminar flow with spiralled rod inserts. *Int. J. Nanopart.*, *5*(1), 37–55.

Sverdrup, H., Nihlgård, B., Svensson, M. G., & Thelin, G. (2002). Principles of sustainable forest management. In *Developing principles and models for sustainable forestry in Sweden* (pp. 33–56). Springer.

Swetha, T. A., Bora, A., Mohanrasu, K., Balaji, P., Raja, R., Ponnuchamy, K., Muthusamy, G., & Arun, A. (2023). A comprehensive review on polylactic acid (PLA)–Synthesis, processing and application in food packaging. *International Journal of Biological Macromolecules*, *234*, 123715.

Szekeres, A., & Jeswiet, J. (2013). Defining Sustainability: Critical Factors in Sustainable Material Selection. in Re-engineering Manufacturing for Sustainability: Proceedings of the 20th CIRP International Conference on Life Cycle Engineering, Singapore. Springer.

Tagliaferri, S., Gaspard, L., Au, H., Mattevi, C., Titirici, M. M., & Crespo-Ribadeneyra, M. (2024). Nature-inspired batteries: From biomaterials to biomimetic design strategies. *Green Chemistry*, *26*(12), 6944–6958.

Tahir, M. B., Sagir, M., & Rafique, M. (2021). Challenges and Future of Nanotechnology. *Nanotechnology*, *141–142*, 141–142. Advance online publication. DOI: 10.1007/978-981-15-9437-3_6

Talaei Khoei, T., & Kaabouch, N. (2023). Machine Learning: Models, Challenges, and Research Directions. *Future Internet*, *15*(10), 332. Advance online publication. DOI: 10.3390/fi15100332

Talreja, N., Chuahan, D., & Ashfaq, M. (2024). Carbon-based two-dimensional (2D) materials: A next generation biocidal agent. *Materials Advances*, *5*(4), 1454–1461. DOI: 10.1039/D3MA00952A

Tan, Q., & Li, J. (2015). Recycling metals from wastes: A novel application of mechanochemistry. *Environmental Science & Technology*, *49*(10), 5849–5861.

Tarasova, L., Gnann, S., Yang, S., Hartmann, A., & Wagener, T. (2024). Catchment characterization: Current descriptors, knowledge gaps and future opportunities. *Earth-Science Reviews*, •••, 104739.

Tariq, A., & Mushtaq, A. (2023). Untreated wastewater reasons and causes: A review of most affected areas and cities. *International Journal of Chemical and Biochemical Science*, *23*(1), 121–143.

Tariq, R., Torres-Aguilar, C. E., Sheikh, N. A., Ahmad, T., Xamán, J., & Bassam, A. (2022). Data Engineering for Digital Twining and Optimization of Naturally Ventilated Solar Façade with Phase Changing Material under Global Projection Scenarios. *Renewable Energy*, *187*, 1184–1203. Advance online publication. DOI: 10.1016/j.renene.2022.01.044

Tawiah, B., Ofori, E. A., & George, S. C. (2024). Nanotechnology in Societal Development. Nanotechnology in Societal Development, 1–64. DOI: 10.1007/978-981-97-6184-5_1

Teferra, K., & Rowenhorst, D. J. (2021). Optimizing the Cellular Automata Finite Element Model for Additive Manufacturing to Simulate Large Microstructures. *Acta Materialia*, *213*, 116930. Advance online publication. DOI: 10.1016/j.actamat.2021.116930

Teixeira, L. V., Bomtempo, J. V., Oroski, F. D. A., & Coutinho, P. L. D. A. (2023). The diffusion of bioplastics: What can we learn from poly (lactic acid)? *Sustainability*, *15*(6), 4699.

Temraz, M., & Keane, M. T. (2022). Solving the Class Imbalance Problem Using a Counterfactual Method for Data Augmentation. *Machine Learning with Applications*, *9*(September), 100375. DOI: 10.1016/j.mlwa.2022.100375

Temuujin, J., Minjigmaa, A., Rickard, W., Lee, M., Williams, I., & Van Riessen, A. (2010). Fly ash based geopolymer thin coatings on metal substrates and its thermal evaluation. *Journal of Hazardous Materials*, *180*(1–3), 748–752. DOI: 10.1016/j.jhazmat.2010.04.121 PMID: 20488615

Terán-Cuadrado, G., Tahir, F., Nurdiawati, A., Almarshoud, M. A., & Al-Ghamdi, S. G. (2024). Current and potential materials for the low-carbon cement production: Life cycle assessment perspective. *Journal of Building Engineering*, *96*, 110528.

Thakkar, H., Lawson, S., Rownaghi, A. A., & Rezaei, F. (2018). Development of 3D-printed polymer-zeolite composite monoliths for gas separation. *Chemical Engineering Journal*, *348*, 109–116. DOI: 10.1016/j.cej.2018.04.178

Theodosiou, A. A., & Read, R. C. (2023). Artificial intelligence, machine learning and deep learning: Potential resources for the infection clinician. *The Journal of Infection*, *87*(4), 287–294. DOI: 10.1016/j.jinf.2023.07.006 PMID: 37468046

Thiruppathi, A. R., Sidhureddy, B., Salverda, M., Wood, P. C., & Chen, A. (2020). Novel three-dimensional N-doped interconnected reduced graphene oxide with superb capacitance for energy storage. *Journal of Electroanalytical Chemistry (Lausanne, Switzerland)*, *875*, 113911. DOI: 10.1016/j.jelechem.2020.113911

Thomas, J., Patil, R. S., Patil, M., & John, J. (2023). Addressing the sustainability conundrums and challenges within the polymer value chain. *Sustainability*, *15*(22), 15758.

Thulasisingh, A., Kumar, K., Yamunadevi, B., & Poojitha, N. (2022). S. SuhailMadharHanif, & S. Kannaiyan, Biodegradable packaging materials. *Polymer Bulletin*, *79*(7), 4467–4496.

Tian, J., Yu, L., Xue, R., Zhuang, S., & Shan, Y. (2022). Global low-carbon energy transition in the post-COVID-19 era. *Applied Energy*, *307*, 118205. DOI: 10.1016/j.apenergy.2021.118205

Tian, S., Arshad, N. I., Toghraie, D., Eftekhari, S. A., & Hekmatifar, M. (2021). Using Perceptron Feed-Forward Artificial Neural Network (ANN) for Predicting the Thermal Conductivity of Graphene Oxide-Al2O3/Water-Ethylene Glycol Hybrid Nanofluid. *Case Studies in Thermal Engineering*, *26*(August), 101055. DOI: 10.1016/j.csite.2021.101055

Tian, X., Ye, C., Zhang, L., Sugumar, M. K., Zhao, Y., McKeown, N. B., Margadonna, S., & Tan, R. (2025). Enhancing membrane materials for efficient Li recycling and recovery. *Advanced Materials*, *37*(5), 2402335. DOI: 10.1002/adma.202402335

Timoshenko, J., Jeon, H. S., Sinev, I., Haase, F. T., Herzog, A., & Cuenya, B. R. (2020). Linking the Evolution of Catalytic Properties and Structural Changes in Copper-Zinc Nanocatalysts Using: Operando EXAFS and Neural-Networks. *Chemical Science (Cambridge)*, *11*(14), 3727–3736. Advance online publication. DOI: 10.1039/D0SC00382D

Tirumanadham, N. S. K. M. K., Thaiyalnayaki, S., & Ganesan, V. (2025). Towards Smarter E-Learning: Real-Time Analytics and Machine Learning for Personalized Education. *International Journal of Computational and Experimental Science and Engineering*, *11*(1). Advance online publication. DOI: 10.22399/ijcesen.786

Tiza, T. M., Kpur, G., Ogunleye, E., Sharma, S., Singh, S. K., & Likassa, D. M. (2023). The potency of functionalized nanomaterials for industrial applications. *Materials Today: Proceedings*. Advance online publication. DOI: 10.1016/j.matpr.2023.03.212

Tiza, T. M., Mogbo, O., Singh, S. K., Shaik, N., & Shettar, M. P. (2022). Bituminous pavement sustainability improvement strategies. *Energy Nexus*, *6*, 100065. DOI: 10.1016/j.nexus.2022.100065

Tiza, T. M., Singh, S. K., Kumar, L., Shettar, M. P., & Singh, S. P. (2021). Assessing the potentials of Bamboo and sheep wool fiber as sustainable construction materials: A review. *Materials Today: Proceedings*, *47*, 4484–4489. DOI: 10.1016/j.matpr.2021.05.322

Tofail, S. (2012). *Biological Interactions with surface charge in biomaterials (No. 21)*. Royal Society of Chemistry.

Tofail, S. A. M., Koumoulos, E. P., Bandyopadhyay, A., Bose, S., O'Donoghue, L., & Charitidis, C. (2018). Additive manufacturing: Scientific and technological challenges, market uptake and opportunities. *Materials Today*, *21*(1), 22–37. DOI: 10.1016/j.mattod.2017.07.001

Toka, E. B., & Olgun, M. (2023). Mechanical and economic evaluation of recycled concrete aggregate as granular road base and subbase material. *Journal of Materials in Civil Engineering*, *35*(5), 04023097. Advance online publication. DOI: 10.1061/JMCEE7.MTENG-14388

Tonelli, D., Scavetta, E., & Gualandi, I. (2019). Electrochemical Deposition of Nanomaterials for Electrochemical Sensing. *Sensors (Basel)*, *19*(5), 1186. DOI: 10.3390/s19051186

Tonnarello, F., Vermiglio, C., Migliardo, C., & Naciti, V. (2025). The Impact of EU Taxonomy for Sustainable Activities on European Utilities' Performance. *Business Strategy and the Environment*, *34*(3), 2848–2862.

Tootkaboni, M. P., Ballarini, I., & Corrado, V. (2025). Towards climate resilient and energy-efficient buildings: A sensitivity analysis on building components and cooling strategies. *Building and Environment*, *270*, 112473. DOI: 10.1016/j.buildenv.2024.112473

Torrisi, C., Carrabino, R., Santonocito, A., Matarazzo, A., & Mertoli, F. (2022). Industrial symbiosis in the production of sustainable textiles: orange fiber srl. P-ESEM, 995.

Trinh, C., Tbatou, Y., Lasala, S., Herbinet, O., & Meimaroglou, D. (2023). On the Development of Descriptor-Based Machine Learning Models for Thermodynamic Properties: Part 1—From Data Collection to Model Construction: Understanding of the Methods and Their Effects. *Processes (Basel, Switzerland)*, *11*(12), 3325. DOI: 10.3390/pr11123325

Tripathy, J., Mishra, A., Pandey, M., Thakur, R. R., Chand, S., Rout, P. R., & Shahid, M. K. (2024). Advances in nanoparticles and nanocomposites for water and wastewater treatment: A review. *Water (Basel), 16*(11), 1481. DOI: 10.3390/w16111481

Tsai, Y. C., Huang, P. C., & Dai, C. L. (2025). Editorial for the Special Issue on MEMS/NEMS Devices and Applications. *Micromachines, 16*(2), 189. DOI: 10.3390/mi16020189

Turku, I., Keskisaari, A., Kärki, T., Puurtinen, A., & Marttila, P. (2017). Characterization of wood plastic composites manufactured from recycled plastic blends. *Composite Structures, 161*, 469–476.

Turner, L. K., & Collins, F. G. (2013). Carbon dioxide equivalent (CO_2-e) emissions: A comparison between geopolymer and OPC cement concrete. *Construction & Building Materials, 43*, 125–130. DOI: 10.1016/j.conbuildmat.2013.01.023

Tzeng, S.-C., Lin, C.-W., & Huang, K. D. (2005). Heat transfer enhancement of nanofluids in rotary blade coupling of four wheel-drive vehicles. *Acta Mechanica, 179*(1-2), 11–23.

Ullah, K., Alam, N., Uddin, S., & Oh, W.-C. (2024). Advanced concept and perspectives toward MXenes based energy storage device: Comprehensive review. *Materialia, 34*, 102089. DOI: 10.1016/j.mtla.2024.102089

Umar, U. M., & Muthusamy, K. (2023). Potential of waste material as coarse aggregates for lightweight concrete production: A sustainable approach. *Construction (Arlington), 3*(1), 87–114. DOI: 10.15282/construction.v3i1.9217

Ungureanu, E. L., Mocanu, A. L., Stroe, C. A., Panciu, C. M., Berca, L., Sionel, R. M., & Mustatea, G. (2023). Agricultural byproducts used as low-cost adsorbents for removal of potentially toxic elements from wastewater: A comprehensive review. *Sustainability (Basel), 15*(7), 5999. DOI: 10.3390/su15075999

Utsev, T., Tiza, T. M., Mogbo, O., Singh, S. K., Chakravarti, A., Shaik, N., & Singh, S. P. (2022). Application of nanomaterials in civil engineering. *Materials Today: Proceedings, 62*, 5140–5146. DOI: 10.1016/j.matpr.2022.02.480

Vagner, C., Finqueneisel, G., Zimny, T., Burg, P., Grzyb, B., Machnikowski, J., & Weber, J. V. (2003). Characterization of the surface properties of nitrogen-enriched carbons by inverse gas chromatography methods. *Carbon, 41*(14), 2847–2853. DOI: 10.1016/S0008-6223(03)00412-3

Väisänen, T., Das, O., & Tomppo, L. (2017). A review on new bio-based constituents for natural fiber-polymer composites. *Journal of Cleaner Production, 149*, 582–596.

Vaishag, P. V., & Noh, J.-S. (2024). A Comparative Review of Graphene and MXene-Based Composites towards Gas Sensing. *Molecules (Basel, Switzerland)*, *29*(19), 4558. DOI: 10.3390/molecules29194558

Van Aert, S., Batenburg, K. J., Rossell, M. D., Erni, R., & Van Tendeloo, G. (2011). Three-dimensional atomic imaging of crystalline nanoparticles. *Nature*, *470*(7334), 374–377. DOI: 10.1038/nature09741

Van Allsburg, K. M., Tan, E. C., Super, J. D., Schaidle, J. A., & Baddour, F. G. (2022). Early-stage evaluation of catalyst manufacturing cost and environmental impact using CatCost. *Nature Catalysis*, *5*(4), 342–353.

Varzakas, T., & Antoniadou, M. (2024). A Holistic Approach for Ethics and Sustainability in the Food Chain: The Gateway to Oral and Systemic Health. *Foods*, *13*(8), 1224. DOI: 10.3390/foods13081224 PMID: 38672896

Vasudevan, R. K., Smith, R., Mehta, A., Hattrick-Simpers, J., Vlcek, L., Kusne, G., Ziatdinov, M., Tavazza, F., Choudhary, K., & Kalinin, S. V. (2019). Materials Science in the AI age: High-throughput library generation, machine learning and a pathway from correlations to the underpinning physics. *MRS Communications*, *9*(3), 821–838. DOI: 10.1557/mrc.2019.95 PMID: 32166045

Venkatesan, J., Qian, Z. J., Ryu, B., Thomas, N. V., & Kim, S. K. (2011). A comparative study of thermal calcination and an alkaline hydrolysis method in the isolation of hydroxyapatite from Thunnus obesus bone. *Biomedical Materials (Bristol, England)*, *6*(3), 1–12. DOI: 10.1088/1748-6041/6/3/035003

Verma, R., Jha, S., Harimurugan, D., Kasthurirengan, S., Shivaprakash, N. C., & Behera, U. (2022). Nanomaterials' Synthesis Approaches for Energy Storage and Electronics Applications. Applications of Nanomaterials in Energy Storage and Electronics, 240–257. DOI: 10.2174/9789815050714122030017

Verma, D., & Senal, I. (2019). Natural fiber-reinforced polymer composites: Feasibiliy study for sustainable automotive industries. In *Biomass, biopolymer-based materials, and bioenergy* (pp. 103–122). Elsevier.

Verma, S. K., Arbind Prasad, S., & Katiyar, V. (2024). State of Art Review on Sustainable Biodegradable Polymers with a Market Overview for Sustainability Packaging. *Materials Today Sustainability*, *26*(June), 100776. DOI: 10.1016/j.mtsust.2024.100776

Verma, S., & Gupta, G. (2017). Natural dyes and its applications: A brief review. *International Journal of Research and Analytical Reviews*, *4*(4), 57–60.

Verma, V. K., & Verma, S. (2022). Machine learning applications in healthcare sector: An overview. *Materials Today: Proceedings*, *57*, 2144–2147. DOI: 10.1016/j.matpr.2021.12.101

Vilén, A., Laurell, P., & Vahala, R. (2022). Comparative life cycle assessment of activated carbon production from various raw materials. *Journal of Environmental Management*, *324*, 116356. Advance online publication. DOI: 10.1016/j.jenvman.2022.116356

Vonarbourg, A., Passirani, C., Saulnier, P., & Benoit, J.-P. (2006). Parameters influencing the steal thiness of colloidal drug delivery systems. *Biomaterials*, *27*(24), 4356–4373.

Vs, N., T, R., Kempahanumakkagari, S., Hr, P., & Bm, P.VS. (2023). Efficient strategies to produce Graphene and functionalized graphene materials: A review. *Applied Surface Science Advances*, *14*, 100386. DOI: 10.1016/j.apsadv.2023.100386

Vyas, A. M., & Kushwah, G. S. (2023). *Sand Battery: An Innovative Solution for Renewable Energy Storage (A Review). in 2023 IEEE Renewable Energy and Sustainable E-Mobility Conference (RESEM)*. IEEE.

Wadhawan, S., Wadhawan, D., Jain, A., & Mehta, S. K. (2024). Toxic implication of nanoparticles: A review of factors, mechanism, exposure and control strategies. *International Journal of Environmental Science and Technology*, *22*(2), 1203–1224. DOI: 10.1007/s13762-024-05810-6

Wadhwa, S. A., Mahapara, A., Akhai, S., Kumar, D., & Kumar, P. (2024). Integrating Taguchi Optimization for Multi-Criteria Decision Making in Engineering Applications. In Recent Theories and Applications for Multi-Criteria Decision-Making (Vol. 1, No. 1, pp. 125-150). IGI Global.

Wagner, M. (2015). The link of environmental and economic performance: Drivers and limitations of sustainability integration. *Journal of Business Research*, *68*(6), 1306–1317.

Wallah, S. E., & Rangan, B. V. (2006). *Low-Calcium fly ash-based geopolymer concrete: Long-term properties*. https://www.geopolymer.org/dl/?get=curtin_flyash_GC-2.pdf

Walls, J., & Smith, M. R. (1998). *Life-cycle cost analysis in pavement design: Interim technical bulletin (No. FHWA-SA-98-079)*. Federal Highway Administration.

Walsh, P. J., Buchanan, J. F., Dring, M., Maggs, C., Bell, S., & Walker, G. M. (2008). Low-pressure synthesis and characterisation of hydroxyapatite derived from mineralise red algae. *Chemical Engineering Journal, 137*(1), 173–179. DOI: 10.1016/j.cej.2007.10.016

Wang, B., Yan, L., Fu, Q., & Kasal, B. (2021). A comprehensive review on recycled aggregate and recycled aggregate concrete. *Resources, Conservation and Recycling*, *171*, 105565.

Wang, C., O'Hagan, M. P., Willner, B., & Willner, I. (2022). Bioinspired artificial photosynthetic systems. *Chemistry (Weinheim an der Bergstrasse, Germany)*, *28*(9), e202103595.

Wang, G., & Ma, Y. (2018). Drying shrinkage of alkali-activated fly ash/slag blended system. *Journal of Sustainable Cement-Based Materials*, *7*(4), 203–213. DOI: 10.1080/21650373.2018.1471424

Wang, H., You, Z., Mills-Beale, J., & Hao, P. (2011). Laboratory evaluation on high temperature viscosity and low temperature stiffness of asphalt binder with high percent scrap tire rubber. *Construction & Building Materials*, *26*(1), 583–590. DOI: 10.1016/j.conbuildmat.2011.06.061

Wang, J., Wang, L., Gardner, D. J., Shaler, S. M., & Cai, Z. (2021). Towards a cellulose-based society: Opportunities and challenges. *Cellulose (London, England)*, *28*, 4511–4543.

Wang, J., Yin, Y., Che, C., & Cui, M. (2025). Research Progress of Thermoelectric Materials—A Review. *Energies*, *18*(8), 2122. DOI: 10.3390/en18082122

Wang, L., & Yang, R. T. (2012). Significantly Increased CO_2 Adsorption Performance of Nanostructured Templated Carbon by Tuning Surface Area and Nitrogen Doping. *The Journal of Physical Chemistry. C, Nanomaterials and Interfaces*, *116*(1), 1099–1106. DOI: 10.1021/jp2100446

Wang, M., Li, Y., Yang, H., Ding, Z., & Liu, R. (2024). Enhancing ion conductivity in polyethylene oxide-based polymers through semi-interpenetrating networks with liquid crystalline polymer for lithium metal batteries. *Journal of Energy Storage*, *100*, 113781.

Wang, W. Y., Lu, J., Li, G., Wang, X., Su, Y., Li, J., Zhang, S., Ren, Y., Gao, X., & Song, H. (2024). Artificial intelligence enabled smart design and manufacturing of advanced materials: The endless Frontier in AI^+ era. *Materials Genome Engineering Advances*, *2*(3), e56. Advance online publication. DOI: 10.1002/mgea.56

Wang, Y., Wang, K., & Zhang, C. (2024). Applications of Artificial Intelligence/ Machine Learning to High-Performance Composites. *Composites. Part B, Engineering, 285*(October), 111740. DOI: 10.1016/j.compositesb.2024.111740

Wang, Y., Zhang, Q., Wang, G.-G., & Cheng, H. (2024). The application of evolutionary computation in generative adversarial networks (GANs): A systematic literature survey. *Artificial Intelligence Review, 57*(7), 182. Advance online publication. DOI: 10.1007/s10462-024-10818-y

Wang, Z., Zhan, L., Ge, M., Xie, F., Wang, Y., Qiao, W., Liang, X., & Ling, L. (2011). Pith based spherical activated carbon for CO2 removal from flue gases. *Chemical Engineering Science, 66*(22), 5504–5511. DOI: 10.1016/j.ces.2011.06.070

Warr, M., & West, R. E. (2020). Bridging Academic Disciplines with Interdisciplinary Project-based Learning. *The Interdisciplinary Journal of Problem-Based Learning, 14*(1). Advance online publication. DOI: 10.14434/ijpbl.v14i1.28590

Wei, H., Deng, S., Hu, B., Chen, Z., Wang, B., Huang, J., & Yu, G. (2012). Granular bamboo-derived activated carbon for high CO2 adsorption: The dominant role of narrow micropores. *ChemSusChem, 5*(12), 2354–2360. DOI: 10.1002/cssc.201200570

Wei, L., Kuai, X., Bao, Y., Wei, J., Yang, L., Song, P., & Wang, X. (2021). The recent progress of MEMS/NEMS resonators. *Micromachines, 12*(6), 724. DOI: 10.3390/mi12060724

Weiss, M., Haufe, J., Carus, M., Brandão, M., Bringezu, S., Hermann, B., & Patel, M. K. (2012). A review of the environmental impacts of biobased materials. *Journal of Industrial Ecology, 16*, S169–S181.

Wenelska, K., Adam, V., Thauer, E., Singer, L., Klingeler, R., Chen, X., & Mijowska, E. (2022). Fabrication of 3D graphene/MoS2 spherical heterostructure as anode material in Li-ion battery. *Frontiers in Energy Research, 10*, 960786. Advance online publication. DOI: 10.3389/fenrg.2022.960786

Werker, A., Pei, R., Kim, K., Moretto, G., Estevez-Alonso, A., Vermeer, C., & de Vries, E. (2023). Thermal pre-processing before extraction of polyhydroxyalkanoates for molecular weight quality control. *Polymer Degradation & Stability, 209*, 110277.

Westlake, J. R., Tran, M. W., Jiang, Y., Zhang, X., Burrows, A. D., & Xie, M. (2023). Biodegradable biopolymers for active packaging: Demand, development and directions. *Sustainable Food Technology, 1*(1), 50–72.

Wheatley, M. C. (2024). Advancements in Renewable Energy Technologies: A Decade in Review. *Science, 1*, 100013.

White, A. D. (2022). Deep Learning for Molecules and Materials. *Living Journal of Computational Molecular Science*, *3*(1), 1499. Advance online publication. DOI: 10.33011/livecoms.3.1.1499

Wickramaratne, N. P., & Jaroniec, M. (2013). Importance of small micropores in CO_2 capture by phenolic resin-based activated carbon spheres. *Journal of Materials Chemistry. A, Materials for Energy and Sustainability*, *1*(1), 112–116. DOI: 10.1039/C2TA00388K

Wijaya, A., Wagner, J., Sartory, B., & Brunner, R. (2024). Analyzing microstructure relationships in porous copper using a multi-method machine learning-based approach. *Communications Materials*, *5*(1), 59. Advance online publication. DOI: 10.1038/s43246-024-00493-5

Wiley. (2015). High-Resolution X-ray diffraction and fluorescence microscopy characterization of Alkali-Activated Slag-Metakaolin binders - White Rose Research online. https://eprints.whiterose.ac.uk/id/eprint/86418/

Wojtuch, A., Jankowski, R., & Podlewska, S. (2021). How Can SHAP Values Help to Shape Metabolic Stability of Chemical Compounds? *Journal of Cheminformatics*, *13*(1), 74. Advance online publication. DOI: 10.1186/s13321-021-00542-y

Woo, S. H., Lee, J., Kim, J. P., Hong, J., Jang, J. M., Ko, T. Y., Kang, Y. C., Kim, D. W., & Kim, S. J. (2024). Electrostatic assembly of charge-modulated MXene and graphene oxide into conductive porous nanohybrids via rapid annealing for EMI shielding. *Applied Materials Today*, *39*, 102271. DOI: 10.1016/j.apmt.2024.102271

Wu, F. C., Tseng, R. L., & Juang, R. S. (2005). Comparisons of porous and adsorption properties of carbons activated by steam and KOH. *Journal of Colloid and Interface Science*, *283*(1), 49–56. DOI: 10.1016/j.jcis.2004.08.037

Wu, J., Ding, Y., Wang, J., Li, T., Lin, H., Wang, J., & Liu, F. (2018). Facile fabrication of nanofiber-and micro/nanosphere-coordinated PVDF membrane with ultrahigh permeability of viscous water-in-oil emulsions. *Journal of Materials Chemistry. A, Materials for Energy and Sustainability*, *6*(16), 7014–7020. DOI: 10.1039/C8TA01539B

Wu, L., Sun, Z., & Cao, Y. (2024). Modification of recycled aggregate and conservation and application of recycled aggregate concrete: A review. *Construction & Building Materials*, *431*, 136567.

Xiao, W., Li, B., Yan, J., Wang, L., Huang, X., & Gao, J. (2023). Three dimensional graphene composites: Preparation, morphology and their multi-functional applications. *Composites. Part A, Applied Science and Manufacturing*, *165*, 107335. DOI: 10.1016/j.compositesa.2022.107335

Xia, Y., Mokaya, R., Walker, G. S., & Zhu, Y. (2011). Superior CO2 Adsorption Capacity on N-doped, High-Surface-Area, Microporous Carbons Templated from Zeolite. *Advanced Energy Materials*, *1*(4), 678–683. DOI: 10.1002/aenm.201100061

Xie, J., Li, X., & Xia, Y. (2008). Putting electrospun nanofibers to work for biomedical research. *Macromolecular Rapid Communications*, *29*(22), 1775–1792. DOI: 10.1002/marc.200800381

Xie, S., Zhang, Y., Song, Y., Ge, F., & Huang, X. (2020). Comparison of the Corrosion Behavior of Brass in TiO2 and Al2O3. Nanofluids. *Nanomaterials (Basel, Switzerland)*, *10*, 1046.

Xie, X., Du, L., Yan, L., Park, S., Qiu, Y., Sokolowski, J., & Shao, Y. (2022). Oxygen evolution reaction in alkaline environment: Material challenges and solutions. *Advanced Functional Materials*, *32*(21), 2110036.

Xie, Y., Sattari, K., Zhang, C., & Lin, J. (2023). Toward Autonomous Laboratories: Convergence of Artificial Intelligence and Experimental Automation. *Progress in Materials Science*, *132*, 101043. Advance online publication. DOI: 10.1016/j.pmatsci.2022.101043

Xie, Z., Ventura, H., & Ardanuy, M. (2024). A Review of the Electrical Conductivity Test Methods for Conductive Fabrics. *Textiles*, *4*(3), 284–308. DOI: 10.3390/textiles4030017

Xing, W., Liu, C., Zhou, Z., Zhang, L., Zhou, J., Zhuo, S., Yan, Z., Gao, H., Wang, G., & Qiao, S. Z. (2012). Superior CO2 uptake of N-doped activated carbon through hydrogen-bonding interaction. *Energy & Environmental Science*, *5*(6), 7323. DOI: 10.1039/c2ee21653a

Xiong, Y., Zhang, D., Ruan, X., Jiang, S., Zou, X., Yuan, W., Liu, X., Zhang, Y., Nie, Z., Wei, D., Zeng, Y., Cao, P., & Zhang, G. (2024). Artificial intelligence in rechargeable battery: Advancements and prospects. *Energy Storage Materials*, *73*, 103860. DOI: 10.1016/j.ensm.2024.103860

Xu, C., Jiang, X., & Coen-Pirani, P. (2023). Empirical Study of Overfitting in Deep Learning for Predicting Breast Cancer Metastasis. *Cancers (Basel)*, *15*(7), 1969. DOI: 10.3390/cancers15071969 PMID: 37046630

Xu, D. (2023, October). Technologies and challenges of hydrogen powered aviation. [). IOP Publishing.]. *Journal of Physics: Conference Series*, *2608*(1), 012003. DOI: 10.1088/1742-6596/2608/1/012003

Xu, D., Karray, M. H., & Archimède, B. (2017). A semantic-based decision support platform to assist products' eco-labeling process. *Industrial Management & Data Systems*, *117*(7), 1340–1361.

Xu, M., Liang, T., Shi, M., & Chen, H. (2013). Graphene-like two-dimensional materials. *Chemical Reviews*, *113*(5), 3766–3798. DOI: 10.1021/cr300263a

Xu, P., Ji, X., Li, M., & Lu, W. (2023). Small Data Machine Learning in Materials Science. *npj Computational Materials*, *9*(1), 42. Advance online publication. DOI: 10.1038/s41524-023-01000-z

Xu, S., Wu, Z., Xu, Y., & Chen, S. (2014). Performance Evaluation of High RAP Asphalt Mixture with Fiber and Warm Mixing Agent. *Conference: 10th Asia Pacific Transportation Development Conference*, 343–351. DOI: 10.1061/9780784413364.042

Xu, X., Chen, Y., Wan, P., Gasem, K., Wang, K., He, T., Adidharma, H., & Fan, M. (2016). Extraction of lithium with functionalized lithium ion-sieves. *Progress in Materials Science*, *84*, 276–313. DOI: 10.1016/j.pmatsci.2016.09.004

Xu, Z., Deng, W., & Wang, X. (2021). 3D hierarchical carbon-rich micro-/nanomaterials for energy storage and catalysis. *Electrochemical Energy Reviews*, *4*, 269–335.

Yadav, K., Bidnyk, S., & Balakrishnan, A. (2024). Artificial intelligence and machine learning in optics: Tutorial. *Journal of the Optical Society of America. B, Optical Physics*, *41*(8), 1739. DOI: 10.1364/JOSAB.525182

Yahia, A. K. M., & Shahjalal, M. (2024). Sustainable materials selection in building design and construction. *International Journal of Science and Engineering*, *1*(4), 10–62304.

Yaman, M. Y., Kalinin, S. V., Guye, K. N., Ginger, D. S., & Ziatdinov, M. (2023). Learning and Predicting Photonic Responses of Plasmonic Nanoparticle Assemblies via Dual Variational Autoencoders. *Small*, *19*(25), 2205893. Advance online publication. DOI: 10.1002/smll.202205893

Yaminsky, I. V. (2024). Scanning Probe Bionanoscopy: Methods and Equipment, Results and Achievements. *Moscow University Physics Bulletin*, *79*(6), 839–855. DOI: 10.3103/S0027134924702473

Yan, Y., Zhang, N., Wang, L., Qin, X., Fan, J., & Wu, Y. (2015). A restricted Boltzmann machine based two-lead electrocardiography classification. 1–9. DOI: 10.1109/BSN.2015.7299399

Yang, H., Fang, L., Yuan, Z., Teng, X., Qin, H., He, Z., Wan, Y., Wu, X., Zhang, Y., Guan, L., Meng, C., Zhou, Q., Wang, C., Ding, P., Hu, H., & Wu, M. (2023). Machine learning guided 3D printing of carbon microlattices with customized performance for supercapacitive energy storage. *Carbon*, *201*, 408–414. DOI: 10.1016/j.carbon.2022.08.083

Yang, J., Bao, W., Jaumaux, P., Zhang, S., Wang, C., & Wang, G. (2019). MXene-Based Composites: Synthesis and Applications in Rechargeable Batteries and Supercapacitors. *Advanced Materials Interfaces*, *6*(8), 1802004. Advance online publication. DOI: 10.1002/admi.201802004

Yang, X. Y., Wei, Y. X., Su, Y. Q., Zhang, Z. W., Tang, X. Y., Chen, Y. E., Yuan, M., & Yuan, S. (2025). The Strategies Microalgae Adopt to Counteract the Toxic Effect of Heavy Metals. *Microorganisms*, *13*(5), 989. DOI: 10.3390/microorganisms13050989

Yang, X., Adair, K. R., Gao, X., & Sun, X. (2021). Recent advances and perspectives on thin electrolytes for high-energy-density solid-state lithium batteries. *Energy & Environmental Science*, *14*(2), 643–671.

Yang, X., Huang, Z., Du, L., Li, Q., & Ye, S. (2024). Active Sites and Stability Study of Fe/N/C Catalyst in PEMFCs: A Decade of Stunning Progress and the Remaining Challenges. *ACS Catalysis*, *14*(20), 15471–15488.

Yang, Y., Yuan, W., Hou, J., & You, Z. (2022). Review on physical and chemical factors affecting fines migration in porous media. *Water Research*, *214*, 118172.

Yang, Z., & Buehler, M. J. (2022). High-Throughput Generation of 3D Graphene Metamaterials and Property Quantification Using Machine Learning. *Small Methods*, *6*(9), 2200537. Advance online publication. DOI: 10.1002/smtd.202200537

Yan, J., Ren, C. E., Maleski, K., Hatter, C. B., Anasori, B., Urbankowski, P., Sarycheva, A., & Gogotsi, Y. (2017). Flexible MXene/Graphene Films for Ultrafast Supercapacitors with Outstanding Volumetric Capacitance. *Advanced Functional Materials*, *27*(30), 1701264. Advance online publication. DOI: 10.1002/adfm.201701264

Yan, X., Xiao, X., Au, C., Mathur, S., Huang, L., Wang, Y., Zhang, Z., Zhu, Z., Kipper, M. J., Tang, J., & Chen, J. (2021). Electrospinning nanofibers and nanomembranes for oil/water separation. *Journal of Materials Chemistry. A, Materials for Energy and Sustainability*, *9*(38), 21659–21684. DOI: 10.1039/D1TA05873H

Yan, Y., Hernando, D., Park, B., Allen, C., & Roque, R. (2021). Understanding asphalt binder cracking characterization at intermediate temperatures: Review and evaluation of two approaches. *Construction & Building Materials*, *312*, 125163. DOI: 10.1016/j.conbuildmat.2021.125163

Yao, H., You, Z., Li, L., Goh, S. W., Lee, C. H., Yap, Y. K., & Shi, X. (2012). Rheological properties and chemical analysis of nanoclay and carbon microfiber modified asphalt with Fourier transform infrared spectroscopy. *Construction & Building Materials*, *38*, 327–337. DOI: 10.1016/j.conbuildmat.2012.08.004

Yao, R., Sun, K., Zhang, K., Wu, Y., Du, Y., Zhao, Q., Liu, G., Chen, C., Sun, Y., & Li, J. (2024). Stable hydrogen evolution reaction at high current densities via designing the Ni single atoms and Ru nanoparticles linked by carbon bridges. *Nature Communications*, *15*(1), 2218. Advance online publication. DOI: 10.1038/s41467-024-46553-9

Yao, Y., Rui, X., Bai, R., Ouyang, Y., Li, G., Zhao, Y., Zhu, Y.-H., Zhao, M., Li, B.-Q., Zhang, X., Li, Z., Ling, F., Ma, C., Ma, J., Zhou, F., Ren, Z., Shi, X., Zhao, Z., Lu, Y., & Yu, Y. (2025). Roadmap for Next-Generation Electrochemical Energy Storage Technologies: Secondary Batteries and Supercapacitors. *ACS Nano*, *19*(34), 30568–30687. DOI: 10.1021/acsnano.4c10091

Yarali, E., Mirzaali, M. J., Ghalayaniesfahani, A., Accardo, A., Diaz-Payno, P. J., & Zadpoor, A. A. (2024). 4D printing for biomedical applications. *Advanced Materials*, *36*(31), 2402301.

Yarin, A. L., Koombhongse, S., & Reneker, D. H. (2001). Taylor cone and jetting from liquid droplets in electrospinning of nanofibers. *Journal of Applied Physics*, *90*(9), 4836–4846. DOI: 10.1063/1.1408260

Ye, Y., Ren, J., Wang, S., Wan, Y., Razzak, I., Hoex, B., & Zhang, W. (2024). Construction and application of materials knowledge graph in multidisciplinary materials science via large language model. *Advances in Neural Information Processing Systems*, *37*, 56878–56897.

Ye, Z., Jiang, Y., Li, L., Wu, F., & Chen, R. (2021). Rational design of MOF-based materials for next-generation rechargeable batteries. *Nano-Micro Letters*, *13*, 1–37.

Yin, J., Wang, H., Pyle, D., Choi, S., Liu, Y., Wen, J., Guest, J. R., Lyding, J. W., & Dong, G. (2025). Synthesis and Self-Assembly of Monodisperse Graphene Nanoribbons: Access to Submicron Architectures with Long-Range Order and Uniform Orientation. *ACS Nano*, *19*(4), 4366–4376. DOI: 10.1021/acsnano.4c12313

Yin, M., Zheng, X., Humphrey, J. D., & Karniadakis, G. E. (2021). Non-Invasive Inference of Thrombus Material Properties with Physics-Informed Neural Networks. *Computer Methods in Applied Mechanics and Engineering*, *375*, 113603. Advance online publication. DOI: 10.1016/j.cma.2020.113603

Yogesh, G. K., Nandi, D., Yeetsorn, R., Wanchan, W., Devi, C., Singh, R. P., Vasistha, A., Kumar, M., Koinkar, P., & Yadav, K. (2025). A machine learning approach for estimating supercapacitor performance of graphene oxide nano-ring based electrode materials. *Energy Advances*, *4*(1), 119–139. DOI: 10.1039/D4YA00577E

Yoo, H. D., Park, Y., Ryu, J. H., & Oh, S. M. (2011). Electrochemical activation behaviors studied with graphitic carbon electrodes of different interlayer distance. *Electrochimica Acta*, *56*(27), 9931–9936. DOI: 10.1016/j.electacta.2011.08.117

Younas, F., Mustafa, A., Farooqi, Z. U. R., Wang, X., Younas, S., Mohy-Ud-Din, W., & Hussain, M. M. (2021). Current and emerging adsorbent technologies for wastewater treatment: Trends, limitations, and environmental implications. *Water (Basel)*, *13*(2), 215. DOI: 10.3390/w13020215

Younis, M., Ahmad, S., Amjad Farooq, M., Atiq, A., Huang, M., & Abbas, M. (2023). Recent Progress in Azobenzene-Based Supramolecular Materials and Applications. *Chemical Record (New York, N.Y.)*, *23*(11), e202300126. Advance online publication. DOI: 10.1002/tcr.202300126 PMID: 37435961

Yousefi, A. A., Haghshenas, H. F., Underwood, B. S., Harvey, J., & Blankenship, P. (2022). Performance of warm asphalt mixtures containing reclaimed asphalt pavement, an anti-stripping agent, and recycling agents: A study using a balanced mix design approach. *Construction & Building Materials*, *363*, 129633. DOI: 10.1016/j.conbuildmat.2022.129633

Yuan, B., Wang, J., Chen, Y., Wu, X., Luo, H., & Deng, S. (2016). Unprecedented performance of N-doped activated hydrothermal carbon towards C_2H_6/CH_4, CO_2/CH_4, and CO_2/H_2 separation. *Journal of Materials Chemistry. A, Materials for Energy and Sustainability*, *4*(6), 2263–2276. DOI: 10.1039/C5TA08436A

Yuan, H., Chen, J., Li, D., Chen, H., & Chen, Y. (2019). 5 Ultramicropore-rich renewable porous carbon from biomass tar with excellent adsorption capacity and selectivity for CO2 capture. *Chemical Engineering Journal*, *373*, 171–178. DOI: 10.1016/j.cej.2019.04.206

Yudaev, P., Chuev, V., Klyukin, B., Kuskov, A., Mezhuev, Y., & Chistyakov, E. (2022). Polymeric Dental Nanomaterials: Antimicrobial Action. *Polymers*, *14*(5), 864. DOI: 10.3390/polym14050864

Yue, L., Xia, Q., Wang, L., Wang, L., DaCosta, H., Yang, J., & Hu, X. (2018). CO2 adsorption at nitrogen-doped carbons prepared by K2CO3 activation of urea-modified coconut shell. *Journal of Colloid and Interface Science*, *511*, 259–267. DOI: 10.1016/j.jcis.2017.09.040

Yuwen, T., Shu, D., Zou, H., Yang, X., Wang, S., Zhang, S., & Zang, G. (2023). Carbon nanotubes: A powerful bridge for conductivity and flexibility in electrochemical glucose sensors. *Journal of Nanobiotechnology*, *21*(1), 320.

Zachary, D., & Herde, R. D. G. S. J. S. T. J. S. (2020). Impact of hydrolysis on surface area and energy storage applications of activated carbons produced from corn fiber and soy hulls. *Carbon Resources Conversion*, *3*, 19–28. DOI: 10.1016/j.crcon.2019.12.002

Zaidi, N., Mir, M. A., Chang, S. K., Abdelli, N., Hasnain, S. M., Ali Khan, M. A., & Andrews, K. (2025). Pharmaceuticals and personal care products as emerging contaminants: Environmental fate, detection, and mitigation strategies. *International Journal of Environmental Analytical Chemistry*, •••, 1–29.

Zain Ul Abidin, M., Ikram, M., Moeen, S., Nazir, G., Kanoun, M. B., & Goumri-Said, S. (2024). A comprehensive review on the synthesis of ferrite nanomaterials via bottom-up and top-down approaches advantages, disadvantages, characterizations and computational insights. *Coordination Chemistry Reviews*, *520*, 216158. DOI: 10.1016/j.ccr.2024.216158

Zalghout, A., Castro, S., Karam, J., & Kaloush, K. (2022). Laboratory and field evaluation of plant produced asphalt mixtures containing RAP in hot climate: A case study from Phoenix, Arizona. *Construction & Building Materials*, *351*, 128322. DOI: 10.1016/j.conbuildmat.2022.128322

Zeenat, L., Zolfagharian, A., Sriya, Y., Sasikumar, S., Bodaghi, M., & Pati, F. (2023). 4D printing for vascular tissue engineering: Progress and challenges. *Advanced Materials Technologies*, *8*(23), 2300200.

Zeng, C., Xie, F., Yang, X., Jaroniec, M., Zhang, L., & Qiao, S. (2018). Ultrathin Titanate Nanosheets/Graphene Films Derived from Confined Transformation for Excellent Na/K Ion Storage. *Angewandte Chemie*, *130*(28), 8676–8680. DOI: 10.1002/ange.201803511

Zeng, N., Liu, Y., Mao, C., & König, M. (2018). Investigating the relationship between construction supply chain integration and sustainable use of material: Evidence from China. *Sustainability*, *10*(10), 3581.

Zhang, D., Xu, Z., Huang, Z., Gutierrez, A. R., Blocker, C. J., Liu, C.-H., Lien, M.-B., Cheng, G., Liu, Z., Chun, I. Y., Fessler, J. A., Zhong, Z., & Norris, T. B. (2021). Neural network based 3D tracking with a graphene transparent focal stack imaging system. *Nature Communications*, *12*(1), 2413. DOI: 10.1038/s41467-021-22696-x

Zhang, E., Dao, M., Karniadakis, G. E., & Suresh, S. (2022). Analyses of Internal Structures and Defects in Materials Using Physics-Informed Neural Networks. *Science Advances*, *8*(7), eabk0644. Advance online publication. DOI: 10.1126/sciadv.abk0644

Zhang, H., (2022). Microstructural degradation of phase change materials under thermal cycling. *Renewable & Sustainable Energy Reviews*, *156*, 111987.

Zhang, J., Zhong, W., Chen, N., & Weng, Y. (2025). Multiobjective Optimization of the Economic Efficiency of Biodegradable Plastic Products: Carbon Emissions and Analysis of Geographical Advantages for Production Capacity. *Sustainability*, *17*(7), 2874.

Zhang, L., Jiang, Y., Ding, Y., Povey, M., & York, D. (2007). Investigation into the antibacterial behaviour of suspensions of ZnO nanoparticles (ZnO nanofluids). *Journal of Nanoparticle Research*, *9*, 479–489.

Zhang, L., Xie, Y., Tian, Z., Liu, Y., Geng, C., & Xu, S. (2021). Thermal conductive encapsulation enables stable high-power perovskite-converted light-emitting diodes. *ACS Applied Materials & Interfaces*, *13*(25), 30076–30085.

Zhang, P., Zhen, Y. Y., Yin, F. J., Chan, T. H. T., & Fu, P. G. (2021). Intelligent Modelling of Clay Compressibility Using Hybrid Meta-Heuristic and Machine Learning Algorithms. *Geoscience Frontiers*, *12*(1), 441–452. DOI: 10.1016/j.gsf.2020.02.014

Zhang, S., Qian, Z., Huang, K., Zhang, R., Xiao, J., He, Y., & Lu, C. (2023). Robust Generative Adversarial Network. *Machine Learning*, *112*(12), 5135–5161. Advance online publication. DOI: 10.1007/s10994-023-06367-0

Zhang, X., Chen, L., Lim, K. H., Gonuguntla, S., Lim, K. W., Pranantyo, D., Yong, W. P., Yam, W. J. T., Low, Z., Teo, W. J., Nien, H. P., Loh, Q. W., & Soh, S. (2019). The pathway to intelligence: Using stimuli-responsive materials as building blocks for constructing smart and functional systems. *Advanced Materials*, *31*(11), 1804540. DOI: 10.1002/adma.201804540

Zhang, Y., Li, X., Luo, X., & Huang, Y. (2021). Multifactor accelerated aging of photovoltaic materials: Protocols and mechanisms. *Progress in Photovoltaics: Research and Applications*, *29*(4), 379–397.

Zhang, Y., & Sun, L. (2022). Research on hydrogen fuel application (inaircraft) based on carbon neutralization. In *E3S Web of Conferences* (Vol. 341, p. 01010). EDP Sciences. DOI: 10.1051/e3sconf/202234101010

Zhao, Y., Zhang, Z., Pan, Z., & Liu, Y. (2021). Advanced bioactive nanomaterials for biomedical applications. Exploration, 1(3). Portico. DOI: 10.1002/EXP.20210089

Zhao, L., Fan, L.-Z., Zhou, M.-Q., Guan, H., Qiao, S., Antonietti, M., & Titirici, M.-M. (2010). Nitrogen-Containing Hydrothermal Carbons with Superior Performance in Supercapacitors. *Advanced Materials*, *22*(45), 5202–5206. DOI: 10.1002/adma.201002647

Zhao, R., Pandyaswargo, A. H., & Onoda, H. (2024). A bottom-up approach for greenhouse gas emission estimation at the community level: A case study in Japan. *Energy*, *307*, 132530. DOI: 10.1016/j.energy.2024.132530

Zhao, X., Boruah, B., Chin, K. F., Đokić, M., Modak, J. M., & Soo, H. S. (2022). Upcycling to sustainably reuse plastics. *Advanced Materials*, *34*(25), 2100843.

Zhao, X., Ruan, C., Sang, Y., Wang, S., & Liu, H. (2024). MXenes and their composites for advanced cathodes in multivalent ion batteries. *Energy Materials*, *4*(6). Advance online publication. DOI: 10.20517/energymater.2024.48

Zheng, B., & Gu, G. X. (2020). Machine Learning-Based Detection of Graphene Defects with Atomic Precision. *Nano-Micro Letters*, *12*(1), 181. DOI: 10.1007/s40820-020-00519-w

Zheng, B., Zheng, Z., & Gu, G. X. (2022). Designing mechanically tough graphene oxide materials using deep reinforcement learning. *npj Computational Materials*, *8*(1), 225. DOI: 10.1038/s41524-022-00919-z

Zheng, L., Gao, L., Huang, Z., Shi, Z., & Zhou, Y. (2025). A systematic meta-analysis of the impacts of group awareness tools on learning achievements, learning behaviors, and learning perceptions from 2010–2023. *Interactive Learning Environments*, *33*(6), 1–18. DOI: 10.1080/10494820.2025.2454440

Zhi, C., Shi, S., Si, Y., Fei, B., Huang, H., & Hu, J. (2023). Recent progress of wearable piezoelectric pressure sensors based on nanofibers, yarns, and their fabrics via electrospinning. *Advanced Materials Technologies*, *8*(5), 2201161. DOI: 10.1002/admt.202201161

Zhi, D., Li, T., Li, J., Ren, H., & Meng, F. (2021). A review of three-dimensional graphene-based aerogels: Synthesis, structure and application for microwave absorption. *Composites. Part B, Engineering*, *211*, 108642. DOI: 10.1016/j.compositesb.2021.108642

Zhou, T., Wu, C., Wang, Y., Tomsia, A. P., Li, M., Saiz, E., Fang, S., Baughman, R. H., Jiang, L., & Cheng, Q. (2020). Super-tough MXene-functionalized graphene sheets. *Nature Communications*, *11*(1), 2077. DOI: 10.1038/s41467-020-15991-6

Zhu, C., Li, C., Di Wu, W. Y., Shi, S., Ming, H., Zhang, X., & Zhou, K. (2021). A Titanium Alloys Design Method Based on High-Throughput Experiments and Machine Learning. *Journal of Materials Research and Technology*, *11*(March), 2336–2353. DOI: 10.1016/j.jmrt.2021.02.055

Zhu, G., & Gao, Y. (2023). Multi-objective optimal scheduling of an integrated energy system under the multi-time scale ladder-type carbon trading mechanism. *Journal of Cleaner Production*, *417*, 137922.

Zhu, J., Dong, Y., Wang, Q., Han, J., Li, Z., Xu, D., Fischer, L., Ulbricht, M., & Ren, Z. (2024). Advancements in magnetic catalysts: Preparation, modification, and applications in photocatalytic and environmental remediation. *The Science of the Total Environment*, *957*, 177595. DOI: 10.1016/j.scitotenv.2024.177595

Zhu, L., Han, Y., Li, Q., Luo, Q., Zhang, Y., Zhang, L., Hu, B., & Chen, Q. (2024). Prediction of ultimate tensile strength of Al-Si alloys based on multimodal fusion learning. *Materials Genome Engineering Advances*, *2*(1), e26. Advance online publication. DOI: 10.1002/mgea.26

Zhuo, Y., Tehrani, A. M., & Brgoch, J. (2018). Predicting the Band Gaps of Inorganic Solids by Machine Learning. *The Journal of Physical Chemistry Letters*, *9*(7), 1668–1673. Advance online publication. DOI: 10.1021/acs.jpclett.8b00124

Zhu, X.-Y., & Wang, B.-R. (2021). Novel Nanofluidic Cells Based on Nanowires and Nanotubes for Advanced Chemical and Bio-Sensing Applications. *Nanomaterials (Basel, Switzerland)*, *11*, 90.

Zimniewska, M. (2022). Hemp fibre properties and processing target textile: A review. *Materials (Basel)*, *15*(5), 1901.

Znidi, F., Morsy, M., & Uddin, M. N. (2024). Navigating challenges and solutions for metal-halide and carbon-based electrodes in perovskite solar cells (NCS-MCEPSC): An environmental approach. *Heliyon*, *10*(12), e32843. DOI: 10.1016/j.heliyon.2024.e32843

Zou, M., Jiang, W.-G., Qin, Q.-H., Liu, Y.-C., & Li, M.-L. (2022). Optimized XGBoost Model with Small Dataset for Predicting Relative Density of Ti-6Al-4V Parts Manufactured by Selective Laser Melting. *Materials (Basel)*, *15*(15), 5298. DOI: 10.3390/ma15155298

Zou, X., Gu, F., Zhang, X., & Zhang, X. (2025). Self-assembly synthesis of FeOOH/Cu2O heterostructure porous microspheres with enhanced photocatalytic activity. *Journal of Alloys and Compounds*, *1010*, 178017. DOI: 10.1016/j.jallcom.2024.178017

About the Contributors

Deepanraj Balakrishnan is currently working as Assistant Research Profesor in the Department of Mechanical Engineering, Prince Mohammad Bin Fahd University, Saudi Arabia. He has 13 years of working experience, which include both teaching and research. Dr. Deepanraj has published more than 200+ research articles in peer reviewed international journals and conferences and holds 4 Indian patents. He is serving as guest editor, editorial board member and reviewer for many international peer reviewed journals including well known publishers like Elsevier, Springer and Inderscience. His area of interest includes Renewable Energy Utilization; Materials Engineering, Waste Management and Sustainability.

Ratna Sunil received his bachelor's degree in mechanical engineering from Andhra University, India, in 2004 and Master's degree in Mechanical Engineering from Osmania University, India in 2007. Dr Ratna Sunil earned his Ph.D. from the Indian Institute of Technology Madras, India, in 2014 in the field of Biomaterials. Dr. Sunil has 15 years of experience in teaching and research in the areas of Materials Engineering, Manufacturing Processes and Surface Engineering. Dr. Sunil has been listed in world top 2% scientists announced by Stanford University, USA, consecutively for five years (from 2020 to 2024). Dr. Sunil has published more than 140 scientific papers in international peer-reviewed journals and conference proceedings and authored a book titled "Surface Engineering by Friction-Assisted Processes: Methods, Materials, and Applications" available with CRC Press, USA. From the research findings, Dr. Sunil has filed 20 Indian patents.

Wattala Rohan Fernando holds the position of Lecturer in Mechanical and Industrial Engineering at the School of Science and Engineering (SSEN), and he is also a lecturer at The Centre for Energy, Petroleum, and Mineral Law and Policy (CEPMLP), where he teaches energy-related modules associated with SSEN. After completing his studies at Northumbria University Newcastle Upon Tyne, UK, in 2011, Fernando gained significant experience through roles with Ericson, Vodafone, and Huawei, contributing to various projects, including the London Olympic Project. In 2015, he completed his MSc in Energy, Water, and Environmental Management at Abertay University Dundee, UK. The practical experience acquired at Ericson, Vodafone, and Huawei proved invaluable, particularly in power and energy engineering, facilitating his MSc studies. In 2020, he successfully earned his PhD in energy engineering from Edinburgh Napier University. Concurrently, during his PhD, Fernando completed six European Credit Transfer and Accumulation System (ECTS) PhD courses at Aalborg University in Aalborg, Denmark. Before joining the University of Dundee in January 2023, Fernando held positions as a research officer at Hydrogen Accelerator, St Andrews University, a research and teaching assistant at Edinburgh Napier University, and a visiting lecturer at the University of Applied Management Studies, Mannheim, Germany, and German Business School, the University of Tunis, for an MBA in Energy Management. With more than 12 years of experience, he actively participated in national and international industry projects in electronics, telecommunications, and power engineering before transitioning to academia.

***

Muhammad Akram is an experienced academic and researcher specializing in Operations and Supply Chain Management, Digital Transformation, AI, and Big Data Analytics. He holds a PhD in Operations Management and an MRes in Management from Cranfield University, UK, along with an MSc and BSc in Industrial and Systems Engineering from King Fahd University of Petroleum & Minerals, KSA. With a strong background in both academia and industry, Dr. Akram has led innovative research projects, secured significant research grants, and published in high-impact journals. His teaching experience spans undergraduate, postgraduate, and executive education, where he has consistently received excellent teaching evaluations. He has also supervised several PhD and MSc research students in areas related to supply chain risk management, AI in business processes, digital transformation, and sustainable operations. Beyond academia, Dr. Akram has been actively involved in consulting projects, working with organizations on process optimization, digitalization, and strategic decision-making. His expertise in integrating AI and analytics into business operations has contributed to improving efficiency and resilience across various industries. He is available for PhD supervision in supply chain management, digital transformation, AI applications in operations, and sustainable business strategies.

Ghida AlZohbi is an assistant professor in the department of mechanical engineering at Prince Mohamed Bin Fahd University in Al Khobar, Saudi Arabia.

She has been a full member of the Renewable Energy Institute since October 2021. Ghida holds a master's degree in energy engineering. In 2011, she began her PhD studies at the Brussels School of Engineering, Université Libre de Bruxelles in Belgium, where she also worked as a lecturer from 2012 to 2015. She received her PhD in Engineering Sciences and Technology from the same institution in 2015. After earning her PhD, Ghida joined the Department of Aero-Thermo-Mechanics at the Brussels School of Engineering as a postdoctoral researcher. Her research interests include renewable energy (solar and wind energy), energy storage (pumped hydro storage systems and compressed air energy storage), hydrogen energy, building-integrated photovoltaics (BIPV), energy efficiency, and retrofit of buildings. In 2017, she contributed to the International Energy Agency (IEA) PVPS Task 15-subtask B, focusing on identifying, benchmarking, and developing business models for BIPV products across different regions and countries. She has also worked as a consultant on numerous projects related to wind energy, photovoltaic (PV) and concentrated photovoltaic (CPV) systems, compressed air energy storage, pumped hydro storage systems, and hydrogen generation and storage. Ghida has served as a technical committee member for many conferences and as a reviewer for numerous journals. She has published more than 50 papers

Bader Eddin Alasali is a Chartered Civil and Transportation Engineer and academic with over 23 years of international experience. He is currently working as an assistant professor in civil and environmental engineering department in National University of Science and Technology, Oman. He is a PhD holder, he specializes in transportation systems, tunnels, and infrastructure design. He combines industry consultancy with university leadership, research supervision, curriculum development, and innovative engineering solutions across global institutions.

Ercan Bursal serves as a "Full Research Professor" at the College of Engineering, Prince Mohammad bin Fahd University. Dr. Ercan Bursal has authored more than 70 articles in SCI-indexed journals on his research area with having more than 5400 citations and 35 h-index. He is the coauthor of 20 book chapters, as well. He supervised one Ph.D. and seven M.S. students. The main topics of his studies are antioxidant activity, bioorganic compounds, medicinal plants, enzyme activity/inhibition, and biochemical parameters in biological samples. Also, his main specializations are UV-Spectrophotometry, HPLC, and LC-MS-MS.

Waqas Choudhary is currently pursuing his Master's degree in Plastic and Packaging Technology at FH Campus Wien – University of Applied Sciences, Austria. He previously earned both a Master's and a Bachelor's degree in Maintenance and Industrial Engineering from the University of Engineering and

Technology (UET), Pakistan. Alongside his academic journey, Waqas has worked in industry where he has successfully led maintenance-related projects, combining his technical expertise with practical problem-solving skills. His interests lie at the intersection of packaging innovation, materials sustainability, and industrial process optimization.

Nancy Elizabeth Dávila Guzmán earned her PhD in Sustainable Processes from the Autonomous University of Nuevo León, where she leads the Chemical Engineering research group. A founding member and current president of the Mexican Adsorption Association, she has helped build a strong national community around adsorption science. Her research centers on adsorption-based separation processes, including water treatment, air purification, biofuel upgrading, green synthesis of metal–organic frameworks (MOFs), and the development of 3D-printed adsorbents. She has published more than 25 peer-reviewed articles and four book chapters, and received the UANL Research Award in 2022. Beyond the lab, Dr. Dávila drives educational innovation through mobile app development and augmented-reality tools for STEM learning. She is also active in science communication: she founded the YouTube channel CienciasIQ UANL to share foundational science content and previously coordinated the Científicas Mexicanas podcast, which highlights women's contributions to science.

Ravikumar Dumpala is an Associate Professor at VNIT Nagpur, specializing in manufacturing and surface engineering. With a Ph.D. from IIT Madras, his research focuses on Micro-arc Oxidation (MAO), wear and corrosion behavior of light alloys, and surface treatments for 3D-printed materials. He is a recipient of the SERB Early Career Research Award and ISTE's National Award for Best M.Tech. Thesis (Supervisor), and actively mentors UG, PG, and Ph.D. students in advanced surface engineering topics.

Rituraj Jain is currently working in the Department of Information Technology, Marwadi University, Rajkot, Gujarat, India. He has published many papers in International Journals and Conferences. He has published 4 books. He is an Oracle Database SQL Certified Expert. He is also associated with professional bodies like the Indian Society of Technical Education (ISTE), Computer Science Teachers Association (CSTA), Academy & Industry Research Collaboration Center (AIRCC), International Association of Computer Science and Information Technology (IACSIT), International Association of Engineers (IAENG). His research interests include Cloud Computing, Machine Learning, Deep Learning, Fuzzy Logic, IoT, Genetic Algorithms, Big Data and Software Engineering.

Ceren Karaman received her B.Sc. in Chemical Engineering from Gazi University (2011), her M.Sc. from Hacettepe University (2013), and her Ph.D. from Ankara University (2020). During her doctoral studies, she was supported by the Turkish Ministry of National Education's scholarship program as a visiting doctoral fellow in the United States (2013–2014). She began her academic career as a Research Assistant at Ankara University (2014–2016) and Pamukkale University (2016–2020). Since 2020, she has been a faculty member in the Department of Electrical and Energy at Akdeniz University, where she has also served as Deputy Head of Department since 2021. In 2023, she was promoted to Associate Professor in Chemical Engineering. Dr. Karaman has been involved in numerous national and international research groups and projects. Her expertise covers the synthesis and characterization of nanomaterials and their applications in energy storage, conversion systems, electrochemical sensors, and environmental protection. She has edited 5 international books, authored 6 book chapters, published over 90 SCI-indexed journal articles, and presented more than 40 papers at international conferences. She also serves as an associate editor, editorial board member, and guest editor for more than 15 SCI-indexed journals, and is a member of the American Chemical Engineering Society and Nature. Dr. Karaman holds a patent titled "Supercapacitor Containing Radioactive Components and Method for Preparing This Supercapacitor". She further expanded her research horizons as a postdoctoral researcher at the Institut Català de Nanociència i Nanotecnologia (ICN2) in Barcelona, Spain (2023–2024), focusing on the application of 3D hybrid graphene structures in energy storage and conversion systems.

Onur Karaman completed his doctoral studies in 2019 and was awarded the title of Associate Professor in Physics in 2022. Since 2014, he has been serving as a faculty member in the Medical Imaging Techniques Program at the Vocational School of Health Services, Akdeniz University. His academic interests focus on material physics, electrochemical sensor technologies, and artificial intelligence–based data analysis. His research brings together interdisciplinary topics such as the synthesis and characterization of nanostructured hybrid materials, their application in biosensor platforms, the optimization of electrochemical measurement techniques, and the processing of biomedical data through artificial intelligence algorithms. Assoc. Prof. Dr. Karaman has published numerous international journal articles, book chapters, and conference papers; he has taken part as a principal investigator and researcher in TÜBİTAK-funded projects and has developed collaborations within international research consortia. His academic endeavors aim to integrate fundamental sciences with engineering and health sciences in order to produce multidisciplinary solutions.

Çağdaş Kizil earned undergraduate degree from Akdeniz University, Faculty of Science, Department of Chemistry in 2010. He Completed master's education at the same university in 2016 and his doctorate in 2022. During graduate education, carried out studies in the field of food toxicology and prepared a master's thesis on the detection of toxic chemicals that harm the environment and human health. In this context, he has carried out researches, seminars, trainings and scientific publications related to sustainable food production that does not harm human health and food analysis in accordance with European Union standards. Since his doctoral education, he carried out studies in the field of energy storage and conversion within the scope of sustainable energy production. In particular, he executed scientific studies and projects on lithium-ion batteries, supercapacitors and fuel cells, which are rapidly becoming widespread in the world in order to reduce the effects of climate change. He also has two international patents and five articles published in high-impact journals in the field of energy storage and conversion.He is now working on the development of qualified and high value-added initiatives in the field of sustainable industrial production and green transformation at the Antalya Chamber of Commerce and Industry. At the same time, he also carries out consultancy and mentoring activities regarding national and international grant supports offered to entrepreneurs and investors.

Ali İhsan Kömür graduated from the Department of Chemistry at Süleyman Demirel University with master's degrees (2016) and bachelor's degrees (2014) and received his PhD from the Department of Chemistry at Akdeniz University (2024). During his master's degree, he took part in interdisciplinary projects focusing on electrochromic smart glasses, nanofibers, thin films, electrochemistry, plasma science, and nanotechnology. During his doctoral studies, he developed projects in the field of energy storage and conversion. During his doctoral studies, he significantly enriched his academic profile in the fields of nanotechnology, advanced materials synthesis (nanomaterials), and electrochemical energy storage technologies (batteries). He completed his doctorate at Akdeniz University on the development and different aspects of sodium-ion batteries (Na-ion batteries) (2024). As of 2024, in addition to his studies (electrochemical energy storage systems such as Na-ion and Li-ion batteries) in which he is an expert at Akdeniz University, he continues to enrich his scientific profile with national and international projects in different branches and fields (artificial intelligence-supported biosensors and cancer detection).

Jacob Salazar has undergraduate (2009) and postgraduate studies in Chemical Engineering (2015). In 2017 he joined as a full professor at the Autonomous University of Nuevo León, in the Faculty of Chemical Sciences, where he develops the following lines of research: Evaluation and development of Surface Physicochemical Processes in the sanitation of impacted aqueous effluents. Synthesis and characterization of materials with environmental applications. Mathematical modeling with a phenomenological foundation of physicochemical processes.

Sitesh Kumar Singh is an academician, researcher, Fellow of Advance HE, Chartered Engineer and currently working as an Assistant Professor at the National University of Science and Technology, Oman. He holds a Ph.D. in civil engineering. He has more than 9 years of teaching and research experience. He has published 4 books, 12 book chapters, 46 research papers in reputed journals, 7 patents, and 15 conference presentations.

Kamal Upreti is currently working as an Associate Professor in Department of Computer Science, CHRIST (Deemed to be University), Delhi NCR, Ghaziabad, India. He completed is B. Tech (Hons) Degree from UPTU, M. Tech (Gold Medalist), PGDM(Executive) from IMT Ghaziabad and PhD in Department of Computer Science & Engineering.He has completed Postdoc from National Taipei University of Business, TAIWAN funded by MHRD. He has published 50+ Patents, 32+Magazine issues and 110+ Research papers in in various reputed Journals and international Conferences. His areas of Interest such as Modern Physics, Data Analytics, Cyber Security, Machine Learning, Health Care, Embedded System and Cloud Computing. He has published more than 45+ authored and edited books under CRC Press, IGI Global, Oxford Press and Arihant Publication. He is having enriched years' experience in corporate and teaching experience in Engineering Colleges.

Index

A

activated carbons 333, 360, 362, 370, 371, 373, 374, 376, 382, 383, 384, 385, 386, 391, 392, 394
Adsorption 61, 65, 72, 132, 154, 155, 157, 158, 159, 161, 166, 167, 172, 182, 193, 328, 333, 361, 362, 366, 367, 368, 369, 370, 371, 372, 373, 374, 376, 379, 381, 382, 383, 384, 385, 386, 387, 388, 389, 390, 391, 392, 393, 394
Advanced Sustainable Materials 305, 321, 323, 325, 352
agro-industrial waste 362
Alginate 329, 330, 331, 354, 355, 357
Applications 2, 3, 4, 6, 7, 12, 14, 16, 19, 21, 22, 25, 26, 28, 30, 32, 34, 37, 39, 41, 43, 51, 56, 57, 58, 60, 61, 62, 64, 65, 69, 70, 71, 72, 73, 74, 76, 83, 87, 90, 91, 93, 97, 98, 100, 101, 102, 106, 108, 111, 112, 113, 115, 119, 124, 127, 128, 129, 130, 131, 132, 133, 134, 135, 136, 137, 138, 139, 140, 141, 142, 143, 144, 145, 148, 150, 151, 152, 153, 154, 155, 156, 157, 158, 160, 161, 163, 164, 165, 167, 169, 170, 172, 173, 175, 176, 178, 179, 181, 183, 184, 185, 186, 187, 188, 189, 190, 191, 192, 194, 195, 196, 197, 201, 202, 204, 207, 208, 209, 210, 211, 212, 214, 215, 218, 219, 220, 222, 227, 228, 234, 236, 240, 241, 242, 243, 244, 245, 246, 247, 248, 251, 252, 253, 255, 258, 259, 260, 262, 263, 264, 265, 268, 269, 277, 278, 279, 280, 281, 282, 283, 284, 289, 294, 297, 298, 299, 301, 303, 309, 310, 311, 313, 314, 316, 317, 319, 325, 328, 330, 331, 332, 333, 336, 338, 342, 345, 351, 354, 355, 356, 357, 358, 384, 385, 387, 389, 390, 394, 397, 402, 409, 417, 424, 433
Artificial Intelligence 1, 2, 7, 25, 26, 27, 28, 30, 31, 32, 33, 34, 49, 74, 95, 98, 105, 106, 108, 109, 138, 170, 173, 194, 217, 227, 229, 230, 242, 243, 245, 248, 425, 430
Atomic Layer Deposition 126, 350

B

Bio-Based Polymers 265, 270, 276, 277, 278, 282, 283, 287, 288, 289, 293, 296, 297, 298, 307, 309, 311, 316, 321, 428
Biochar 102, 104, 158, 159, 166, 167, 180, 333, 334, 360, 361, 362, 367, 368, 369, 384, 385, 388, 389, 411
Bio-inspired Materials 321

C

calcium 251, 252, 253, 254, 255, 256, 257, 258, 259, 260, 262, 263, 264, 280, 407, 409, 414, 415, 431, 432, 434
Carbon-dioxide 384
Carbon Nanotubes 7, 32, 42, 52, 72, 74, 75, 90, 113, 137, 148, 151, 155, 157, 158, 159, 160, 165, 167, 169, 172, 175, 209, 212, 213, 225, 227, 349, 420
Cellulose Nanocrystals 54, 325, 327, 328
Chitosan 289, 297, 322, 325, 329, 330, 345, 346, 355, 356, 357
Circular Economy 58, 192, 265, 267, 268, 276, 282, 283, 288, 290, 292, 295, 299, 300, 305, 307, 310, 311, 312, 313, 314, 315, 316, 317, 321, 355, 396, 416, 417, 419, 425, 428

D

Durability 43, 44, 51, 52, 69, 73, 151, 189, 237, 280, 281, 293, 294, 298, 299, 300, 303, 309, 344, 350, 397, 398, 405, 409, 410, 412, 414, 426, 427, 428

E

Eco-Friendly Materials 265
Energy 2, 5, 7, 9, 14, 16, 19, 22, 23, 24,

26, 27, 30, 31, 35, 36, 37, 38, 40, 41, 44, 45, 46, 47, 50, 51, 52, 53, 55, 56, 57, 58, 59, 60, 62, 63, 64, 65, 66, 68, 69, 70, 71, 72, 73, 74, 75, 76, 78, 80, 81, 90, 93, 96, 101, 102, 104, 105, 106, 107, 111, 112, 113, 117, 119, 120, 121, 123, 125, 130, 132, 134, 138, 139, 140, 142, 144, 145, 147, 148, 149, 150, 151, 152, 153, 154, 155, 156, 157, 160, 161, 163, 164, 165, 166, 167, 169, 170, 171, 172, 173, 174, 175, 176, 177, 178, 179, 180, 181, 182, 183, 184, 185, 188, 192, 194, 195, 196, 197, 198, 201, 209, 210, 214, 217, 218, 219, 220, 221, 222, 223, 224, 225, 226, 227, 228, 229, 230, 232, 234, 235, 236, 239, 240, 241, 242, 243, 244, 245, 246, 247, 248, 249, 257, 267, 268, 271, 274, 275, 276, 278, 279, 281, 282, 284, 285, 286, 287, 288, 289, 290, 291, 293, 294, 298, 299, 300, 301, 312, 314, 315, 316, 317, 319, 321, 322, 323, 324, 327, 331, 332, 333, 334, 336, 338, 340, 341, 342, 343, 344, 345, 349, 350, 351, 352, 354, 355, 357, 361, 367, 368, 369, 371, 373, 374, 376, 385, 386, 388, 389, 390, 391, 393, 394, 402, 404, 405, 406, 415, 420, 421, 422, 428, 433
Energy Materials 2, 31, 35, 36, 37, 40, 41, 44, 45, 46, 53, 62, 63, 64, 66, 68, 69, 70, 73, 101, 182, 230, 245, 249, 393
Engineering 3, 6, 7, 12, 18, 21, 22, 23, 24, 25, 26, 27, 28, 29, 30, 31, 32, 33, 34, 35, 37, 38, 41, 47, 48, 50, 51, 52, 54, 55, 56, 57, 58, 59, 60, 61, 63, 65, 66, 70, 71, 72, 73, 74, 78, 80, 86, 91, 98, 99, 100, 101, 104, 106, 107, 108, 109, 111, 113, 124, 128, 130, 131, 132, 134, 135, 138, 139, 140, 141, 148, 153, 154, 157, 163, 164, 172, 174, 176, 179, 180, 181, 184, 187, 188, 190, 192, 193, 195, 201, 208, 212, 213, 214, 218, 219, 223, 225, 226, 227, 229, 231, 236, 237, 238, 240, 241, 243, 244, 246, 249, 258, 259, 262, 263, 264, 297, 309, 310, 312, 313, 315, 316, 318, 319, 320, 321, 323, 325, 327, 332, 334, 335, 341, 344, 350, 352, 354, 355, 356, 358, 376, 384, 385, 386, 388, 389, 390, 392, 393, 394, 395, 396, 399, 415, 416, 420, 422, 424, 425, 428, 430, 431, 432, 433, 434
Environment 4, 8, 10, 13, 22, 75, 83, 96, 104, 112, 117, 123, 131, 133, 134, 147, 148, 157, 158, 159, 160, 161, 163, 164, 169, 171, 172, 173, 176, 180, 184, 186, 210, 211, 252, 259, 260, 266, 267, 274, 275, 280, 282, 283, 284, 288, 296, 304, 309, 312, 314, 317, 319, 322, 355, 362, 396, 399, 402, 405, 410, 412, 415, 416, 417, 420, 422, 428
Environmental Engineering 131, 264, 431

G

Generative Models 2, 6, 24, 80, 88, 89, 95, 100
global warming 149, 152, 169, 291, 361, 368, 369, 382

H

Heat 7, 9, 32, 46, 74, 89, 90, 106, 107, 154, 155, 156, 163, 164, 187, 188, 189, 190, 192, 193, 194, 195, 196, 197, 198, 199, 200, 201, 202, 203, 204, 208, 209, 210, 212, 213, 214, 257, 264, 361, 376, 379, 386, 387, 397, 398, 399, 400, 401, 404, 409, 420
High-Throughput Screening 1, 83, 239
Hybrid Modeling 21, 192
hydroxyapatite 103, 136, 252, 253, 254, 255, 256, 257, 258, 262, 263, 264

I

Ionic Conductivity 36, 38, 44, 53, 55, 59, 60, 64, 67, 70

L

Lifecycle Evaluation 265, 290

M

Machine Learning 1, 2, 4, 5, 9, 11, 14, 15, 16, 18, 21, 23, 25, 26, 27, 28, 29, 30, 31, 33, 34, 40, 48, 49, 57, 63, 66, 70, 73, 77, 78, 80, 89, 98, 99, 100, 101, 102, 103, 104, 105, 107, 108, 109, 170, 185, 192, 194, 219, 227, 228, 231, 232, 234, 238, 239, 241, 242, 243, 244, 245, 246, 247, 248, 249, 340, 346, 427
Materials 1, 2, 3, 4, 5, 6, 7, 8, 9, 10, 11, 12, 13, 14, 18, 20, 21, 22, 23, 24, 26, 27, 28, 29, 30, 31, 32, 33, 34, 35, 36, 37, 38, 39, 40, 41, 42, 43, 44, 45, 46, 48, 52, 53, 54, 57, 58, 59, 60, 61, 62, 63, 64, 65, 66, 68, 69, 70, 71, 72, 73, 74, 75, 76, 77, 78, 79, 80, 81, 82, 83, 84, 85, 86, 87, 88, 89, 90, 91, 92, 93, 94, 95, 96, 97, 98, 99, 100, 101, 102, 103, 104, 105, 106, 107, 108, 109, 110, 111, 112, 113, 114, 115, 117, 120, 122, 123, 125, 126, 127, 131, 132, 133, 134, 136, 137, 138, 140, 141, 142, 144, 147, 148, 149, 150, 152, 153, 154, 155, 156, 157, 158, 159, 160, 161, 163, 164, 165, 166, 167, 169, 170, 172, 173, 174, 175, 176, 177, 178, 179, 180, 181, 182, 183, 184, 185, 186, 187, 188, 192, 194, 196, 198, 202, 207, 208, 211, 212, 213, 217, 218, 219, 220, 222, 223, 224, 225, 226, 227, 228, 230, 231, 234, 236, 237, 238, 239, 240, 241, 242, 243, 244, 245, 246, 247, 248, 249, 251, 252, 253, 254, 255, 257, 258, 259, 260, 262, 263, 264, 265, 266, 267, 268, 269, 270, 271, 272, 273, 274, 275, 276, 277, 278, 279, 280, 281, 282, 283, 284, 285, 286, 287, 288, 289, 290, 291, 292, 293, 294, 295, 296, 297, 298, 299, 300, 301, 303, 304, 305, 306, 307, 308, 309, 310, 311, 312, 313, 314, 315, 316, 317, 318, 319, 320, 321, 322, 323, 324, 325, 331, 333, 335, 336, 343, 344, 345, 347, 348, 350, 351, 352, 354, 355, 356, 357, 358, 360, 361, 362, 363, 367, 370, 372, 373, 374, 376, 378, 379, 380, 383, 384, 385, 386, 387, 388, 389, 390, 391, 392, 393, 394, 395, 396, 397, 398, 399, 402, 404, 405, 406, 407, 411, 412, 414, 415, 416, 417, 418, 419, 420, 422, 423, 424, 425, 426, 427, 428, 429, 430, 431, 432, 433, 434, 435
Material Science 9, 163, 245, 298, 334, 339, 417
Materials Discovery 30, 71, 73, 87, 94, 100, 103, 104, 105
medical implants 61, 289, 297, 325
Metal–Organic Frameworks 37, 169, 239
Metal-oxide 121, 182
MXene 100, 217, 218, 219, 220, 222, 223, 224, 225, 226, 230, 231, 232, 234, 235, 236, 237, 238, 239, 240, 241, 242, 243, 244, 245, 246, 247, 248, 249

N

Nanobiotechnology 75, 136, 139
Nanocomposites 44, 56, 57, 58, 70, 71, 73, 78, 112, 122, 125, 127, 130, 131, 138, 157, 158, 164, 183, 184, 217, 218, 219, 220, 221, 222, 223, 225, 226, 239, 241, 244, 316, 330
Nanomaterials 2, 6, 7, 9, 25, 36, 37, 39, 54, 72, 74, 75, 77, 87, 89, 90, 91, 99, 102, 111, 113, 114, 115, 116, 117, 119, 120, 121, 122, 123, 127, 130, 131, 132, 135, 136, 137, 138, 139, 140, 141, 143, 144, 145, 148, 150, 159, 166, 167, 169, 170, 172, 175, 176, 178, 179, 180, 181, 183, 211, 212, 213, 214, 215, 218, 219, 220, 223, 241, 242, 244, 246, 338, 339, 354, 355, 356, 387, 393, 397, 399, 400, 401, 402, 433, 434
Nanoparticles 14, 28, 42, 47, 71, 72, 89, 103, 112, 117, 119, 120, 126, 127,

131, 133, 136, 137, 138, 139, 140, 141, 143, 144, 148, 151, 152, 157, 158, 178, 182, 183, 184, 187, 188, 189, 190, 191, 192, 193, 194, 195, 196, 197, 199, 200, 201, 202, 203, 204, 205, 206, 207, 208, 209, 210, 211, 212, 213, 214, 215, 289, 323, 336, 337, 339, 349, 350, 355, 397, 400, 401, 402, 432
Nanoscience 111, 114, 117, 142, 213
Nanotechnology 22, 72, 105, 111, 112, 113, 114, 115, 129, 130, 131, 132, 133, 135, 136, 137, 138, 139, 140, 141, 142, 143, 147, 148, 150, 151, 157, 158, 173, 179, 212, 213, 269, 336, 397, 399, 402
New Materials 1, 5, 23, 81, 83, 87, 91, 96, 97, 169, 170, 279, 322, 397, 416, 426

P

Pavement Materials 395, 396, 397, 419, 422
Polyhydroxyalkanoates 75, 270, 278, 279, 282, 323, 326
Polylactic 7, 270, 277, 278, 279, 282, 289, 294, 317, 318, 326, 328, 341, 357
Polymers 1, 6, 7, 26, 36, 44, 45, 54, 57, 59, 61, 62, 64, 68, 71, 73, 75, 77, 83, 84, 87, 89, 90, 91, 96, 102, 108, 122, 128, 139, 144, 148, 161, 164, 175, 193, 207, 223, 239, 265, 270, 276, 277, 278, 279, 282, 283, 287, 288, 289, 293, 296, 297, 298, 307, 309, 310, 311, 313, 316, 317, 321, 328, 335, 339, 344, 346, 347, 349, 354, 356, 357, 358, 374, 379, 396, 397, 399, 411, 414, 418, 421, 428, 432
Property Prediction 2, 4, 8, 9, 10, 24, 87, 89, 92, 100, 239

R

Recycled PET 272, 331

S

Single-atom Catalyst 47
Smart Manufacturing 18, 19, 23, 24, 214
Smart Materials 7, 22, 131, 133, 156, 163, 164, 180, 187, 314, 345
Sustainability 16, 19, 22, 33, 40, 41, 45, 72, 73, 74, 78, 96, 97, 104, 105, 108, 111, 114, 131, 134, 138, 139, 140, 141, 149, 150, 152, 156, 160, 169, 170, 173, 174, 175, 176, 177, 178, 179, 180, 181, 182, 183, 184, 185, 192, 199, 267, 268, 269, 275, 286, 290, 291, 292, 293, 295, 297, 299, 300, 301, 302, 303, 304, 305, 306, 307, 308, 309, 310, 311, 313, 314, 315, 317, 318, 319, 320, 322, 324, 329, 331, 332, 335, 336, 341, 345, 351, 352, 354, 357, 358, 388, 393, 394, 396, 397, 410, 415, 416, 417, 422, 425, 433
Sustainable Energy 16, 37, 40, 41, 46, 51, 53, 75, 104, 106, 111, 130, 147, 148, 152, 153, 154, 155, 170, 179, 182, 195, 243, 312, 314, 321, 341, 343
Sustainable Materials 91, 181, 265, 266, 267, 268, 269, 270, 276, 277, 279, 280, 282, 284, 286, 288, 289, 290, 292, 293, 294, 295, 303, 304, 305, 306, 307, 308, 309, 311, 313, 314, 316, 320, 321, 323, 325, 336, 350, 352, 415, 426

T

Thermostability 61

www.ingramcontent.com/pod-product-compliance
Lightning Source LLC
LaVergne TN
LVHW081317300926
849161LV00008B/1787